Sustainable food planning

evolving theory and practice

edited by:
André Viljoen
and
Johannes S.C. Wiskerke

Wageningen Academic
P u b l i s h e r s

Wageningen Academic Publishers,
P.O. Box 220, 6700 AE Wageningen,
the Netherlands.
www.WageningenAcademic.com
copyright@WageningenAcademic.com

ISBN: 978-90-8686-187-3

Cover image by: Bohn&Viljoen Architects

First published, 2012

© Wageningen Academic Publishers
The Netherlands, 2012

Note from the editors

One overriding aim has directed our efforts in putting this book together: how to get the many people from diverse disciplinary backgrounds who are developing ideas about sustainable food planning to talk to each other, so as to better understand where common themes exist, where there is difference, and above all how we might work together.

It is true to say that most contributors come from what is termed 'the alternative food movement', but it is also correct that many of these alternatives are becoming mainstream. We hope that this book is a step on the way to building the critical mass that helps to give voice to this evolving practice.

We would like to acknowledge the inspiration provided by Jerry Kaufmann, Professor Emeritus of Urban and Regional Planning at the University of Wisconsin, whose work has established food systems as a legitimate topic for the planning community in the USA and increasingly within Europe. His work and its recognition within the Association of European Schools of Planning (AESOP) food group, under the direction of Professor Kevin Morgan, has helped to bring together the authors in this collection.

Looking ahead, we hope and anticipate that more voices and new perspectives will develop the ideas on food, planning, and communities presented here. After all, the subject is vast, challenging, and nothing short of fundamental to our shared futures.

André Viljoen and Han Wiskerke

Table of contents

Part 2. Integrating health, environment and society

Part 3. Urban agriculture

Part 4. Planning and design

Foreword

Tom Bliss
www.urbal.tv

How long have we got? That is the question which most occurs to me when contemplating the Herculean task of adapting our cities to make them sustainably and equitably resilient in times of peaking resources, economic chaos and ecologically catastrophic climate change. Ignoring for the moment those with closed minds, opinions generally range from people who think in terms of responsibility to future generations, to those who, like Asterix's Chief Abraracourcix, are expecting the sky to fall at any moment. The latter tend to be desperate for quick, revolutionary fixes capable of saving society from itself in the nick of time, while the former prefer a more methodical, evolutionary approach, which minimises transitional damage and, who knows, maybe even finally delivers that elusive Utopian world order. Of course; the revolutionaries risk getting it disastrously wrong in their haste, while the evolutionaries risk achieving too little too late.

But actually, we have no way of knowing how long we have got – so perhaps in a way it does not matter. Maybe all that matters is that all of us who are committed to finding and promoting potential solutions do so with alacrity, while fostering a sense of urgency – and, crucially, achievability – at every turn. It is this achievability that is the nub, though, because (apart from the immutable need to safeguard global ecological viability) the two things we know for sure are that, one, the task is almost impossibly complex, and, two, we have never been here before, so the known/unknown paradox is at its most opaque.

We do have some useful social models from history though, not least (as my film 'The Urbal Fix' suggests) my forebear Ebenezer Howard's pre-combustion-engine Social City, with its pedestrian ergonomics, productive green spaces and self-governing commonwealths. But even better, we now also have a growing number of inspirational contemporary experiments from cities all over the world – some of the most exciting of which are analysed fruitfully in this book.

That said, finding a reliable way of unpicking the status quo, of assessing and then adapting whatever constituent parts we hope may still work, while introducing new ideas (some of them radical and very probably unpopular), before somehow promoting a functioning, just, resilient metropolis, presents a daunting challenge. In The Urbal Fix, Professor Lord Tony Giddens (former Director of the London School of Economics) comments, 'I think we're at the outer edge of some of the biggest changes in our history across the world, Not just another industrial revolution – bigger than the industrial revolution.' And I think he is right.

So where are the drivers (and road map) for such a major change? I am not, myself, a fan of the 'bottom up'/'top down' paradigm. I feel it forces an unrealistic triangular structure and an over-simplistic positive/negative charge on a society which is in reality vastly complex, presenting, as it does, a set of competing pyramids (politics, commerce, academia, celebrity etc.) each with conflicting forces driving in diverse directions. ('Top down' is anyway only 'bottom up' fed very imperfectly – thanks, largely, to commercial interference – through the political machine, which is why progress to date has been so painfully slow). I prefer to think in terms of 'centre out' change, by which I refer to an energising of the expertise within every citizen (specially those who consider themselves to be 'experts' and have accepted what they feel are appropriate responsibilities, and rewards, within society); be it food growing, entrepreneurship, social organising, communication, education, academic analysis, professional practice, political power, or any number of other offerings.

Most people, when approached at the right moment, will admit to disquiet at the current situation, along with, often, an acceptance that major change is probably inevitable – and even, when pressed, a willingness to play their part. The problem is that social habits, systems, patterns and forces (laws, prices, ownerships, debts, etc.) make it easier for them (us, actually, if we are honest) to continue with 'business as usual' – for now, anyway. This is why we continue to hear that economic growth is the solution to the current financial/ecological crisis, when actually it is the cause of the problem, and the solution must lie in the internalisation of the externalities which caused the system to fail. Certainly, an economy subservient to society, and a society subservient to the environment does currently look like our safest bet for the permanent safeguarding of ecosystem services (without which, no food) and an equitable civilisation.

But even the most recalcitrant casino-capitalist or denialist will have ideas, and the best ideas often come from the least likely places, so we would do well to keep our eyes and ears wide open, and make sure the debate is as open as possible. After all, Howard himself was a court stenographer (and inveterate inventor, mainly of typewriters, as it happens), not a writer or town planner (a profession he effectively created). This is why, although a twice-times landscape student (and now lecturer), when I first considered my own contribution, I chose initially to create an on-line film, then a blog and an open source mapping project, rather than to write academic papers. I felt I could reach more people more quickly, and promote a wider debate through the internet, than by pouring my heart into the hushed shadows of the Leeds Met library.

So does that mean I feel we do not need books like this? Of course not. Analytical and creative thought is the best tool we have – possibly the only one – so it must be encouraged in all forms, and at all levels. Many of the people we need to engage are themselves academics and professionals – or other opinion and policy-makers who are used to relying on a solid body of nuanced, peer-reviewed research to buttress their opinions and decisions. They will find this in spades (and, as it happens, forks) here. But I hope this book will be widely read in the broader community too. Drawing as it does on the best available thinking and practice around localised food planning and production across the globe, it provides an invaluable practical reference resource for anyone working in this (increasingly small, urban) field. And by the same token, I hope also that those who have contributed to these pages will not rest upon their laurels (and, preferably, harvestable bushes), but will continually seek out ever more fluid and powerful ways to publish and debate – through networks both virtual and real – this absolutely vital topic.

In the last clip of The Urbal Fix, Hilary Benn (then British Secretary of State for the Department of Food and Rural Affairs) says:

The thing about humankind's existence on the earth is; we have shown an astonishing capacity to adapt. And it seems to me that all of the building blocks that we need, in terms of science, technology, inventiveness, resourcefulness, politics and so on, are there. And if we use that, then I think the challenge can be overcome, because in the end we're quite rational. I'm optimistic that we will make the changes that are needed. Why would we not want to?

Why indeed?

Tom Bliss teaches landscape architecture and sustainability part time at Leeds Metropolitan University, is a professional songwriter and musician, and runs the media company turnstone.tv.

'Urbal' is the polar opposite of 'Rurban.' The Urbal Fix can be viewed online at www.urbal.tv.

Chapter 1

Sustainable urban food provisioning: challenges for scientists, policymakers, planners and designers

Johannes S.C. Wiskerke[1] and André Viljoen[2]
[1]Wageningen University, Rural Sociology Group, Hollandseweg 1, 6706 KN Wageningen, the Netherlands; [2]University of Brighton, School of Architecture and Design, Grand Parade, Brighton, BN2 0JY, United Kingdom; han.wiskerke@wur.nl

1.1 Introduction

Since 2007 more people live in urban than in rural areas. It is expected that the current world population of 7 billion will grow to (at least) 9 billion by 2050, of which 6.5 billion will be living in urban areas. Roughly speaking this means that in the next four decades the urban population will grow by 3 billion people. Although this implies that each day an additional 205 thousand urban mouths have to be fed, food is, or at least was until recently, often not an issue on the urban planning, development and/or policy agenda.

For a long time food has been, as Pothukuchi and Kaufman (2000) rightfully state, 'a stranger to the field of urban planning'. This is first of all due to the fact that many urban residents, especially in most Western countries, are taking food for granted:

> *Food arrives on our plates as if by magic, and we rarely stop to wonder how it might have got there. But when you think that every day for a city the size of London, enough food for thirty million meals must be produced, imported, sold, cooked, eaten and disposed of again, and that something similar must happen every day for every city on earth, it is remarkable that those of us living in cities get to eat at all* (Steel, 2008).

A second reason for food being a stranger to urban planning has been the rapid industrialisation of food production and processing and growing geographical distance between the place of production and the place of consumption. Although this radical restructuring of the urban food provision system has resulted in the disconnection of producers and consumers and the territorial disembedding of food (Wiskerke, 2009), 'the 'thereness' of food in cities for the majority of urban residents has not changed' (Pothukuchi and Kaufman, 1999). Third, food is often considered not to be part of the urban public domain. This is rooted in the historical process of urbanisation, which led to certain issues being defined as essentially urban and other issues as essentially rural. Food and agriculture are generally considered to be typical rural issues. This persistent dichotomy in public policy between urban and rural policy has, according to Sonnino (2009a) resulted in three shortcomings in urban food research and policy:

- The study of chains and networks of food provisioning is confined to rural and regional development studies, thereby missing that the city is the space, place and scale where demand is greatest for food products including the 'alternative' products (e.g. organics, local, origin labelled, etc…).

- Urban food security failure is seen as a production failure instead of a distribution and access failure and this has constrained much needed interventions in the realm of urban food security.
- It has promoted the view of food policy as a non-urban strategy, delaying research on the role of food in sustainable urban development as well as on the role of cities as food system innovators.

This book is an attempt to address these shortcomings by providing a wide range of empirical and theoretical explorations about the role of food in urban development, planning and design. In particular, this book shows how food can play a central role in sustainable urban and regional development.

1.2 Planners and designers, strangers in the same room

If food is thought of as a stranger to the field of urban planning, we might also think that planners and designers, by and large, are strangers to one another. Although both engaged with questions related to food systems, when they do meet, language and disciplinary practice often get in the way. This is a typical 'silo problem', as well as touching on a more fundamental schism between the arts and sciences. Yet, as both acknowledge, food systems are complex, overlapping and intertwined. The driving forces and arguments that challenge the status quo may be scientific, quantifiable and policy related, but the consequences are often spatial and physical. Spatial and physical environments need to be designed, and they have a significant impact on quality of life, beyond the satisfaction of hunger. Designers and planners need to speak to and understand each other. If we doubt this, then just consider the impact that twentieth century urban design has had on shaping the car-reliant city, and the abandonment of shared public space within the urban realm. Irrespective of food, if cities are to expand as predicted, and residents are to remain connected to seasonal cycles, the outdoors, and a sensual experience of the world (*urban organoleptics)*, then the spatial implications of new food paradigms need to be considered. The intriguing challenge of imagining the city, in part at least, as a farm and of rethinking 'the countryside', is already engaging architects, designers, landscape architects and artists, and has been well articulated in two seminal exhibitions, the 2007 *Edible City* exhibition hosted by the Netherlands Architecture Institute and the 2009 *Foodprint* exhibition curated by the Den Haag based arts organisation Stroom. The argument has been made that accommodating food related spaces within urban design should be thought of as 'essential infrastructure' (Viljoen and Bohn, 2009) and the publication in 2011 of *Carrot City, creating places for urban agriculture* (Gorgolewski *et al.*) presents many practical examples that can be thought of as components for this new infrastructure.

Notwithstanding the ongoing work of planners and designers, it is still the case that policy makers need convincing if food systems are to change, and the evidence that food systems do need to change is mounting.

1.3 Contemporary urban food provisioning: the dark side of urban dwelling?

Examining how food can play a central role in sustainable urban development requires us to understand why food should be an integral part of any urban development strategy. The importance and significance of food in sustainable urban development encompasses much more than the fact of feeding 2.5 to 3 billion city dwellers more than we have now. The key challenge for the decades to come is how to feed the growing urban world population in a way that can be defined as socially, economically and environmentally sustainable and ethically sound. Living and eating in cities is nowadays inextricably linked to globalised chains of food production, processing and distribution (Murdoch *et al.*, 2000; Steel, 2008). This globalised food system has brought many benefits to the urban population (in particular to urban citizens in the Western world): food is usually constantly available at relatively low prices and many food products have a year round supply. However, these benefits have also come at a cost, or more accurate a series of costs (Lang, 2010; Wiskerke, 2009). These costs constitute the challenges for developing sustainable and ethically sound modes of urban food provisioning.

- Downward pressure on farm family incomes
  In the last decades the mainstream system of food provision has changed from a supply to a demand driven food supply chain. Concomitantly a shift in power has occurred within the food supply chain from primary production to the retail sector, which has become the main outlet for processed as well as fresh food products. With concentration processes in the food processing industry and retail sector, price competition between and within food supply chains has become vigorous (Kirwan *et al.*, forthcoming). The impact of these dynamics on farm family incomes can be illustrated by two aspects. First, the cost prize squeeze on agriculture: a stagnating or even declining Gross Value of Production (GVP) combined with increasing costs for primary production (Van der Ploeg *et al.*, 2000). Second the subordinate economic position of primary producers in the food supply chain, illustrated by the uneven distribution of value added in the food supply chain (Kirwan *et al.*, forthcoming). One of the most dramatic expressions of this is the Dutch pork supply chain. Per euro market value of fresh pork in the supermarket, the supply sector (such as animal feed) has a share of 30%, the pig farmer retains 6% and the slaughterhouse, de-boner, pre-packer and retailer have a respective share of 4%, 8%, 20% and 24% in the sales price (Hoste *et al.*, 2004). These two aspects, embedded in the overall process of primary agriculture being integrated in an agro-industrial complex (Goodman and Redclift, 1990), have accumulated into a 'treadmill effect' (Morgan and Murdoch, 2000), i.e. farmers feeling compelled to continue along the path of increasing production levels and enlarging the scale of operation in order to reduce the costs of production per unit of product and/or per unit of labour.

- Loss of labour, skills, competences and knowledge
  Globally we are witnessing a rapid decline in agricultural labour and concomitantly a loss of skills, competences and knowledge related to food production and on-farm processing of food products. In his farewell address upon his retirement as professor of rural sociology

at the Institute of Social Studies in the Hague, Ben White (2011) warned that the lack of interest in farming among the youth (in the global South as well as the global North) may result in a disappearance of traditional family farms. Given the importance of the peasantry in global food production (Van der Ploeg, 2008), this may seriously threaten global food security.

- Environmental pollution and degradation
  The intensive nature of food production has taken place (and still does) at the expense of contributing to environmental pollution, such as emission of nitrate to groundwater, of ammonia to the air, phosphate saturation of soils and emission of pesticide residues to the air and to ground and surface water, depending the mode of production and production practices (Goodlass et al., 2003; Parris, 1998; Van Eerd and Fong, 1998). Furthermore, due to sectoral specialisation (e.g. on dairy, beef, pork, cereals, apples or potatoes) of farm enterprises and agricultural regions (Van der Ploeg, 2010) combined with a separation of urban and rural nutrient and waste flows (Steel, 2008), nutrient cycles are not closed resulting in enormous spatial differences in nutrient surpluses and shortages (Lang, 2010).

- Waste
  A large part of the food that urban dwellers buy is not consumed but ends up as waste (Steel, 2008). Lang (2010) states that in 2007 approximately 33% of all food purchased was thrown away. In addition to food ending up as waste, the increasing use of processed food products has also resulted in a rapid increase of food packaging that has to be disposed of after food consumption. According to Pothukuchi and Kaufman (1999) food waste (including food packaging) makes up to one third of the urban household, commercial, and institutional wastebasket.

- Fossil fuel dependency
  Food production, processing, distribution, storing and sales have become heavily dependent on fossil fuels and as a result the globalised food system contributes significantly to greenhouse gas emissions and hence to climate change (Carlsson-Kanyama et al., 2003; Carlsson-Kanyama and Gonzalez, 2009; Lang, 2010). Life cycle analyses of Western diets indicate that it takes on average seven calories of fossil fuel energy to produce one calorie of food energy (Heller and Keoleain, 2000). Although different elements of the global food supply chain contribute to this energy inefficiency, the 'heavy fossil fuel users' are pesticides and chemical fertiliser, food processing and packaging, food transport (depending on the means of transport) and cooling (during transport, storage and sales) (Pimentel et al., 2008).

- Climate change
  Climate change is expected to have a large impact on the productive capacity of agricultural regions across the globe (Garnett, 2008). Some regions are expected to benefit from global warming as this will create a more productive environment (longer growing season, sufficient rainfall), while many other regions are likely to suffer from

global warming due to severe droughts and floods and will hence be confronted with food shortages. According to the FAO (2008) 'climate change will affect all four dimensions of food security: food availability, food accessibility, food utilization and food systems stability'. Agriculture is not only affected by climate change, but also contributes to it by emitting greenhouse gasses. This implies that agriculture can also 'contribute to climate change mitigation through reducing greenhouse gas emissions by changing agricultural practices' (FAO, 2008).

- Water stress
  Most of the world's fresh water is used for the production of food. It takes, for instance, 2,400 litres of water to produce one beef-burger of 150 grammes (Lang, 2010). Of the daily amount of water consumed in the UK only 0.2% is consumed as drinking water, while 65% is consumed as water embedded in food. It has been estimated that if the entire world population were to adopt a Western-style diet, 75% more water would be necessary for agriculture and this could imply that the world runs out of fresh water (Wiskerke, 2009).

- Loss of (agro)biodiversity
  Intensification of production has also resulted in a dramatic reduction in agro-biodiversity as food production systems have increasingly been based on a few high productive plant varieties or animal breeds: 'Modern agricultural practices, stemming from the rise of a modern breeding industry and from the Green Revolution, have caused massive genetic erosion, the disappearance of many diverse populations of crops maintained by farmers and adapted to local circumstances' (Visser, 1998).
  In addition to genetic erosion in farm crops and animals, the modernisation of farms and the countryside has also resulted in the loss of non-agricultural biodiversity. Land reconsolidation measures that were implemented to make the countryside suitable for modern farming, have led to the destruction of natural habitats and historico-cultural landscapes (Wiskerke, 2009). Similarly large parts of the Amazon rainforest disappear annually as it is converted into agricultural land for the production of soy or biofuels.

- Decline in organoleptic quality and diversity
  The loss of agro-biodiversity due to the focus on high productive plant varieties and animal breeds combined with the standardisation of food production and processing techniques have also resulted in a dramatic loss of organoleptic quality and diversity of fresh produce (Cayot, 2007; Nosi and Zanni, 2004). Simultaneously the introduction of strict food hygiene rules and regulations combined with upscaling in the food processing industry (and concomitantly the disappearance of small scale processing units and artisanal processing techniques) have further contributed to the loss of specific organoleptic qualities (Kirwan *et al.*, forthcoming). Organoleptic diversity is increasingly becoming an end-of-chain issue, created by the food processing industry by adding colorants and (artificial) flavours to a standardised primary product.

- Agricultural land

  As a result of the growing world population the competition over land use is fierce (Lang, 2010). Agricultural land is needed for the expansion of cities (or construction of new cities), for industrial development and for infrastructure. In many European countries we also witness a growing demand for alternative forms of land use in rural areas, such as land for recreation, nature and rural dwelling (Van Dam *et al.*, 2006). A final competing claim regarding agricultural land use is the competition between food production and the production of biofuels. With an increase in the price of oil, the production of biofuels becomes an economically interesting alternative for food production. With regards to agricultural land, Lang (2010) also rightfully mentions the disproportionate use of land by cities in the global North:

  > London ... *actually uses 48,868,000 global hectares (gha) of land to keep its consumers; that is, 6.63 gha per person living in the city. London's footprint – its land use – thus far exceeds its actual geography. To make London's land use more equitable, that land use ought to drop to 0.16 gha per capita.*

- Soil degradation

  Soil is a vital resource to produce the food that we consume and to produce the feed and fodder for the animals that we consume. Maintaining or improving the productive capacity of the soil requires good management of the soil (Van der Ploeg, 2008). However, many soils across the globe are not well management leading to soil degradation (Lang, 2010). Ye and Van Ranst (2009) simulated the effect of soil degradation, on long-term food security in China using a web-based land evaluation system. Their scenario study predicts that 'food crops may experience a 9% loss in productivity by 2030 if the soil continues to be degraded at the current rate .... Productivity losses will increase to the unbearable level of 30% by 2050 should the soil be degraded at twice the present rate'.

- Public health

  As mentioned in the *European strategy for child and adolescent health and development* of the World Health Organization 'the growing obesity epidemic is one of the most worrying emerging health concerns in many European countries' (WHO, 2005: 5). Obesity rates in Europe range from 10% to 38% of the population. In particular the rapidly rising prevalence of overweight children is alarming (Lobstein *et al.*, 2005). Obesity costs society tens to hundreds of Euros per person per year (Van Baal *et al.*, 2006) and is responsible for approximately 25% of the annual increase in medical spending (Thorpe *et al.*, 2004). Simultaneously, and this seems paradoxical at first sight, malnutrition is also a growing health concern which, like obesity, is more prevalent among the socially and economically disadvantaged sections of the urban population. Surveys in the United States in the 1990s revealed that up to 80% of elderly people in homes were suffering from malnutrition (Pothukuchi and Kaufman, 1999). Research carried out by the charity *Age Concern* in the UK show that 40% of people aged over 65 admitted to a NHS hospital are malnourished, while an additional 20% may develop malnutrition during their hospital stay (Age Concern, 2006). Another expression of malnutrition is the increase in so-called 'food deserts'

(Cummins and Macintyre, 2006; Wrigley, 2002; Wrigley *et al.*, 2002), i.e. impoverished urban neighbourhoods that lack supermarkets and grocery stores, but boast dozens of fast food and snack shops. With supermarkets and grocery stores moving to the outskirts of cities for logistical reasons, ownership of a car becomes more or less a prerequisite to have access to fresh food for home preparation and consumption (Pothukuchi and Kaufman, 1999). If public transport facilities to these outskirts are underdeveloped or simply lacking, then disadvantaged people are deprived of access, or at least easy access to nutritious foodstuffs.

The aforementioned challenges cannot be addressed as single issues, but need to be dealt with as an interrelated and mutually reinforcing set of challenges. All together the prevailing system of urban food provisioning seems to be heading for a catastrophe. This may indeed be true if we stick to business as usual or keep focussing on single challenges, as solutions for one problem may well lead to a worsening of other issues (Lang, 2010). Despite the gloomy picture, there is also reason to be hopeful: across the globe there are many initiatives emerging (initiated by farmers, consumers, retailers or NGOs) that (attempt to) address (several of) the aforementioned challenges (Sonnino, 2009; Wiskerke, 2009). Together these initiatives contribute to the development of a new alternative food geography.

1.4 Addressing the key challenges: the emergence of an integrated territorial food geography

As a response to the multitude of food-related health and sustainability concerns a new food geography (or collection of new food geographies) is forcing itself onto the scientific, political and planning agenda (Watts *et al.*, 2005). This new food geography is grounded in a different logic and incorporating different values to the industrial global food geography. Central to this new geography of food is a sustainability discourse that no longer accepts the externalisation of environmental, social and even economic costs (Morgan *et al.*, 2006).

Driven as it is by new concerns about food quality and safety, nutrition, food security and carbon food prints, the emerging new food geography is developing along three partly interrelated and mutually reinforcing societal axes (see also Figure 1.1):
1. Short producer to consumer food chains – new relations between civil society and the chain of food provision.
2. Re-valuing public food procurement – new relations between the public sector (as buyer and consumer of food) and the chain of food provision.
3. Urban food strategies – the rise of municipalities and city-regions as food policy makers, pointing to new relations between the (local/regional) government and civil society.

1.4.1 Short P2C food chains

The development of short food supply chains (sFSCs) or alternative food networks (AFNs) (Renting *et al.*, 2003) is generally considered to be the first sign of the emerging new food

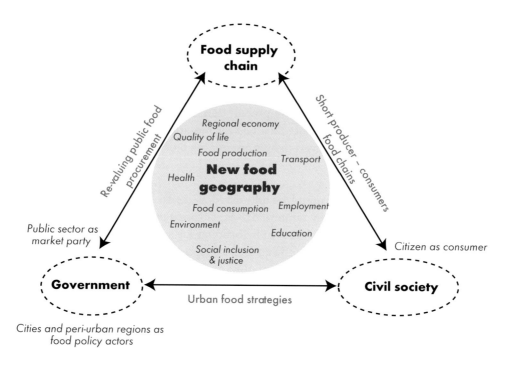

Figure 1.1. The integrated territorial food geography (Wiskerke, 2009).

geography (Watts *et al.*, 2005). In the last decade, sociologists, economists and geographers have provided ample evidence that alternative food networks are steadily gaining ground (Morgan *et al.*, 2006; Sonnino and Marsden, 2006; Watts *et al.*, 2005), as

> concerns about food safety and nutrition are leading many consumers in advanced capitalist countries to exercise more caution in their consumption habits. A growing number of discerning consumers are demanding "quality" products... Moreover, quality is coming to be seen as inherent in more "local" and more "natural" foods... Thus, quality food production systems are being re-embedded in local ecologies (Murdoch *et al.*, 2000: 107).

As a result of increasing scholarly interest a steadily maturing body of socio-spatial food theories, concomitant with a rapid growing number of well elaborated cases, has been developed under the umbrella of the notion of 'alternative food networks' (Renting *et al.*, 2003; Watts *et al.*, 2005). Alternative food networks represent spatially bound relations between consumers and the food market; they are considered to be the outcome of

> the deliberate intention to create alterity (or otherness) in the food system and to produce change in the 'modes of connectivity' between the production and consumption of food, generally through reconnecting food to the social, cultural and environmental context of its production (Kirwan, 2004: 395).

1. Sustainable urban food provisioning

Many case studies have been published about spatially bound food supply chains such as farm shops (Holloway *et al.*, 2007; Ilbery and May, 2005; Ventura and Milone, 2000), farmers' markets (Kirwan, 2004), box schemes (Seyfang, 2006) and community supported agriculture (Hinrichs, 2000). But also the more spatially extended food supply chains with regionally specific food products such as those protected by PDO and PGI regulations (Barham, 2003; De Roest and Menghi, 2000) as well as alternative modes of food production (and processing) such as organics and 'quality' foods are usually considered to be expressions of alternative food networks (Renting *et al.*, 2003). Building on the distinction made by Watts *et al.* (2005) between weak alternative *food* networks (e.g. PDO, PGI, organics, 'quality' foods) and strong alternative food *networks* (i.e. predominantly spatially bound networks of production, processing, distribution and consumption) it are particularly the alternative networks that are part and parcel of the alternative food geography, yet it are very often also alternative foods that are produced, processed, distributed and consumed in many of these alternative networks. Due to the burgeoning literature on AFNs a rich database of AFN cases has been built (see Watts *et al.* (2005) for an excellent literature overview) concomitant with a lively theoretical debate about network dynamics, conventions, producer-consumer linkages, power relations and paradigmatic change (see e.g. Goodman, 2004; Morgan *et al.*, 2006).

4.1.2 Re-valuing public food procurement

Another dimension of the new food geography is the rising awareness of the power of the (semi-) public sector to enhance sustainable food production and consumption patterns by changing its food procurement strategies. Recent studies about hospital catering (Kirwan and Foster, 2006) and school meals (Morgan and Sonnino, 2008) show that the

> *public sector is emerging as a powerful actor in the food chain – one that has the capacity to reconnect producers and consumers through a process of qualification that extends beyond the market and the food products alone. By also acting upon the less visible aspects of the food system – including service, transport, labor, eating practices – procurement policies ... are designing an 'economy of quality' that has the potential to deliver the environmental, economic, and social benefits of sustainable development – in and beyond the food system* (Sonnino, 2009b: 426).

Both developed and developing countries are using school food reform as a tool to develop new supply chains that set a high premium on the use of 'quality' food, which is generally equated with fresh, locally produced food (Morgan and Sonnino, 2008). Much more could be achieved if the power of purchase were to be harnessed across the entire spectrum of the public sector – in hospitals, nursing care homes, colleges, universities, prisons, government offices, and the like. As planners and policy-makers throughout the world become increasingly engaged with this new politics of the public plate, different challenges continue to arise in the realms of infrastructural development, transport, land use and citizens' education, to name just a few. The EU has identified the greening of public procurement as a key policy instrument in achieving more sustainable consumption and production (Council of the European Union, 2006). In the UK, public sector bodies serve around 3.5 million meals per weekday, spending

Sustainable food planning: evolving theory and practice   **27** approximately £2 billion each year with 50% of that amount spent in schools (Strategy Unit, 2007). There has been an ongoing policy tension at national level over the priority for public procurement between the priority of achieving far greater savings in costs, and using it as a policy instrument for environmental and sustainability objectives. This has led to a wider debate over what constitutes 'best value' where longer term sustainability gains are costed-in to the equation (Morgan, 2007). Public sector – hospitals, care homes, schools, universities, prisons and canteens in government buildings – represents a significant part of any national food economy and so their potential in delivering healthy and sustainable communities is large. However, the story of public procurement is largely a tale of untapped potential (Morgan, 2006), despite the enormous potential power in driving behavioural change in economy and society.

1.4.3 Urban food strategies

The third dimension of the new food geography regards urban food strategies, i.e. the active role of cities and metropolitan regions as food policy makers. For decades food policy was considered to be the responsibility of the nation state (or even the supranational state) if we consider the EU's Common Agricultural Policy as a form of food policy), implying that cities are becoming a new actor with regard to food policy design and implementation. Although urban food strategies differ from city to city, the common denominator is the intention to connect and create synergies between different public domains (environment, spatial planning, public health, education, employment, social cohesion, et cetera) that are in one way or the other related to food (Wiskerke, 2009). These connections have not escaped the attention of academics. Indeed, much has been written in the last decade about urban food sustainability, especially in relation to the long-standing issue of food security. In this context, the attention of researchers has focused primarily on the 'production' dimension of the food security problem, as evidenced by the emergence of a large body of literature on urban agriculture and its real and potential contributions to improving the quality of the living environment for urban residents as well as their individual and collective health and well-being (see Halweil and Nierenberg, 2007; Koc et al., 1999; Mougeout, 1999, 2006; Redwood, 2009; Smit et al., 1996). In the context of the 'new food equation' – fashioned by rapid urbanisation, food price hikes, dwindling natural resources and looming climate change (Morgan and Sonnino, 2010), urban agriculture is likely to play an increasingly important role in meeting the most basic food needs of urban residents. However, research insights that have been provided so far are not sufficient to help policy-makers to address the most compelling questions that are emerging in relation to urban food provisioning and land-use planning. In fact, the lack of comprehensive and comparative studies on urban agriculture makes it difficult to understand under what specific conditions this activity can deliver its alleged public health, social, economic and environmental benefits, and to whom (Redwood, 2009: 154). In addition, there is an urgent need to develop new conceptual frameworks that integrate the vast amount of literature on urban food production with studies that focus on the most fundamental dimension of food security: access to food. Socially and economically, increased marketplace activities of corporate chains have displaced local food retailers

(Dixon *et al.*, 2007: 124-125), creating urban 'food deserts' where people – especially low-income people – have little or no access to fresh, nutritious and healthy food (Guy *et al.*, 2004; Wrigley, 2002). In this context, there is an urgent need for integrated urban food policies that create new linkages and new relationships between different stages and actors of the food chain to improve urban food provisioning and to create positive connections between food, the environment, health, the economy, and culture. Pioneering city-governments all over the world have begun to address this need by devising food strategies that aim to calibrate demand and supply. Social scientists from different disciplinary backgrounds have documented the emergence of these strategies (in Rome, London, Toronto and Amsterdam) and their sustainable development potential (see, for example, Donald and Blay-Palmer, 2006; Morgan and Sonnino, 2010; Sonnino, 2009; Wiskerke, 2009). However, so far these pioneering efforts have remained too often confined to their local context, be it place-based or disciplinary-based (Healy and Morgan, forthcoming). In other words, it can be stated that the new research literature on urban food systems has not yet been disseminated to, or integrated with, policy-making communities, especially those that are responsible for forging sustainable production-consumption linkages.

1.5 Towards new modes of sustainable urban food provisioning: challenges for science, policy, planning and design

The new food geography, as visualised in Figure 1.1, not only reflects a territorial (as opposed to global) approach to food production and consumption, but also an integrated conceptualisation of food. In the new food geography, food is more than a commodity or a substance containing calories, vitamins, proteins, nutrients, et cetera that we need to eat in order to survive; it is a product and a process that links environmental pollution (e.g. pesticides and fertiliser), transport (e.g. food miles), environmental degradation (e.g. loss of biodiversity), environmental quality (e.g. productive green spaces in cities) social (in)equality (e.g. differences in access to food), public health (e.g. obesity and malnutrition), employment (e.g. food stores, restaurants, urban farmers), education (e.g. food lessons at primary schools, farmer-to-school initiatives), et cetera. The new food geography can thus be characterised as an integrated territorial geography.

This also has considerable implications for scientific research, policymaking, planning and design: if the sectoral approach to food (i.e. food production = agriculture, food processing = industry, food distribution = transport, and food selling = retail) is making way for an integrated approach to food (i.e. food = environment + public health + social justice + employment + education + quality of life), scientific research, policymaking, planning and design have to change accordingly. Hence, scientists have to cross their disciplinary borders, policymakers have to cross their departmental borders and planners and designers will have to cross sectoral borders. While interdisciplinary work is, also within the domain of food studies, gradually becoming a habit in many scientific communities (MacMynowski, 2007), interdepartmental policymaking and integrated planning and design, especially regarding food, is just starting to take place or, very often, not occurring at all (Wiskerke, 2009).

Moreover, linking *interdisciplinary* food studies to *interdepartmental* food policymaking and *integrated* food systems planning and design is, at present time, the real challenge. It demands that we abandon 'the continuing primacy of academic disciplines in designing and conducting research about food' (Hinrichs, 2008). Hence, the definition of problems, articulation of questions, design of research activities, collection and analysis of data, interpretation of results, formulation of recommendations, implementation of activities, monitoring and evaluation of activities, definition of new problems, et cetera is a collective and interactive process of scientists, policymakers, planners and other stakeholders, albeit with clearly defined roles and responsibilities (resulting from specific capacities and capabilities) for the different actors involved. And finally the territorial approach to food implies the need to abandon scientific universalism and instead depart from the understanding of place-specificity. Challenges with regard to creating more sustainable food production and consumption patterns will differ from place to place and so will the solutions.

1.6 About this book

With this book we aim to contribute to the challenge of linking science, policymaking, planning and design in the realm of an integrated territorial food geography. This book explores the emerging fields of urban food studies, urban food policies and governance and urban food planning and design. It combines accounts of evolving practices around the world, albeit it with an emphasis on the European situation, with theoretical explorations and explanations of these evolving practices. Placing the aforementioned twelve challenges for developing sustainable and ethically sound modes of urban food provisioning in the context of the newly emerging food geography, we have identified four themes for further theoretical and empirical elaboration:

- *Urban food governance.* This theme aims to better understand the roles that cities can play in fashioning sustainable food systems and, second, to document what cities and metropolitan regions are already doing with respect to urban food policies and planning.
- *The integration of health, environment and society.* This theme explores how the integration of health and environment might add to the development of sustainable food policies in modern societies and how this could be addressed without widening inequalities.
- *Urban agriculture.* This theme examines the impact of urban agriculture on city life, with some contributions discussing the effect of urban agriculture on the economy of cities, while others try to map its effects on urban ecology or social structures.
- *Food systems planning and design.* This theme focusses on the infrastructure for food systems in towns and cities and explores how places for food production, processing and trading can be re-introduced in urban settings and the spatial, social and environmental opportunities this represents.

Each thematic part of this book consists of a selection of chapters[1]. Although each chapter is linked to one particular part, many chapters also relate to one or more other thematic parts of the book. For instance, some of the chapters about urban agriculture (Part 3) also discuss issues of urban planning (Part 4) and/or the incorporation of urban agriculture in urban policies (Part 1). Likewise, many chapters about food planning and design (Part 4) are built upon urban agriculture cases (Part 3). This again illustrates the integrated nature of the urban food geography. Some of the chapters in this volume are primarily empirical, describing a specific case in detail, while others combine empirical descriptions and analyses with theoretical reflections. Each thematic part starts with an introductory chapter by the editors of the theme in which they set the thematic scene and position the chapters that follow within that scene.

This introduction to the book is followed by two context-setting chapters. First a chapter by Carolyn Steel in which she discusses the relationship between food and cities and argues why we need to rethink the contemporary relationship between food and cities. This is followed by a chapter by Joe Nasr and June Komisar, in which they discuss the integration of food and agriculture into urban planning and design practices, thereby mainly drawing on the North American experience.

In this introductory chapter we have tried to make a case for the fact that new paradigms for urban and regional planning capable of supporting sustainable and equitable food systems are urgently needed. This edited volume addresses this urgent need. By working at a range of scales and with a variety of practical and theoretical models, this book reviews and elaborates definitions of sustainable food systems, and begins to define ways of achieving them.

References

Age Concern, 2006. Hungry to be heard: The scandal of malnourished older people in hospital. Age Concern England, London, UK, 26 pp.

Barham, E., 2003. Translating terroir: the global challenge of French AOC labeling. Journal of Rural Studies 19: 127-138.

Carlsson-Kanyama, A., Ekström, M.P. and Shanahan, H., 2003. Food and life cycle energy inputs: Consequences of diet and ways to increase efficiency. Ecological Economics 44: 293-307.

Carlsson-Kanyama, A. and González, A.D., 2009. Potential contributions of food consumption patterns to climate change. American Journal of Clinical Nutrition 89: 1704S-709S.

Cayot, N., 2007. Sensory quality of traditional foods. Food Chemistry 102: 445-453.

Council of the European Union, 2006. Review of the EU Sustainable Development Strategy (EU SDS) – Renewed Strategy 10917/06. Council of the European Union, Brussels, Belgium.

---

[1] Each chapter has gone through a double peer-review process. Draft chapters have been reviewed by the theme editors. Revised versions have been reviewed by one of the theme editors as well as by the author(s) of another chapter within the same thematic part.

Cummins, S. and Macintyre, S., 2006. Food environments and obesity: Neighbourhood or nation? International Journal of Epidemiology 35: 100-104.

De Roest, K. and Menghi, A., 2000. Reconsidering 'traditional' food: The case of Parmigiano Reggiano cheese. Sociologia Ruralis 40: 439-451.

Dixon, J., Omwega, A.M., Friel, S., Burns, C., Donati, K. and Carlisle, R., 2007. The health equity dimensions of urban food systems. Journal of Urban Health 84: 118-129.

Donald, B. and Blay-Palmer, A., 2006. The urban creative-food economy: producing food for the urban elite or social inclusion opportunity? Environment and Planning A 38: 1910-1920.

Garnett, T., 2008. Cooking up a storm: Food, greenhouse gas emissions and our changing climate. Food Climate Research Network, Centre for Environmental Strategy, University of Surrey, UK, 156 pp.

(FAO) Food and Agriculture Organization, 2008. Climate change and food security: A framework document. Food and Agriculture Organization of the United Nations, Rome, Italy, 107 pp.

Goodlass, G., Halberg, N. and Verschuur, G., 2003. Input output accounting systems in the European community: An appraisal of their usefulness in raising awareness of environmental problems. European Journal of Agronomy 20: 17-24.

Goodman, D., 2004. Rural Europe redux? Reflections on alternative agro-food networks and paradigm change. Sociologia Ruralis 44: 3-16.

Goodman, D. and Redclift, M., 1990. The farm crisis and the food system: some reflections on the new agenda. In: Marsden T. and Little, J. (Eds.) Political, Social and Economic Perspectives on the International Food System. Avebury, Aldershot, UK, pp. 19-35.

Gorgolewski, M., Komisar, J. and Nasr J., 2011. Carrot city: Creating places for urban agriculture. Monacelli Press, New York, NY, USA, 240 pp.

Guy, C., Clarke, G. and Eyre, H., 2004. Food retail change and the growth of food deserts: a case study of Cardiff. International Journal of Retail and Distribution Management 32: 72-88.

Halweil, B. and Nierenberg, D., 2007. Farming the cities. In: Starke, L. (Ed.) The World Watch Institute, State of the World 2007: Our Urban Future. The World Watch Institute Washington D.C., USA, pp. 48-63.

Heller, M.C. and Keoleian, G.A., 2000. Life Cycle-Based Sustainability Indicators for Assessment of the U.S. Food System. Center for Sustainable Systems, University of Michigan, Ann Arbor, MI, USA, 61 pp.

Hinrichs, C.C., 2000. Embeddedness and local food systems: notes on two types of direct agricultural market. Journal of Rural Studies 16: 295-303.

Hinrichs, C.C., 2008. Interdisciplinarity and boundary work: challenges and opportunities for agrifood studies. Agriculture and human values 25: 209-213.

Holloway, L., Kneafsey, M., Venn, L., Cox, R., Dowler, E. and Tuomainen, H., 2007. Possible food economies: a methodological framework for exploring food production–consumption relationships. Sociologia Ruralis 47: 1-19.

Hoste, R., Bondt, N. and Ingenbleek, P.T.M., 2004, Visie op de varkenskolom, Wetenschapswinkel Wageningen UR Rapport 207, Wageningen, the Netherlands, 83 pp.

Ilbery, B. and Maye, D., 2005. Food supply chains and sustainability: evidence from specialist food producers in the Scottish-English borders. Land Use Policy 22: 331–344.

Kirwan, J., 2004. Alternative strategies in the UK agro-food system: interrogating the alterity of farmers' markets. Sociologia Ruralis 44: 395-415.

Kirwan, J. and Foster, C., 2006. Public sector food procurement through partnerships: the Cornwall Food Programme. In: Roep, D. and Wiskerke, J.S.C. (Eds.) Nourishing Networks: fourteen lessons about creating sustainable food supply chains. Reed Business Information, Doetinchem, the Netherlands, pp. 155-164.

Kirwan, J., Slee, R.W., Foster, C. and Wiskerke, J.S.C., (forthcoming). Dynamics and diversity in food supply chains across Europe. In: Wiskerke, J.S.C., Van Huylenbroeck, G. and Kirwan, J. (Eds) Sustaining food supply chains: grounded perspectives on the dynamics and impact of new modes of food provision. Ashgate, London, UK.

Koc, M., MacRae, R., Mougeot, L.J.A. and Welsh, J., 1999. For hunger-proof cities: sustainable urban food systems. International Development Research Centre, Ottawa, Canada, 294 pp.

Lang, T., 2010. Crisis? What Crisis? The Normality of the Current Food Crisis. Journal of Agrarian Change 10: 87-97.

Lobstein, T., Rigby, N. and Leach, R., 2005. Obesity in Europe – Briefing Paper for the EU Platform on Diet, Physical Activity and Health. International Obesity Task Force, London, UK.

MacMynowski, D.P., 2007. Pausing at the brink of interdisciplinarity: power and knowledge at the meeting of social and biophysical sciences. Ecology and Society 12: 20

Morgan, K., 2006. School food and the public domain: the politics of the public plate. Political Quarterly 77: 379-387.

Morgan, K., 2007. Greening the public realm: sustainable food chains and the public plate. BRASS working paper series. 43. The Centre for Business Relationships, Accountability, Sustainability and Society (BRASS), Cardiff University, Cardiff, UK.

Morgan, K. and Murdoch, J., 2000. Organic vs. conventional agriculture: knowledge, power and innovation in the food chain. Geoforum 31: 159-173

Morgan, K., Marsden, T.K. and Murdoch, J., 2006. Worlds of food: Place, power, and provenance in the food chain. Oxford University Press, Oxford, UK, 225 pp.

Morgan, K. and Sonnino, R., 2008. The school food revolution: public food and the challenge of sustainable development. Earthscan, London, UK, 256 pp.

Morgan, K.J. and Sonnino, R., 2010. The urban foodscape: world cities and the new food equation. Cambridge Journal of Regions, Economy and Society 3: 209-224.

Mougeot, L.J.A., 1999. For self-reliant cities: urban food production in a globalizing south. In: Koc, M., MacRae, R., Mougeot, L.J.A. and Welsh, J. (Eds.) For hunger-proof cities: sustainable urban food systems. International Development Research Centre, Ottawa, Canada, 294 pp.

Mougeot, L.J.A., 2006. Growing better cities: urban agriculture for sustainable development. International Development Research Centre, Ottawa, Canada, 106 pp.

Murdoch, J., Marsden, T.K. and Banks, J., 2000. Quality, nature and embeddedness: some theoretical considerations in the context of the food sector. Economic Geography 76: 107-125.

Nosi, C. and Zanni, L., 2004. Moving from "typical products" to "food-related services": the Slow Food case as a new business paradigm. British Food Journal 106: 779-792.

Parris, K., 1998. Agricultural nutrient balances as agri-environmental indicators: An OECD perspective. Environmental Pollution 102 – Supplement 1: 219-225.

Pimentel, D., Williamson, S., Alexander, C.E., Gonzalez-Pagan, O., Kontak, C. and Mulkey, S.E., 2008. Reducing energy inputs in the US food system. Human Ecology 36: 459-471.

Pothukuchi, K. and Kaufman, J.L., 1999. Placing the food system on the urban agenda: the role of municipal institutions in food systems planning. Agriculture and Human Values 16: 213-224.

Pothukuchi, K. and Kaufman, J.L., 2000. The food system: A stranger to the planning field. Journal of the American Planning Association 66: 113-124.

Redwood, M., 2009. Agriculture in urban planning. Generating livelihoods and food security. Earthscan, London, UK, 272 pp.

Renting, H., Marsden, T.K. and Banks, J., 2003. Understanding alternative food networks: exploring the role of short food supply chains in rural development. Environment and Planning A 35: 393-411.

Seyfang, G., 2006. Ecological citizenship and sustainable consumption: Examining local organic food networks. Journal of Rural Studies 22: 383-395.

Smit, J., Ratta, A. and Nasr, J., 1996. Urban agriculture: food, jobs and sustainable cities. UNDP, New York, NY, USA, 328 pp.

Sonnino, R., 2009a. Feeding the city: Towards a new research and planning agenda. International Planning Studies 14: 425-435.

Sonnino, R., 2009b. Quality food, public procurement and sustainable development: the school meal revolution in Rome. Environment and Planning A 41: 425-440.

Sonnino, R. and Marsden, T.K., 2006. Beyond the divide: rethinking relationships between alternative and conventional food networks in Europe. Journal of Economic Geography 6: 181-199.

Steel, C., 2008. Hungry city: How food shapes our lives. Random House, London, UK, 400 pp.

Strategy Unit, 2007. Food: an analysis of the issues. Cabinet Office, London, UK, 113 pp.

Thorpe, K.E., Florence, C.S., Howard, D.H. and Joski, P., 2004. The impact of obesity on rising medical spending. Health Affairs W4: 480-486.

Van Baal, P.H.M., Heijink, R., Hoogenveen, R.T. and Van Polder, J.J., 2006. Zorgkosten van Ongezond Gedrag, Zorg voor Euro's 3. Rijksinstituut voor Volksgezondheid en Milieu, Bilthoven, the Netherlands, 62 pp.

Van Dam, F., De Groot, C. and Verwest, F., 2006. Krimp en Ruimte: bevolkingsafname, ruimtelijke gevolgen en beleid. NAI Publishers, Rotterdam, the Netherlands, 109 pp.

Van der Ploeg, J.D., 2008. The new peasantries: Struggles for autonomy and sustainability in an era of empire and globalization. Earthscan, London, UK, 352 pp.

Van der Ploeg, J.D., 2010. The food crisis, industrialized farming and the imperial regime. Journal of Agrarian Change 10: 98-106.

Van der Ploeg, J.D., Renting, H., Brunori, G., Knickel, K., Mannion, J., Marsden, T., De Roest, K., Sevilla-Guzmán, E. and Ventura, F., 2000. Rural development: from practices and policies towards theory. Sociologia Ruralis 40: 391-408.

Van Eerdt, M.M. and Fong, P.K.N., 1998. The monitoring of nitrogen surpluses from agriculture. Environmental Pollution 102 – Supplement 1: 227-233.

Ventura, F. and Milone, P., 2000. Theory and practice of multi-product farms: farm butcheries in Umbria. Sociologia Ruralis 40: 452-465.

Viljoen, A.M. and Bohn, K., 2009. Continuous Productive Urban Landscape (CPUL) Essential Infrastructure and Edible Ornament. Open House International 34: 50-60.

Visser, B., 1998. Effects of biotechnology on agro-biodiversity. Biotechnology and Development Monitor 35: 2-7.

Watts, D.C.H., Ilbery B. and Maye, D., 2005. Making reconnections in agro-food geography: alternative systems of food provision. Progress in Human Geography 29: 22-40.

White, B., 2011. Who will own the countryside? Dispossession, rural youth and the future of farming. Valedictory Address delivered on 13 October 2011 on the occasion of the 59th Dies Natalis of the International Institute of Social Studies, The Hague, the Netherlands.

Wiskerke, J.S.C., 2009. On place lost and places regained: reflections on the alternative food geography and sustainable regional development. International Planning Studies 14: 361-379.

WHO (World Health Organization), 2005. European Strategy for Child and Adolescent Health and Development. World Health Organization – Regional Office for Europe, Copenhagen, Denmark.

Wrigley, N., 2002, Food Deserts' in British cities: Policy context and research priorities. Urban Studies 39: 2029-2040.

Wrigley, N., Warm, D., Margetts, B. and Whelan, A., 2002. Assessing the impact of improved retail access on diet in a 'Food Desert': A preliminary report. Urban Studies 39: 2061-2082.

Ye, L. and Van Ranst, E, 2009. Production scenarios and the effect of soil degradation on long-term food security in China. Global Environmental Change 19: 464-481.

Chapter 2

Sitopia – harnessing the power of food

Carolyn Steel
Kilburn Nightingale Architects, 26 Harrison Street, London WC1H 8JW, UK;
cs@kilburnnightingale.com

Abstract

The relationship between city and country is fundamental to civilisation, and set to dominate socio-political, economic and ecological agendas for years to come. The idea that cities can endlessly extract the means of their subsistence from the natural world is the product of our predominantly rural past. For architects and planners, the question of how to design and build must now be weighed against conditions in which finite global resources will play an ever greater role. If we are to create an equitable and sustainable future, new dwelling models must be sought. The scale and complexity of the task demands a broadening of the architectural and planning discourse to embrace fields not traditionally considered relevant. New tools are needed, both in order to comprehend the issues at hand, and to make effective use of the creative capacity of spatial imagination. Food is one such tool. The primary agent of urban-rural relationships, food is embedded in our lives at every level. Our most vital shared commodity, food is embedded in our lives socially, physically, and symbolically. Using food as a lens, we can interpret human civilisations as various forms of sitopia: 'food-places' whose diverse characteristics share a common thread. By embracing food in this way, we can use it as a practical and conceptual tool with which to think, act, and create. Food already shapes our world. By harnessing its power, we can shape the world better.

Keywords: cities, agriculture, networks, design, utopia

2.1 The urban paradox

Our need to eat shapes our daily existence; yet for those of us living in cities, it also creates a paradox. Our urban lifestyles depend upon sustenance from elsewhere: a place we persist in calling 'the countryside', although the images conjured by such a term often bear little resemblance to the realities of modern food production.

Of all the resources needed to sustain a city, none is more vital than food. Before industrialisation, this was obvious, since the physical challenges of producing and transporting it made food the dominant priority of every urban authority. No city was ever built without first considering where its food was to come from, and perishables such as fruit and vegetables were grown as locally as possible, often in the city fringes. Meat and fish were consumed seasonally, the excess preserved by salting, drying or pickling. Nothing was wasted: kitchen scraps were fed to pigs and chickens, and human and animal waste was collected and spread on the outfields as manure. In the pre-industrial city, the sights and smells of food were inescapable (Steel, 2008).

Things are very different now. The advent of railways in the 19th century emancipated cities from geography, making it possible for the first time to build them any size, shape, and place. As cities sprawled, food systems industrialised, and the two began to grow apart. While architects and planners dreamt of cities free of mess and smell, the nascent food industry sought ever-greater 'efficiencies' in the pursuit of profit. As the relative costs of transport shrank, food production was increasingly located, not close to cities, but far afield, wherever natural resources and cheap labour could be most readily exploited.

Our very concept of a city, inherited from a distant, predominantly rural past, assumes that the means of supporting urbanity can be endlessly extracted from the natural world. But can it? So few people once lived in cities (just three percent in 1800) that their ecological impact was limited. Today, with over half the global population living in cities and the number expected to double by 2050, the opposite is true. Food riots are increasingly common as failed harvests, soaring oil prices, bio-fuels and commodity speculation push food prices to record levels. Despite our technical ability, we are no closer to solving the urban paradox than were our ancient ancestors.

Current trends suggest that nothing short of a complete review of our way of life is required if we are to avoid ecological calamity. Yet our social, political and economic systems are set against such a change. Like the bones of an ancient body, they have become brittle. Faced with sudden shocks, they have two responses: to carry on as normal, or break. In rapidly evolving and uncertain times, this lack of flexibility arguably poses a greater threat to our future than do dwindling physical resources.

The global response to the food and banking crises of 2008 illustrates the problem. Despite the obvious flaws in an economic system that relies on unsustainable, permanent 'growth', the international response to its collapse has been to restore business as usual. Proposals to start a 'Green Revolution' in Africa are similarly myopic. The increased crop yields achieved during the early years of the original 'Green Revolution' in the Indian Punjab were followed by declining harvests, loss of soil fertility, water stress and multiple farmer suicides; yet these tragic outcomes are tacitly ignored (Shiva, 1991).

2.2 Sitopia

In order to cope with complex, interconnected, global problems, we need new ways of thinking and acting. We need instruments better attuned to the conditions of modernity, better able to respond to uncertainty, complex enough to reflect reality, yet simple enough to be grasped. But where are we to find such tools?

Food provides an answer. Our landscapes and cities were shaped by food. Our daily routines revolve around it, our politics and economies are driven by it, our identities are inseparable from it, our survival depends on it. What better tool, then, with which to shape the world?

Of course, food already shapes the world; it just does so in ways of which we are largely unaware. We might notice its most obvious effects – on our waistlines after Christmas, for instance – but how many of us notice the way in which food shapes our political and economic systems, the landscapes we create, our use of public and private space, or the social bonds we form? It takes a special kind of 'seeing' to notice food's all-pervasive influence; yet seeing through food is precisely how we can use it as a tool.

Our greatest need in response to global threats is not more technology, money, or physical resources, but philosophy. We need a vision of the life we are trying to create, save or adapt. Only then will we be able to act effectively; for only then will we see climate change and peak oil for what they are: physical threats to our material existence, not existential ones that threaten life itself.

Creating such vision is where food can help us most. Since we must all eat, the question of *how* we should eat approximates to that of how we should live. Through food, we can judge whether or not the life we lead is 'good' in every sense. By acting upon such judgements, we can start to build sitopia (food-place): a society in which food-based values are commonly shared and practiced (Steel, 2008).

2.3 The value of food

What, exactly, is a food-based value? Nothing less than the acknowledgement that, in a world of temporally finite resources, our most vital common necessity is a good measure by which to live. Societies around the world differ in the habits, beliefs and skills they apply to food, and the values they attach to it. Significantly, however, every pre-industrial society has placed food at its social and spatial heart, while every post-industrial one has placed food at the periphery. Why is this? The short answer is that food represents power; a fact all too evident in pre-industrial societies, yet relatively obscure in ours. Lulled by the prospect of year-round, 'cheap' food, we have allowed our most precious common resource to slip beyond our grasp.

The results, from a social and ecological point of view, have been disastrous. Food and agriculture today account for one third of global greenhouse gas emissions. Thirteen million hectares of forest are lost each year to logging and agriculture, and an estimated 24% of the world's arable land is already degraded (FAO, 2011; IFPR1, 2011). Eighty-five percent of global fish-stocks are either depleted or fully exploited (FAO, 2010). Seventy percent of the world's freshwater is used for farming, while 1.2 billion people worldwide live with water scarcity (FAO, 2007). Each calorie of food we consume in the West takes an average of ten to produce, yet one half of the food produced in the USA is wasted (Jones and Andy, 2001;

Stuart, 2009). A billion people worldwide go hungry, while a further billion are overweight, one third of those obese.

The costs of industrial food are great, yet almost all are excluded from the price we pay for it in shops. Factor in the externalities, however, and it becomes clear that 'cheap food' is an illusion, a ruinous one at that. One recent estimate by the Centre for Science and the Environment in India put the true cost of a burger made from beef raised on recently cleared forest land to be US \$200 (Patel, 2010).

Our failure to value food properly affects more than the cost of a burger: it destabilises our value systems as a whole. By treating food as though it were cheap, we live in denial of our lives' true cost, which in turn blinds us to the value of life itself. We strive for 'prosperity' and 'growth', ignoring that the foundations upon which such concepts are based are unsound.

2.4 A mundane life

If our future society is to be well-founded, we must seek alternative ways of defining terms such as 'prosperity' and 'growth'. Food can help us here, too. The relationship between food and sun is perhaps too obvious to pay it much heed in the modern world, yet it is fundamental to our innate senses of space, time and wellbeing. It follows that a good life must be lived according to an earthly rhythm. Happily, the sun's enduring energy means that, in human terms at least, the earth's resources are not *absolutely* finite. Our challenge is not therefore that of meting out dwindling resources, but of managing renewable ones; a feat well within our reach.

Living according to the seasons, as we must, links the everyday rhythm of our lives to the universal: a dual meaning captured whenever we speak of 'mundane' existence. Although we use the word dismissively, 'mundane' is defined in the Oxford English Dictionary as 'of this world, worldly, of the universe, cosmic, dull, routine' (Skyes, 1983). By paying greater attention to the mundane, we can regain our sense of place in the world, and gain insight into better ways of living.

Thinking and acting through food orientates us and shows us our boundaries. Take the example of a shared meal. Table manners are rituals inherited from ancient times, designed to preserve a sense of fairness and strengthen social bonds. How one ate with others was of supreme importance in every past culture: in ancient Athens, for example, to be greedy at table was taken as a sure sign of political untrustworthiness (Davidson, 1995). Even today, we intuitively understand the importance of table manners, and behave accordingly whenever we eat formally with others. In a globalised world, the sharing of food takes on new significance. We must extend our table manners to those whom we have never met, and to species whose existence we can only guess at. Sharing food in this way becomes a means of living well, both directly and metaphorically.

2.5 The urban-rural divide

Sitopia has many uses, but perhaps its most obvious is as a means of addressing the urban paradox. Ten thousand years ago, when urbanity and agriculture first co-evolved, the relationship between cities and their hinterlands was clear. Early city-states were communities in which city and country were socially, physically and conceptually bound together. In contrast, most of us today live hundreds, if not thousands of miles from the sources of our sustenance, know little about the industry that feeds us, and exert little or no control over it. Cities, whose growth was once limited by geography, currently gain 1.3 million new rural migrants every week. A billion people live in informal shanty-towns and slums with no access to fresh water, power, or sanitation.

Opinion is divided as to whether or not this headlong rush to cities is a good thing. Commentators including Stewart Brand and Edward Glaeser argue that cities are the 'greenest' way to live, and that informal shanty-towns are fertile breeding-grounds for the sort of human ingenuity that will lift their new inhabitants out of poverty (Brand, 2010; Glaeser, 2011). Others, including Raj Patel and Jules Pretty, find the mass abandonment of rural life, with its concomitant loss of local knowledge, skills, communities and sovereignty, tragic (Patel, 2007; Pretty, 2002). Either way, what is clear is that the relationship between non-food-producing communities (aka cities) and food-producing ones (aka countryside) is dangerously out of kilter, and that billions of people in urban and rural areas alike lead lives of crushing poverty and deprivation.

What lies at the root of the current mass exodus to cities? The pursuit of social opportunity, for sure, but also the fact that small and medium-scale farmers are being driven off the land by the spread of global agri-business, a system in which only the largest, most 'efficient' farms can survive. Cities, and the industrial food systems that have evolved to feed them, are making rural life untenable. In order to create equitable, stable societies worldwide, it follows that we must address the power structures governing food. Only then can we hope to restore the balance between city and country, the relationship upon which civilisation depends.

2.6 Social food networks

Connecting producers to consumers is where all food-based power lies, and as Tim Lang, Michael Heasman and others have shown, that power is increasingly consolidated in fewer, corporate hands (Lang and Heasman, 2004). Represented as a diagram, the global food system now resembles a tree, in which many roots (producers) channel nutrients through a narrow trunk (supermarkets) to feed many branches (consumers) (Grievink, 2003). In such a system, the trunk exerts a stranglehold over the entire food chain. But what if consumers were to forge direct relationships with those who grew their food? Entirely different systems would then emerge: complex networks of flexible, personal, equable connections.

Park Slope Food Coop in Brooklyn NYC is one such system. Established in 1973, the Coop has 14,000 members, each of whom works a few hours' shift every month in exchange for between 20-40% savings on their groceries. The Coop maintains long-term relationships with 40 small-scale local farms within a 100-mile radius of the city, giving farmers a level of security rare in the modern food industry. The Coop also has a strong ethical code, enforced through monthly members' meetings. Its members are effectively 'co-producers': a term coined by Slow Food founder Carlo Petrini to describe knowledgeable consumers who actively promote ethical food networks through their actions (Petrini, 2007).

Democratising food trade in this way opens up new possibilities for rural communities. Once the monopolistic corporate 'trunk' is bypassed, the social potential inherent in urban food systems can reassert itself. Instead of a tree, the system can resemble a complex series of interconnections, in which all the tree's 'branches' bend down to touch its 'roots'. That is what a democratic food system looks like. In such a case, the trunk represents nothing more than regulatory order: the support role once played by urban authorities.

Opportunities to create complex food systems are greatly increased by modern communication technology. For the first time, the advantages of social proximity are not limited to those living in cities. Access to markets, news and knowledge are now available online. In industrial nations, farmers are now able to connect to consumers via the internet in order to supply them with boxes of organic vegetables and so on. In Kenya, Masai cattle-ranchers can share information about market trends before deciding when and where to sell their animals. Such communicative networks are vital, because they loosen the corporate stranglehold over markets and food distribution.

The critical outcome of establishing more equitable food networks is that is humanises the entire food chain. The social and ecological benefits of having *more*, rather than fewer, people working in food – particularly as producers on land and sea – are profound. In contrast to the slash-and-burn approach of corporate agri-food, small and medium-scale farmers and fishermen take the long-term view of food production. As Raj Patel, Jules Pretty and others have pointed out, traditional producers have inherited knowledge of specific territory, as well as family continuity in mind, so stewardship of land and sea is inherent in what they do (Patel, 2007; Pretty, 2002). With good access to markets, life as a small producer can also be a rich and rewarding one. But in order to make such ways of life possible, the rest of us must invest food with its proper value, by becoming co-producers.

Acting together, city dwellers can develop what Julie Brown, founding director of London organic box scheme Growing Communities, calls 'community-led trade'. In building up her business, Brown realised that her greatest challenge was not that of finding farmers willing to supply Londoners with organic produce, but of finding Londoners who cared enough about food to pay the necessary premium. In order to develop an ethical, sustainable food system, she reasoned, people need to understand the full effects of their diet. Only then will they be prepared to adjust the way they eat.

Brown's efforts shifted away from the specifics of her business to an educational programme aimed at changing people's perceptions and food choices. Her 'Food Zones' model, still in evolution, is one outcome of this work (Brown, 2011). The model, inspired by Johann von Thünen's land use theory, shows how Londoners could feed themselves from a global patchwork of farms, working out from the city itself (where Brown has found it only makes sense to grow salad leaves), to the UK, Europe, and the rest of the world (no-one, even Julie Brown, expects Londoners to give up coffee or bananas).

2.7 The problem of scale

Pioneering models such as Park Slope Coop and Growing Communities demonstrate food's potential to shape a better world. But we are still far from understanding the true scope of food's transformative power. Placing food back at the heart of society implies a major cultural shift. It suggests new political and economic structures, new planning models, a new social order: away from neo-liberalism towards community-based trade; away from unchecked urbanisation towards urban-rural regionalism.

As Michael Pollan and others have noted, a new 'Food Movement' is underway (Pollan, 2010). Inspirational food projects abound; yet, compared to the juggernaut of global agri-business, their combined effect remains puny indeed. Like separate currents in the ocean, they flow more or less independently, which is why it is possible in the UK for example, to see the recent growth of farmers' markets, artisanal food producers and TV cookery programmes (aka 'food porn'), accompanied by a continued decline in cookery skills, rising childhood obesity levels, and dairy farms going bankrupt. The two sets of phenomena are in direct opposition, yet are deeply connected.

For those leading the Food Movement, the question of how to scale up good practice is uppermost. Many issues stand in the way: not least, the inherent un-scalability of food itself. Since food is linked to life in myriad ways, any attempt to upscale its production, transportation, trading or consumption, leads inevitably to radical shifts in its meanings and effects. Mixed farming, for example, has much to recommend it when practiced at a small scale: synergies through local waste management and crop rotation, resilience through variety, and a mixture of jobs that make farm-work more enjoyable. However, when scaled up to industrial proportions, mixed farming raises serious issues over pollution, worker conditions and animal welfare. Crucially, the difference between the two is the degree of skilled, concerned, and caring human involvement: the very thing that industrial 'scaling up' seeks to eradicate in the name of 'efficiency'.

2.8 Sitopian evolution

A truly food-based society would look radically different from ours; so much so, that getting there will require a revolution. The scale of the task is immense, making it tempting to shy away from it altogether. Yet face it we must. This is the sort of problem that sitopia can help us address. Food transcends scale, so by thinking and acting through it, we give ourselves permission to think 'big' (without, of course, forgetting that 'small is beautiful').

'Revolution' is a troublesome term, suggesting violence and destruction; yet it can also be gentle, as the 1989 Czech 'Velvet Revolution' demonstrated. The key to such peaceful transition is preparation: the gradual shifting of conditions to the point that, when change comes, people, and infrastructures, are ready. This is how sitopia works. Sitopia is not utopia; it is not an ideal place, but rather a transitional tool that works *towards* an ideal. It is an evolutionary, not a revolutionary movement: always in the process of becoming, never complete. In sitopia, small gains are significant, because they help create conditions in which life lived according to food-based values can flourish.

Global agri-food will doubtless be with us for years to come. That need not, however, discourage us. On the contrary, we must prepare the way for a time when the agri-food juggernaut finally hits the buffers, as it surely will. By acting now, we can make sure that the transition from a wasteful, unethical, inequitable food system to an ethical, sustainable, sitopian one is velvety. Many paths of action are open to us: informing ourselves about food, teaching others, designing and planning with food in mind, demanding action from politicians, becoming co-producers, growing our own, cooking more, eating better, and so on. We may not be able to tackle the juggernaut overnight, but by working together, we can hasten its end.

According to Plato, the art of asking proper questions was the purpose of civilisation. Anyone not directly engaged in philosophy was simply making up the numbers. Ours is a different time and society, but our fundamental human question remains the same. Our greatest task is not to confront climate change, peak oil, population or poverty, but life itself. Since food is life, it can help us get closer to what that means.

Sitopia, in essence, is a vision of life that embraces the mundane in its truest sense. It shows how we can enjoy real prosperity and lasting growth: through the accumulation, not of material goods, but of richness through wisdom, experience, diversity and imagination. We can celebrate the changing seasons, the uniqueness of place, the pleasure of company, respect for others, the taste of good food, the smell of earth, the joy of learning and teaching, of working with our hands, of appreciating, loving and sharing. Such simple pleasures cost us little, other than the willingness to seek a life lived well.

If that sounds idealistic, it is because sitopia, fully realised, is utopia. The difference between them is that sitopia can be bad as well as good. The way we shape the world through food is up to us.

References

Brand, S., 2010. Whole earth discipline: why dense cities, nuclear power, transgenic crops, restored wildlands, radical science, and geoengineering are necessary. Atlantic Books, London, UK, 323 pp.

Brown, J., 2011. Growing communities food zone manifesto. Available at http://www.growingcommunities.org/about-us/food-zone/manifesto/.

Davidson, J., 1995. Opsophagia: revolutionary eating at Athens. In: Wilkins, J., Harvey, D. and Dobson, M. (Eds.) Food in Antiquity. University of Exeter Press, Exeter, London, UK, pp. 204-213.

FAO (Food and Agriculture Organization of the United Nations), 2007. Coping with water scarcity. p. 4. Available at http://www.fao.org/nr/water/docs/escarcity.pdf.

FAO, 2010. The state of world fisheries and aquaculture. p. 8. Available at http://www.fao.org/docrep/013/i1820e/i1820e00.htm.

FAO, 2011. The state of the world's forests. Available at http://www.fao.org/docrep/013/i2000e/i2000e00.htm.

Glaeser, E., 2011. Triumph of the city: how our greatest invention makes us richer, smarter, greener, healthier and happier. MacMillan, London, UK, 270 pp.

Grievink, J-W., 2003. The changing face of the global food industry. OECD Food Conference Proceedings, The Hague, the Netherlands.

IFPRI (International Food Policy Research Institute), 2011. The economics of desertification, land degradation, and drought. Discussion Paper, p. 57. Available at http://www.ifpri.org/publication/economics-desertification-land-degradation-and-drought.

Jones, A., 2001. Eating oil: food supply in a changing climate. Sustain and Elm Farm Research Centre, London, UK, 5 pp.

Lang, T. and Heasman, M., 2004. Food wars: the global battle for mouths, minds and markets. Earthscan, London, UK, 307 pp.

Patel, R., 2007. Stuffed and starved: markets, power, and the hidden battle for the world food system. Portobello, London, UK, 319 pp.

Patel, R., 2010. The value of nothing: how to reshape market society and redefine democracy. Portobello, London, UK, p. 44.

Petrini, C., 2007. Slow food nation: why our food should be good, clean and fair. Rizzoli Ex Libris, New York, NY, USA, 304 pp.

Pollan, M., 2010. The food movement, rising. New York Review of Books, June 10, 2010. Available at http://www.nybooks.com/articles/archives/2010/jun/10/food-movement-rising/.

Pretty, J., 2002. Agri-culture: reconnecting people, land and nature. Earthscan, London, UK, 264 pp.

Shiva, V., 1991. The Green Revolution in the Punjab. The Ecologist 21: 57-60.

Skyes, J.B. (Ed.), 1983. Concise Oxford Dictionary of Current English. 7th Edition. Clarendon Press, Oxford, 1983, 1255 pp.

Steel, C., 2008. Hungry city: how food shapes our lives. Chatto&Windus, London, UK, 324 pp.

Stuart, T., 2009. Waste: uncovering the global food scandal. Penguin, London, UK. p. 188.

Chapter 3

The integration of food and agriculture into urban planning and design practices

Joe L. Nasr[1] and June D. Komisar[2]
[1]*Centre for Studies in Food Security, KHS 348 C, Ryerson University, Toronto, ON M5B 2K3, Canada; The Urban Agriculture Network, Washington, DC, USA;* [2]*Department of Architectural Science, Ryerson University, 325 Church Street, Toronto, ON M5B 2K3, Canada; and member of the Toronto Food Policy Council, Toronto, Canada; jnasr@ryerson.ca*

Abstract

A host of urban problems that have not always been central to the work of planners and designers are now transforming the understanding of typical urban systems and expanding areas of concern, as well as becoming catalysts to innovative responses. The disconnection between food systems and city dwellers is one of these problems. Food and agriculture are fast gaining recognition as legitimate areas of both planning and design through research, university teaching, design competitions, and awareness by professions – but their integration into the everyday practice of planners and designers is lagging. This chapter analyses the emergence of this new area of practice within the built-environment professions, with a focus on North American case studies at both the urban and building scales that incorporate food and agriculture systems within the larger agenda of sustainable design and planning practices. Although their number is increasing, just a few individuals and agencies are integrating food issues within their daily work as they seek to give shape to the place of food systems within the urban context; however, these can show how planning and architecture can embrace food issues robustly, as part of what is addressed in urban design and planning agendas.

Keywords: food systems, planning practice. architectural practice, architectural design

3.1 Introduction

A host of urban problems that have not always been central to the work of planners and designers are now transforming the understanding of typical urban systems and expanding the areas of concern for urban environments, as well as becoming catalysts to innovative responses to these concerns. These problems include many challenges that are resulting from the current food and agriculture systems, which have disconnected residents of countless cities from food production as well as from the whole system that processes and delivers this food.

As people are learning about the relationships between the food system and health, and as they are being educated about a range of issues such as urban agriculture, food security, food miles and the reduction of greenhouse gases, their awareness of these challenges is growing. It is increasingly understood that local food production and distribution enables a more sustainable use of scarce resources. The smaller ecological footprint that this provides is an important part of a sustainable agenda for planning and design.

While planners have begun to address challenges in the food system for over a decade,[2] designers (from architects and landscape architects to interior designers and industrial designers) have also become increasingly interested in urban food and agriculture issues.[3] Food and agriculture are thus fast gaining recognition as legitimate areas of both planning and design through research, university teaching, design competitions, continuing education, and awareness-raising by professional organisations – but their integration into the everyday *practice* of planners and designers does not result automatically from such growing recognition.

This chapter analyses the emergence of this new area of practice within the built-environment professions. A review of some North American case studies that incorporate food and agriculture systems within the larger agenda of sustainable design[4] reveals a number of different strategies that designers and planners have used. We present here a number of scenarios associated with this emergence, illustrating some of these scenarios with concrete examples of practitioners who have encountered or actively sought responses to food-system challenges from within their place in society as professionals who are specialised in shaping the physical environments in cities or their sociocultural, economic and ecological settings. We will start with scenarios that can be found in the planning world, before suggesting scenarios of designers (and more specifically, architects) who have recently addressed matters of food and agriculture within their practice. Other studies have recently considered the question of the education of such professionals,[5] or the specific content of their work.[6] This chapter is one of the first ones to focus on questions related to professional practice.

3.2 Integration of food systems into planning practice

Food and agriculture are increasingly acknowledged as legitimate planning issues that are of interest to planners[7] – and they are gradually getting integrated into common planning practice. The following section is a reflection on this process of the emergence of the food system as a new field of planning practice (rather than just a new area of interest) – and its gradual assimilation into the work of planners, moving from a position of marginality to slowly becoming integral to this work. A few entry points can evoke the range of ways in which food system planning has started to be practiced by planners.

---

[2] For instance, key publications by Pothukuchi and Kaufman (1999 and 2000) are now over a decade old.
[3] The key publication that formalised the interest by architects in urban agriculture is Viljoen, 2005.
[4] Many case studies have been identified through the Carrot City initiative at Ryerson University, coordinated by this article's authors along with Mark Gorgolewski. See www.carrotcity.org and Gorgolewski *et al.*, 2011.
[5] For planning, see Mendes and Nasr (2011). For architecture, see Komisar *et al.*, 2009.
[6] The American Planning Association (APA) and the Ontario Professional Planners Institute (OPPI) undertook surveys that included questions related to aspects of involvement by planners in food system issues.
[7] See for instance, the adoption by the APA of the *Policy Guide on Community and Regional Food Planning* in 2007 (APA, 2007), or the 'call to action' following an entire symposium on food system planning that the OPPI held in October 2010 (OPPI, 2011).

In recent years, as concerns around the current food system have taken hold across many countries, growing numbers of people have started to wish to concentrate their energy, time and knowledge into a life path aiming to contribute to a transformation of the food system. Some of these individuals (particularly the youth), wanting to specialise in the fields that have emerged around such a desire for transformation, have been looking for a good base to anchor their interest professionally – with planning offering one of the best opportunities. So the planning profession itself is not an end in itself in such cases, but rather a means to an end. We have been contacted on occasion by university students and other young people interested in urban agriculture as a career, trying to figure out which professional degree is best for getting there – and whether planning is well suited to a career that concentrates on food-system challenges. Even secondary students have contacted us to provide such advice.

The demand for food system planning is also emanating from without. Non-planners are thus looking for a role for planners to help address food-system challenges. Many people within the alternate food advocacy movement are nowadays expecting planners to have a role in transforming food systems. These external pressures are leading many planning departments that had never addressed food-system issues to start recognising them and offering certain solutions from within their sphere of intervention – and to assign some existing planners to integrate the food system into their area of work.

In some cases, specific controversies have forced planners to step in. Perhaps the most common catalyst has been the debates around the legalisation of keeping chickens in urban and suburban areas. Such zoning challenges can be seen as classic planning areas of intervention – as such, they are not unusual matters for the planners to deal with. The difference in recent years is that such challenges have emerged out of a direct concern around the food system and its problems – and around the constraints that land use regulations can impose on this system. The 'chicken laws' and similar areas of disagreement (urban beekeeping, front-yard planting, community composting...) have thrust planners – whether they like it or not – into food-system debates.[8]

As a result of this increased attention to food-system challenges, planning studies have been commissioned, involving a concentrated but temporary focus on food systems or some aspects of them. A good example is the Delaware Valley Regional Planning Commission (DVRPC) study of the Philadelphia area's food system. This regional agency had typically dealt regularly with issues such as transportation planning, but they conducted one study on food issues (DVRPC, 2010). This one-time study was a completely new field for the planning commission. Many planning agencies are now tackling food-system planning, but as with the DVRPC initially, the studies are often a special, one-time task rather than an ongoing activity.

---

[8] The City Farmer website (www.cityfarmer.info) features a number of stories from across North America (literally, from San Diego to Halifax) that focus on urban agriculture controversies, particularly the raising of chickens. Some of these involve debates between candidates for local elections.

Food has become an area where the work of planners sometimes intersects that of other areas of public interventions. An exemplary case of food as a connective space for public policy is the collaboration that has been ongoing between the Planning and Health Departments of the Waterloo Region in southern Ontario, a cluster of mid-size towns and surrounding peri-urban belts. This collaboration emerged when these two departments found that a number of challenges that they were both confronting overlapped around the food system. They started by jointly undertaking some cutting-edge studies,[9] which led later to planning and public health interventions based on the knowledge and awareness generated by this work.

Planners are starting to deal with food-system issues not only in special circumstances such as those described so far, but also on a more continuous basis. Very recently, food-system jobs have become commonplace. As such jobs are getting created, planners are stepping in to fill some of them. Examples of food-related non-planning jobs that 'foodie' planners are holding outside planning departments include Food Policy Council staffers, community food animators, farmers' market coordinators, etc. Planners are among the professionals whose knowledge base and training prepares them quite well to respond to the range of needs required by such jobs. Thus it is not the specific technical knowledge, but rather the breadth of knowledge, the system thinking and such less tangible factors that make them well adapted and thus employable in food-system jobs.

Besides non-planning opportunities, new jobs are emerging within planning departments where food-system planning is at least part of the job description. These may be social planning positions that include a focus on food issues, or public health planners for whom addressing the local food system is within their set of tasks, or community planners who are expected to include food matters routinely in their work. [10]

Finally, food-system questions are starting to become routine matters for planners working across the established areas of planning. Moreover, the portfolio of interventions of planning departments or divisions within such departments is ending up with a continuous (rather than one-time) focus on food-system issues. For a good example, we can return to the aforementioned DVRPC, the Philadelphia region's planning agency. The significant work of studying the region's food system eventually led the DVRPC to develop a food-system plan for the region (DVRPC, 2011), prepared by a staff person who was granted the title of 'food system planner'. This title – until very recently, rarely used – is becoming less and less unusual. In fact, we are aware of several professionals bearing this title, from Vancouver to Baltimore. This can be seen as the consecration of the emergence of food systems as a legitimate area of practice for planners.

---

[9] Numerous reports can be found under the 'Food & Healthy Eating' section of the Research Studies area of the Region of Waterloo Public Health website, available at http://chd.region.waterloo.on.ca/en/researchResourcesPublications/researchstudies.asp#FOOD.

[10] In an interesting variant, Baltimore recently hired its first Food Policy Director – housing this position within the City's planning department. See Cohn, 2010.

3.3 Integration of food systems into architectural practice

Increasingly, design students from architecture, landscape architecture and urban design programmes who are interested in sustainable design and sustainable cities have seized upon urban agriculture and food system transformations as part of a sustainable, integrated design strategy. While early on, such interest was largely restricted to the imagination of university students, a recent trend that appears to be taking hold is the integration of food-system thinking into design strategies after young designers leave school to become practitioners. Instead of merely academic projects, these young designers (and some more established ones too) are incorporating urban agriculture into landscape strategies, green building technologies, and adaptive reuse projects.

Case studies showing the transition of young designers from school to practice illustrate the close relationships that can be found between student explorations and later professional work, highlighting the value of teaching about design for urban food and agriculture to the development of more sustainable building practices. Some students interested in this emerging area are fortunate enough to find a like-minded firm to work for when they begin their careers. Others strike out on their own to develop sustainable projects that incorporate elements of the food system, and still others find ways to educate a community about such strategies through community or professional organisations. Three Canadian cases are presented for illustration.

One former Ryerson University architectural science student in Toronto, Micah Vernon, is contributing to bringing urban agriculture and design to the centre of the discussion about sustainable design. As an undergraduate he integrated a productive greenhouse within his urban adaptive-reuse co-housing thesis project (Figure 3.1). After graduation, Vernon has not worked at a firm that integrates growing food with building design. However, he has managed to maintain his interest through activities outside his 'day job'. Notably, he founded an urban food group within Toronto's Architecture for Humanity chapter, helping to disseminate information and enthusiasm for sustainable initiatives as his group organises bi-monthly forums; eventually, this led him to heading the local chapter.

Another recent graduate of the same programme, Jordan Edmonds, focused on the relationships between community development and food with his thesis. This project, a community food centre, functioned as a transitional bridge between housing and parkland through the design of a greenhouse, market space, community gathering spaces, and a community garden. After graduation, Edmonds joined the Toronto firm of Hilditch Architects, a socially and environmentally conscious firm. Their current project '40 Oaks Community Centre' is part of Regent Park, Toronto's largest neighbourhood redevelopment project. The new Centre includes space for food-focused activities from teaching kitchens to demonstration gardens, below four floors of affordable housing.[11] Jordan also contributed to

---

[11] See http://www.tcrc.ca/index.cfm/14511/40-oaks.

Figure 3.1. Adaptive-reuse co-housing in Toronto. Section through the greenhouse and kitchen from Micah Vernon's thesis project, Ryerson University. Image courtesy of Micah Vernon.

the effort, now underway, to provide additional community garden space in the redeveloped Regent Park. Edmonds' experience led to other work that involves food-related design, including a new project to create a community food centre in Stratford, Ontario – mirroring in reality his original aim while a student.

When Rune Kongshaug was a student at McGill University's School of Architecture in Montreal, he was a participant in the Edible Landscape Project of the Minimum Cost Housing Group.[12] The group's action research project involving urban agriculture took place in three developing cities around the world as well as Montreal. These researchers sought to find strategies that would alleviate food insecurity by using participatory community design practices to develop strategies for residents to use for increasing food production. After this experience, Kongshaug began practicing architecture independently. An early project, Maison Productive House, is a multi-family structure that he realised in Montreal (Figures 3.2 and 3.3). The building incorporates adaptive reuse, active and passive solar systems as well as productive ground-level gardens, roof terraces for growing, a greenhouse, vertical surface gardening and areas for composting.[13] He is now planning a follow-up project that scales up his first project.

Of these three young practitioners, two are paving their own way in this emerging field, through their own practice or through non-profit organisations, and one is working for a like-minded firm. In all these cases, the knowledge acquired during the university years provided the foundation for a continued involvement in food issues in later years. In other instances, an interest in food-system issues emerged outside any academic setting. A number

---

[12] See http://www.mcgill.ca/mchg/pastproject/edible-landscape.
[13] See http://productivehouse.com/en. See also Chartrand, 2009.

Figure 3.2. Exterior, Maison Productive House (image courtesy of Rune Kongshaug).

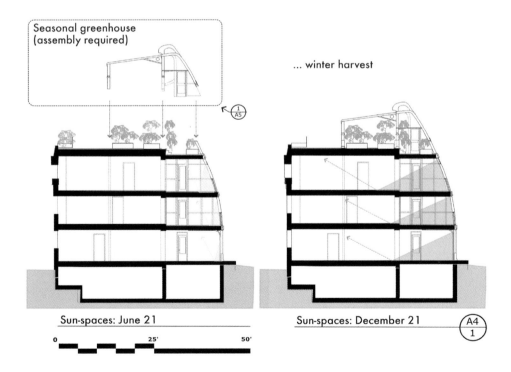

Figure 3.3. House Sections, Maison Productive House (image courtesy of Rune Kongshaug).

of established designers and design firms have thus been incorporating solutions for food-related challenges into their sustainable design work.

The firm of Teeple Architects recently designed an award-winning social housing apartment building for Toronto Community Housing (60 Richmond Street, 2007). Called simply '60 Richmond Street East,' it incorporates spaces for urban agriculture as part of sustainable strategies that include an extensive green roof, passive ventilation, passive and active solar systems, a grow wall that cools and filters the air, and rainwater collection. Kitchen gardens are located six stories above ground as part of outdoor social spaces. They are made viable through vertical and horizontal slices through the almost cubic volume, allowing light to stream in. Food production complements other food components in the building, including a teaching kitchen. This project shows how urban food and agriculture elements can be incorporated into sustainable design strategies – in this case, for affordable housing in tight urban sites. This is an example of a firm that sees food-system pieces as one important component of an emerging focus on sustainable design within a particular project.

In contrast, another Toronto-based architect, Joe Lobko, and his firm du Toit Architects Ltd/ du Toit Allsopp Hillier (DTAH), is becoming known for expertise in designing for urban food and agriculture through sustainable adaptive reuse. Their design practice began to develop this expertise through the rehabilitation of historic streetcar repair barns resulting in the Artscape Wychwood Barns, a community space that includes offices for not-for-profit organisations, artist housing and studios, and a children's theatre space.[14] It was through the design work for the largest tenant, The Stop, that DTAH developed its expertise in designing for urban food and agriculture. The Stop is a community food centre that encourages urban agriculture, runs a farmers' market, teaches about nutrition and productive gardening, and promotes community nutrition, among others. DTAH adapted one of the century-old repair barns for The Stop to become a vast greenhouse as well as introducing an outdoor bake oven and teaching kitchen (Figure 3.4).

In parallel, DTAH was the lead architect on a larger project called Evergreen Brick Works.[15] Located at an abandoned quarry with historic brick-making facilities, the factory buildings and landscaping encompass a 2,500-square metre area with space for a farmers' market, festivals, plant nurseries, and a sheltered garden with raised garden beds planted in a formerly paved area. The architects designed an education component that includes demonstration gardens with a variety of edible plants, native trees, shrubs, wildflowers, and marsh plants. The site has become a hub for teaching about food and agriculture, thanks to careful planning as well as the vision of the many players, including the main tenant, Evergreen, a not-for-profit environmental organisation (Figure 3.5).

---

[14] See http://www.torontoartscape.on.ca/places-spaces/artscape-wychwood-barns.
[15] Other designers working on this large and complex undertaking include Claude Cormier Architectes Paysagistes, Diamond + Schmitt Architects Inc., and E.R.A. Architects. See Hume, 2010.

Figure 3.4. The Stop Community Food Centre Greenhouse and Community Gardens at Wychwood Barns, Joe Lobko, du Toit Architects Ltd./du Toit Allsopp Hillier (DTAH) Architects.

Figure 3.5. Evergreen Brick Works Demonstration Garden, Toronto. Joe Lobko, du Toit Architects Ltd/ du Toit Allsopp Hillier (DTAH) Architects.

The designers of Brickworks and the Wychwood Barns have shown the potential of old abandoned industrial heritage buildings when they are repurposed for community development and education, with food serving as a central element that pervades the site's multiple functions. Along the way, the expertise acquired at DTAH in developing such facilities has grown to fill a niche in this new area of design. More recently, DTAH was asked to undertake a 'Sustainable Neighborhood Action Plan' for the Jane and Finch neighbourhood of Toronto. While food was not explicitly stated in the terms of reference for this study, it emerged quickly as a major element in the Plan. This illustrates well how food-system concerns can become a significant slot within the design profession, as well as an area that is routinely considered within design processes.

Architects can also act as facilitators. Mole Hill, a whole block in Vancouver, is a city-owned project that many activists worked to save from demolition.[16] Two architecture firms and a landscape architect (DIALOG, Sean McEwen and Durante Kreuk) played a critical role as mediators between the residents and the city in a very contested case, where the city wanted to demolish the downtown neighbourhood. Instead this became a model of historic preservation, adaptive reuse, and innovative sustainable systems, resulting in affordable housing combined with community garden plots as well as a naturalised garden that diverts rainwater into an ornamental pond, and innovative pedestrianised paths. The meandering laneway that runs through the site also features a community laundry, workshop, and recycling area. In this way, the designers facilitated the placement of urban agriculture as part of a 'living lane,' creating a community focal point for the residents of Mole Hill (Castrejon Violante, 2010).

The examples above show the great variety that exists in the roles of architects in relation to food-system design. In some cases, particular client needs generated a firm's development of urban agriculture design expertise, whereas in other cases, it was a logical part of a larger sustainability agenda. The contribution of designers may be technical, artistic, procedural or advisory. The knowledge base of this emerging area of design always increases with each subsequent project.

3.4 Conclusion

The scenarios outlined and the cases presented above paint a picture of professions that are starting to find a role for themselves in relation to the growing understanding of the importance of a well-functioning food system. While professional bodies like the American Planning Association have stepped forward institutionally in this regard, it is largely through the interest and commitment of individuals that food systems have emerged as an area of professional practice for planners and designers.

---

[16] See www.mole-hill.ca.

Planners and designers have been discussed separately in this chapter, but the emergence that has taken place is not fully distinct between the two professions. For one, evolutions of the two professions have long been interconnected, and the question of food systems is no different in this regard. Moreover, the practice in one profession impacts the other. For instance, some of the innovative developments described here could not be undertaken without challenging current land use and zoning, such as the laneway adaptation in Mole Hill and the transformation of the repair facilities at the Artscape Wychwood barns. At the same time, such developments had to relate to existing frameworks, including the planning framework.

Bruce Darrell, an architect from Toronto now living in Ireland, argues that food relates to the economic, political, social and cultural environments of a city, yet the complex relationship between food and cities continues to be ignored outside of small circles. He therefore offers the label of 'Food Urbanism' to help conceive of how food imprints the built environment. For Darrell, Food Urbanism is a theory that 'positions food as a primary transforming force capable of organising the city and enhancing the urban experience' (Darrell, 2007). This can help lay the ground for multiple roles for certain professions to take on explicitly and work actively on shaping the urban built (and unbuilt) environments. As the examples above show, individuals, organisations and firms in the planning and design professions are already working on transforming the urban environment as they put into practice emerging aspects of food urbanism, enhancing our landscape, our built form, and the fundamental ways we relate to food.

References

60 Richmond Street East Housing Co-Operative, 2007. Canadian Architect. Available at http://www.canadianarchitect.com/news/60-richmond-street-east-housing-co-operative/1000218352/.

American Planning Association, 2007. Policy guide on community and regional food planning. Available at http://www.planning.org/policy/guides/adopted/food.htm.

Castrejon Violante, L., 2010. Agricultural roads in Vancouver. University of British Columbia, Masters Graduating Project. Available at http://hdl.handle.net/2429/29573.

Chartrand, I., 2009. Vers le vert: Les apprentissages de deux PME en montage et gestion de projets résidentiels verts, Masters thesis, University of Montreal, pp. 33-51. Available at http://www.arclab.umontreal.ca/GRIF/ARTICLES/00028/00028_DOC_1.pdf.

Cohn, M., 2010. Baltimore names its first food czar. Baltimore Sun May 11. Available at http://articles.baltimoresun.com/2010-05-11/health/bs-hs-food-policy-director-20100511_1_food-czar-healthful-ebt-machines.

Darrell, B., 2007. What is food urbanism? 8 January. Available at http://foodurbanism.blogspot.com/2007/01/what-is-food-urbanism.html.

Delaware Valley Regional Planning Commission, 2011. Eating here: The Greater Philadelphia food system plan. Available at http://www.dvrpc.org/Food/SustainableFoodSystems.htm.

Delaware Valley Regional Planning Commission, 2010. Greater Philadelphia food system study. Available at http://www.dvrpc.org/food/FoodSystemStudy.htm.

Gorgolewski, M., Komisar, J. and Nasr, J., 2011. Carrot City: Creating places for urban agriculture. Monacelli Press, New York, 240 pp.

Hume, C., 2010, Brick Works gives Toronto something to build on, Toronto Star, September 26.

Komisar, J., Nasr, J. and Gorgolewski, M., 2009. Designing for food and agriculture: Recent explorations at Ryerson University. Special issue on "Designing edible landscapes," guest ed. Vikram Bhatt and Leila Marie Farah. Open House International 34, no. 2: 61-70.

Mendes, W. and Nasr, J., 2011. Preparing future food system planning professionals and scholars: Reflections on teaching experiences. Journal of Agriculture, Food Systems, and Community Development, 2(1): 15-52.

Ontario Professional Planning Institute, 2011. Healthy communities and planning for food: Planning for food systems in Ontario – a call to action. Available at http://www.ontarioplanners.on.ca/pdf/a_call_to_action_from_oppi_june_24_2011.pdf.

Pothukuchi, K. and Kaufman, J., 1999. Placing the food system on the urban agenda: The role of municipal institutions in food systems planning. Agriculture and Human Values 16: 213-244.

Pothukuchi, K. and Kaufman, J., 2000. The food system: A stranger to the planning field. Journal of the American Planning Association 66 (2): 112-124.

Viljoen, A. (Ed.), 2005, CPULs – Continuous productive urban landscapes: Designing urban agriculture for sustainable cities. Oxford and Burlington, Mass.: Architectural Press/Elsevier.

Part 1. Urban food governance

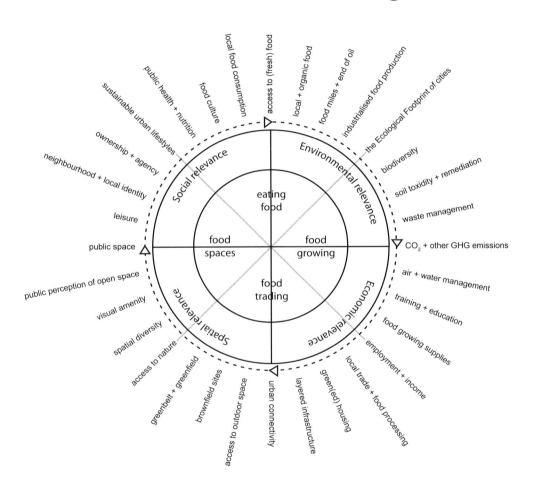

Chapter 4

Food and the city: the challenge of urban food governance

Petra Derkzen[1] and Kevin Morgan[2]
[1]Rural Sociology, Wageningen University, Hollandsweg 1, 6706 KN, Wageningen, the Netherlands; [2]Cardiff School of City and Regional Planning, Cardiff University, Glamorcan building, King Edward VII Avenue, Cardiff, Wales, CF10, United Kingdom; petra.derkzen@wur.nl

4.1 The emergence of planning for food

With over half the world's population now deemed to be urbanised, cities are assuming a larger role in political debates about the security and sustainability of the global food system. The *Urban Food Governance* theme is designed with two aims in mind: first, to help us to better understand the roles that cities can play in fashioning sustainable food systems and, second, to document what they are already doing with respect to food planning, arguably one of the fastest growing social movements in the world today. The rapid growth of the food planning movement – which in a narrow sense embraces planners, policy-makers, politicians and citizen action groups, and more broadly includes anyone seeking to fashion a more sustainable food system – owes much to the fact that food, because of its unique, multi-functional character, has a 'convening power' that helps to bring people together from all walks of life (Morgan, 2009).

4.2 Food Policy Councils

One of the most visible signs of the burgeoning food planning movement is the formation of Food Policy Councils in North America, where some 150 councils have been created in various jurisdictions, be they counties, cities or states. Food Policy Councils can help local states to address two governance challenges that bedevil public policy-makers around the globe, namely: (a) how to engage with, and tap the energy of civil society and (b) how to overcome the fragmented bureaucratic silos in which public policy tends to be designed and delivered. Astute policy-makers know that they need to enlist the creative energy of civil society if they are to seriously address key societal challenges, like climate change, health inequalities and social justice for example, and Food Policy Councils are one of the governance mechanisms for doing so because they create spaces of deliberation in which local state and local civil society can collaborate for mutually beneficial purposes and fashion joint solutions to common problems. Creative engagement with civil society interlocutors is a learning-by-doing exercise in which food is the prism through which local state officials come to better understand the deeper linkages between their formally separate functional departments, many of which have a food policy dimension even if they were not hitherto aware of the fact. This brings us to a fundamentally important point about food governance, which is that *all* public institutions have a food policy whether or not they know it: while some institutions practice this policy by design, and consciously strive to enhance the quality of the food that is served on their premises in the name of public health and ecological integrity; others practice it by default, and remain oblivious to the quality of the food that is served to their staff, a stance that clearly compromises their duty of care if it results in unhealthy diets at work.

As the chapters in this theme will show, food governance is still something of a novelty for most cities, where many urban politicians tend to see the food system as beyond their remit. Indeed, in his contribution on Food Policy Councils, Philipp Stierand goes even further and argues that urban food systems have been rendered virtually obsolete by the industrialised food system. The food supply chain operates on a national and global scale in which the geography of particular cities has no significant role. Indeed, according to Stierand, it has led to the passiveness of the urban citizen and the redundancy of the urban scale. Stierand distinguishes 'new urban food needs' and describes the bottom up projects which try to influence the urban food system from two orientations: problem and opportunity oriented. A Food Policy Council, which is the main theme of his chapter, is part of an opportunity oriented project to highlight the urban development potential of the food system.

However, this potential only becomes apparent, as Plantinga and Derkzen argue, if city governments attach significance to a sustainable urban food system and to the role of collaboration in bringing it about. A case study of the Dutch city of Tilburg showed little willingness on the part of the local government to engage in food policy at city level. Although respondents from the city government and other stakeholders acknowledged the existence of food production and consumption problems, this was not deemed to be sufficiently urgent for the city government to assume responsibility for food policy. Food was seen as an 'add on' to a political agenda dominated by budget cuts. The chapter by Plantinga and Derkzen neatly illustrates the ambiguity that exists in some city governments about the multi-functional character of food and its implications for urban policy and planning.

4.3 Agenda setting for food policy

Clearly, from an urban governance perspective, the multi-functional character of food presents problems as well as opportunities. As Melanie Bedore argues, 'small buzz-less cities' can pose some of the biggest challenges for food planning. Drawing on her research in Kingston, in eastern Canada, she suggests that food system planning is hampered by factors such as preoccupation with land values, limited human resources in the city departments, and therefore limited policy learning capacity. An equally important constraint, she discovers, is 'an ideological resistance to 'interfering' in the free market'.

The neo-liberal conviction that food provisioning should be decided by the market, because food choice is alleged to be a wholly private matter, is the mantra of the food industry everywhere. This mantra has been deployed so successfully that it has persuaded many social democratic governments to tread carefully when it comes to food planning regulations (Morgan *et al.*, 2006). Notwithstanding this mantra, local and central governments are increasingly experimenting with food planning regulations, not least to promote public health and regulate the growth of urban agriculture. This amounts to a new politicisation of food as Nevin Cohen shows for a number of cities in the United States, where urban agriculture is being championed for many different reasons. He shows that in cities where zoning does not allow for food production, the growth of the agricultural activity demanded a city government response. Activists and 'policy entrepreneurs' inside government institutions often work together and this collaboration between local state and local civil society can create a 'policy window' for the articulation of new city policies on food production and consumption.

A similar point is made by Ségolène Darly on urban food procurement in the peri-urban area of Paris. Agricultural development and extension agents played a key role in connecting farmers in the peri-urban fringe of Paris to school meal systems run by local town councils. The new procurement networks that were fashioned in the two cases described were unusual constellations of actors working together. Both chapters attest to the compelling convening power of food: urban food strategies have the potential to bring people together from very different professional, social and ethnic backgrounds, and such cultural diversity nurtures the cosmopolitan localism that typifies the liveliest and most congenial urban environments.

However, as Ségolène Darly also demonstrates, the new networks of food governance are not necessarily aimed at improving the sustainability of the food system in the holistic or multi-dimensional sense of the term. Promoting links between organic farmers and neighbouring schools, for example, was largely driven by the goal of creating more commercial opportunities for local organic produce, and local food is not synonymous with sustainable food.

4.4 New and old fresh markets

The potential tension in the values that constitute 'sustainability' is highlighted in two chapters on the role of municipal markets in the UK. Municipal markets have recently been rediscovered as potential spaces for fresh, healthy and affordable food by cities with renewed interest in the urban food system. However, Machell and Caraher call for a more cautious and indeed a more critical approach to the health claims made on behalf of these markets because of the latter's image of fresh fruit and vegetables. At their case study site, the Leeds Kirkgate Market, the ratio of fast-food eateries to fresh-produce stalls was 3 to 1. The economic survival strategies of these markets seem at odds with policies to engage people in healthier eating habits. Hence they conclude by saying that 'it is not realistic to assume that markets can return to their original function and role in society as the sole provider of a healthy diet to the urban poor'.

In their chapter, Machell and Caraher are largely concerned with the relationship between the municipal market and its working class customers. By contrast, the chapter of Julie Smith, focusing on the traditional municipal markets of the North East and Eastern regions of the

UK, presents a more diverse picture of social class within these markets. Among the shoppers in her case studies were those on low or more restricted incomes, as well as other groups such as office workers, students and ethnic minority groups. She argues that 'the traditional market shopper has been part of a tradition that values affordable quality for far longer than the "new" farmers' market shoppers'. Customers at the farmers' markets confirmed that they would also shop at the municipal markets.

While the farmers' market customers seemed happy to buy at both farmers' and municipal markets, the reverse is more problematic argues Jessica Paddock. The struggle of the traditional municipal markets has paradoxically coincided with the recent growth of new farmers' markets in the UK. According to Paddock, these new and higher priced farmers' markets have become 'middle class consumer institutions'. By means of a discourse analysis, she analyses how social class is subtly inscribed in the language of customers at the farmers' markets. As part of an alternative consumption movement, she argues that farmers' markets have become an instrument in class distinction, where some people feel that they belong, while others are implicitly informed that such markets are above their standard.

As a new phenomenon, farmers' markets have been extensively researched over the last ten years, whereas research on traditional/ municipal markets has been conspicuous by its absence. The two chapters on municipal and one on farmers' markets and their social class analysis illustrate the widening of the discussion about sustainable food systems. We are slowly beginning to recognise that sustainable food needs to be readily available everywhere and not confined to exclusive or alternative outlets. This is one of the reasons why the government needs to step into the garden, as Barmeier and Morin argue so convincingly in their chapter. Demonstrating the transformative power of collaboration, their analysis shows that the resilience of urban community gardening programmes in the United States is due in large part to their 'co-governance relationships'. However, they also show that city council members are more often than not the respondents and the facilitators, rather than the real initiators of community gardening programmes, thus highlighting the crucial role of local civil society in shaping the local state policy.

4.5 Food pivotal in sustainable development

One of the main implications of the chapters in this strand of the book is that public bodies in general, and city governments in particular, begin to see food in a new light when they view it through the prism of sustainable development. Although the food planning movement is still in its infancy, and many cities are unsure about what if anything they should do about the food system, as the Tilburg case illustrates, this uncertainty is itself a sign of progress because, until recently, planners and policy-makers were certain that they had no role whatsoever in shaping the food system. Food is a public policy issue because it is deeply implicated in the most compelling questions of the 21st century – questions like climate change, public health, natural resource management and social justice for example. Because of its centrality to human health and wellbeing, food can never be reduced to being 'just another business', which is what neo-liberals would have us believe.

Another important implication of this stand of the book is that conventional food policy will not be rendered more sustainable without the active involvement of the food planning movement. More often than not the stimulus for change comes not from politicians, policy-makers and planners, but rather from grass roots initiatives which draw on the energy, the talent and the creativity of civil society. To be successful, however, such grass roots initiatives need to elicit political support because they are a necessary but not sufficient ingredient in the recipe for change. Collaboration between local civil society and the local state may be essential to fuel the process of reform, but some form of 'co-governance' arrangement is ultimately necessary to ensure that the gains of sustainability are capable to being sustained, otherwise they could be reversed by an incoming government of a different political persuasion, which is precisely what is now happening to school food reform in Rome.

The food planning movement comes together from many walks of life and for many reasons – the most important of which are to enhance human health, to secure social justice and to promote ecological integrity – but such diversity needs to become an asset rather than a liability and the best way to do this is to ensure that these diverse values, the values at the heart of sustainable development, are so thoroughly embedded in the principles and practices of civil society that incoming politicians need to learn to underline and not undermine these values.

References

Morgan, K., 2009. Feeding the city: the challenge of urban food planning. International Planning Studies 14: 429-436.
Morgan, K., Marsden, T. and Murdoch, J., 2006. Worlds of food: place, power and provenance in the food chain. Oxford University Press, Oxford, UK, 225 pp.

Chapter 5

Food Policy Councils: recovering the local level in food policy

Philipp Stierand
Speiseraeume.de, Große Heimstraße 32, 44137 Dortmund, Germany; ps@speiseraeume.de

Abstract

This chapter looks at Food Policy Councils as a possibility to shape urban food systems. The urban food system lost through up scaling and delocalisation its importance for food supply. With this development the scope for influencing food systems on the local level shrunk and food policy moved to national, European and global level. Recently the need for urban food policies is seen again. The case study of the Food Partnership in Brighton and Hove shows that Food Policy Councils represent a viable possibility to recover the local level in food policy.

Keywords: delocalisation, urban food system, Food Policy Council, Brighton and Hove

5.1 Redundant urban food systems

Urbanisation and industrialisation were accompanied by a revolution of the urban food system. Until industrialisation cities had to be supplied with food from the nearby hinterland and the urban area itself. Due to the limited transport and preservation possibilities crop and stock farming had to be urban. Later the needs for the urban supply system changed rapidly: scores of urban dwellers without time and space for self-supply had to be supplied. The urban food system opened up with growing technical possibilities for regional connections. Perishable food was still being produced within the city but also in a type of Thünen's circles around the city (Atkins, 2003; Steel, 2008; Teuteberg, 1987). For durable food a national market developed while products such as corn came from the colonies (Friedmann *et al.*, 1989). Today trade and consumption in cities has lost almost all boundaries of time/season and space. The spatial scale of the urban food supply changed from subsistence on an urban level towards regional connections, operating today on a national and global scale (see Figure 5.1).

Focusing the analysis on consumption shows that the shift in spatial scale loosened the links between food production and the place of living. As one consequence of this *delocalisation* hunger was defeated in Europe through the equalisation of seasons and the restructuring of the worldwide food production towards the needs of Europe and the US. The second consequence is a large uniformity of food due to the industrial modes of production, supported by a growing mobility of consumers and a decreasing ritualisation of food. The third effect is the urbanisation of food systems, not only because of the growing amount of urban dwellers but through the establishment of the urban diet as a norm (Montanari, 1993).

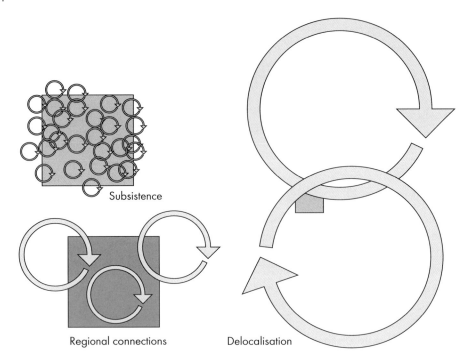

Subsistence

Regional connections   Delocalisation

Figure 5.1. Shift in spatial scale.

Focusing the analysis on the aspect of production and supply chain shows that the main reasons for the shift in spatial scale are cost-effective production and economies of scale. The key consequences of this are: Firstly, a *disconnection* between customers and producers. The relations along the food production chain are characterised by anonymity. Secondly, the *disembedding* of food and its place of origin, meaning that in the dominate food system the places of food production are exchangeable. Place has for the vast majority of products no influence on their characteristics anymore. And thirdly, the specialisation of producers and traders, which goes along with a *disentwining* of food supply chains and the creation of different spheres of activities for specific products and services (Wiskerke, 2009).

With an additional focus on the relation between stakeholders and the urban food system, the following consequences of the shift in spatial scale can be identified:

- *Passiveness.* The role of the urban citizen changed from active participant of the food system to an inactive consumer. Furthermore the entire city is today a passive user of food systems without a significant input to its own nutrition and without significant potency to shape the urban food system.
- *Redundancy.* The stakeholders of the food systems operate on a national to global level. The local farmer is producing for the world market, the local supermarket is supplied by national distribution centres. There is no trade or exchange on the local level anymore. For the functioning of food supply the urban food system is unimportant.

To summarise, today's dominant urban food system could be characterised as delocalised, disconnected and redundant. The urban food system lost not only any traditional confederate but also the feasibility for conventional policy. Together with the local food system the food policies disappeared and they have 'been understood as expressions of "higher" national or global impulses (e.g. agricultural policies, food aid or food safety)' (Mendes, 2008: 943).

5.2 New urban food policies

Consumption seems to be the only function, which is not possible to relocate but bound to the specific urban food system. Even retail lost the monopoly of the closeness (Spannagel and Bange, 1999) as the development towards out of town supermarkets in the last decades and the decline of rural food retail shows. On the other side the urban food system represents the spatial scale where many problems of the entire food system are becoming visible or are being caused. Many problems are in the end borne by individual consumers, other problems are caused by consumption habits. 'Cities find themselves at the forefront of the NFE [new food equation] for both ecological and political reasons' (Morgan *et al.*, 2010: 210).

There are new urban food needs, which are beyond the requirement of a constant supply. These needs are especially:
- *Confidence*: anonym production chains combined with the food scandals of the last decades created the need for trust and closeness within food supply chains.
- *Sustainability*: consumers and politicians getting aware that food production makes massive use of clima relevant resources.
- *Health*: malnutrition and overweight are getting widespread disease and a massive burden for the health system.
- *Fairness*: the fair trade market is booming and there is an active discussion for example about the appropriateness of milk prices.

Local organisations and authorities are getting aware of these new urban food needs and the multifunctional character of the food system. They are starting to develop policies and projects to influence the food system 'bottom up'. Urban food systems are becoming a concern of urban governance again, food policy is more and more becoming an urban policy issue. A great variety of food projects is emerging. Especially in North America and the UK these could be already seen as a social movement backed by activities of local governments and scientific research. In continental Europe the appearance is not as strong but developing.

Food is deeply connected with many parts of the human live. Consequently the food projects are multifaceted, targeting several issues and problems. They are as well adapted to local needs, finding different arrangements in every community. A comparison off the different approaches is therefore difficult.

The food projects vary in spatial scale from a single lot and an urban neighbourhood to city wide projects. Some of them concentrate on specific issues such as health or retailing, others have a more holistic approach (see Figure 5.2). In general these projects could be distinct in two groups:

1. Problem orientated: the projects are focused on a single target and are often a reaction on a specific problem, whether the targeted problem(s) or approaches for solution are connected to the food system.
2. Opportunity orientated: these projects use the food system as a potential for urban development. They expand a problem centred view and are using a more strategic approach.

In my point of view there are only two concepts with a holistic approach and a district-wide covering: Food Strategies and Food Policy Councils. Food strategies are expressing aims and guidelines for the development of the urban food system. Food Policy Councils (FPC) counsel politics and administration on food issues. In the following this chapter will focus on Food Policy Councils – as the instrument with the most extensive approach, often including food strategies – and on the question if and how they can affect the urban food system.

5.3 Food Policy Councils

The concept of Food Policy Councils was developed in the US 'Community Food Security Movement' (Borron, 2003: 4) in the 1980s. The first FPC was established 1982 in Knoxville as

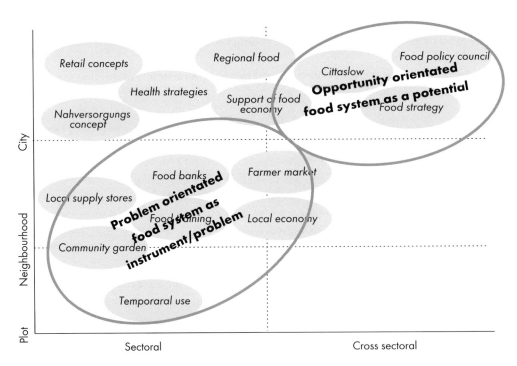

Figure 5.2. Projects in the food system.

a response to studies which found a food desert in the inner city of Knoxville. The definitions of the term Food Policy Council vary (Hamilton, 2002; Harper *et al.*, 2009; Schiff, 2007). The range of definitions could be seen as a lack of consensus but also as the flexibility of the concept according to different local conditions.

There is consent about the objectives of the councils: 'The central aim of most of today's Food Policy Councils is to identify and propose innovative solutions to improve local or state food systems, making them more environmentally sustainable and socially just' (Harper *et al.*, 2009: 16; Pothukuchi *et al.*, 1999). In the 1980s many FPC focused on food access and hunger issues. This was a limitation which 'negatively affects the longer-term success' (Dahlberg, 1994: 10) as early evaluations of FPC already stated. Today a broad approach is accepted, often with an additional focus on health and economic aspects.

The descriptions of the function of a Food Policy Council vary (Lang *et al.*, 2005: 15; Pothukuchi *et al.*, 1999) but the views expressed seem to be similar.

> *Food Policy Councils bring together people engaged in a wide variety of food organizations and activities to share ideas about and help initiate projects that advance community food security and food system sustainability and to develop public understanding that a sustainable and secure food system generates a wide mix of community benefits....* (Roberts, 2010: 173).

Harper (2009: 2) sees four general functions for a Food Policy Council (FPC):
- 'To serve as forums for discussing food issues,
- To foster coordination between sectors in the food system,
- To evaluate and influence policy and
- To launch or support programmes and services that address local needs.'

The food system approach of the councils makes networking one of the central tasks. The councils have to accomplish a wide diversity of views, knowledge and experiences to further develop the interests and ideas in the food system context. Another function is education, in the meaning of communication of ideas and exchange of information. It is targeting on the members of the Food Policy Councils, the stakeholder of the food system and the public. The internal education is fulfilled by exchange between the different members and involving of external speakers. The external public is reached trough brochures and leaflets (e.g. food guides) and trough hosting conferences and events.

Typical projects of FPC are concerned with:
- food system assessment;
- municipal food policies and charters;
- urban gardens and farms;
- community kitchens;
- local food;
- food access;

- institutional food purchasing; and,
- conferences and events (Schiff, 2007).

The described functions and concerns emphasise the underlying principle of Food Policy Councils: They are very much about networking among stakeholders and enabling and empowering of stakeholders of an urban food system.

In terms of membership Schiff (2007) sees a variation in the different definitions of FPC: Some definitions claim a membership of all components of the food system as necessary. Hamilton (2002: 447) accentuated: 'The goal is to have as broad representation of issues and interest and people and institution as possible from across the food system.' Members of a FPC are therefore stakeholders of the food system like farmers, processors, wholesaler, retailer, restaurant owners and consumer, activists from anti hunger or environmental initiates, officials from trade unions, farm organisation and gardeners as well as academics involved in food issues. 'In addition, the state or local government officials involved with the council typically include representatives from the state departments of agriculture, economic development, inspections, education, health, human services, and transportation'(Hamilton, 2002: 447). Members involved in waste management and in food processing are less common (Harper *et al.*, 2009).

Wayne Roberts (2010) disagrees with these criteria for membership. He stresses the importance of membership recruitment on the basis of individual talent. Members that are chosen because they represent a specific stakeholder group would contradict the intention of FPC as public purpose groups. 'The purpose of membership diversity is not to provide representation, but practical knowledge of all sectors to strengthen the value of the council's holistic perspective' (Roberts, 2010: 180).

Most of the Food Policy Councils are not part of the public administration but are serving as an advisory board. Other councils are autonomous from the administration working as a NGO (Borron, 2003; Pothukuchi and Kaufman, 1999; SSAWG, 2005). In general there are three models of organising the relation between a Food Policy Council and the government: a government organisation, a non government organisation and a hybrid model (Schiff, 2007, 2008), each with specific benefits and drawbacks.

A council which is organised as a government organisation has the largest legitimacy within the city and its administration. Such an organisation form is seen as a strong indication for a commitment in the field of food. A government organisation can also avoid competition with NGOs about funding. However, NGOs can establish themselves as an alternative to governmental policy; an external reviewer has much more room to move. The work of a non governmental FPC can be much more flexible and unbureaucratic. On the other hand such a Food Policy Council is only one of many NGOs and could have more difficulties to find its concerns heard. The 'hybrid model, including some formal relationship with government through funding, resources, or otherwise while maintaining some NGO or non-profit status, can offer some of the advantages of status as both a non-profit and a government entity.' (Schiff, 2007). Dahlberg (1994) saw in his evaluations a clear advantage in institutionalised and governmental FPC, Schiff (2008) agrees here.

5.4 Brighton and Hove Food Partnership

The Community Food Security Coalition estimates that there are about 40 FPC on the local level and about 60 FPC on state, regional or county level in North America (Community Food Security Coalition, 2010). In Europe there are some cities with food strategies with London and Amsterdam being pioneers (Morgan *et al.*, 2010). However, continuous working Food Policy Councils are rare. Brighton is an exception with a Food Policy Council founded in 2005 named Brighton and Hove Food Partnership. The description of the Food Partnership is mainly based on research conducted in the context of my PhD (Stierand, 2008, 2007a, 2007b).

The aim of the Brighton and Hove Food Partnership is to develop a local food system which is socially just, economically active, environmentally sustainable and supporting the health and well being of the residents (Brighton and Hove Food Partnership, 2006). The membership in the partnership is open to all residents and organisations. Today there are about 600 members. The Partnership is managed by a board of directors. It consists of elected and appointed members. Nine elected directors are supported by representatives from primary care trust and the city council and appointed experts.

The initiative starting the founding process of the partnership came from the sustainability team of the Brighton and Hove City Council. The team recognised in its work a strong connection between the issues of sustainability and food. On a first meeting of food interested parties (e.g. Primary Care Trust, Food and Health Partnership East Sussex) in October 2002 it was decided to approach the topic food system.

The non-profit consultancy Food Matters obtained the assignment to survey the food system of Brighton and Hove. One conclusion of the resulting report 'The Brighton and Hove Foodshed' is that the variety of activities of food related institutions and organisations is a good basis for further work. Brighton and Hove would be ripe for alternative and innovative food solutions.

> *Although there is an understanding and acknowledgement of the need to address issues relating to the impact of the current global food system among key players in the city, this does not necessarily translate into action. This is most apparent among decision makers and those responsible for public policy* (Food Matters, 2003: 28).

Food Matters was recommending the development of a strategy which considers the connections between food, health, economy, and sustainability. The general information about the food system should be improved and made accessible.

The Brighton and Hove Foodshed was presented at the conference 'Spade to Spoon' in autumn 2003. The Conference was open for the public and had 120 local as well as national visitors from different backgrounds. The conference recommended the establishment of a cross-sector food organisation. This was the kick off of the Food Partnership Brighton and Hove. The Partnership was finally registered as a not-for-profit company in November 2005.

As one of the first key projects the Food Partnership developed a food strategy for Brighton and Hove. The process started with workshops for the development of aims and measures. These drafts were discussed between the sustainability team and the respective administrative officer in one-to-one meetings.

> The final consultation process for the draft strategy and action plan was intended to reach as wide an audience as possible, from individual residents to neighbourhood community groups, to local catering businesses and statutory agencies. This approach reflects the relevance of food and food work across the whole community, and the importance of inspiring action at both the individual and the institutional level (Food Matters, 2003: 2).

This consultation process started with a conference on the 14. June 2005 attended by over 100 invited key stakeholders. In addition to this event some smaller meetings with social and food activists were held. Additionally the Food Partnership had stalls at different events relating to food. A summary of the food strategy complemented with a questionnaire was widely distributed. The returning proposals and requests were used to adapt the food strategy. In a first step the final version of the food strategy was adopted by the Sustainability Commission, in second step by the city council. A review of the Food Strategy started 2009.

The key projects of the Brighton and Hove Food Partnership include:
- *Good food grants.* Since 2006 the Partnership offers financial support to projects aimed towards healthy food and a sustainable food system via a small grand scheme. In the year 2006 24 projects of schools and social initiatives could be supported with a total amount of £ 15,000, in 2009 £ 25,000 were available to support 30 projects.
- *Local food festivals.* In the years 2004 to 2006 three local food festivals helped to bring local food in the minds of the people and to promote local producers.
- *Harvest Brighton and Hove.* This joined project is lead by the partnership and awarded by the lottery fund with £ 500,000. The project aims to make available space and the skills necessary to grow food. It will improve the access to local food and raise the awareness for it.

With the Food Partnership Brighton and Hove created an instrument which deals with the food system on an urban wide and cross-sector level. It is open to the public and all interest groups but has also a strong connection to the municipality. With that combination it has the possibility to mirror and foster the manifoldness of the food system.

The Brighton and Hove Food Partnership could benefit from two circumstances during its founding process. Brighton already had a population with good food awareness. A diverse and critical consumer ship seems to be more important for the current development of the food system than a food cluster in the economic sense. Itself a pioneer in establishing the Food Partnership it could profit from earlier examples of FPC. The Toronto FPC for example, founded in 1990, was initiated after decades of food activism and by a focus on health (Blay-Palmer, 2009). Brighton focused already in early stages of the discussion on the instrument FPC and did choose a very cooperative way to establish a FPC involving the public from the early beginning.

The Food Partnership is organised in a hybrid model avoiding some of the drawbacks and ensuring many of the advantages of the 'government organisation' model. Trough the Food Strategy the Food Partnership established food as a cross-sector issue in the public administration despite the partnership itself is not part of the government. The partnership is competing with other NGO's for public funding but also found a way to be a sponsor itself by the Good Food Grant.

The establishment of the food partnership was resulting from a discussion about sustainability and health, issues in which the main stakeholders are rooted in. This lead to strong and well-defined aims of the partnership but implies as well a restriction. The economic development of the food system is one of the targets of the food strategy but the concentration on local food and the neglect of other important economic functions is a constraint here.

5.5 Recovering a new local level

The shift in the spatial scale of the food systems had not only implications on the provision of food but as well on urban food policies. The function of cities within food systems is reduced to a replaceable consumption and market place. The urban food system seemed to be redundant for food supply. The city widely accepted its passive role as consumer and the responsibility of the national, European and global level for food policies. Due to health and social problems, vigilant consumers and growing social movement cities are confronted with food matters anew. The awareness is rising that the urban food system is the level where a lot of food problems are becoming visible and that the reasons for many of these problems are urban consumption habits. There is a need for a new local level within food policy.

A functioning urban food system may not be important for assuring food supply nowadays. But the city and the well being of its residents depend on the food system and its functionality. This imbalance of dependency is the main dilemma urban food policies have to deal with. The city has a need to affect the food system on the local level. Parallel to the multifunctional character of the food system urban interests are manifold and diverse. One reaction to the dilemma of the imbalance of dependency is the attempt to downsize food systems to the local level, to create alternative food systems that depend on the local. But these attempts run the risk of getting lost in the vacuum that the dominant food system has created on the local level: The benefits of local food supply are not inherent anymore; the traditional confederate are orientated to higher levels and towards the dominant food system.

The other approach is to bundle the remaining urban stakeholders, to gather local interests and needs to create a new level of exchange and activity within food systems. The case study of Brighton showed how the creation of such a new level could work. Food Policy Councils (as the Food Partnership in Brighton) aim to develop the entire food system. They are engaged with all (or many) issues in the food system. They are able to identify links, to develop synergies and to work with the systematic interrelations of the food system. Food Policy Councils start with the development of visions and continue with a conceptual framework for action – as Brighton demonstrated in an almost ideal manner. This ensures the benefit for the food system and the city. With their approach of relying on co-operation of stakeholders and bundling of interests Food Policy Councils have the potential to create a constitutive element in urban food systems which was lost through the up scaling of food systems. Food Policy Councils are undertaking efforts to localise food policies. They attempt to widen the scope for local action and to use it for urban interests. This could be seen as precondition for localising food systems and any other local food projects.

References

Atkins, P. 2003. Is it urban? The relationship between food production and urban space in Britain 1800-1950. In: Hietala, M. and Vahtikari, T. (Eds.) The landscape of food. The food relationship of town and country in modern times. Finnish Literature Society, Helsinki, Finland, pp. 133-144.

Blay-Palmer, A., 2009. The Canadian pioneer. The Genesis of urban food policy in Toronto. International Planning Studies 14: 401-416.

Borron, S.M., 2003. Food Policy Councils. Practice and possibility. Eugene, OR, USA.

Brighton and Hove Food Partnership (Ed.), 2006. Spade to spoon: making the connections, a food strategy and action plan for Brighton and Hove. Brighton and Hove, UK.

Community Food Security Coalition, 2010. Council list. Portland. Available at http://www.foodsecurity.org/FPC/council.html.

Dahlberg, K.A., 1994. Food policy councils. The experience of five cities and one county. Tucson, AZ, USA. Available at http://unix.cc.wmich.edu/~dahlberg/F4.pdf.

Food Matters (Ed.), 2003. The Brighton and Hove foodshed. Mapping the local food system. Brighton and Hove, UK.

Friedmann, H. and McMichael, 1989. Agriculture and the state system: the rise and decline of national agricultures, 1870 to present. Sociologia Ruralis: 93-117.

Hamilton, N.D., 2002. Putting a face on our food: how state and local food policies can promote the new agriculture. Drake Journal of Agricultural Law 7: 408-452.

Harper, A., Shattuck, A., Holt-Giménez. E., Alkon, A. and Lambrick, F., 2009. Food Policy Councils: lessons learned. Institute for Food and Development Policy, Oakland, CA, USA, 66 pp.

Lang, T., Rayner, G., Rayner, M., Millstone, E. and Barling, D., 2005. Policy Councils on food, nutrition and physical activity. The UK as a case study. Public Health Nutrition 8: 11-19.

Mendes, W., 2008. Implementing social and environmental policies in cities: The case of food policy in Vancouver, Canada. International Journal of Urban and Regional Research 32: 942-967.

Montanari, M., 1993. Der Hunger und der Überfluss: Kulturgeschichte der Ernährung in Europa. Beck (Europa bauen), München, Germany, 261 pp.

Morgan, K. and Sonnino, R., 2010. The urban foodscape: world cities and the new food equation. Cambridge Journal of Regions, Economy and Society 3: 209-224.

Pothukuchi, K. and Kaufman, J.L., 1999. Placing the food system on the urban agenda: the role of municipal institutions in food systems planning. Agriculture and Human Values 16: 213-224.

Roberts, W., 2010. Food policy encounters of a third kind: how the Toronto Food Policy Council socializes for sustain-ability. In: Blay-Palmer, A. (Ed.) Imagining sustainable food systems. Theory and practice. Ashgate, Aldershot, UK, pp. 173-200.

Schiff, R., 2007. Food Policy Councils: an examination of organisational structure, process, and contribution to alternative food movements. Murdoch University, Perth, Autralia, 454 pp.

Schiff, R., 2008. The role of food policy councils in developing sustainable food systems. Journal of Hunger & Environmental Nutrition 3: 206-228.

Spannagel, R. and Bange, H., 1999. Standortpolitik und Strukturwandel im Einzelhandel. In: Beisheim, O. (Ed.) Distribution im Aufbruch. Bestandsaufnahme und Perspektiven. Vahlen, München, Germany, pp. 564-580.

SSAWG (Southern Sustainable Agriculture Working Group), 2005. Food security begins at home. Creating Community Food Coalitions in the South. Fayettville, AR, USA.

Steel, C., 2008: Hungry city: How food shapes our lives. Random House UK, London, UK, 400p.

Stierand, P., 2007a. Food partnership Brighton and Hove. Interview with Francesca Iliffe. Brighton and Hove. mp3 file.

Stierand, P., 2007b. Food partnership Brighton and Hove. Interview with Ann Baldrige. Brighton and Hove. mp3 file.

Stierand, P., 2008. Stadt und Lebensmittel. Die Bedeutung des städtischen Ernährungssystems für die Stadtentwicklung. Dissertation, Technical University of Dortmund, Germany. Available at http://hdl.handle.net/2003/25789.

Teuteberg, H.-J., 1987. Zum Problemfeld Urbanisierung und Ernährung im 19. Jahrhundert. In: Teuteberg, H.-J. (Ed.) Durchbruch zum modernen Massenkonsum. Lebensmittelmärkte und Lebensmittelqualität im Städtewachstum des Industriezeitalters. Coppenrath, Münster, Germany, pp. 1-36.

Wiskerke, J.S.C., 2009. On places lost and places regained. Reflections on the alternative food geography and sustainable regional development. International Planning Studies 14: 369-387.

Chapter 6

How food travels to the public agenda

Simone Plantinga and Petra Derkzen
Rural Sociology, Wageningen University, Hollandseweg 1, 6706 KN Wageningen, Netherlands;
simone.plantinga@gmail.com

Abstract

Our daily bread is entering the public domain again after decades of neo-liberal market policies and consumer choice doctrine in food provenance. A combination of factors, described as the 'new food equation' (Morgan and Sonnino, 2010) make that food supply matters again as a political issue. Increasingly, awareness grows of the relations of various (urban) problems with food (Wiskerke, 2009). And equally there is more attention towards the nature of the problem; a systemic problem related to the functioning of modern agriculture, agri-business and retail rather than a problem which can be solved by individual choices. Food is emerging as a policy concept by which it is possible to look at problems in a new and connecting way. Food policy can imply a shift from sectoral thinking and acting towards a more integrated and territorial way of thinking about policy. One of the governance instruments to come to such food policy at the level of a city is by forming a Food Policy Council (FPC). An FPC could fit with the Dutch tradition of consensus and network governance. So far a 'Food Policy Council' does not exist in The Netherlands. Research carried out in the city of Tilburg (the Netherlands) aimed to find the different possibilities and restrictions related to the establishment of a Dutch version of an FPC in Tilburg. This chapter analyses the possibilities and restrictions of a FPC using the following concepts: urgency, representation, role of the government, responsibility and scale. The results show how food as a public issue is a gradual process in which sense of urgency is a key indicator. Lack of this and therefore of responsibility show that many stakeholders are halfway.

Keywords: Food Policy Councils, governance, food policy

6.1 Introduction

Our daily bread is entering the public domain after decades of neo-liberal market policies and consumer choice doctrine in food provenance. A combination of factors, described as the 'new food equation' (Morgan and Sonnino, 2010) are placing food supply matters on the political agenda. With rapid urbanisation the number of people dependent on an external food supply increases while the food price surge of 2007/2008 shows how vulnerable the globalised food supply is. Environmental and socio-political influences of land-use conflicts, 'new' colonialism, climate change and changing eating habits all indicate that food is re-emerging as a political issue after decades of being a taken for granted commodity in at least the developed world (Morgan and Sonnino, 2010). Hence, there is a growing awareness for the relations of various (urban) problems with food (Steel, 2008).

Equally there is more attention towards the nature of the problem; a systemic problem related to the functioning of modern agriculture, global agri-business and retail rather than a problem which can be solved by individual choices.

Through the ongoing modernisation of agriculture and consequently of food production, the place of production has become irrelevant. Wiskerke (2007) mentions three concepts to describe this process:

1. *Disconnecting*: production and consumption have become separated, through physical distance.
2. *Disembedding*: the place of production is losing its influence on the quality and nature of products.
3. *Disentwining*: production and supply chains become more and more specialised and separated.

These three processes have caused products as well as regions to be exchangeable, enforced by economic competition: principles of cost-effective production and economies of scale (Wiskerke, 2007). Other problems regarding food production are: environmental pollution and ecological degradation, which is closely linked to the intensive nature of current farming practices (Wiskerke, 2009) and the loss of organoleptic quality and diversity, due to the use of high productive animal breeds and plant varieties and the standardisation of food production and processing techniques (*ibid.*)

Problems regarding food consumption are related to *trust* and *health*. After various animal disease outbreaks, consumers are no longer blindly able to trust the safety and quality of their food. This gap in trust is palpable both physically as well as in terms of consumer perception, and is also related to the marketing strategies of the retail sector. Product marketing is increasingly focussed on consumption; and the dirty business of food production is hidden.

The top three causes of disease in the developed world are all food-related (diabetes, obesity and heart disease). It is now common knowledge that obesity poses a major health threat, in Europe obesity affects 10% to 38% of the population. With obesity comes with an increased risk of diabetes, cardiovascular diseases and certain forms of cancer, creating public health costs (Wiskerke, 2009). On the other side of the story, worldwide hunger and malnutrition is on the rise. In Western societies, it is mainly the elderly that are at risk of suffering malnutrition, either in care facilities or hospitals as a result of the lack of time and resources. of healthcare staff, but mainly by the limited budgets for food in the healthcare sector. Another group suffering from malnutrition is the inhabitants of the so-called 'food deserts', areas in which little or no fresh products are readily available, and where people with low incomes are essentially forced to buy cheap pre-packaged and industrially processed food (Wigley *et al.*, 2003).

Government is therefore finding its way back to the dinner table, as reflected in the recent 'Sustainable Food Policy' of the Dutch Ministry of Agriculture (Nota Duurzaam Voedsel, 2009). Food is emerging as a policy concept by which it is possible to look at (urban) problems in a new and connected way (Steel, 2008). Food policy can imply a shift from sectoral thinking because 'it is integrated as it connects and creates synergies (or has the potential to do so) between a wide variety of issues, sustainability objectives, and public domains that are directly or indirectly related to food' (Wiskerke, 2009: 377). 'It is territorial in a two-fold way', Wiskerke argues (*ibid.*). On the one hand food connects producers and consumers in a specific region, and thereby supports the local food economy. On the other hand it is about the mode of food governance, which is determined by local specifics like cultural traditions, food related problems and the characteristics of local agriculture.

One way to address these problems and to find solutions is to organise collaboration, joint thinking and synergy among the different stakeholders. One of the governance instruments to merge such joint forces and to address food policy at the level of a city is by forming a Food Policy Council, an instrument originally developed in Canada and the US (Blay-Palmer, 2010). Research carried out in the city of Tilburg (the Netherlands) aimed to find the different possibilities and restrictions related to the establishment of a Dutch version of a Food Policy Council in Tilburg from the point of view of the stakeholders. The remainder of this chapter first analyses the literature on Food Policy Councils in relation to the Dutch context (Section 2), subsequently introduces the empirical study's methods (Section 3) and then follows to describe the results of the interviews with stakeholders in the city of Tilburg (Section 4). The chapter's final section concludes and puts the results in a wider discussion.

6.2 Food Policy Councils

Food Policy Councils, at city or state level, are found mainly in the US, Canada and Australia. A well-known and well-documented example is the Toronto Food Policy Council (TFPC) (Blay-Palmer, 2010). FPCs are networks consisting of various actors represented in the food system. The food system is comprised of private and public actors, e.g. farmers, processing industry, retailers, health care organisations, environmental NGO's as well as government representatives from various fields of policy. FPCs can be official bodies of government, but may exist outside government structures as well. There are cogent arguments both for and against governmental and non-governmental forms. Non-governmental FPCs are expected to be more critical towards the food system but are less likely to have access to (structural) government funding. Governmental FPCs, such as the TFPC, are more likely to receive support (financially and politically) for their ideas and activities (Borron, 2003; Pothukuchi and Kaufman, 1999). The goal of the FPCs is to map and improve the food system, in order to obtain a sustainable food system in an ecological and socially responsible way (Hamilton, 2002; Pothukuchi and Kaufman, 1999; World Hunger Year, 2004 in Schiff, 2007). The main tasks of FPCs are: advising, bridging, networking and providing leadership, which is expressed through various activities, such as: conducting research and writing reports, developing projects, education, supporting other organisations and facilitating network meetings (Hamilton, 2002; Lang *et al.*, 2004; Pothukuchi and Kaufman, 1999; Schiff, 2008).

In The Netherlands, so far, there are no similar examples of FPCs that compare to the well established Toronto FPC. Some cities and municipalities are developing so-called 'urban food strategies', with, to date, a low degree of embedding in the political organisation and policies of these cities. However, as a governance instrument, a Food Policy Council could fit the Dutch socio-political tradition of consensus and network governance. The Dutch political arena is characterised by multiple political parties, coalition governments and a high degree of sectorialisation due to largely autonomous ministerial departments (Andeweg and Irwin, 2002; 139, 148). The horizontal fragmentation is matched with a high degree of sensitivity by sectoral departments to lobbying by interest organisations in their field. This vertical cooperation in policy sectors is referred to as '(neo) corporatism' and has been particularly evident in agricultural policy during the period of modernisation, known as the Green Front (Frouws, 1994). Today, there is a much lower degree of institutionalised incorporation of organised interests but this informal form of corporatism has become known internationally as the Dutch 'Polder Model' in the late nineties of the previous century (Derkzen, 2008). Various cities and civic organisations have begun looking at food as an issue in their own local context. An example is Tilburg, where an environmental NGO 'BMF' is actively promoting food and food policy as an important issue for public debate and action. The empirical study underlying this chapter has been conducted in Tilburg in close collaboration with the environmental NGO BMF.

6.3 Methods

The conducted empirical research can best be defined as explorative field research. This type of research is suitable for describing, interpreting and clarifying behaviour, opinions and 'products' of actors in a usually limited, existing research situation (a field) ('t Hart *et al.*, 1996). The research area comprises the municipality of Tilburg, which consists of the city of Tilburg and two villages. Tilburg is located in the Province of Noord-Brabant, in the southern part of the Netherlands, a province characterised by its dense population and high rate of urbanisation. Of the ten largest Dutch municipalities, three cities in Noord-Brabant are represented. This area was chosen after conducting some explorative meetings with the environmental NGO and a representative of the Province, who were both key informants during the research. Data was collected through explorative topic interviews. The topics are derived from the central research question: What are the preconditions, possibilities, and restrictions for setting up a Food Policy Council?

The respondents were selected from within the network of the two key informants and included stakeholders in and around Tilburg. The selection was based on individual, local involvement and activity of the stakeholders in the local food system of Tilburg. A total of twelve interviews were carried out. Table 6.1 shows the respondents, arranged on the scale of their main activities. The transcribed data was sent to the respondents by email for feedback.

Table 6.1. Respondents, arranged on the scale of their main activities.

Scale	Respondent
Neighbourhood	• Social welfare organisation (MvdZ and KV)
	• Sustainable Tilburg (LB)
City	• Municipality of Tilburg – city councillor (TvdW)
	• Muncipality of Tilburg – Policy makers (WC and JV)
	• Muncipality of Tilburg – alderman (MM)
	• Provincial healthcare organisation (JvdS)
	• Transition Town Tilburg (HA)
	• Estate where retail, recreation and nature are combined – De Groene Kamer (TK)
Province	• Environmental NGO (JV and PR)
	• Province of Noord-Brabant – Department of ecology (EvdB)
	• Social and cultural planning agency (SC and GG)
	• Agricultural innovation organisation (GW)

6.4 A Food Policy Council in Tilburg: the results

Based on interviews with stakeholders in Tilburg this chapter analyses the possibilities and restrictions of an FPC according to the respondents, using the following criteria: urgency, scale, role of the government, responsibility and representation.

6.4.1 Urgency

The environmental NGO BMF pitched the subject of food as an integrating concept and the notion of the FPC within its network by setting up activities related to food and by seeking cooperation with other organisations in the area. A successful example is the participatory urban development of an area close to Tilburg. The urban planning was produced by a consortium of farmers, citizens and supported by architects. And the design is called Nieuwe Warande. In this plan agriculture, retail and education are combined within a new neighbourhood. This plan was handed over to the Municipality of Tilburg, as an inspirational document. Through this and other activities conducted by BMF, several respondents in the research became aware of how the subject of food connects to other fields of policy. Initially only a few respondents were familiar with the concept FPC.

The respondents were asked about their perception of the problems and issues regarding food production and consumption. All respondents acknowledged the high impact of the food related problems. Summarised, the main problems, as mentioned by the respondents, are:

- the large scale of food production (in the primary industry as well as the processing industry), results in a psychological distance between producer and consumer, and contributes to the anonymity of food provenance;
- the place of production – the countryside is under pressure (especially in the Province of Noord-Brabant, where there is a large debate going on about Q fever and mega barns), and;
- obesity and malnutrition, caused by poverty and a lack of knowledge.

The respondents agreed upon the need to address the current food production and consumption problems. However, most respondents did not relate this need to their own organisation and their potential role in food policy or food governance. Respondents working on 'food' at neighbourhood level (see both Table 6.1 and 6.2) had little affinity with 'policy' in general but saw the advantage of sharing knowledge and experience. Others questioned the usefulness and necessity of 'new structures' such as an FPC, or in renewed food policy. According to the representative of 'De Groene Kamer' 'new structures are not needed per se, but they could speed-up the process' (TK, interview March 2010). Additionally a commonly held opinion amongst city government officials and politicians was the idea that an FPC or food-system related governance should 'come from society'. The city government is willing to support bottom-up initiatives, but will not put food on its own agenda. 'If no one brings it up, then it will not happen, it [food] is not a primary issue. At this point we have to cut our expenses, there is only attention for the key tasks' (aldermen MM, interview April 2010).

Table 6.2. Food related activities, arranged on the scale.

Scale	Activities
Neighbourhood	• Cooking workshops for children
	• Neighbourhood restaurant 'Van harte resto'
	• Children's farm
	• Allotment/children's gardens
	• Comic book about sustainable food for children 'Eetje Eerlijk'
	• Selling vegetables on a barrow pulled by a donkey
City	• Climate program
	• Specific developments in the rural area
	• Green roofs subsidy
	• Development of an estate where retail, recreation and nature are combined – De Groene Kamer
	• Consumer network – food box scheme
	• Subsidies for children's gardens
	• Food teams
	• Nieuwe Warande
Province	• Rural – urban development forum
	• Development of provincial landscapes
	• Support, development and innovations in the primary sector

However, there was an initial acknowledgement of the inherent relationship between 'food' and 'sustainability' made by the city councillor who said; 'If we want to keep up our position as the most sustainable city [of the Netherlands], we have to pay attention [to food]' (city councillor TvdW, Interview March 2010).

A conclusion, posited at a feedback meeting with respondents and other interested parties, was that the urgency for an FPC appears to be lacking in Tilburg. When this conclusion was presented to the different stakeholders, the response contrasted slightly with the results of the interviews. There is a sense of urgency related to the acknowledgement of how big the food related problems are but these phenomena are not yet perceived as personal and local organisational matters. The need for a translation of the systemic problems to lower levels of organisation such as at the municipal level was felt only by a small group of stakeholders.

6.4.2 Scale

The Province of Noord-Brabant is very densely populated (about 500 inhabitants per square kilometre) and highly urbanised. In the middle of the Province four of the main cities (two cities of over 200,000 and two cities of over 100,000 inhabitants), are surrounded by a number of smaller cities (over 25,000 and over 50,000 inhabitants). As a result of the geographical situation, the municipalities have a history of cooperation on various issues. Due to the population density and high level of urbanisation, combined with the intensive form of agriculture in this province, various problems arise. Several outbreaks of animal disease has occurred here in recent years, leading to an intense debate about the scale and intensity of agriculture. In such a situation the question is at which scale an FPC could be effective. Most respondents in the research preferred the scale of the city of Tilburg for an FPC, followed by the provincial level. At the level of the province one argument for an FPC is that 'there is no long-term vision for Noord-Brabant at the provincial level, the time is right for political attention' (JvdS, interview April 2010). During the research, the FPC concept was clarified using examples of urban food policy and urban food strategies in the Netherlands and elsewhere. Those opposed to a FPC on a city level were afraid it would lead to policy cacophony if every municipality would have an FPC (GW, interview March 2010) and therefore pleaded for an FPC on the provincial level. Other arguments, referred to the need for an even larger scale, to be able to function as a foil to the food industry (GW, interview March 2010). Since the respondents were selected on the basis of their involvement with 'food' in some way, an overview of all their food related activities is shown in Table 6.2. The most concrete activities are all focused on consumer awareness at the neighbourhood level, contributing to knowledge and practices that stimulate sustainable and healthy consumption. While these activities are both valuable and necessary, it was remarkable that the respondents felt that 'once the consumer knows better, he/she will consume more responsibly', and hence the system would change.

The contrast in scale is stark; on the one hand there exists a systemic problem, on the other hand the solution is defined at the individual level. The middle ground of joint action on a municipal or provincial level has yet to be perceived. Moreover neither the processing nor the retail industry were mentioned by the respondents as potential partners for action towards a more sustainable production and consumption model.

6.4.3 Role of government

As food begins to enter the public domain, due to different reasons mentioned in the introduction of this chapter, government is expected to assume responsibility for what has been (or still is) seen as a private business and consumer issue. An FPC or more general, the development of food policy demands an active role by the public sector, a role for which the government of Tilburg is not prepared at this moment. Four respondents working at different levels within the city government did not see initiating an FPC as their task. Other respondents, equally not keen on their own involvement did however maintain the expectation that the city government should initiate, facilitate, and stimulate the process of setting up an FPC. The city government countered by stating that it is only willing to support 'bottom-up' initiatives. Their unwillingness is related to its current priorities: of which 'food' is not one, absent as it is from political or policy agendas. At this point in time, the municipality has to cut expenses, and apparently food sounds like an expensive issue. Moreover, it could well be that the city government, confronted with a new issue or 'demand', perceives food as an additional burden on the budget rather than as an integrative instrument to explore new synergies or alternative city planning and policy targets. Whatever the reason, food is rejected as an issue to which the city government of Tilburg feels responsibility.

6.4.4 Participation or representation

The concept of an FPC fits well with the Dutch governance and consensus tradition, 'the Polder Model'. Governance networks and other types of participatory or bottom-up projects are found throughout the country. So too in Noord-Brabant where so-called 'reconstruction committees' are part of the mandatory provincial countryside development plan; seven regional committees were established in 2005 to agree on zoning and relocation of farms and nature preservation. The committees are comprised of (local) representatives from organisations such as the environmental NGO 'BMF', the province, the farmers' interest group ZLTO, municipalities, the water board and other regional interest groups. The committees have worked on a legally binding spatial planning with far reaching consequences for farmers in particular. If consensus could not be reached, the national government would take over the role of the committees.

In forming an FPC, other governance networks could be taken as an example, in so far as participation and representation are key factors (Derkzen and Bock, 2009). Questions to be asked are: Who participates? And: Which role or interest is at stake? A characteristic of governance in the Netherlands, and more specifically of corporatism, is that governance networks have a high degree of co-decision but therefore, these networks are usually comprised of the 'known' sectoral interest groups best equipped to deal with the complicated negotiation format (Derkzen, 2008). This is true for the reconstruction committees in Noord-Brabant as well as for local governance networks in the city of Tilburg. The city of Tilburg works with 'De Groene Mal' a nature conservation and development network which consists of the exact same stakeholders as those in the reconstruction committees.

The high stakes make these committees more focused on representation than on voluntary and passion-based participation. However, pure representation of sectoral interests can lead to sectoring, 'sectoring' is oriented towards protecting, if not advancing the differentiation of one sector from another'(Degeling, 1995: 294; Derkzen *et al.*, 2009) and in which the boundaries of the sector remain beyond question.

A high risk for an FPC in the Dutch context therefore would be to repeat the existings format in which all of the well-equipped interest organisations meet each other for the next negation issue; food. It is important that participation in a governance network such as an FPC, is based on voluntarism, personal involvement, and passion to achieve the creativity needed for future solutions (see Blay-Palmer, 2010). In order to avoid 'sectoring' in this research, respondents were chosen according to their individual activity within the Tilburg food system/network instead of on their representation within the sector, resulting in interviews with relative 'outsiders'. However, during the interviews most respondents mentioned the 'known' sectoral interest groups as possible participants in an FPC. Furthermore, only one interest group per sector was mentioned. This could result in further strengthening the possibility of sectoring (see Derkzen *et al.*, 2009).

6.5 Conclusion and discussion

In order to conclude this chapter, the different restrictions and possibilities for forming an FPC will be further analysed. Given the results of the research, it is clear that the actors in the Tilburg food system are aware of the various food related problems. However, this does not lead to a sense of urgency to tackle these problems, nor does it stimulate the actors to take responsibility in doing so. Furthermore, the stakeholders are not clear about their own role in a possible FPC or even willing to have a chair in it (such as the NGO BMF who pitched the idea). The various organisations are clear about the role that other organisations could take, especially about the role of the municipal government, which they expect to initiate and stimulate an FPC. The government on the other hand does not see this action as their role, further amplifying the pervasive lack of responsibility.

One key factor for this might be the unfamiliarity with and 'newness' of food as a policy issue for the municipal government. Lack of overview on how food is connected to many of the problems the city is facing with regard to its ambitious sustainability policy might influence the response of the city respondents.

Indeed, the concept of an FPC required a lot of explanation during the interviews. Even though 'food' plays a great role in the respondents' current activities, using it as a tool for integrating and connecting different fields of policy, or as a lens to view problems in a broader perspective is not routinely considered. To the city government the establishment of an FPC may sound like the creation of policy and consequently budget to the existing fields/sectors of policy.

A further possible explanation for the lack of willingness to undertake action may be that the concept of an FPC itself is 'foreign' and therefore hard to engage. It is not in use in the Netherlands yet, and comparable concepts, like urban food strategies, are still a developing practice. Another possible reason for the reluctance of some respondents to embrace the concept could be the existence of the word 'policy' in 'FPC'. To a number of neighbourhood practitioners on the neighbourhood-level, 'policy' is the opposite of 'practise' and policy usually leads to nothing good.

In any case, a clear conclusion is that the respondents in this research perceived no urgency for an FPC in Tilburg at this point. The systemic nature of the problems is widely acknowledged but solutions are being sought at the individual consumer-level. An often repeated response was that consumers need to become aware and knowledgeable about food and the food system.

We might say in response that a similar awareness is needed at the level of established interest organisations and government. An indication that food is 'becoming public' in a gradual process can be found in the earlier mentioned feedback meeting organised as a discussion of the results of this research. Much to our surprise, over 30 people were present from diverse organisations ranging from local to national levels. During the meeting various people expressed their willingness to share knowledge and exchange experiences resulting in an at-hoc email-database and a follow-up meeting in September 2010 organised by the NGO 'BMF' and Transition Town Tilburg.

The geographical situation of Noord-Brabant, densely populated and highly urbanised, raises questions about the desired level of scale for a Dutch FPC and in a more general sense, about the level of government at which food should be present on the agenda. Several cities are situated within a small area, pointing to cooperation between these cities as more desirable than the situation in which individual cities all form a (partly overlapping) FPC. In this sense, a regional approach to food would be wise to consider, as several respondents mentioned.

Whichever path is taken, a future FPC could both build on the Dutch tradition of collaboration and consensus or be constrained by it. There is a danger that an FPC begins negotiating such that all parties have a stake in it, rather than creatively re-thinking our urban relationship with food. It is therefore important that whoever takes the lead in forming an FPC is aware of the tendency of thinking within policy sectors and their representatives. This phenomenon frequently leads to solutions of the 'lowest common denominator' by which none of those involved are threatened. The vast and integrated nature of urban food problems, however, beg for equally nuanced and holistic actions. Bringing together people who are passionate about the subject of food, health, and sustainability in a 'think tank' context or other inspirational setting is a likely first step.

References

Andeweg, R.B. and Irwin, G.A., 2002. Governance and politics of the Netherlands. Palgrave, New York, NY, USA, 304 pp.

Blay-Palmer, A., 2010. The Canadian Pioneer; the genesis of urban food policy in Toronto. International Planning Studies 14: 401-416.

Borron, S.M., 2003. Food Policy Councils: practice and possibility. Bill Emerson National Hunger Fellow Congressional Hunger Center Hunger-Free Community Report, Eugene, OR, USA, available at http://www.edibleaustin.com/content/sustainable-food-policy-board-resources-178/246?task=view&69b07cc833e0fcb9f5e80e74070ef07f=6a7069d.

Degeling, P., 1995. The significance of 'sectors' in calls for urban public health intersectoralism: an Australian perspective. Policy and Politics 23: 289-301.

Derkzen, P. and Bock, B.B., 2009. Partnership and role perception, three case studies on the meaning of being a representative in rural partnerships. Environment and Planning C: Government and Policy 27: 75-89.

Derkzen, P., 2008. The politics of rural governance. Phd thesis, Wageningen University, Wageningen, the Netherlands, 170 pp.

Derkzen, P., Bock, B.B. and Wiskerke, J.S.C., 2009. Integrated rural policy in context: A case study on the meaning of 'integration' and the politics of 'sectoring'. Journal of Environmental Policy & Planning 11: 143-163.

Frouws, J. 1994. Mest en macht. Een politiek-sociologische studie inzake de mestproblematiek in Nederland vanaf 1970. Phd thesis, Landbouwuniversiteit Wageningen, Wageningen, the Netherlands.

Hamilton, N.D., 2002 Putting a face on our food: how state and local food policies can promote the new agriculture. Drake Journal of Agricultural Law 7: 407-412.

't Hart H., J. Van Dijk, M., De Goede, Jansen W. and Teunissen, J., 1996. Onderzoeksmethoden. Boom, Amsterdam, the Netherlands, 381 pp.

Lang, T., Rayner, G., Rayner, M., Barling D. and Millstone, E., 2004. Policy Councils on food, nutrition and physical activity: the UK as a case study. Public Health Nutrition: 8: 11-19.

Ministerie van Landbouw, Natuur en Voedselkwaliteit, 2009. Nota duurzaam voedsel. Available at http://www.rijksoverheid.nl/bestanden/documenten-en-publicaties/notas/2009/08/11/nota-duurzaam-voedsel/nota-duurzaam-voedsel.pdf.

Morgan, K. and Sonnino, R. 2010. The urban foodscape: world cities and the new food equation. Journal of Regions, Economy and Society 3: 209-224.

Pothukuchi, K. and Kaufman, J.L., 1999. Placing the food system on the urban agenda: the role of municipal institutions in food systems planning. Agriculture and Human Values 16: 213-224.

Schiff, R., 2007. Food Policy Councils: an examination of organization structure, process and contribution to alternative food movements. Available at http://wwwlib.murdoch.edu.au/adt/browse/view/adt-MU20070906.103640.

Schiff, R., 2008. The role of food Policy Councils in developing sustainable food systems. Journal of Hunger & Environmental Nutrition 3: 206-228.

Steel, C., 2008. Hungry city. Chatto & Windus, London, UK, 400 pp.

Wiskerke, J.S.C., 2007. Robuuste regio's: dynamiek, samenhang en diversiteit in het metropolitane landschap. Inaugural speech, Wageningen Universiteit, Wageningen, the Netherlands.

Wiskerke, J.S.C., 2009. On places lost and places regained: reflections on the alternative food geography and sustainable regional development. International Planning Studies 14: 369-387.

Wrigley, N., Warm, D. and Margetts, B., 2003. Deprivation, diet, and food-retail access: findings from the Leeds 'food deserts' study. Environment and Planning A 35: 151-188.

Interviews (carried out by S. Plantinga for Master thesis research)

8 March 2010 – LIB – Geert Wilms (Agricultural Innovation Platform).

6 March 2010 – Gemeente Tilburg – Raadslid Tine van de Weyer (City Councillor).

29 March 2010 – De Groene Kamer – Thieu Kessels (Retail manager).

8 April 2010 – Gemeente Tilburg – Wethouder Marieke Moorman (Alderman).

29 April 2010 – GGD Hart voor Brabant – Jos van de Sande (Health organisation).

Chapter 7

Food system planning in small, buzz-less cities: challenges and opportunities

Melanie Bedore
University of Ottawa, 310 College Street, Kingston, ON, K7L4M4, Canada;
mbedore@uottawa.ca

Abstract

For over a decade, planning scholars have promoted measures that cities can take to incorporate food systems into the urban planning agendas. Large cities such as Toronto, Rome, London, and others are hubs of these activities such as municipal Food Policy Councils, institutional procurement programmes, and formal municipal food documents. In contrast, there is significantly less policy activity in many smaller Canadian cities. In this chapter, I focus on the challenges and opportunities facing cities that are distinct only because they are small, ordinary – even mundane – and lacking the 'buzz' of large, fast-growing cities. I draw from my doctoral research and anecdotal experience based in Kingston, Ontario, Canada. Kingston is a small city with 117,000 residents, located about two hours from larger cities like Montreal, Ottawa, and Toronto. Between 2006 and 2010, I conducted my doctoral fieldwork by exploring the urban retail food landscape of Kingston's lower-income neighbourhoods, including the closure of two full-service grocery stores. Specifically, I draw from forty-two interviews with local actors like elected officials, urban planners, real estate brokers and local food activists. I intend for this presentation to initiate a lively discussion about food systems planning in the context of the small, buzz-less city. Set within the theoretical context of Logan and Molotch's (1987) growth machine theory, I suggest that food systems planning is hampered by factors such as preoccupation with land values, limited human resources, policy learning capacity, or inter-sectoral networks. Most importantly, an ideological resistance to 'interfering' in the free market is pervasive. Policy ideas may include but should not be limited to encouraging better inter-sectoral communication, low-cost ideas for municipal food networks and programming, capitalising on local knowledge and human capital, and incorporating food into other municipal agendas.

Keywords: food planning, small cities, urban planning, growth machine theory

7.1 Introduction

For over a decade, planning scholars have promoted measures that cities can take to incorporate food systems into the urban planning agendas. Scholars have, for example, promoted urban food planning in special issues of the *Journal of Planning Education and Research* (2004) and *International Planning Studies* (2009). Some of the most exciting food system planning in the developed world is emerging in major metropolitan centres. World cities like Toronto, Rome,

London and many others are increasingly the subject of scholarly research on innovative food policy tools and programmes, including institutional procurement policies, Food Policy Councils, and food charters (e.g. Blay-Palmer, 2009; Morgan and Sonnino, 2008; Welsh and McRae, 1998). These cities are establishing important precedence, and their ideas stand to be adopted by other cities as they become increasingly aware of food as an issue for urban planning (e.g. Morgan, 2009; Pothukuchi and Kaufman, 1999).

Less well addressed in the scholarly literature, however, is the degree to which food system planning lags or is virtually non-existent in small cities. In this chapter, I draw from my recent doctoral project in Kingston, Ontario, Canada. I situate this chapter theoretically within Logan and Molotch's growth machine theory of urban development and land use politics. I use interview testimonies and anecdotal experience to illustrate the significant challenges that small 'buzz-less' cities face with respect to generating action on food planning. I group these challenges (affecting both urban planning departments and the local organisations doing local 'food work') into four categories: most significantly, ideology and pervasive faith in the 'free market'; the financial politics of urban development and land values; resource constraints; and leadership and jurisdictional challenges. In my concluding section, I suggest several opportunities and assets that small cities can seize to enhance their current level of food system planning.

7.2 Case study research

The results that I present in this chapter are based on my doctoral dissertation project, titled *The just urban food system: exploring the geographies of social justice and urban retail food access in Kingston, Ontario*. Kingston is a small city in South-Eastern Ontario with a population of about 117,000 people. It is located approximately two to three hours from major cities like Toronto, Ottawa and Montreal, and it is the largest city within its immediate region, giving it some importance as a regional commercial and administrative centre. Kingston boasts a vibrant and historic downtown that fronts Lake Ontario, as well as newer amalgamated communities to the east and west. Nevertheless, Kingston is slow-growing: much of the city's employment comes from the public sector, including local penitentiaries, post-secondary institutions, a Canadian Forces military base, and three hospitals. Due to a relatively weak private sector, the city is neither failing nor fast-growing: Kingston fails to attract immigrant populations at the provincial rate, and local economic development plods along at a slow pace, without the 'buzz' of larger cities.

While Kingston has a sizable middle class owing to its institutional employment base, for my doctoral project I explored two grocery store closures in Kingston's North End, it's lowest-income area with a mixture of gentrifying areas, public housing complexes, seniors, people with disabilities, people receiving social assistance, single-parent families, and people with compounding social problems like poverty, addiction, and chronic unemployment. Between 2006 and 2009, two small-scale, older grocery stores serving the North End closed and one new large-scale grocery store opened (see Figure 7.1). Some of Kingston's poorest residents, who live in the Rideau Heights neighbourhood, were without a nearby walkable grocery store for three years.

I studied this changing urban retail food landscape using a normative, critical political economy lens. I developed research questions asking why communities with poor retail food access (also called the 'food desert' problem) are emerging, how this situation constitutes a

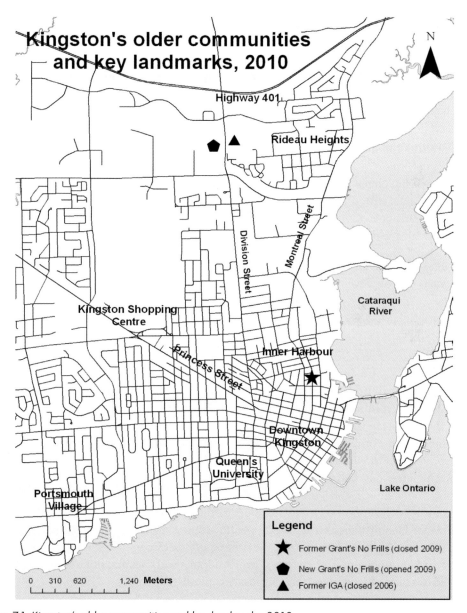

Figure 7.1. Kingston's older communities and key landmarks, 2010.

social injustice, and what alternative models and programmes might lead us to a more just urban food system. My research methods included focus groups with North End residents, door-to-door surveys in Rideau Heights, and archival research. For this chapter, however, I draw from forty-two qualitative interviews with key informants conducted between 2008 and 2009. These people consisted of local food and poverty activists and non-profit workers, food retailers, commercial real estate and property development informants, and elected city councilors and city bureaucrats. The 'foodie' community of Kingston alone could number in the hundreds, including small-scale/artisanal growers and producers, a strong farming/ agricultural community, post-secondary researchers, volunteers and consumers from a range of socio-economic backgrounds. Interviews typically lasted between one and two hours and were immediately transcribed and analysed. While I never directly asked interviewees about the challenges and opportunities of food systems planning in a small city, I observed many themes and ideas that sparked my interest in this topic. Insights came primarily – although not exclusively – from city bureaucrats, councilors, and urban planners.

7.3 Theoretical and municipal context

Many of the challenges that small cities face within respect to engaging in food systems planning can be situated theoretically within Logan and Molotch's (1987) growth machine theory. In their book, *Urban fortunes: The political economy of place*, the authors propose this concept as a way of understanding key questions of urban politics, 'who governs?' but also 'for what?'. They suggest that

> For those who count, the city is a growth machine, one that can increase aggregate rents and trap related wealth for those in the right position to benefit. The desire for growth creates consensus among a wide range elite groups, no matter how split they might be on other issues.... Although they may differ on which particular strategy will best succeed, elites use their growth consensus to eliminate any alternative vision of the purpose of local government or the meaning of community. (p. 50-51)

According to Logan and Molotch, members of the growth coalitions are those who can benefit strongly from local economic development, namely increasing land and property values. They dedicate time and resources to local affairs because they stand to gain or lose disproportionately compared to the average citizen. Elected politicians, for example, have a complex relationship with local business elites and professionals like lawyers and property brokers; the latter can be influential in ensuring that politicians are elected to office, and the former, once in office, have systemic power to create a favourable policy environment for the business community. Other members of the growth coalition who stand to benefit from increased land values and aggregate wealth – and a role in promoting and maintaining growth – include local media, utilities, and other quasi-public agencies, universities, cultural institutions like museums and theatres, professional sports teams, organised labour, self-employed or corporate retailers, and corporate capitalists.

For this project, Logan and Molotch's contribution to urban political theory is perhaps most significant in terms of the inevitable centrality of land and property values to planning debates and decisions. In Ontario (and Canada more broadly), municipalities rely heavily on property taxes to fund the delivery of services such as police, fire and ambulance services, waste management, sewage and utilities, and other key functions (Association of Municipalities of Ontario, n.d.). This creates an incentive for municipal governments to approve land use and development projects that increase property values and the local tax base. This financing scheme has become more complicated in Ontario since the late 1990s, however. In the late 1990s, the provincial Progressive Conservative party, led by Premier Mike Harris, restructured the provincial-municipal spending burden for social services. As part of his Common Sense Revolution platform, Harris mandated that the Ontario provincial government assume financial responsibility for education, while social services were downloaded onto municipalities (Keil, 2002; McMullin *et al.*, 2002). To this day, Ontario municipalities struggle with this funding arrangement: first, there is division and debate as to the extent to which municipalities must take up social agendas and social problems that stretch their role as traditional service providers (B. Donald, personal communication, June 15, 2010). Second, and most relevant to this chapter, food sits awkwardly within both the traditional realms of land use planning, retail development, tourism and business development, and many problems like poverty, food insecurity, homelessness, and food access. As I illustrate in the remainder of this chapter, municipal actors continue to wrestle with these questions of the role of food in urban and regional planning.

7.4 Challenges to food system planning in small, 'buzz-less' cities

7.4.1 Ideological resistance to food systems planning

Perhaps most significantly to this project, there are many ideological reasons for cities to be cautious or even opposed to engaging in food system planning – especially 'interfering' in the retail food landscape of a city. With respect to low-income communities like the North End that, for a time, were underserved by retail food capital, several elected councilors expressed their faith in the private market for food provision, as well as caution about interfering using taxpayers' money:

> *So we live in Canada, in a free market society. It's not a society where the government can dictate where you and I work and what we do for a living, and the government doesn't get into the grocery store business in Canada. I sympathize with what people would like to see and people feel that certain areas, where they don't have transportation, that people would like to see more grocery stores, but there isn't anything the government— the municipal government can do—to force a business into place.... Well, I might think that if the reality was that the market wasn't taking care of those needs, but I don't think that's the case, I think that the market is taking care of the needs of citizens in terms of providing them choice and locations.... You know, I don't think there's any cause for government intervention if the government could intervene and I don't think that in the present legislation that we can* (elected official).

Well, I've already made the point that I think that the food market is your ultimate democracy. So if you, in a way, then, try to direct that market rather than letting the individual consumers make their own choice by either restricting the choice... I'm not sure what this sort of 'distributive' system that you're suggesting does.

At the margin, you're sort of taking public revenues in a way, and you're making decisions about the best way—and you're sort of saying 'it's better for us to do that, than say to redo the sewers,' or 'it's better to do that, than to pave the road that needs paving,' or 'it's better to do that than offer a swimming pool.' I mean, you know, you're making these choices (local business educator).

Indeed, it was not uncommon to encounter similar attitudes toward cities engaging in planning their retail food environment and intervening to improve access for underserved communities. At times, the tone implies that to do so would be tantamount to state socialism, or a gross perversion of free market logic. At other times, it seems that interviewees are simply not aware of the many cases where (often larger) North American cities are engaging in the retail food environment through zoning and subsidies, incentives, and transportation. This lack of awareness may speak to small cities' time and resource constraints, as discussed below. In any case, this resistance or caution by influential local figures suggests that food has yet to be understood as a commodity that deserves and requires distinct treatment in the field of urban planning.

7.4.2 Land values and property tax base

In many ways, interurban competition for commercial investment and the drive to increase land and property values is at the heart of food system planning challenges for small Canadian cities. The urge to accept new residential development proposals to expand the city's housing supply and property tax base, for example, has lead to poor decision-making with respect to communities without accessible retail, according to one interviewee:

Well, I could say in Greenwood [there will be future access problems], it's because the planning let it happen. Previous councils weren't more forceful about saying 'no, we want commercial [-zoned buildings] to go in there and make it work.' I think maybe it's something to do with our sprawl, we've allowed our city to just grow out, it's sort of a phenomenon of suburban culture where you drive to the grocery store in your minivan and buy twenty bags' worth of groceries for the week (elected councilor).

Development corporations, she explains, can be very argumentative about the profitability or viability of commercial space in newer-style planned subdivision communities. She explains that an identical situation will likely occur with a new community currently under construction:

So we're building a whole new 3,000 unit, 9,000-person subdivision at Bayridge and Princess St. North. And they would be happy to just put in houses. And no commercial, no institutional, like the institutional—they have to leave room for the schools, and the school board will eventually decide whether they go in there or not. But you HAVE to leave the commercial in there. And what they'll do is they'll say 'we'll have the commercial', and then they'll say 'well, the commercial isn't working, we're putting in housing'. But we're saying you cannot just go do that... (elected councilor).

Other slow-growing cities likely face similar challenges: they are compelled to grow their tax base and are nearly desperate to attract a greater portion of migrants to the city as 'evidence' of a vibrant, growing local economy. Small cities in this predicament would be hard-pressed to reject such proposals as they stand to increase housing volume and tax revenues, even if it means a lack of commercial space or retail food planning in these communities. In a way, this situation reinforces the dominance of free market values described above, as small city officials (and growth coalition members) are likely to feel that property tax, as a taxation instrument, best serves the city when left to the 'free market' for housing and residential development.

In the same line of thought, small and slow-growing cities may also be especially eager to approve proposed large-scale retail franchises and other commercial developments because they are assumed to boost jobs, customer choice, property tax income, and the city's profile. Such large-scale developments, however, can have questionable consequences: aside from external costs of 'big box' retailing like environmental damage, the net benefit to the community is debatable, as one urban planner explains using the example of a controversial proposal for a large-scale home improvement store in Kingston,

they talk about 'oh, the economy's bad, we need new jobs'. Well, okay, first of all, they're gonna pay an average of about $11 an hour. And some of those jobs aren't going to be new jobs, some of those guys are going to be Colour Your World, who already told us, if they're approved and they come to town, let me know 'cause I'm selling my franchise and getting out. So all of those employees, laid off, done.... And so you're not, really, it's not like you're getting new stores, stores moved because they want to be in a better location; you're losing jobs, that's why you heard the phrase 'net jobs', so you may get 125 new jobs, you may lose seventy-five, so you're only netting fifty new jobs, which isn't as grand as 125 (urban planner).

Approving whatever new retailer is willing to come is likely very common amongst smaller cities that are desperate to increase or maintain growth. The problem, however is that by catering to the wishes of large-scale food retailers (through land use changes and special allowances, for example), planners and officials in small cities may neglect other qualities of strong cities, such as economic diversity, beautification, well-paying employment, accessibility, or the viability of their central business district. Again, 'free market' trends in retailing may win out over quality of life interventions that are within planners' jurisdiction.

Finally, interviews with elected municipal councilors suggested that they may be reluctant to take on new food planning initiatives that would impose financial costs on the city. The limitations of the property tax base surfaces again as a reason why city bureaucracy would avoid food initiatives to increase food access in underserviced neighbourhoods like North Kingston, for example,

council is working to increase [public transit] ridership and has some significant success in the last four years, in the midst of the governments, both provincially or federally, are providing gas money to the municipalities specifically to increase ridership on mass transit. But we're already taxed in terms of the services that we provide, off the property tax base, and the municipality wants to be pretty careful in terms of adding what services they want to provide, and adding the delivery of food would be a pretty significant service (elected councilor).

This reluctance may reflect the degree to which popular beliefs about government's role in the provision of services is affected by neoliberal thinking. On a more tangible level, municipal funding arrangements are the framework within which municipal managers and officials might consider – and dismiss – more unconventional food planning initiatives. Concerned about the public reaction to the spending of public money, it is not unexpected that they would shy away from committing to provide (costly) services through a food planning framework, whatever the less tangible long-term social, health or economic benefits might be.

7.4.3 Resource constraints

Planning departments and city bureaucracies in small cities may face disproportionate constraints in their resources, limiting the potential for food system planning. One interviewee, an employee for a local non-profit food organisation, gave an extensive account of trying to work with Kingston's elected officials and planning administration with regards to urban agriculture in Kingston's recent new Official Plan. They observe that:

The planning department, they're good people, they really are, and they just want what they think is best for Kingston. I think they have very archaic views of what a planning department can do. In the US, there's a network of Mayors' Innovation something, it's a network of mayors from across the US from Chicago and LA to Scranton, PA, who meet to discuss innovative policy that they can promote and to share ideas around this, right? And the research and the committee I'm on—the Kingston Urban Agriculture Action Committee—KUAAC—the more we research, the more we see that Kingston is really far behind … it's frustrating; they're not prioritizing it, they're not reading the research we give them, nobody's really taking on, exploring, all the links in the papers and things that we can provide them with, to help to assuage their fears. And there was a planning committee meeting on December 4th, and … it was just very clear that most of the committee members hadn't read the stuff that we'd sent. And the planning department, the staff, had read some, but were really, really stuck on their own notions of what things were, right?... Maybe we just didn't get on this quick enough around the official plan, we haven't had the opportunity really make them learn what we need them to learn, so it's been frustrating (non-profit worker and food activist).

In this account, several issues emerge with respect to time and resource constraints. While it again reinforces the low priority that food issues are assigned within governments and the strength of the 'free market' mentality over active planning, this testimony also suggests that without strong networks with other municipalities or mayors, cities may not enjoy the benefits of learning about other cities' experiences, initiatives and best practices. With only so much time in each work day, planners in these departments likely do not locate and read food planning literature easily; even literature provided by local groups in not well-read. It is impossible to say whether this is because of disinterest or lack of time on the part of the planner or local politician, however it is important to assume that city officials receive this kind of literature and information from all kinds of interest groups on a variety of issues (a lot of other 'things on their plate', as this interviewee said), so it is understandable that a relatively novel – and potentially unsettling – idea like urban agriculture would fail to command the attention that it deserves in small cities.

7.4.4 Leadership and jurisdictional issues

Finally, because food system planning is new and relatively unexplored by the small city of Kingston, lack of leadership and jurisdictional cohesion is a barrier. Many aspects of local economic development can be hampered by a poor working relationship between different sectors in a small city like Kingston. Economic growth and progressive politics can be stifled when different urban actors with social agendas or business agendas fail to break down 'silos' and work together (Bedore and Donald, 2010). As one interviewee notes,

> *Social services and business have to work closer together, because of the programs cross over. Something that I talk about when I meet with the social service groups is, as much as they find it difficult to integrate into business meetings, business people, why do I walk into a room with social service advocates and feel like I have three heads on my shoulders?* (economic development officer).

Food is unavoidably a public and private entity, and virtually all aspects of urban food systems necessarily involve public and private actors. If there are poor working relationships between municipal actors, the business sector, and the social services sector, it can only mean foregone opportunities for cross-sectoral collaboration and serious mobilisation around local food-related issues.

A second jurisdictional challenge relates to the multidisciplinarity of food. Because food is inherently relevant to the mandates of municipal departments such as social services, public health, transportation, land use planning, waste and environment, recreation and culture, economic development and many others, it can be an enormous logistical challenge for these departments to collaborate, and even more so for food activists to target the appropriate municipal actors, 'One of our initial struggles was to find out which department we should even be dealing with: was it parks? Was it planning? Culture? Who would do urban agriculture, who would take that on? And the planning department is where we kept getting referred to' (non-profit worker and food activist). Moreover, some public agencies may be charged with economic development, however retail food does not fit neatly into their mandate:

> *We don't do a lot of work on the retail [attraction] side. Our corporation's broken down into two parts: Tourism Kingston and Business Development, that's under the umbrella of [Kingston Economic Development Corporation]. Within the Business Development group, we have people who work and specialize with small business clients, so that would include retail, but those us of working on larger accounts and larger companies, we generally don't get involved in retail development* (economic development officer).

While larger cities also inevitably struggle with the question of where and how food 'fits' into multiple departmental mandates, they may also have more resources and foresight to appoint key departments to take on food issues. In some cases, North American cities have been partnering members of coalitions that have actively recruited food retail into underserved areas (Pothukuchi, 2005); cities with more progressive social agendas may be open to broader thinking about the role of the public service in service delivery as it relates to food. For smaller cities that are less familiar with food planning, there may still be outstanding issues around (1) whether there is political will to address issues for smaller populations living in smaller communities; (2) who in local government should do what, and (3) whether there is the capacity to formally take on additional food planning work. Too often, neoliberal ideas around the proper restricted role of government may dominate popular thought when it comes to municipal food issues.

7.5 Conclusion: opportunities for food systems planning small cities

In this short essay, I have suggested four key challenges that affect small, 'buzz-less' cities like Kingston, Ontario that act as barriers to greater engagement in food systems planning. Related to Logan and Molotch's (1987) growth machine theory, small cities may make land-use planning decisions that enhance property values and economic development in the city, however these may be at the expense of other goals (such as communities that can incorporate accessible food retail), while causing some disincentive to take on food planning projects that generate use value but do not contribute to land use values. I also suggested that smaller cities may suffer from scant human, time and financial resources that are needed to learn about what is happening in other jurisdictions, or to become more familiar with terms like urban agriculture. There may also be ideological resistance to food systems planning of the basis of 'interfering' with the free market, or leadership and jurisdictional challenges that prevent smaller cities from taking on food systems planning. While planning departments and related organisations in larger Canadian cities are generating tremendous excitement with their policy innovations and programming, these barriers may be somewhat unique to the case of small, slow-growing or 'buzz-less' cities like Kingston, Ontario.

Despite – and in resistance to – municipal leaders' adherence to neoliberal thinking about 'free market' dynamics and food issues, there is a role for planning, and small cities may have certain assets on their side to address these challenges I have described. Short of taking on levels of service delivery requiring a municipal budget for additional salaries, for example, small cities could make excellent use of volunteer and subsidised labour to conduct food planning and policy work. Canadian cities far smaller than Kingston have universities and colleges, meaning that students might be recruited to do *pro bono* policy and planning work. Organisations may also have ample experience applying for grants and salary subsidies from funders like the United Way, providing more opportunities for affordably hiring. As well, community leaders in small cities may have an easier time assembling the local 'foodie' community, simply because this may be a tighter-knit group of people who are familiar with each other and their work. This local or student labour could be an important part of lower-cost municipal food planning initiatives, such as charters, 'friendly' policy language and participation in multi-sectoral networks.

Even though I suggested that local leaders may be resistant to deploying certain policy tools, that is not to say that much is not already being done to stimulate underserved areas. In this project, for example, interviewees mentioned that the 'food desert' problem can be addressed through initiatives such as addressing public transportation, business attraction and retention, zoning as a tool to facilitate neighbourhood revitalisation, cleanup of brownfields, and various other tools and incentives to stimulate development in Kingston's Old Industrial Area, located in the North End. Even without a clear vision of food planning in the city, reactions to the North Kingston grocery store closures suggest that there is action occurring on food issues as they emerge. Such initiatives are likely to continue as small cities like Kingston attempt to modernise their planning vision: Kingston has recently adopted its Four Pillars of Sustainability (economic health, environmental sustainability, cultural vitality, and social equity) (Sustainable Kingston, 2009), giving local activists and organisations a means of holding city leaders accountable, as well as providing planners and other officials new grounds for progress on urban food issues.

Reference

Association of Municipalities of Ontario. n.d. Government in Ontario. Available at www.amo.on.ca/YLG/ylg/govinont.htm.

Blay-Palmer, A., 2009. The Canadian pioneer: The genesis of urban food policy in Toronto. International Planning Studies 14: 401-416.

Donald, B. and Bedore, M., 2010. Revisiting the politics of class in urban development: Evidence from the study of the social dynamics of economic performance. Urban Affairs Review 47: 183-217.

Keil, R. 2002. "Common-sense" neoliberalism: Progressive conservative urbanism in Toronto, Canada. Antipode 34: 578-601.

Logan, J.R. and Molotch, H.L. 1987[2007]. Urban fortunes: The political economy of place (20th anniversary edition). University of California Press, Berkley, Los Angeles A, London, 383 pp.

McMullin, J.A., Davies, L. and Cassidy, G., 2002. Welfare reform in Ontario: Tough times in mothers' lives. Canadian Public Policy 28: 297-314.

Morgan, K., 2009. Feeding the city: The challenge of urban food planning. International Planning Studies 14: 341-348.

Morgan, K. and Sonnino, R., 2008. The school food revolution: Public food and the challenge of sustainable development. Earthscan, London, UK, 256 pp.

Pothukuchi, K., 2005. Attracting supermarkets to inner-city neighborhoods: Economic development outside the box. Economic Development Quarterly 19: 232-244.

Pothukuchi, K. and Kaufman, J.L., 1999. Placing the food system on the urban agenda: The role of municipal institutions in food systems planning. Agriculture and Human Values 16: 213-224.

Welsh, J. and McRae, R., 1998. Food citizenship and community food security: Lessons from Toronto, Canada. Canadian Journal of Development Studies 19: 237-255.

Chapter 8

Planning for urban agriculture: problem recognition, policy formation, and politics

Nevin Cohen

The New School, 72 Fifth Avenue, room 518, New York, NY 10011, USA; cohenn@newschool.edu

Abstract

Using John Kingdon's multiple streams policy framework, this chapter discusses the nature of urban agriculture planning and policymaking in the US and the conditions necessary for policy implementation. Case studies of five US cities illustrate how problem recognition and policy formation, in the presence of supportive political conditions, creates a window of opportunity for implementing urban agriculture policy. By identifying the factors that enable urban agriculture policy development, food system planners will be better able to advance new initiatives.

Keywords: policy entrepreneurs, food policy, urban farms, community gardens

8.1 Introduction

Despite the recent flurry of urban agriculture activity, many US cities have lacked policy tools to deal with these diverse new practices. In the past few years, however, recognising that urban food production can produce social, economic, and environmental benefits, a number of cities have begun to plan for the growth of urban agriculture, reassessing local zoning codes, ordinances, and development policies to accommodate, regulate, and support urban agriculture efforts (Hodgson *et al.*, 2011). These activities, together with increased attention to urban agriculture by the media, professional organisations, and activists, are leading to a profusion of policy innovations and the transfer of these policies from city to city. A key question is what accounts for the recent surge in urban agriculture policymaking, particularly after urban planners and other city officials had neglected the food system for decades (Pothukuchi and Kaufman, 2000).

8.2 Policy streams framework

To understand the nature of urban agriculture policymaking, this study uses Kingdon's (2002) multiple streams framework of policy development. According to Kingdon, three interacting but distinct process streams – problem recognition, the formation and refining of policy proposals, and politics – run through the policy development process (Henstra, 2010). Policy is not arrived at through comprehensive, rational decision-making, but rather through the efforts of individual actors, and networks of actors, who work to secure particular policy outcomes (Zahariadis, 2007).

8.2.1 Problem recognition

Problem recognition is the process by which agendas are set, in which certain problems emerge as salient while others are ignored. The problem recognition stream is defined by changes in social and economic conditions, focusing events, and symbols that galvanise action and make concrete for people the issues that they feel are important (Birkland, 1998; Kingdon, 2002). The media, interest groups and other stakeholders work to categorise problems and place them in a value context as part of the problem recognition process.

The recent surging interest in urban agriculture is just one manifestation of a broader concern among policy makers about the food systems that cities rely on. Morgan and Sonnino (2010) contend that this is the result of a 'new food equation,' a collection of global challenges, including food price surges, food insecurity, climate change effects, land conflicts, and rapid urbanisation, that have created substantial vulnerabilities for cities. In many cities, these vulnerabilities have prompted planning and policy responses, including support for urban agriculture (Pothukuchi, 2009).

8.2.2 Policy formation

The policy formation stream is the process by which the communities of policy specialists in and around government, including legislative staffs, planners, academics, consultants, and interest groups propose and innovate ideas for consideration (Kingdon, 2002). These policy entrepreneurs help to define problems, mobilise public opinion, and formulate and promote policy solutions and therefore also influence problem recognition. Often, the agents of policy formation are organised in networks, as members of advocacy organisations, interest groups, or think tanks (Godwin and Schroedel, 2000; Haas, 1992). For example, Pothukuchi (2009) describes the creation of a food planning policy guide by the American Planning Association and the growth of national networks such as the Community Food Security Coalition as critical to the adoption and diffusion of municipal food and agriculture policies.

8.2.3 Politics

The political stream is composed of swings in national and local public opinion, changes of administration, shifts in the ideology of political leaders, and interest group pressure (Kingdon, 2002). Support for or opposition to a policy hinges on the assessment of the political power of proponents and the political impacts of policy adoption, which are in turn dependent on the underlying political landscape. Morgan and Sonnino's (2010) new food equation explains some of the political dynamics influencing the ability of planners to advance food policy. Urban agriculture policy also represents a reaction to globalisation, with individuals attempting to re-scale the food system to counter the ecological, social, and individual disconnections they experience as a result of their alienation from food production (McClintock, 2010). Accompanying the new food equation is a neoliberal shift in planning over the past few decades towards competition and entrepreneurialism (c.f., Harvey, 1989) that has encouraged cities to view agriculture as a way to provide services, such as community economic development, education, and food production that cities facing fiscal crisis cannot or choose not to provide.

8.2.4 Policy windows

The confluence of the three streams is a necessary condition for policy implementation, which Kingdon (2002) describes as the brief opening of a 'policy window.' If the convergence of the problem, policy, and politics streams, and the opening of a policy window, is not followed expeditiously by implementation, the opportunity for policy development often passes, as governments move on to other issues (*ibid.*).

8.3 Methodology

This chapter reports the results of exploratory research on urban agriculture policy development in five US cities. The research involved semi-structured personal interviews with urban agriculture advocates, practitioners, and policy makers in each city, as well as reviews of relevant policy documents. Interviewees were identified by contacting urban agriculture leaders and, through snowballing other key informants were selected. Using a common interview protocol, 30 in-person interviews varying from 30 to 90 minutes were conducted during summer 2010. Interview notes were organised to highlight important concepts raised by respondents, and themes and patterns in the interview responses were examined to uncover the extent to which the process of urban agriculture policy development fit the policy stream framework.

8.4 Case profiles

The following cases follow Kingdon's multiple streams framework in describing the formulation of urban agriculture policy. Each profile discusses the problem recognition, policy formation, and politics streams, and the extent to which the streams have converged to facilitate urban agriculture policy implementation.

8.4.1 New York City

Until recently, New York City's food policy has been defined by its mayor, who has emphasised three issues: improving the nutritional quality of the food served by city agencies; increasing the flow of federal food dollars to low-income New Yorkers; and increasing the availability of fresh fruits and vegetables in low-income neighbourhoods (Morgan and Sonnino, 2010; NYC Department of City Planning, 2008; Tester *et al.*, 2010). Within the last few years, urban agriculture policy has become much more salient in New York City, in part because of an on-going debate about the tenure rights of community gardens (Elder, 2005; Moynihan, 2010) and the emergence of entrepreneurial for-profit urban agriculture ventures, like rooftop farms that have opened up in Brooklyn and Queens, and non-profit urban farm programmes. These two core issues, the protection of existing agriculture and the community development and entrepreneurial opportunities of urban farms, have focused the public's attention on urban agriculture and raised questions about whether existing laws are sufficiently supportive.

Policy entrepreneurs inside and outside of government have developed new proposed urban agriculture policies. The Manhattan Borough President convened meetings of food advocates to produce a blueprint for food policy in New York (Stringer, 2010). The Speaker of the New York City Council released a food policy report called 'FoodWorks', developed with input from advocates, business leaders, NGOs, and academics, that also focused on the job development possibilities of sustainable food (Brannen, 2010). The mayor released a revised sustainability plan ('PlaNYC 2030') that includes a section on food policy, which suggests the need to address a broader range of food system concerns than nutrition and food access (New York City Office of Long Term Planning and Sustainability, 2011).

Networks of citizens have also advocated for urban agriculture policies. The New York City Community Gardening Coalition, for example, has worked to secure supportive policies for gardens (NYC Community Garden Coalition, 2010). The NGO Just Food rallied members to lobby successfully to legalise beekeeping, and recently created The New York City School of Urban Agriculture to teach urban farming and food policy (Just Food, 2011). The non-profit Design Trust for Public Space is preparing a citywide urban agriculture plan with substantial input from practitioners, policymakers, and funders (Design Trust for Public Space, 2011).

The political milieu in New York is supportive of urban agriculture, particularly low- or no-cost policy initiatives that support community development, entrepreneurship, asset building, and food production, as evidenced by the proliferation of the above policy proposals and urban agriculture activities. Potential mayoral candidates such as the current Borough President and City Council Speaker have tried to distinguish themselves by promoting citywide food policies, including urban agriculture. The addition of food and agriculture as a topic in the city's updated sustainability plan signals support by the Mayor.

The confluence of policy streams in New York City has opened up a window of opportunity for the adoption of new food legislation, including new legislation that will be introduced in the City Council to implement the Council Speaker's FoodWorks initiative (Brannen, 2010). However, New York's elected officials are limited to two terms of office, so the policy window could close if new political leaders have different priorities.

8.4.2 Seattle

Seattle has had an extensive community garden programme, called 'P-patch', since 1973 (Hou *et al.*, 2009) but several recent actions have made urban agriculture even more recognisable as an issue within the city. In April, 2008, the Seattle City Council adopted Resolution Number 31019, which established food system goals and a policy framework, including proposed policies to support the expansion of urban agriculture (City of Seattle, 2011a). In November

2008, 59% of Seattle voters enacted a Parks and Green Spaces Levy that provided $2 million in funding for the development of new P-Patch community gardens (Seattle Parks and Recreation, 2011). In February 2010, the Mayor and the City Council declared 2010 as 'the year of urban agriculture,' launching a campaign to promote urban food production and access to locally grown food.

Seattle's strong network of community gardeners, environmental activists, and policy entrepreneurs has worked to support the formation of urban agriculture policy. In addition, the City Council's food system resolution fostered policy entrepreneurship within city government by identifying the roles of multiple city agencies in supporting a sustainable food system, and encouraging collaboration among them. Seattle established a food systems interdepartmental team comprising eight city agencies that provide interagency coordination, with the Department of Planning and Development taking the lead in integrating food systems analysis in all major land use decisions.

Integrating urban agriculture in the mission of different agencies has led to innovative policy development. For example, the Seattle Public Utilities department, whose primary mission is waste management, created a backyard composting programme to support gardens, and a food recovery programme to distribute food to organisations that would otherwise become waste (C. Woestwin, personal communications). The Parks and Recreation department is working to integrate its community centres and gardens into the P-patch program, providing space for the Department of Neighborhoods to run the programme and community kitchens to neighbourhood groups (R. Harris-White, personal communication). The City's sustainable building initiative, a joint Seattle Public Utilities and Department of Planning and Development project, is exploring how its green roof programme can promote rooftop farms (J. Banslaben, personal communication; McIntosh, 2010). In August 2010, the Seattle City Council adopted land use code changes recommended by the Planning Department to legalise five different urban agriculture uses (City of Seattle, 2011b).

Seattle's political environment has been supportive of urban agriculture policy. Mayor McGinn, a former Sierra Club activist, and council members who are concerned about environmental issues, have actively promoted urban agriculture. The passage of the 2008 garden levy was described as clear evidence that Seattle voters are fully supportive of expanding local food production (A. Petzel, personal communication).

The salience of urban food production, policy entrepreneurs within and outside of government, and a political environment supportive of urban agriculture have combined to create a large window of opportunity to advance urban agriculture policy in Seattle that has been open at least since the 2008 enactment of Resolution 31019. One factor that may close the window is the city's budget crisis, which has required budget cuts.

8.4.3 Chicago

In Chicago, private foundations have helped to shape food system problem recognition through their support for food planning initiatives. In 1998, the W.K. Kellogg Foundation funded a citywide food security workshop. In 2001 and 2002, the Chicago Community Trust funded an Illinois Food Security Summit. As an outgrowth of the 2002 summit, activists and government officials formed a food policy advisory council, which has sponsored annual food policy conferences and published studies about Chicago's food system (Allen, *et al.*, 2008). Building on these efforts, an organisation called Advocates for Urban Agriculture developed a plan in 2006 (updated in 2010) that called for secure land for urban farming, food incubators, urban agriculture zoning, composting, agriculture training, and food distribution systems (Advocates for Urban Agriculture, 2010). This led to the Chicago Plan Commission's adoption of a comprehensive food systems plan on June 21, 2007. The themes repeated throughout Chicago's various food plans and policy reports is that food access remains a critical issue in low income neighbourhoods, and that a robust food system, including urban agriculture, can be a strategy for economic development.

Many policy entrepreneurs inside and outside of Chicago government continue to influence urban agriculture policy. They include networks of activists like Advocates for Urban Agriculture, food policy advisory council members, and government officials, such as those in the Chicago Metropolitan Agency for Planning who created a comprehensive food element in the region's strategic plan (Chicago Metropolitan Agency for Planning, 2010). Within city government, the Department of Zoning and Land Use Planning has been working on the creation of urban agriculture zoning changes to legalise urban agriculture in most zones (Chicago Department of Zoning and Land Use Planning, 2010). To develop its zoning initiative, the department assembled a working group with other departments, including the health department and the department of the environment, and members of Advocates for Urban Agriculture (K. Dickhut, personal communication).

Highly innovative urban agriculture planning has also occurred at the community level, where agriculture has been incorporated in several neighbourhood plans. In the low-income Englewood community, for example, the city's planning department worked with the residents to create a plan for an urban agriculture district, which would include a neighbourhood food hub developed by community-based organisations, such as Growing Home, along with local financial institutions, the local community college, and city planning officials (H. Rhodes, personal communication).

Chicago's mayoral administration, which installed a green roof on City Hall to demonstrate the city's commitment to the environment, has been supportive of environmental initiatives (Portney, 2009). The economic recession, however, has focused attention on economic development and job creation needs, which may be consistent with community-based urban agriculture plans that have explicit economic development and job training components, like the Englewood initiative.

A window of opportunity exists to move urban agriculture forward in Chicago as a result of advocacy for food production as an economic development strategy, but urban agriculture is only one of a number of food system issues that food policy advocates are concerned about. Other problems, like food access, may take precedence over food production in the future. Moreover, if the political priorities of the current Mayor differ from those of Daley, the urban agriculture policy window may not remain open very long.

8.4.4 San Francisco

The San Francisco Bay Area, where food activists coined the term 'locavore' (Cohen, 2010), has a tradition of community gardening and support for urban agriculture (Lawson 2005). San Francisco's 1996 sustainability plan, which was adopted as a non-binding city policy, has a chapter on food (SF Environment, 1996). Ordinances requiring farmers markets to take government benefits cards, banning agencies from buying bottled water, and resolutions supporting cage-free chickens and opposing *foie gras* have been passed in recent years (City and County of San Francisco, 2011). In 2009, Mayor Gavin Newsom helped to make urban agriculture even more salient by issuing an executive directive that required, among other things, an assessment of publicly available land that could be put into food production (Newsom, 2009). The executive directive also articulated a vision of a food system with greater availability of nutritious food, shorter distances between consumers and food producers, protections for worker health and welfare, reduced environmental impacts of food, and strengthened connections between urban and rural communities (Newsom, 2009).

Newsom's directive engaged a broad range of policy entrepreneurs through the creation of a Food Policy Council (Newsom, 2009). In addition, urban agriculture activists formed the San Francisco Urban Agriculture Alliance to work on policy development (P. Jones, personal communication). San Francisco also has a vibrant group of entrepreneurs who have raised the visibility of urban agriculture through their attempts to build new ventures around innovative disaggregated models of urban farming, including backyard farms. The desire to support these entrepreneurial farms prompted San Francisco planning officials to amend the planning code to define allowable urban agriculture uses within the city.

San Francisco is a politically progressive city, and is supportive of policies to expand urban agriculture. It might seem like a city with a window of opportunity for policymaking that is indefinitely open. However, the city faces a $480 million deficit for fiscal year 2012-2013, and a $642 million deficit for fiscal year 2013-2014 (Sabatini, 2011). The city and state's fiscal woes may shift attention to financial and fiscal concerns, diverting attention from urban agriculture and food system reforms and possibly closing the window on new policies, particularly those with costs to the city.

8.4.5 Detroit

Detroit's economic woes and its 40 square miles (10,360 hectares) of vacant property within the city limits have helped to define urban agriculture as a potential economic development opportunity, yet diverse stakeholders have, as policy entrepreneurs, attempted to advance very different visions of how it might be implemented. Community based organisations like the Detroit Black Community Food Security Coalition, Earthworks Urban Farm, and the non-profit Greening of Detroit have proposed the expansion of community gardens and neighbourhood-based urban farms. The non-profit organisation Self-Help Addiction Rehabilitation (SHAR) has developed a comprehensive proposal for a food production, processing, and retail cluster of businesses, called Recovery Park, which would create jobs for SHAR's clients and provide revenue for the organisation. (D. Pitera, personal communication). Another proposal has been advanced by Hantz Farms, a company that wishes to build a large-scale commercial farm in Detroit (Gallagher, 2010). Dan Carmody, the CEO of Eastern Market, Detroit's historic wholesale and retail food market, has advanced yet another proposal to create a food hub that incorporates a three-acre teaching farm and value added food processing capacity (D. Carmody, personal communication).

Despite these proposals from Hantz, SHAR, Eastern Market, and numerous smaller organisations, the city continues to grapple with whether, and to what extent, urban agriculture is an appropriate use of the city's vacant land. A new comprehensive plan for Detroit ('Detroit Works') is currently in preparation that will help, in part, to define the city's priorities in the area of urban agriculture. A background report for the Detroit Works plan indicates that residents and some urban agriculture stakeholders are concerned with large-scale farming within the city (AECOM, 2010). Mayor Bing has expressed reservations about large-scale urban agriculture, stating that the city was 'not going to go full steam ahead' (Heinze, 2010). Part of the reason for this hesitation is that Detroit faces extraordinary fiscal challenges: a $325 million deficit, nearly 30% unemployment, and 50,000 homeowners facing foreclosure (Bing, 2010).

There is a great deal of interest in possibilities for large-scale urban agriculture in Detroit, with a small window of opportunity for one or more of the proposed large-scale schemes to come to fruition. Much will be determined in the next year as individual project proponents negotiate with city officials for land access. Depending on the public's input to the Detroit Works planning process, agriculture may remain on the agenda and land for agriculture may be incorporated into the city's general plan, but the policy window may close if projects do not materialise or start and then fail financially.

8.5 Analysis and conclusions

The cases illustrate that Kingdon's multiple streams – problem recognition, policy formation, and politics – can help to organise our understanding of the complex process of urban agriculture policy development. However, Kingdon provides a general framework, not a predictive model. Notwithstanding the commonalities among the cases, each of the five cities profiled has unique political environments, different types of stakeholders involved in policy development, and distinctly articulated problems for which urban agriculture is posed as a potential solution.

Urban agriculture has been identified as a policy 'problem' for a number of reasons in all of the cases. The recession, and the desire of cities to find low- or no-cost options to use undeveloped land, support neighbourhood improvements, and enable low-income residents to grow vegetables to enhance their diets are common issues that prompt the development of urban agriculture policy. Even in relatively economically robust cities, like New York and Chicago, urban agriculture has been discussed as a potential solution to inequitable food access and the need for community development. In Detroit, economic development is the key impetus for urban agriculture. Urban agriculture policy is generally part of a broader consideration of city or metropolitan food policy, and often the issue of urban agriculture is recognised and articulated through overall food plans or policy statements. Where pre-existing zoning conflicts with urban food production, as in Detroit, Chicago, or San Francisco, the growth and increased visibility of agricultural activity has forced a municipal response, opening up an opportunity for a more fundamental discussion of the role of agriculture in the city.

The development of plans or policies comes from a variety of thought leaders, stakeholders, and organised networks of advocates from the public, non-profit, and private sectors. Political officials in New York, Seattle and San Francisco have largely helped to coalesce the ideas of policy entrepreneurs, while private foundations have taken the initiative in Chicago and Detroit. In all of the cases, advocates from the non-profit sector and government policy entrepreneurs have tended to work together in networks on policy councils or advisory boards, or in ad-hoc committees or conferences. Indeed, the entanglements among policy entrepreneurs make it difficult to identify the source of policy innovation, though one or more individuals may become associated with particular policies because they are principally responsible for shepherding them through the policy process. In the case of Seattle, for example, a great deal of policy creativity exists within government agencies, even among those not explicitly connected to food production, as agency staff collaborate to advance the city administration's goal of expanding urban agriculture. In cities like Chicago, foundation-funded planning efforts have brought together policy entrepreneurs from the non-profit and government sectors to plan for urban agriculture. Policy generation also comes from for- and not-for-profit entrepreneurs as they begin to create new ventures that require policy support, such as SHAR in Detroit, Growing Home in Chicago, or the rooftop Brooklyn Grange in New York, and as entrepreneurs lobby for policy changes to enable them to farm in various places and at different scales in their respective cities.

Given the current nationwide popularity of urban agriculture, from the White House (White House, 2011) to grassroots community gardens, the underlying political environment has been supportive of urban agriculture policy. In cities with long traditions of pro-environment policies, like San Francisco and Seattle, there is little reason to believe that the political stream will change course dramatically in the foreseeable future. In cities with more recent green initiatives, like Chicago and New York, politics may indeed change with new administrations. In Detroit, there is little formal political support for urban agriculture, though private and non-profit sector entrepreneurs may be able to bring their projects to fruition, which if successful and at the large scale currently envisioned, may help to build support among political leaders.

The relatively recent turn to urban agriculture by planners and policymakers makes it difficult to assess Kingdon's theory that policy windows stay open only for a limited time. In cities like San Francisco, Seattle, New York and Chicago, and even Detroit, a robust advocacy community is likely to continue to keep urban agriculture policy issues on the political agenda. However, attention to urban agriculture may wane if the problem recognition stream shifts, if policy entrepreneurs and their agency or foundation supporters turn their attention to other aspects of the urban food system, or if the underlying political climate changes with shifting administrations or changing economic conditions. Shifts in any of the three streams could cause the window of opportunity for urban agriculture policy to close. Tracking the policy development process longitudinally using Kingdon's framework would enable planners and policymakers to more clearly understand the role of each policy stream and assess how their confluence determines opportunities for urban agriculture policy implementation.

References

Advocates for Urban Agriculture, 2010. Plan for sustainable urban agriculture in Chicago. Available at http://auachicago.files.wordpress.com/2010/03/aua-plan-updated-3-4-10.pdf.

AECOM, 2010. Policy audit topic: urban agriculture and food security. The Detroit Works Project. December 17, 2010. Available at http://detroitworksproject.com/?s=agriculture.

Allen, E., Clawson, L., Cooley, R., De Coriolis, A. and Fiser, D., 2008. Building Chicago's community food systems. Chicago Food Policy Advisory Council, Chicago, IL, USA.

Bing, D., 2010. State of the city. The city of Detroit. March 23, 2010. Available at http://www.clickondetroit.com/download/2010/0323/22924611.pdf.

Birkland, T.A., 1998. Focusing events, mobilization, and agenda setting. Journal of Public Policy 18: 53-74.

Brannen, S., 2010. FoodWorks: A vision to improve NYC's food system. Speaker Christine C. Quinn, The New York City Council, New York, NY, USA.

Chicago Department of Zoning and Land Use Planning, 2010. An ordinance amending various provisions of the zoning code regarding urban agriculture uses. Available at http://www.cityofchicago.org/city/en/ordinances/2010/december_08__2010/an_ordinance_amending0.html.

Chicago Metropolitan Agency for Planning, 2010. Go to 2040 plan. Available at http://www.cmap.illinois.gov/2040/download-the-full-plan.

City and County of San Francisco, 2011. SF Food Policy and Reports. Available at http://www.sfgov3.org/index.aspx?page=760.

City of Seattle, 2011a. Resolution No. 31019. City of Seattle legislative information service. Available at http://tinyurl.com/29ywrt7.

City of Seattle, 2011b. Council bill number: 116907, ordinance number: 123378. City of Seattle Legislative Information Service. Available at http://tinyurl.com/26z5fzq.

Cohen, N., 2010. Locavore. In: Mulvaney, D. (Ed.) Green Food: an A-Z guide. Sage Publications, Washington DC, USA.

Design Trust for Public Space, 2011. Five Borough Farm. Available at http://www.designtrust.org/projects/project_09farm.html.

Elder, R.F., 2005. Protecting New York City's community gardens. New York University Environmental Law Journal 13: 769.

Gallagher, J., 2010. Reimagining Detroit: opportunities for redefining an american city. Wayne State University Press, Detroit, MI, USA.

Godwin, M.L. and Schroedel, J.R., 2000. Policy diffusion and strategies for promoting policy change: Evidence from California local gun control ordinances. Policy Studies Journal 28: 760-776.

Haas, P.M., 1992. Epistemic communities and international policy coordination. International Organization 46: 1-35.

Harvey, D., 1989. From managerialism to entrepreneurialism: the transformation in urban governance in late capitalism. Geografiska Annaler. Series B. Human Geography 71: 3-17.

Heinze, K., 2010. Greening of the great lakes. Detroit Mayor Dave Bing hopes golden opportunities will help grow Detroit green. Available at http://tinyurl.com/42zkznh.

Henstra, D., 2010. Explaining local policy choices: a multiple streams analysis of municipal emergency management. Canadian Public Administration 53: 241-258.

Hodgson, K., Campbell, M.C. and Bailkey, M., 2011. Urban agriculture: growing healthy, sustainable places (PAS 563). APA Planning Advisory Service. APA Planners Press, Chicago, IL, USA, 144 pp.

Hou, J., Johnson, J.M. and Lawson, L.J., 2009. Greening cities growing communities: Learning from Seattle's urban community gardens. University of Washington Press, Seattle, USA, 232 pp.

Just Food, 2011. Farm School NYC. Available at http://www.justfood.org/farmschoolnyc.

Kingdon, J., 2002. Agendas, alternatives, and public policies. (Longman Classics Edition) (2nd Edition) Longman, New York, NY, USA, 304 pp.

Lawson, L.J., 2005. City bountiful: A century of community gardening in America. University of California Press, Berkely, CA, USA, 382 pp.

Mcclintock, N., 2010. Why farm the city? Theorizing urban agriculture through a lens of metabolic rift. Economy and Society. 1-17.

McIntosh, A., 2010. Green roofs in Seattle: A study of vegetated roofs and rooftop gardens. City of Seattle and University of Washington, Seattle, USA.

Morgan, K. and Sonnino, R., 2010. The urban foodscape: world cities and the new food equation. Cambridge Journal of Regions, Economy and Society 3: 209-224.

Moynihan, C., 2010. Taking over a tree to save community gardens. City Room, The New York Times, August 10, 2010.

New York City Community Garden Coalition, 2010. Response to NYC Dept. of parks and recreation's rules for community gardens. Available at http://www.nyccgc.org/.

New York City Department of City Planning, 2008. Going to market: New York City's neighborhood grocery store and supermarket shortage. Presentation – October 29, 2008. Available at http://www.nyc.gov/html/dcp/html/supermarket/index.shtml.

New York City Office of Long Term Planning and Sustainability, 2011. PlaNYC: a greener, greater New York. Available at http://www.nyc.gov/html/planyc2030/html/theplan/the-plan.shtml.

Newsom, G., 2009. Executive Directive 09-03. healthy and sustainable food for San Francisco. Office of the Mayor, City and County of San Francisco, USA.

Portney, K.E., 2009. Sustainability in American cities: A comprehensive look at what cities are doing and why. In: Mazmanian, D.A. and Kraft, M.E. (Eds.) Toward sustainable communities: transition and transformations in environmental policy. MIT Press, Cambridge, MA, USA, pp. 227-254.

Pothukuchi, K. and Kaufman J.L., 2000. The food system: A stranger to the planning field. Journal of the American Planning Association 66: 113-124.

Pothukuchi, K., 2009. Community and regional food planning: building institutional support in the United States. International planning studies: 349-367.

Sabatini, J., 2011. San Francisco's growing budget deficits threaten basic services. San Francisco Examiner. April 7, 2011.

Seattle Parks and Recreation, 2011. Parks and green spaces levy. Available at http://www.cityofseattle.net/parks/levy/.

SF Environment, 1996. Sustainability plan for San Francisco. Available at http://www.sfenvironment.org/downloads/library/sustainabilityplan.pdf.

Stringer, S.M., 2010. FoodNYC: A blueprint for a sustainable food system. Press release, Office of Manhattan Borough President Scott M. Stringer, New York, NY, USA.

Tester, J.M., Stevens, S.A., Yen, I.H. and Laraia, B.L., 2010. An analysis of public health policy and legal issues relevant to mobile food vending. American journal of public health 100: 2038-46

White House, 2011. Replanting the White House garden. The White House Blog. Available at http://www.whitehouse.gov/blog/2011/03/17/replanting-white-house-garden.

Zahariadis, N., 2007. The multiple streams framework: Structure, limitations, prospects. In: Sabatier, P.A. (Ed.) Theories of the Policy Process, 2nd ed. Westview Press, Boulder, CO, USA, 65-92.

Chapter 9

Urban food procurement governance: a new playground for farm development networks in the peri-urban area of greater Paris region?

Ségolène Darly
University of Paris 8, 2 rue de la Liberté, 93526 Saint-Denis, France;
segolene.darly@univ-paris8.fr

Abstract

In France, where state governance is still highly centralised, the main objectives of urban food policies have been determined after a nationwide consultation (the 'Grenelle de l'environnement'). It prescribes that, by the end of 2012, 20% of the food served in schools must be cooked from organic products. In the greater Paris region (or Ile-de-France), this objective has been integrated into various regional policies (from school food to the development of organic farming) and has been implemented by a small number of municipalities without any attempt to structure the governance of these multi-level and multi-site initiatives. This lack of city governance is especially visible in peri-urban areas, where the territorial governance is fragmented between a great numbers of independent local authorities. In this context, the urban food governance, in the peri-urban area of Ile-de-France, tends to be driven by agricultural development agents, rather than cities government. In order to evaluate the pro-active role of farming development agent in constructing and orienting the objectives new city food network, we will here use the *translation* cycle framework described by Callon. Our research is based on two case studies located in the peri-urban countryside of Paris. The first one is the case of a school food project initiated by the mayor of the town of Nemours that has been managed, and reshaped, by the local comity of the Chamber of agriculture. The second case is also a school food project, managed at a greater scale than the previous by the agricultural development agent of the Natural Regional Park (Parc Naturel Régional) of Vexin français. The overall objective of this contribution is to extend the understanding of urban food governance by focusing on the original situation of peri-urban areas in France.

Keywords: school food procurement, local farm development, alternative food networks, Ile-de-France

9.1 Introduction

In France, the school food system is composed by a total of 12,000 cooking structures (2,000 for primary schools, 6,000 for secondary schools and 4,000 for private schools)[17] that serve

---

[17] Figures updated in 2002 or 2004, extracted from a public rapport of the french National Food Council (Conseil National de l'Alimentation, Avis n°47, 2004).

more than a billion meals per year: 330 millions in primary, 500 millions in secondary where 50% of the students eat in the school restaurant, 200 millions in private schools (CNA, 2004). The extremely large procurement budget associated to this system, known as part of the 'power of the public plate' (Morgan and Sonnino, 2008), has been recently identified by the national authorities as a tool to influence agricultural development and develop alternative and more sustainable farming systems (Assemblée nationale, 2010). In 2007, at the end of a nationwide consultation (the 'Grenelle de l'environnement'), the government took the engagement to foster more sustainable farming by promoting the use of organic products in the school food cooking process (20% of the food served have to be organic, MEDDTL, 2009).

Despite this nationwide commitment, the link between the food policy of local collectivities and the promotion of alternative local farming require the coordination of actors throughout cooperation networks that remain to be created, with collectivities, agro-food service industry and farmers. In France, municipalities, which are legally responsible of the primary school food market, are most of the time very small structures that do not always have the resources to achieve these coordination tasks, especially in rural areas. What we have seen in the rural belt of the Ile-de-France region (the most inhabited and urbanised region of France), where more than 80% of the areas are dedicated to farming, is that farm development actors tend to overcome this constrain by assuming themselves the task of creating new procurement networks (Murdoch, 2000).

The national political commitment of the 'Grenelle de l'environnement' presented above is largely based on the specific sustainability concept of organic farming. The implicit hypothesis which is made here is that the sustainability concept of organic farming will be embodied into initiatives that will change existing food procurement networks (Foster and Kirwan, 2004). The point developed here is that when initiatives of alternative school food procurement are taken up by the local farm development agent, the sustainable concept embodied by these initiatives is not the organic farming but the 'proximity confidence' (a mutual confidence shared by producers and consumers on the bases of geographic and social proximity).

In this communication, we will describe two case studies and interpret the role of farm development agents in building new alternative city food networks using the Actor Network Theory (ANT) framework developed by Latour (1999), which focuses on the relations constructed between actors, how they emerge, extend and stabilise. According to this perspective, 'building networks depends on actors' capacities to direct the movement of intermediaries such as texts, technologies, materials and money' (Burgess *et al.*, 2000: 123). This building process is conceptualised as a *translation* cycle which follows four consecutive stages distinguished by Callon (1986). The first stage is the *problematisation* phase where 'an actor analyses a situation, identifies and defines the problem and proposes a solution'(Roep *et al.*, 2006: 5). It is followed by three other stages where other actors are gathering (Interessment phase), generate a new network of interest (enrolment phase) that will work to implement the solution promoted (mobilisation phase). Our hypothesis is that, by being pro-active in the rising of alternative food procurement initiatives, farm development agents are engaging in the translation cycle that lead to the formation of new city food networks. While doing so, they bring city food procurement actors to acknowledge the fact that local food procurement is the best solution to develop alternative food sector.

Using this theoretical framework, we perform an empirical analysis of two school food projects driven by agricultural development agents. Our objective is to describe the processes of network creation that lead to new forms of procurement governance and to show how the sustainability concept of 'proximity confidence', carried on through short supply chains, substitute the organic farming one.

9.2 Cases presentation and data collection methods

The two case studies presented in this chapter are located in the greater Paris region, within the boarder of the administrative region of Ile-de-France (Figure 9.1). These areas are part of the 'rural belt' that surrounds the urban core of Paris. This rural belt is composed by more than 700 municipalities inhabited by about 2 million people (the total population of Ile-de-France is 11.5 million, or 18% of the national population for 2% of the territory). Open spaces (agriculture, forest and nature) still occupies 80% of this territory, thanks to a highly productive agriculture, which products are exchanged in global food market. Geographical proximity between urban populations and agricultural spaces is therefore very strong.

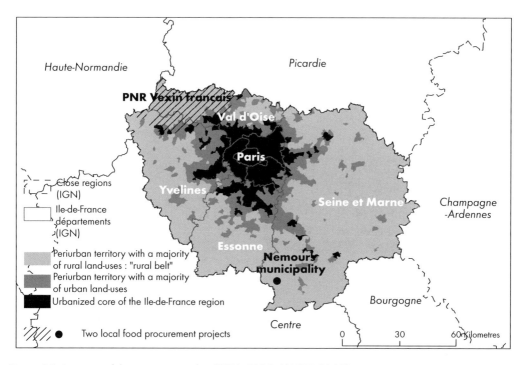

Figure 9.1. Location of the two case studies (IGN, 2000; IAURIF, 2003).

During the year 2009, the Regional Direction of Food and Agriculture (the public body in charge of implementing state policy at a territorial scale) formed a regional working group to analyse the procurement strategies of collective restaurants (school, administrative and enterprises) and their capacity to absorb local farming products. The group is composed of experts from INRA-Paris, representatives of Chambers of agriculture and of environment and local development department of the regional Council. Its members could identify a serie of initiatives dedicated to the development of local procurement in collective restaurants. As we participated to these meetings, we were able to select the following two cases.

The first case is a school food project initiated by both the mayor of the town of Nemours and the local elected representative of the Chamber of Agriculture. Nemours is a town of almost 13,000 inhabitants, located in the south-west part of the region, at 80 km from the centre of Paris, half an hour away from the international airport of Orly. The territory of the municipality (1,082 hectares) is entirely occupied by urban and forest zones but is surrounded by rural municipalities where agricultural land is predominant. The landscape in this area, as in the majority of the region, reflects the high proportion of productivist cereal cropping systems in the agricultural economy.

The economy was essentially dedicated to the agricultural sector until the fifties, when the glass industry development transforms the small rural burg into an industrial pole. With the departure of these industrial activities in the 1990s, the local economy is now driven by small industrial and business parks.

In march 2008, the political orientation of the municipality council changed with the election of a new mayor from the 'UMP' party (Union pour la Majorité Parlementaire), the party of most of the members of the current government (including the president) in France. As the former mayor, from the Socialist Party (PS), was retired from industry sector, the new one, a woman, is an administrative manager for the professional organisation of private collective transports of the region.

From September 2009 to June 2010, one meal '100% terroir' (which means that 100% of the products used to cook that meal is grown 'locally') has been monthly served to 600 kids from the primary schools of Nemours. That day, small presentations on farm and crop techniques were performed by one or two of the farmers who procured the food. This project, initially experimental, has been extending for one more year. It is today, from what we know, the first case in Ile-de-France where, thanks to the coordination between the cook and the farmers, an entire meal, from starter to desert, is cooked from locally produced food on regular bases for more than one communicative action.

To understand how this project was initiated, how it works and who where the different actors that were involved in it, we run two interviews (semi-directive) of the local agent of the Chamber of Agriculture, the first one in January 2010 and the second one in July 2010.

We also had access to internal documents related to the project (list and location of farmers, menus, evaluation synthesis etc).

The second initiative that we present is developed by the farm development agent of the Regional Natural Park (PNR in french) of Vexin français and is still at the early stage of project designing. Nevertheless, we argue that it would be a good case to see how farm development networks invest the field of food procurement governance at the scale of a larger territory than just one municipality, which implies also the coordination of several municipalities in addition to the coordination of the local farmers.

A few miles north of Paris, the PNR of Vexin français has been created in 1995. It spreads on a large rural land area of 66,000 hectares, and counts today 79,000 inhabitants on 94 municipalities. Most of the municipalities have less than 1000 inhabitants and the population of the most populated one does not exceed 6,000 inhabitants.

The two major economic sectors are building and agriculture and the modern cereal crop systems of the farms located inside the park are known to be of the most productive of France. In order to welcome other economic activities inside the park, such as agro-food industries and services (transport, cleaning), several business and commercial zones have been determined near the main burgs (Magny-en-Vexin, Marines, Vigny, Ennery, Génicourt, Boissy-l'Aillerie).

Its luminous landscapes have remained as they were a century ago, when they attracted impressionist painters. However, it is faced with serious problems including pressures from urbanisation, tourism and large infrastructure projects, which could destroy landscapes in only a few years' time. That is why local authorities have joined with the national government in a contractual programme to halt and reverse the process: in France, a Regional natural park (French: *Parc naturel régional*) is indeed a public establishment covering a rural area of outstanding beauty, in order to protect the scenery and heritage as well as setting up sustainable economic development in the area. In order to implement this contractual programme, financial resources are allocated (by local authorities and the French government) to support the functioning of specific decisional and operational structures.

In the Vexin, the PNR actions are influenced by and dedicated to farm interests and development. One of the missions of the park agents and representatives, defined by the contractual charter signed, is to *develop agriculture with respect of environment and by promoting rational land use planning*. An open 'Committee for Agriculture' gathers the elected representatives of the municipalities (a majority of farmers elected in their own residential locality), associations and public partners to propose and validate the operational actions of the technical agent specifically in charge of *farm development*. The vice-president of the committee, who leads the group and played a major role in the park creation process is a former agro-industrial entrepreneur and keeps strong relationships with the private agro-industrial sector through his personal and professional networks.

At the beginning of the year, the technical agent in charge of farm development programmes engaged a project for the development of local procurement in the school restaurants of the park. An initial study has been driven and 17 municipalities have been involved into the project.

Data about the project have been collected from several presentation materials, performed during meetings we assisted to at the Regional Direction of Food and Agriculture. To complete this first approach, we interviewed the technical agent of the park who initiated and managed the project.

9.3 Results: local food procurement networks built by agricultural development actors

9.3.1 The '100% terroir' meal project in Nemours' school restaurants

The idea of introducing locally grown food products in the school restaurants of Nemours has been proposed to the city mayor by the local elected representative of the Chamber of agriculture at the very beginning of 2009. It was at the town hall and the mayor just mentioned in her new year's speech her wish to introduce organic food in the school restaurants menus. Later during the reception, the representative of the Chamber of agriculture suggested to prefer locally grown products, which might encourage the development of organic practices in local farms, instead of paying for organic products grown in farms that can be located thousands of kilometres away. For him, marketing local production throughout school food supply chain is a way of enhancing diversification and reducing economical weakness of local farm systems. It also strengthens its electoral base (farmers of the locality). He then proposed the help of the local development agent of the Chamber to evaluate the feasibility of such a project with the municipality.

The first study presented to the mayor by the Chamber of agriculture demonstrated that most of the organic products used by the traditional private suppliers were mostly imported from other countries, and costs the same or higher prices than similar locally grown products. These figures, and the perspective of promoting at the same time alternative food procurement and a local economical sector (and communicate on it), were, from the local development agent point of view, decisive in the decision of the mayor to prefer locally grown products to organic products.

During the first part of the year 2009, the local development agent of the chamber listed and contacted all the farmers in the area, studied the menus composition and evaluated the prices that could be afforded by the municipality and the producers in order to initiate business negotiations. She led coordination meetings with the fifteen producers that were willing to participate, elected representatives of the municipality and operational manager of the public central kitchen where the meals are cooked for primary schools. The composition of the 10 menus (Figure 9.2), and the name of the farmers in charge of the procurement for each menu, was collectively planned during the coordination meeting of august 2009, and

Thursday October 22nd	Thursday November 26th	Thursday December 17th
Chicken paté	Leek with vinaigrette	Salad with ewe's cheese
Hamburger steak	Beef Bourguignon	Guinea-fowl cooked with
Carrots with cream	Cooked courgette	apples
Plain yoghurt	Goat's Cheese	Gingerbread
Pear	Apple	Apple Juice
Bread form Gâtinais	Bread from Gâtinais	Bread from Gâtinais

Thursday January 28th	Thursday February 18th	Thursday March 25th
Grated carrots	Potato salad	Grated red beetroots
Roasted chicken	Sheep sausage	Duck shepherd's pie
French fries	Roasted cabbage	Ewe yoghurt
Flavoured yoghurt	Plain yoghurt	Apple compote
Pear	Apple	Bread form Gâtinais
Bread form Gâtinais	Bread form Gâtinais	

Thursday April 15th	Thursday May 27th	Thursday June 24th
Duck paté	Cucumber salad	Cauliflower with vinaigrette
Hamburger steak	Sheep tajine with vegetable	Basque chicken
Mashed parsnips	Goat's cheese	Season vegetable
Ewe cheese with apple	Gingerbread	Plain yoghurt
compote or honey	Bread form Gâtinais	Cherry
Bread form Gâtinais		Bread form Gâtinais

Figure 9.2. Composition of the '100% terroir' menus served in the school restaurants in Nemours.

validated by the school menus Council of the municipality before the beginning of school year. Eleven farms are located at less than 20 km from the centre of Nemours (Figure 9.3). Four others, located further in the Seine-et-Marne *département* (up to a hundred kilometres) were contacted thanks to the Chamber of agriculture network to provide products that were not available locally (essentially bread and beef).

During the ten following months, in order to cook the '100% terroir' meal, the central kitchen manager directly contacted the identified farmers several days in advanced. The products were delivered by the producers. The breeders took in charge to bring the animals to the slaughterhouse and butchers before delivering the meat. The goat cheeses were made at the farm. The vegetables and fruits were delivered without any primary processing.

By June 2010, as the ten meals had been served, an evaluation was run by the chamber of agriculture. It showed that both municipality's representatives and farmers were highly satisfied of the experience. It also revealed the obstacles to the scaling up of the local procurement.

For the mayor, the additional costs of these special menus were affordable for the city budget and were justified, for the municipal council, by the positive image that was transmitted to the

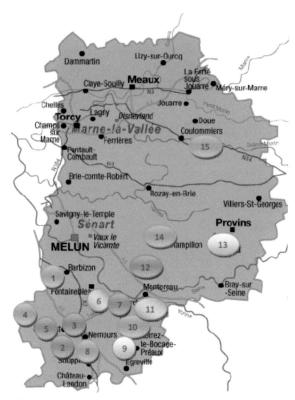

Figure 9.3. Location of the farmers that procured the food products for the '100% terroir' meals in Nemours.

children's families and to the community in general. Despite the success of this experimental initiative, the idea of extending it to two days a month was rejected by the mayor as she could not see what more it could bring in order to justify the costs.

For the farmers, as they were all already marketing their products through alternative and direct selling, it was easy for them to respond to that demand even for a single year engagement: the quantities needed for the meals were not important enough to impose specific investments. As it was not a condition to participate, most of the farmers engaged in the project were not converted to organic techniques. There were organic products included in the food (coming from the two or three organic producers participating) but it was not mentioned in the menus. As the project was just experimental for one year, no farmer took it as an opportunity to engage a transition process toward organic farming techniques (several years are necessary to complete the process of transition). The adults in charge of the kids during lunch time reported that the kids didn't notice any change of taste of quality in the food, expect for bread and cheese.

Scaling up this local supply chain initiative will be favored in the near future. Indeed, as the public central kitchen has been saved from closing down thanks to the new municipal council, a new public call for tender will be open for food procurement for four years.

This call will include a special clause for locally grown products. Responding will certainly implicate a stronger coordination between farmers and the participation of other private partners capable of procuring more quantities of less expensive products specifically grown for school food purpose (some cheeses or cooked meats were very high quality products). The local development agent of the chamber of agriculture is planning on helping the interested farmers in this up-scaling process.

Apart from this Nemours initiative, this agent is now playing a central role in promoting local supply chain in other collectivities located in her professional territory or in other parts of the region.

9.3.2 The 'PNR' supply chain project in the Vexin français

The project of developing local supply chain for school restaurants in the territory of the Vexin PNR has been initiated by the farm development agent of the park in order to promote diversification of activities and marketing opportunities for the farms located in the park. As for most of the farm development projects, the committee did not propose the initiative but validated it.

The farmers and the local municipalities that would potentially participate were identified by the PNR agent. She also ran a survey in order to identify the main difficulties that such a project would have to deal with. The first obstacles seemed to come from the fact that almost the majority of the municipality are delegating to private sector the procurement and cooking process, and therefore argued that they could not experiment direct local procurement. But mostly, it seemed that, expect for the 17 municipalities that responded positively to the project proposition (the majority is delegating cooking process to private actors), there was no interest for local rather than conventional procurement.

Despite the low number of participants, the fact that food committees of several municipalities accepted to participate to coordination meetings is seen as a positive and innovative result. For the development agent, bringing local committees to work together for farm development is rare.

To coordinate and enhance the capacity of local farmers to respond to the need of school food supply chains, she built a partnership with a research agency and a social integration working centre in order to support investments and infrastructures for collective primary processes and selling platform. The project is currently emerging and the first coordination meeting was held by the end of august 2010[18].

---

[18] By the end of year 2010, the operational study was reformatted for the procurement of only two municipalities (which were able to cook the meals in their own collective kitchen). With the departure of the two development agents that were promoting the project within the park administration, the project was not implemented as planned. It is now another development agency (funded by a LEADER project) which took the lead on local food procurement project in a more extended area.

9.4 Conclusions

For agricultural development actors, the *power of the public plate* can impulse more sustainable local farming because it is seen as a risk reduction system. From a financial point of view, it allows the farmers to invest into alternative processes such as organic farming as the volumes of products ordered are important enough to secure the financial risks they are taking. From an agronomic perspective, as the menus are known long in advance, the field works can be planned with less uncertainty and more efficiency. One argument frequently mentioned by development agents is the fact that the deliveries to the school kitchen require less work than other direct marketing systems and therefore improve the work conditions at the farm. In the Nemours example, these objectives were only partially achieved as the small size of the project did not require enough volumes to truly reduce the risks of new investments into alternative framing. Nevertheless, the project of up-scaling such initiatives is still considered as a way to achieve more sustainable alternative farming in peri-urban areas.

In France, the school food system remains highly fragmented. The farm development actors that operate in rural and peri-urban areas are able to connect each local municipality to a network of farmers located in close neighbourhood (but still outside its territorial limits) which can procure local food. In this sense, they can play a central role in building new development networks in the school food system. As the municipalities of the PNR are invited by the development agent to participate to a local school food supply chain, the mayor of Nemours is convinced by the agent of the Chamber of agriculture. In both cases, they initiate the project and support the coordination between public or private actors. Doing so, they connect two independent networks: public school food networks and private local farming sector. In both cases, the development agents are working for organisations where conventional farming representatives have a strong influence. In order to initiate a concrete experiment, they work mainly with farmers of more alternative sector, engaged in alternative marketing (direct selling) or techniques (organic farming). While enhancing local food procurement governance, agricultural development agents achieve also other network connections between the conventional and alternative farming sectors.

If we refer here to the Actor Network Theory developed by Latour (1999) and to the *translation process* by which networks are evolving, we can assume that, in the two projects presented, agricultural development agents initiate a *problematisation* phase (Callon, 1986) where they *analyse a situation, identify and define the problem and propose a solution* (Roep *et al.*, 2006: 5). Doing so, they carry forward their own conception of sustainability, which is developed around the geographic and social proximity between producers and consumers and the confidence that grows thanks to this proximity.

The stability of these new networks, and their ability to up-scale alternative school food projects in order to influence the development of more sustainable farming systems is today uncertain. The becoming of such initiatives constitute great research material that we intent to follow in a short and long-term perspective.

References

Assemblée nationale, 2010. Loi n° 2010-874 du 27 juillet 2010 de modernisation de l'agriculture et de la pêche. Available at http://www.legifrance.gouv.fr/.

Burgess, J., Clark, J. and Harrison, C., 2000. Knowledges in action: an actor-network analysis of a wetland agri-environment scheme. Ecological Economics 35: 119-132.

Callon, M., 1986. Some elements of a sociology of translation: domestication of the scallops and the fishermen of St. Brieuc Bay. In: J. Law (Ed.) Power, action and belief: a new sociology of knowledge? Routledge and Kegan Paul, London, UK, pp. 196-233.

CNA (Conseil National de l'Alimentation), 2004. Avis n°47. 29 pp. Available at http://agriculture.gouv.fr/IMG/pdf/avis47.pdf.

Latour, B., 1999. On recalling ANT. In: Law, J. and Hassard, J. (Eds.) Actor Network Theory and after. Blackwell, Oxford, UK, pp. 15-25.

MEDDTL (Ministère de l'écologie, du développement durable, des transports et du Logement), 2010. Les engagements numérotés du Grenelle. Available at http://www.legrenelle-environnement.fr/

Morgan, K. and Sonnino, R., 2008. Sustainable development and the public realm: The power of the public plate. In: Morgan, K. and Sonnino, R. (Eds.) The school food revolution, public food and the challenge of sustainable development. Earthscan, Oxford, UK, 165-222.

Murdoch, J., 2000. Networks – A new paradigm of rural development? Journal of Rural Studies 16-4: 407-419.

Roep, D., Oostindie, H., Brandsma, J.P. and Wiskerke, H., 2006. Constructing sustainable food supply chains: trajectories, lessons and recommendations, SUS-CHAIN Practical and Policy Recommendations. Wageningen University, Wageningen, Netherlands, 44 pp.

Chapter 10

The role of municipal markets in urban food strategies: a case study

Georgia Machell and Martin Caraher
Centre for Food Policy, City University London, Northampton Square, London EC1V 0H, United Kingdom; georgia.machell.2@city.ac.uk

Abstract

Municipal markets have been recognised by both government and industry as valuable social spaces which can address growing public health issues in urban areas such as obesity. Historically, traditional British markets have been considered as important municipal bodies that ensured urban dwellers have access to an adequate and affordable diet. Yet, there is a lack of evidence linking markets to the public health impacts that are claimed. This chapter will provide indicatory measures of food access at a large municipal market in Leeds and extrapolate the findings into a discussion on the wider potential role of municipal markets in urban food strategies. Specific focus is given to the role on markets in addressing food access for low-income urban communities. As urban food strategies develop in towns and cities across Britain, steering groups and urban planners need to take a realistic look at potential existing food strategies.

Keywords: food access, food planning, municipal markets

10.1 Introduction

The recent developments of urban food strategies across the UK should be both celebrated and analysed. Their existence and development indicates a growing awareness of urban food issues from local government, however the potential impact of the various instruments within strategies need to be realistically considered. One such instrument that has been referred to in urban food strategies is the municipal market. Historically, traditional British markets have been important municipal bodies that ensured urban dwellers have access to an adequate and affordable diet (Schmeichen and Carls, 1999). In the UK municipal markets have been recognised by both government and industry as valuable social spaces that can address growing public health issues in urban areas such as obesity (The Cabinet Office, 2008). Yet, there is a lack of evidence linking markets to the public health impacts that are claimed due to a historic lack of research on markets (Zasada, 2009). As urban food strategies develop in towns and cities across Britain, food strategy steering groups need to take a realistic look at the current role of municipal markets in urban food systems.

Leeds Kirkgate Market and the Leeds Food Strategy 2006-2010 provide a framework for a case study that helps us investigate market development with more rigour. A common theme across food strategies is increasing access to healthy and affordable food to low income consumers. This chapter will specifically focus on the role of markets in addressing food access for low-income urban communities. The discussion is framed by the following questions: what actions do municipal markets currently take to contribute to urban food strategies? What further actions could municipal markets take to contribute to urban food strategies and what forms of governance are necessary to ensure actions occur?

10.2 Context

10.2.1 Food strategies

Food strategies are elevating urban food issues to new levels of importance. Most major cities in the UK now have an urban food strategy. Pothukuchi and Kaufman (1999) suggests urban food issues should be viewed with the same magnitude as other urban issues such as transport and crime and that urban food strategies have the capacity to do this, by linking food systems to urban environments and local government structures.

10.2.2 Markets in the UK today

Markets have existed in towns and cities across the UK for hundreds of years. In recent years there has been a drive from central government to utilise markets as policy vehicles for delivering wider agendas pertaining to food policy, thus suggesting that markets are valuable assets to urban food strategies (The Cabinet Office, 2008). Such agendas include social exclusion and public health. These issues have been strongly linked to food access (McEntee, 2008; White *et al.*, 2004; Wrigley *et al.*, 2002). Statistics on the British market industry present a mixed picture. Currently the British market industry employs 96,000 people, has an average annual turnover of £125 million (Rhodes, 2005). This figure accounts for all goods sold at markets, not only food. A lack of research in the area of food access and markets is making it challenging to accurately predict the potential impact that municipal markets could have on urban food strategies.

10.2.3 Food access

In the area of food access, there is a large literature. Wrigley *et al.'s* (2002) multi-method triangulation of food access research in an area of Leeds offers different insights into the social and cultural barriers preventing everyone from making healthy food choices. Wrigley *et al.'s* research is framed by studies across Britain investigating the relationship between food access and socio-economic status (White *et al.*, 2004). McEntee (2008) suggests that food access and inequality constitute a threefold issue comprising access to knowledge of food and nutrition, economic access and physical access.

Leeds is in the north of England and has a population of 715,404 (Office of National Statistics, 2010). Leeds has been the focus of in depth food access studies (Whelan *et al.*, 2002; Wrigley *et al.*, 2002, 2004). The studies recognise the range of food access issues that exist in Leeds to different groups living in low income settings as well as recognising the multifaceted nature of food access issues recognising that financial access, physical access and access to information need to be addressed simultaneously to have meaning impacts. The wider literature on food access further remarks that engaging consumers in the concept of changing behaviour to access a healthier diet is the most challenging aspect of addressing food access (Dowler, 2008; McEntee, 2008).

Leeds City Council's goal is to assist and encourage everyone in Leeds to have the opportunity to eat a healthy diet (Healthy Leeds Partnership, 2006). Leeds Food Matters – A Food Strategy for Leeds: 2006 – 2010 Healthy, Affordable, Safe and Sustainable food for all (Healthy Leeds Partnership, 2006) is an ambitious strategy to address myriad food issues in the city of Leeds and place them on the urban agenda. The plan sets out both the range of issues and goals to be achieved between 2006 and 2010. The strategy is currently being evaluated. Goals include increasing the amount of fruit and vegetables consumed by all residents of Leeds, increase accessibility to fresh produce and educate people about the importance of a healthy diet. The Leeds Food Strategy also provides a context to measure the policy approach taken by local government to address the issue of food access, public health and market retailers. The Leeds Food Strategy does not explicitly discuss a role for the city's markets', however a market trader is quoted as saying: 'people always ask me if what they have bought makes up 5 portions of fruit and vegetables. It can be confusing but very popular with customers' (Healthy Leeds Partnership, 2006: 8).

10.3 Methodology

We identified key barriers to food access from academic discourse (Dibsdall *et al.*, 2002; McEntee, 2008; Pothukuchi and Kaufman, 1999; Whelan *et al.*, 2002; White *et al.*, 2004; Wrigley *et al.*, 2002) and created food access indicator questions to gauge the level of food access at the market. Three areas of food access were identified: physical access, financial access and access to information about food and health issues. Within these three areas, questions were developed that would indicate the level of food access at the market. Indicatory data was collected through observations and document analysis in July 2009.

10.4 Case study

The case study indicates the level of accessibility to a healthy and affordable diet at Leeds Kirkgate Market and examines the markets role within the Leeds Food Strategy. The case study is based on Leeds Kirkgate Market, a traditional municipal market in the north of England. The market was a central point of early urban planning in Leeds and originally reflected historic market values, primarily: a duty of care and responsibility for the townspeople (Schmeichen and Carls, 1999).

Profile: Leeds Kirkgate Market

- *Footfall:* estimated footfall of 10,000 visitors per day.
- *Demographic of shoppers:* elderly, young mothers, low income from surrounding housing estates.
- *Number of traders:* 635 trading stall including 434 permanent indoor stalls and 2001 temporary outdoor (open) stalls.
- *Opening times*: Indoor trading Monday – Saturday 9 am-5 pm. Outdoor trading Monday – Saturday 9 am-5 pm with a half day on Wednesdays.
- Location: South East Central Leeds. Adjacent to City Bus Station (Metro, 2009).
- Governance: Leeds City Council Markets Division.

Food access findings at Leeds Kirkgate Market

The data collected though observations at Leeds Kirkgate Market indicate indicate that the contribution of the municipal market to the Leeds food strategy is unclear. For many low income consumers, accessing healthy and affordable food at Leeds Kirkgate Market is made challenging by the high ratio of fast-food eateries to fresh-produce stalls (3:1), creating an obesogenic environment. There was little evidence of the market being explicitly involved with the Leeds Food strategy or its core principles and values. This raises questions of governance as both the food strategy and the market are under the purview of the local authority, but are clearly disconnected.

The single link to the Food Strategy was the existence of an All Being Well stall within Leeds Kirkgate Market. The main function of the All Being Well Stall was to be a 'health point' or drop in centre where shoppers at the market could receive information on health and cooking from a range of agencies. Different agencies host activities of their choice. When observations were undertaken, the Leeds Vegan Society was manning the stall providing vegan recipes to passersby. Although the All Being well stall exists to provide a public service and received funding from the Food Standards Agency, a local health charity and East Leeds PCT, within the market it essentially operates as any other trader, paying rent with little voice in terms of market development.

The All Being Well stall presents a practical example of the lack of joined-up working to address food access across local government departments. Since the research was undertaken in July 2009, the All Being Well Stall has evolved into a Jamie's Ministry of Food[19]. The new focus is on developing cooking skills in the shoppers at Leeds Kirkgate Market as well as continuing to offer a 'health point'. As the food projects pay full rent on their space, it appears that the market itself has not made any changes to engage with the strategy.

---

[19] Jamie Oliver's Ministry of Food shops are drop-in centres aimed to teach people cooking skills. See http://www.jamieoliver.com/jamies-ministry-of-food/.

10.5 Discussion

Historically, municipal markets demonstrated a duty of care to vulnerable groups of people in urban settings. It is reasonable to expect that the output from any municipal institution should have the best interest of the local residents in mind. There is currently pressure on other municipal institutions such as schools, hospitals and prisons to understand the risks of serving poor quality and unhealthy food. It is therefore worrying that a municipal market can have a ratio of 3:1 unhealthy eateries to fresh food stalls. A main challenge to accessing healthy and affordable food at LKM is the number of stalls selling cheap, unhealthy food essentially creating a local authority-endorsed obesogenic environment (Lake and Townsend, 2006). Lake and Townsend (2006) discuss the impact that disproportionate access to unhealthy food can have and how there is potential for positive impact in the design of urban public spaces to help to encourage healthier choices; 'shaping the environment to better support healthful decisions has the potential to be a key aspect of obesity prevention intervention' (Lake and Townsend, 2006). This presents a clear issue in looking to the market as a tool in a wider food strategy.

The placement of the All Being Well stall and now a Jamie Oliver's Ministry of Food outlet in Leeds Kirkgate Market reflects a growing trend in urban food policy development. This is the value of both voluntary and funded 'food projects' to effectively respond to structural failures of urban food systems (Dowler and Caharer, 2003). The existence of these projects acknowledges that the local authority is aware of urban food issues; however it also underestimates the scale and complexity of the issues. Dowler *et al.* (2007) suggests that this is a typical response to issues of food access in that it is a minor intervention, avoiding any explicit up-stream intervention, and in practice maintaining a primarily volunteer-led community initiative. Thus externally it appears that Leeds Kirkgate Market actually is addressing the issue of food access, which has gained much attention from other local authority departments. However further examination in the case study revealed how in practice the All Being Well stall is not having the impact that it has the potential to achieve.

Further questions arise over who will use the new Ministry of Food outlet. Cooking sessions are £2 for participants on any form of social benefit and £4 for anyone else. The cooking sessions are 2 hours long; the goal is for customers to walk away with new skills and a healthy meal. The market are enthused by the Ministry of Food as there are hopes it will bring in new customers and help revitalise parts of the market increase the foot traffic which has been dwindling in recent years. However, the reason why the Ministry of Food exists is for the people already using the market – low income consumers. So what will happen if the Ministry of Food attracts new customers who are interested in learning new cooking skills but are not the people most at risk of diet related illness? Many urban food strategies focus on developing new initiatives and projects to address issues that have occurred not because of a lack of food projects but because of larger structural and cultural issues such as the rise in fast-food culture, the dominance of the food industries and new technological developments that encourage minimal cooking. Food access literature stresses the value of integrated and joined-up responses to food access issues. It's clear that the Leeds Food Strategy reflects this; however responses must adjoin from both sides and do more than acknowledge a resource.

The reported clientele of the market are mainly from low income areas of Leeds. One action that could have a potential food access impact would be the acceptance of Healthy Start Vouchers[20] at stalls that sell fresh produce or even the Ministry of Food. Healthy Start is a government food welfare scheme for eligible low income mothers and/or pregnant women to purchase fruits, vegetables, milk, infant formula and vitamins. Currently no stalls accept Healthy Start, however the market management is encouraging the market stall holders to accept Healthy Start vouchers. This cannot be enforced by the market management as each market stall owner is technically an independent business. This classification of market stalls being independent businesses presents a number of challenges to the market engaging with the Leeds Food Strategy. The stalls cannot be effectively regulated, as the council's role is first and foremost as a landlord and effectively providing a service to the traders. This raises questions regarding how food strategies can be effective. The food access literature is reflected in the Leeds Food strategy as it stresses the coordinated approach to increasing people's access to healthy food in Leeds. The action plan in the Leeds Food Strategy includes measures to promote the strategy to public and private sector groups including the retail sector.

There appears to be two opposing agendas; the market needs to ensure financial survival and thus increase the amount of foot traffic through it – the food strategy steering group want to use the market as a central location to engage people in healthier eating habits. The crux of the issue appears to be that the market management are not yet engaged with healthier eating agendas and the primary focus leans towards keeping the market in business and preserving the architectural heritage of the market. The area in which local authorities have invested is primarily in preserving the Edwardian architecture of the main market hall, which the case study illustrates is under-utilised. It is arguable that this type of investment is not primarily in the best interest of the public's needs. Lang, Caraher and Barling (2009) would ascertain that this indicates a weakness in the food system in which markets are an actor. This type of investment creates a tension between local authority departments, specifically as it neglects to address public health consumer benefits, disregarding the potential role of LKM in Leeds Food Matters (Healthy Leeds Partnership, 2006). This is an example of investing in the past, winning out over investing in the future. Tension stems from how markets are being defined and questions are being raised as to whether markets are a public service, a business or heritage sites that needs to be preserved. In general markets are aiming to reflect their communities, whether or not it is in the best interest of the consumer.

As the case study illustrates and the government reports recognise, markets do have the potential to respond to aspects of the needs of low income urban populations. However, doing this would mean taking a risk in terms of the economic value of a market. Rhode's survey estimated that markets in the UK are a £125 million industry employing over 96,000

---

[20] See http://www.healthystart.nhs.uk/.

people (Rhodes, 2005). Shifting the role of markets from economic entities to service oriented entities would inevitably compromise trade bodies such as the National Association of British Market Authorities. The findings at Leeds Kirkgate Market found that for every trader providing fruit and vegetables there were almost three fast food eateries. Given the current market situation, where the stall occupancy rate is 75% and falling (Rhodes, 2005), shifting the ratio of fast-food eateries to fruit and vegetable shops would essentially mean closing or replacing stalls and losing jobs. Thus trade bodies and market managers would be compromised. Not to mention, the other agendas markets are being encouraged to respond to such as tourism, environment and regeneration present further areas of tension. These tensions and pressures to effectively blend issues in a collective response present in essence a key challenge to food policy in general (Lang *et al.*, 2009).

Markets are more complex then the mythic images we see of fresh and fruit and vegetables being sold. Markets are rife with politics and competition for both customers and space. Assumptions about what markets provide have been constructed throughout history. Images of fresh produce are often associated with provision of food from markets. Observations from the case study suggest how areas of Leeds Kirkgate Market are not immediately recognisable as a market due to the presence of retail chains that are more often associated with the high street, for example; Greggs the Baker, a national chain of bakeries and Jack Fulton a national chain of frozen food shop. Even the media corporation Sky had a stall where they were trying to sell TV and broadband packages to shoppers at Leeds Kirkgate Market. The market's competition is reflected in the many of the changes that have occurred in the last few decades. This perhaps explains the rise in commercial developments at markets such as Leeds Kirkgate Market which deconstruct the archetypal market image.

10.6 Implications

The case study highlights key areas of tension within food governance; on a local level the key tensions are between healthy and unhealthy foods being sold within markets, between the role of the local authority as both a landlord and as a health promoter and between the various functions of a municipal market i.e. is it primarily a public service, a business or a heritage site? Wider tensions are presented through the unfounded claims being made about the impact markets could have to enhance urban food systems and the statistics that show markets as dwindling in size and popularity. If the tensions explored in the case study are not addressed, then the potential for markets to be influential in food access will not be realised and their longevity will continue to be vulnerable as they exist precariously and undefined within urban food systems.

10.7 Recommendations

To address the aforementioned tensions, we recommend actions at both a local and national level. On a local level, recommendations include explicitly connecting the market and the food strategy through greater engagement with the wider goals and issues of the strategy. This could include connecting the food strategy directly to market traders to encourage healthier options to be sold. The case study does suggest that the number of fast food eateries needs to be regulated in a sensitive and constructive way, i.e. the market management support traders to change what they sell as opposed to shutting down trading stalls completely.

The mandatory acceptance of Healthy Start vouchers at all stalls that sell fresh fruits and vegetables would potentially have a large impact on providing food access to vulnerable groups and possibly drive the demand for more fruit and vegetable stalls to exist in the market.

On a national level further research needs to be commissioned to explore the range of markets operating in Britain today and the range of issues they face. This could inform the scale of impact that markets could have. Before linking markets to ambitious government agendas such as food access, clear strategies for the survival of the market industry need to be clearly developed taking into account the range of markets and range of issues.

10.8 Conclusion

As Schmiechen and Carls (1999) describe, in the past markets have been key in responding to the dietary needs of urban low income consumers throughout history. Today the picture is far more complex. It is not realistic to assume that markets can return to their original function and role in society as the sole provider of a healthy diet to the urban poor (Schmeichen and Carls, 1999), although there is an argument that markets be judged for their contribution to healthy eating and public health. The basic need to provide urban low income consumers with healthy food still exists, the myriad of actors in the process of healthy food provision have become more complex. Outside factors have changed and providing food access is more multifaceted due to diverse populations, increased barriers to choice, the development of supermarkets, shifting priorities of the industry and tensions between policy actors at a local level.

The case study indicates that there is potential for the market to take a more active role in the Leeds Food Strategy, however the market management would need to fully engage with both the key principles of the strategy and the challenges the market is faced with. In theory many municipal markets do present a central point that food strategies could act through, however there are clearly some barriers that need to be crossed before this can happen.

References

Dibsdall, L., Lambert, N., Bobbin, R. and Frewer, L., 2002. Low-income consumers' attitudes and behaviour towards access, availability and motivation to eat fruit and vegetables. Public health nutrition 6: 159-168.

Dowler, E., 2008. Policy initiatives to address low-income households' nutritional needs in the UK. Proceedings of the Nutrition Society 3: 289-300.

Dowler, E. and Caraher, M., 2003. Local food projects: the new philanthropy? The Political Quarterly 74: 57-65.

Dowler, E., Caraher, M. and Lincoln, P., 2007. Inequalities in food and nutrition: challenging 'lifestyles'. In: Dowler, E. and Spencer, N., (Eds.) Challenging health inequalities: from Acheson to 'choosing health'. The Policy Press, Bristol, UK, 127-155.

Healthy Leeds Partnership, 2006. Leeds Food Matters. A food strategy for Leeds: 2006-2010. Healthy, affordable, safe and sustainable food for all. The Leeds Initiative, Leeds, UK, 17 pp.

Lake, A. and Townsend, T., 2006. Obesogenic environments: exploring the built and food environments. The Journal of the Royal society for the Promotion of Health 126: 262.

Lang, T., Barling, D. and Caraher, M., 2009. Food policy:integrating health, environment society. Oxford University Press, Oxford, UK, 336 pp.

McEntee, J., 2008. Food deserts: contexts and critiques of contemporary food access assessments. Working paper series No. 46, BRASS Centre, Cardiff, UK, 37 pp.

Metro, 2009. Your where and when market guide for West Yorkshire. Available at http://www.wymetro.com/NR/rdonlyres/83BEC06B-6218-4B54-92A0-9175CDCC776E/0/MarketGuide2009_infoleaflet.pdf.

Office for National Statistics, 2010. Population Estimates: Leeds. 2001 Census. Available at http://www.statistics.gov.uk/census2001/pop2001/Leeds.asp.

Pothukuchi, K. and Kaufman, J.L., 1999. Placing the food system on the urban agenda: The role of municipal institutions in food systems planning. Agriculture and Human Values 6: 213-224.

Rhodes, N., 2005. National retail market survey. Retail Enterprise Network, Manchester Metropolitan University Business School, Manchester, UK.

Schmiechen, J. and Carls, K., 1999. The British market hall: a social and architectural history. Yale Univ Press, New Haven, CT, USA, 326 pp.

The Cabinet Office, 2008. Food matters: Strategy for the 21st Centur. The Cabinet Office, London, UK.

Townsend, T. and Lake, A.A., 2009. Obesogenic urban form: Theory, policy and practice. Health and Place 15: 909-916.

Whelan, A., Wrigley, N., Warm, D. and Cannings, E., 2002. Life in a'food desert'. Urban Studies 39: 2083-2100.

White, M., Bunting, J., Raybould, S., Adamson, A., Williams, L. and Mathers, J., 2004. Do food deserts exist? A multi-level, geographical analysis of the relationship between retail food access, socio-economic position and dietary intake. The Food Standards Agency, University of Newcastle upon Tyne, Newcastle upon Tyne, UK.

Wrigley, N., Warm, D., Margetts, B. and Lowe, M., 2004. The Leeds'food deserts' intervention study: what the focus groups reveal. International Journal of Retail & Distribution Management 32: 123-136.

Wrigley, N., Warm, D., Margetts, B. and Whelan, A., 2002. Assessing the impact of improved retail access on diet in a'food desert': a preliminary report. Urban Studies 39: 2061-2082.

Zasada, K., 2009. Markets 21: A policy & research review of UK retail and, wholesale markets in the 21st century. The Retail Markets Alliance, UK, 84 pp.

Chapter 11

Traditional food markets: re-assessing their role in food provisioning

Julie Smith

Countryside and Community Research Institute, University of Gloucestershire, Oxstalls Campus, Gloucester GL2 9HW, United Kingdom; juliesmith.juke@gmail.com

Abstract

Rapid transformation in the food retail supply system accompanied by rational economic efficiency has marginalised the role that traditional markets play in the UK food distribution system. Yet these markets survive, some even thrive, implying that traditional food markets cannot be defined simply in terms of their distribution function. Traditional markets occupy a contested space in food provisioning, seen as part of the conventional food system but with a history and geography that has evolved in response to social and cultural norms and market forces; thus providing an important 'hidden' retail source for fresh and affordable food that is both globally and locally sourced. This chapter presents the first detailed assessment of traditional food markets in England. It maps and identifies patterns of concentration at different geographical scales using database research on traditional markets, wholesale markets and more specialist niche markets, including farmers' markets. The chapter goes on to present some early fieldwork findings from case study research on traditional markets in the North East and Eastern regions of the UK. These findings provide new insights into where this affordable fresh food comes from, what the people operating, trading and shopping on traditional markets think about the food on sale, about fresh food shopping habits, and the influences affecting fresh food choices. The findings also raise questions about perceptions of 'value' and 'quality', and about distinctions between 'alternative' and 'conventionally' produced food. The conclusions suggest that there should be wider recognition within the public policy agenda of the role that traditional food markets and their supply chains play in delivering sustainable food provisioning.

Keywords: fresh food, value, public policy

11.1 Introduction

There is little published data on the traditional food market sector in the UK. Some refer to it as a 'hidden' retail sector, although recent figures reveal there are over 1,100 traditional food markets operating with an estimated annual turnover of £3.5 billion (Zasada, 2009). Rapid transformation in the food retail supply system accompanied by rational economic efficiency has marginalised the role that traditional markets play in the UK food distribution system. Yet these markets survive, some even thrive, implying that traditional food markets cannot be defined simply in terms of their distribution function.

In 2009, the Select Committee for Communities and Local Government (SCCLG, 2009) conducted a wide-ranging inquiry into traditional markets. It concluded that competition from supermarkets and other low-cost discounters, neglect by local authorities, the difficulty of attracting new traders, a struggling wholesale market sector and a restrictive regulatory context had all left a feeling of market decline.

However, it also acknowledged that the social benefits generated by markets were as important as the economic benefits, including the role they play in town centre regeneration, in supplying fresh and affordable food and in reducing the environmental impacts of the retail sector.

This chapter presents the first detailed assessment of traditional food markets in England, including how they have been affected by retail restructuring, by mapping patterns of concentration, and by presenting some preliminary case study results about their fresh food supply chains – i.e. fresh fruit and vegetables, meat and fish – shopping habits and other factors influencing fresh food choice on these markets.

11.2 Retail restructuring

Agri-food, retail and consumption geographies have emerged to capture new interpretations and approaches to the rapid transformations in the food retail supply system and the interest in the cultural processes that shape it. However, there has been a tendency to under-theorise how these transformations and processes have been absorbed by *traditional* food systems. This includes consideration of the uneven and differentiated development of spaces and places, and how lifestyles and identities relate to traditional food distribution and access in both urban and rural environments.

Boundaries between conventional and alternative food systems are blurred (see for e.g. Maye *et al.*, 2007; Whatmore and Thorne, 1997). Traditional markets, although often ignored in this burgeoning literature, operate as an urban food system, evolving in response to market forces and changing consumer preferences, and providing an important retail source for fresh and affordable food that is both globally and locally sourced. Plattner's (1982) US study demonstrates how public markets offer an outlet for inferior quality produce not acceptable for sale in US chain stores. The market in the study acted as a 'shock absorber for the modern, vertically integrated mass-distribution produce economy'. Its informal economy was seen as 'an integral, positively functional part of the formal fresh-produce industry. The market did not exist in spite of the corporate-farm chainstore industry, but in positive integration with it' (*ibid.* p. 401).

In parallel, the cultural diversity of world cities suggests the need for 'transformative boundary crossings' (Tolia-Kelly, 2010) between local and global food systems – a 'cosmopolitan localism' (Morgan and Sonnino, 2010). Again, traditional markets sit well with such claims. For example, Imbruce's (2006) work on Chinese food markets in New York City illustrates how the Chinatown system may operate outside the industrial, corporately controlled food system with produce grown, sourced and sold in a contained system within the South East Asian communities, but it does not do so by consciously resisting the industrial system (as some alternative food networks supposedly do); instead, 'it is a result of new spatial arrangements made by individuals' – an alternative globalisation, with businesses thriving at competitive prices (*ibid.* p. 176).

Urban strategies for sustainable regional economies are also emerging that reconnect cities with their surrounding regions. Sonnino (2009) uses the concept of food as a prism to understand the complex web of connections that tie cities to wider relations, places and processes. Techouèyres (2007) discusses how contemporary consensual attachments to markets could infer a craving for the past and idealised notions of traditional societies. She says, 'markets retail the illusion of sociability, the pleasure of urban life on a human scale, where the crowd is friendly, in what one might call the anti-modern city' (*ibid.* p. 248). As the popularity of rural places as leisure destinations, second homes and increasingly as 'suburbs' of the cities, these conceptualisations are closely linked with redefinitions of local rural identity.

11.3 The impact of retail restructuring on traditional food markets

Retail restructuring has taken its toll on traditional importers, wholesalers and market traders. Freidberg (2007: 322-323) refers, for instance, to how the standards implemented by supermarkets to assure quality and food safety are a kind of 'imperial knowledge … that threatens the practical knowledge traditionally employed in the production and trade of food, and, in particular, perishable produce'. She suggests that these schemes 'may undermine the very expertise, social relations and livelihoods that have historically helped to both define and provision quality food'.

These modes of retail restructuring have notable impacts on traditional food markets and their supply chains. Wholesale markets have been left with a declining role in food distribution in Western Europe (Cadilhon *et al.*, 2003; Dolan and Humphrey, 2004; Saphir, 2002) and have frequently been moved to the periphery of towns and cities and away from the centrally placed traditional food retail markets. De la Pradelle (2006: 1) suggests, 'Market society has no need of its street or stallholder markets. It has developed other forms of distribution that better satisfy its demands for rational efficiency and profit'.

A recent empirical survey of markets in the UK (Rhodes, 2005) found that competition from supermarkets and other food retail outlets, lack of investment, and changes in food shopping habits, have put many traditional markets under pressure. Others comment on the difficulties small retailers face without the obvious advantages presented by the economies of scale operated by supermarkets (Burt and Sparks, 2003; Hallsworth *et al.*, 2006).

Social and spatial health gaps, the inequality of social class, and the economic fall-out from industrial decline, all add to the complex combination of factors that affect how food is both distributed and retailed in the UK. The next section of the chapter maps the current geographic spread of markets in the UK and considers some of the reasons for their current locations.

11.4 Mapping traditional food markets

The empirical study begins with a look at the current state of traditional food markets in geographic and historical context and provides the first detailed assessment of traditional food markets in England. It maps and identifies patterns of concentration at different geographical scales using database research on traditional markets, wholesale markets and more specialist niche markets, including farmers' markets. This is supported by more detailed analysis of survey data collected from a sampling frame of traditional food markets in England.

Between September 2008 and February 2009 a national database of traditional food markets, wholesale, markets, and more specialist niche markets, including farmers' markets, was constructed from secondary data sources. The research found 2,105 markets of all types, including 1,124 traditional food retail markets (60% run by the public sector), 26 primary wholesale markets and 605 farmers' markets.

The database was geocoded to provide the means to carry out a geographical analysis of the current spread and distribution of markets at postcode level. Geocoded data was then used to map the geographic spread of these markets throughout the UK. Because the principal interest of this study is English markets, data on markets in Wales, Scotland and Northern Ireland was not so thoroughly researched.

Figure 11.1 reveals a concentration of traditional food markets through the central corridor, with a swathe down through the North West, the Midlands, the South East and London. The explanation could in part be historical. There was increased market hall building in the nineteenth century where there was the greatest population and industrial growth (except for London) and market halls were best suited to regions with large working populations whose food demands dictated large new market spaces. Thus market halls were built in the Midlands and North West as regions with the most pressing needs for improved food supply. Also the industrial north, with its growing wealth, underwent a shift in consumer demand from bread to meat at the very time the west began to increase its meat, dairy products, and vegetables that supported the market system. This, in turn, was supported by the evolution of better systems of canals and roads in these regions that encouraged centralised marketing.

In parallel, Figure 11.2 maps the distribution of primary wholesale markets established by government at the end of the First World War to encourage more efficient distribution. This coincided with the development of the railway system. The central swathe mirrors the development of the transport network (traditional retail markets clustered in close

Retail markets ·

Figure 11.1. Spatial distribution of traditional food retail markets by postcode (National database of UK markets, 2009).

proximity). The lack of a primary wholesale market serving the Eastern and South West regions is marked. An explanation could be the predominantly rural nature of these regions and the corresponding low population densities.

Farmers' markets show heavier clustering in the South East and South West regions and clustering in the Eastern region, although there is a reasonable spread throughout England. Just as the geographic spread of traditional food markets could be seen to reflect the process of urban growth and the accompanying transport networks, so farmers' markets (which have only developed in this form over the past 20 years) have clustered more significantly where small producers are located, or where the population demographics are able to support these more specialised niche markets (like in the South East) (Figure 11.3).

Traditional markets – although often ignored in geographical and sociological accounts of food retailing – have been at the centre of the UK food retail system for much of history and have mirrored the social and economic development. The current spatial distribution of traditional food markets (and wholesale and farmers' markets) reveals much about this previous history and association with place and place-making. The more recent and rapid growth of niche markets, including farmers' markets (see Kirwan, 2004), reveals a new set of

Wholesale markets ·

Figure 11.2. Spatial distribution of wholesale markets by postcode (National database of UK markets, 2009).

relationships. The next section of the chapter considers the fresh food sold on these markets in more detail.

11.5 Placing traditional food markets: some preliminary case study findings

Case study research carried out on a selection of different types of traditional food markets shifts the research to a local level. Two case studies of: (1) a large, local authority operated, indoor urban market hall in the North East of England, open 6 days a week (the Grainger Market) and (2) an outdoor, urban, local authority operated traditional market in Eastern England, open 6 days a week (Cambridge) were conducted. The case studies also included farmers' markets operated by both local authorities on either the same site or an adjoining site. In order to consider how these cities reconnect with their surrounding regions, traditional and farmers' markets operated in nearby market towns were also included in the research studies.

These preliminary findings provide new insights into how fresh food is sourced on these markets, about shopping habits, and about how the fresh food is valued by those involved in this type of market exchange.

Farmer's markets •

Figure 11.3. Spatial distribution of farmers' markets by postcode (National database of UK markets, 2009).

11.5.1 Sourcing fresh food

Most of the traders selling fresh food on the Grainger market in Newcastle do not immediately make a distinction between food that is industrially produced and imported and food that is locally grown, albeit on an industrial scale. When asked a bit more about this, it is obvious that a reasonable proportion of what they buy at the wholesale market to sell on the market depends on the seasons and this is likely to be produced locally or regionally (or sometimes from further a-field in the UK) but for them, this is implicit in the market offer, it's what they have always sold. The same also appears to be true of the more traditional (older) market shoppers – they seem to see no need to immediately make distinctions about where the food is from – they just know. That being said, traders have also adapted to the change in peoples' shopping habits and to the cultural diversity in the population. This has changed what they source for sale on their stalls and they now offer a wider choice of imported fruit and vegetables and, to a lesser extent, imported meat and fish.

Fruit and vegetable traders in Cambridge and Ely (a market town 15 miles to the north) were more mixed in their response to questions about sourcing. Most commented on the decline in the numbers of small growers. In Cambridge some had adapted by offering good quality produce (exotics, fresh herbs, a range of mushrooms etc.) which was labelled as Class 1 rather than by provenance. Another targeted tourists and students by offering freshly squeezed fruit and vegetable juices and exotics rather than 'the basics' (potatoes, cabbages etc.). One fruit and vegetable trader had always sourced from local producers (although he acknowledged that there would be nobody to take their place when they were gone) and his customers tended to be middle-class and looking for seasonal, local food. He said, 'Some of the old things are coming back – rhubarb is now quite fashionable. In the past you couldn't give it away because everybody used to grow it.' He felt the market survived well because it had similarities with a farmers' market and that his younger customers had been brought up on farmers' markets. In Ely, one fruit and vegetable trader also continued to source directly from farmers and used local wholesalers. Another sourced most things from Spitalfields in London. Traders pointed out that the bigger local producers were sending their produce down to the London wholesale markets and market traders were going down to bring it back up.

Fish traders tended to favour UK sourcing. In fact many went to great lengths to ensure this was the case. It was pointed out that it was difficult to buy from one port because of the quotas changing all the time. One trader (also running a business supplying restaurants) had prices and availability faxed and emailed to him every morning. Another said he would not countenance foreign fish. Some butchers said everything they sold was local – one farmed, slaughtered and butchered all his own meat, another said he did not buy from anywhere more than 40 miles away and operated on the principle that if meat was supplied locally, it tended to be reared locally.

11.5.2 Shopping habits

The Cambridge market operates outdoors in a smaller, university city, with a more middle-class population. Many of the traders commented on how people now shop a lot later (mostly between 11.30 and 15.00). Some put this down to the bus pass not offering the elderly free transport before 9.30. Even on a six-day week market there was still a feeling that trade built up as the week progressed. Thursday was often considered 'market day' and Saturdays had got a lot busier in the last two or three years and brought very different customers. The fish trade had changed dramatically, partly because of farmed fish making the price more manageable, awareness of diet and health, negative attitudes to supermarket fish, and the influence of cookery programmes. In addition, the butcher on the market was doing well, combining the market trade with his own farm and shop, but commented on how Sunday opening has killed trade because 'people aren't at home cooking a roast'.

Low-income shoppers were not obvious on Cambridge market although older shoppers, office workers, students and tourists were. The Saturday market was very busy and younger couples and families were more in evidence and even more so on the Sunday farmers' market (which was really just one large fruit and vegetable trader) who was swamped with customers. In Cambridge, shoppers' comments reflected their desire for fresh and preferably local produce, and a tendency to look for quality over price. Most also expressed a preference for shopping outdoors because of the atmosphere.

In Newcastle apart from the traditional market shoppers, who tend to be profiled as older and often on low or more restricted incomes, there are other groups of shoppers representing core elements of the customer base including office workers, students, and individual ethnic groups. Some of these groups appear to have more complex relationships with the decisions they make about food choice, tied up with cultural traditions, values and beliefs. In the students' cases this may be about anti-supermarket feeling, but decisions to shop on the traditional market were also to do with the cooking advice offered by traders, the ability to buy small quantities, and the fresh food available at cheap prices. For the immigrant communities in the city it was often about cultural preferences and freshness, and increasingly for some ethnic restaurants, the cheapest source of fresh ingredients available.

11.5.3 Valuing fresh food

The growth of farmers' markets in the UK has paradoxically coincided with many traditional markets struggling. For example, local authorities often establish a farmers' market on either the same site or an adjoining site to the traditional market, a situation that has resulted in various tensions between the two types of markets.

Early analysis of the fieldwork conducted in Newcastle shows that traders and some shoppers on the traditional market are suspicious of the monthly farmers' market held on an adjoining square. Prices are too high, traders and customers are not the same ('more money than sense'), the council privileges these traders with low rents, and traditional traders see these farmers' market traders as being able to operate with far fewer health and safety restrictions. But, it seems, those shopping on the traditional market are often looking for a similar shopping experience as those shopping on the farmers' market, yet the 'value' placed here is different. This is 'proper' shopping.

Traditional market shoppers may be associated with a different profile – although early observations indicate it is more varied than this – yet they are appreciating the same things those on higher incomes, or with supposedly different values, appreciate on a farmers' market. The traditional market shopper has been part of a tradition that values affordable quality for far longer than the 'new' farmers' market shoppers. Ask traditional market customers about the produce in the traditional market and this becomes the 'special, fresh' produce, not like the tired, packaged produce available in the supermarket. Ask shoppers on the farmers' market if they shop on the traditional Grainger market and in this small sample, all of them said yes. They perceive the quality of the traditional market produce as superior to the supermarket offer. Initial impressions also confirm a Northern commercial notion of 'value' (Sonnino and Marsden, 2006). Traditional market traders and shoppers are aware that the seasonal fresh fruit and vegetables are likely to have been grown in the UK (and often locally or regionally), meat is often reared locally or regionally, and fish has been sourced from nearby fishing ports and is often caught or farmed within the UK, but they often voice other priorities concerned with freshness, taste and price. Quality is therefore constructed and negotiated as Sonnino and Marsden (2006) point out, and as these early findings begin to demonstrate, it is also determined by the social and spatial context.

11.6 Conclusions

Following Everts and Jackson (2009), it is important to point out that academic and colloquial notions of 'traditional' are complex; they suggest the term 'pre-modern' as more appropriate because the contrast between 'modern' and 'premodern' retail forms is 'a discursive construction rather than a simple description of historical changes in consumer practices' and that,'(c)onsumption practices vary not just over time and space but also at the same time and in the same place' (*ibid.* p. 918-921). In other words, references to 'traditional' food retailing need to be seen within the contemporary world, rather than as a linear, historical shift and cannot be seen in isolation from the wider retail landscape. Although the spatial distribution of local, regional and national market networks is linked to history and to recent social, economic and cultural change, traditional markets continue to act as central nodes in towns and cities where commodities and cultures intersect as they adapt to changing market forces, population demographics and consumer preferences.

In 2006, Defra (2006), noted that centralised supermarket distribution systems could not cope with increased consumer demand for local produce and were beginning to source this produce from regional and smaller suppliers as demand continued to grow. In addition, the recent Cabinet Office Strategy Unit (2008) report, *Food Matters: Towards a strategy for the 21st century,* published in July 2008, whilst acknowledging the apparent decline of street markets across the UK, notes their importance as a source of 'affordable, good quality food including fresh fruit and vegetables' (para 56, p. 65). The report goes on to say that the success of farmers' and specialist markets and revitalised large city markets 'provide models for greater local engagement with fresh, affordable food and highlight an opportunity to modernise or develop new food retail markets' (para 57, p. 66), and that cities and towns can use their planning policies and food strategies to support markets as part of a sustainable food strategy. However, traditional food markets are frequently not viewed as part of the local retail economy by local authorities and have suffered from a lack of investment, often being administered across a number of departments (2008; Rhodes, 2005; Watson and Studdert, 2006).

As the early findings from this study begin to demonstrate, traditional food markets continue to play an important, although often 'hidden', role in food provisioning offering a source of fresh and affordable food that is often sourced regionally or locally within the UK and valued by shoppers from across the social spectrum. Shaw *et al.* (2002) express frustration with how policy relevant work within human geography is rarely used to enter directly into policy debates. The traditional food market could present just such a case for making engagement with public policy debate more meaningful by encouraging local authorities to adopt planning policies and food strategies that recognise the value, and long-term sustainability, of their traditional food markets.

References

Burt, S. and Sparks, L., 2003. Competitive analysis of the retail sector in the UK. Department of Trade and Industry, London, UK.

Cabinet Office, 2008. Food Matters: Towards a strategy for the 21st century. Strategy Unit, London, UK.

Cadilhon, J.J., Fearne, A., Hughes, D.R. and Mouse, P., 2003. Wholesale markets and food distribution in Europe: new strategies for old functions. Discussion paper No. 2, Centre for Food Chain Research, London, UK.

Communities and Local Government Select Committee, 2009. Market failure?: Can the traditional market survive? House of Commons, London, UK.

De la Pradelle, M. 2006 [1996]. Market Day in Provence. University of Chicago, London, UK, 243 pp.

Department for Environment, Food and Rural Affairs, 2006. Food security and the UK: an evidence and analysis paper. Food Chain Analysis Group, London, UK.

Dolan, C. and Humphrey, J., 2004. Changing governance patterns in the trade of fresh vegetables between Africa and the United Kingdom. Environment and Planning A 36: 491-509.

Everts, J. and Jackson, P., 2009. Modernisation and the practices of contemporary food shopping. Environment and Planning D: Society and Space 27: 917-935.

Freidberg, S., 2007. Supermarkets and imperial knowledge. Cultural Geography 14: 321-342.

Hallsworth, A., Parker, C. and Rhodes, N., 2006. Retail markets: present status, future prospects. Retail Enterprise Network, Metropolitan University Business School, Manchester, UK.

Imbruce, V., 2006. From the bottom up: the global expansion of Chinese vegetable trade for New York City markets. In: Wilks, R., (Ed.) Fast food, slow food: the cultural economy of the global food system. AltaMira Press, Plymouth, MA, USA, pp. 163-179.

Kirwan, J., 2004. Alternative strategies in the UK agro-food system: Interrogating the alterity of farmers' markets. Sociologia Ruralis 44: 395-415.

Maye, D., Kneafsey, M. and Holloway, L., 2007. Introducing Alternative Food Geographies. In: Maye, D., Holloway, L. and Kneafsey, M. (Eds.) Alternative Food Geographies. Elsevier Ltd, Oxford, UK, pp. 1-22.

Morgan, K. and Sonnino, R., 2010. The urban foodscape: World cities and the new food equation. Cambridge Journal of Regions, Economy and Society 3: 209-224.

Plattner, S., 1982. Economic decision-making in a public marketplace. American Ethnologist 9: 399-420.

Rhodes, N., 2005. National Retail Market Survey. Retail Enterprise Network, Metropolitan University Business School, Manchester, UK.

Saphir, N., 2002. Review of London wholesale markets: a review for the department for environment, food and rural affairs and the corporation of London. Defra, London, UK, 119 pp.

Shaw, M., Dorling, D. and Mitchell, R., 2002. Health, place and society. Pearson, London, UK, 207 pp.

Sonnino, R., 2009. Urban food and public spaces: planning for security and sustainability. AESOP sustainable food planning conference. October 9-10, 2009. Almere, the Netherlands.

Sonnino, R. and Marsden, T., 2006. Beyond the divide: rethinking relationships between alternative and conventional food networks in Europe. Journal of Economic Geography 6: 181-199.

Techouèyres, I., 2007. Food markets in the city of Bordeaux – from the 1960s until today: historical evolution and anthropological aspects. In: Atkins, P.J., Lummel, P. and Oddy, D.J. (Eds.) Food and the city in Europe since 1800. Ashgate Publishing, Aldershot, UK, pp. 239-249.

Tolia-Kelly, D.P., 2010. The geographies of cultural geography 1: identities, bodies and race. Progress in Human Geography 34: 358-367.

Watson, S. and Studdert, D., 2006. Markets as sites for social interaction. Open University report, The Policy Press, Bristol, UK, 64 pp.

Whatmore, S. and Thorne, L., 1997. Nourishing networks: alternative geographies of food. In: Goodman, D. and Watts, M.J. (Eds.) Globalising food: agrarian questions and global restructuring, Routledge, London, UK, pp. 287-304.

Zasada, K., 2009. Markets 21: A policy & research review of UK retail and, wholesale markets in the 21st century. The Retail Markets Alliance, UK, 84 pp.

Chapter 12

Marking the boundaries: position taking in the field of 'alternative' food consumption

Jessica Paddock
Cardiff University, School of Social Sciences, Climate Change Consortium of Wales (C3W),
King Edward VII Avenue, Cardiff CF10 3WT, United Kingdom; paddockjr@cardiff.ac.uk

Abstract

The goal of realising environmental sustainability has long been interwoven with the ambition of achieving the objective of *equitable* sustainability whilst adhering to principles of distributive social justice. Through an analysis of food consumption practices, this chapter demonstrates that such an ambition demands a critical and *classed* approach to environmental social science. To this end, this chapter presents consumer narratives regarding their relationship with food, and their experiences of engaging with alternative food networks in South Wales at both a farmer's market and a community food co-operative that each promotes a form of 'alternative' food consumption. Consumers from various socio-economic backgrounds at each research site articulate their experience through the descriptive frame of class. By exploring such accounts, we come to understand the extent to which reluctance to engage with 'alternative' food networks on behalf of the consumer hinge upon relations of social class. Following a discourse analysis of interview data, this chapter then concludes that efforts of sustainable food planning, albeit within the context of this research, do little to promote *equitable* sustainable consumption. Crucially, it becomes clear that in order to realise environmental sustainability based upon sound principles of distributive social justice, we must first accomplish a nuanced understanding of contemporary class relations.

Keywords: social class, sustainable consumption, distinction

12.1 The reinstatement of class analysis: a provocation

This chapter advocates a return to class analysis for three reasons. Firstly, class analysis may enrich our understanding of social inequalities in the face of claims that individual agency and choice govern one's ability to succeed in 'high modernity' (Beck, 1992; Giddens, 1990). Secondly, a return to class analysis seems fundamental to the realisation of sustainability within the context of global environmental change with respect to climate change and food security in particular. Lastly, is seems reasonable to suggest that sustainable consumption – as part of the normative agenda of sustainable development – faces some difficulty when applied within high consumption societies. Such difficulties may be seen to arise given the international focus given to issues of equity in relation to sustainable consumption. A contention of this chapter is that we require a nuanced understanding of sustainable consumption initiatives as implemented within Western contexts. In turn, it is argued that such initiatives may benefit from recognition of the social inequalities that persist at a local, national and global level.

This first supposition; that an understanding of [is imperative, is exacerbated by the urgency required to meet the challenge of adapting to and mitigating the effects of environmental degradation. Given the complex social, ecological, political, cultural and institutional context within which food systems operate, those seeking develop sustainable food systems are faced with numerous interlocking challenges. It is in dealing with these interlocking challenges that an appreciation of the inequalities experienced through food become necessary. For example, inequalities in access to food can have profound impact on one's health. It is in the study of health that the commitment to social class as a classificatory category (Glass, 1954; Goldthorpe, 1980) for understanding inequality has remained steadfast. However, the use of social class as a category for exploring the subjective and cultural elements of inequality has experienced some demise. To some extent this demise can be traced to the summation that explanatory command has been lost to categories of race and gender as well as postmodern conceptions of lifestyle choice as directing individual identity projects. This perception that social class offers little in explaining everyday life (Pakulski and Waters, 1996) is however, contested. In this way, sociologists have asked why an interest in social class has declined even in the context of increasing economic polarisation between rich and poor (Savage, 2000). In response, strong defences have been mounted in claiming that social class is embodied and experienced through culture (Skeggs 1997, 2004) and even account for the discursive means through which class is reproduced through *unspoken* and thus *apparently* invisible identity politics (Lyle, 2008; Walkerdine, 1990). Further defence of social class as a subjective and cultural category is provided by Sayer (2005) wherein social actors are seen to have 'sensitive antennae' with regards to social class. To have 'sensitive antennae' in relation to class is not to relegate the categories of race, gender and lifestyle choice as redundant. Rather, social class is seen as a means of underpinning subjective identities as well as a means of economic categorisation. Here, it seems reasonable to suggest that the 'death' of class has been prematurely declared on account of newly acknowledged categories of description. The classed baby, it seems, was thrown out with the proverbial bathwater.

The application of social class analysis in the case of food consumption has a long tradition in, for example, the social theories of Elias (1939, 1969) in relation to 'The Civilising Process' and in the theories of Engels (1844) as he charts the conditions of death by starvation faced by the working classes in London and Manchester. Furthermore, theories that pertain to food and class can be found in the work of Veblen (1899) where food is seen as an arena for competitive display. Douglas and Isherwood (1979) however, take this notion of food as competitive display and indeed as an essential need for human subsistence to suggest that commodities are necessary for 'making visible and stable the categories of culture' (*ibid*. p. 38) therefore proposing, albeit implicitly, that food is analogous to the social system. In this way, it would appear that social class can be interpreted not only through demographic data, but can be read and understood through the practices of consumption that embody the relations of our culture.

More recently Bourdieu (1984) charts the relationship between food, class and distinction. Indeed, theorising food as a resource of the social field that is inextricably bound with processes of distinction has a great tradition in the social sciences, as briefly summarised above. However, since the 'death of class' thesis (Pakulski and Waters, 1996) and the 'omnivorousness' thesis (Peterson and Kern, 1996) gained favour on the scholarly stage, it seemed that food is no longer seen as a terrain for the battles of distinction as was noted by Bourdieu (1984) in his study of French consumers. Such a trend is refuted by Johnston and Baumann (2007, 2010) in their suggestion that contemporary food consumption practices continue to reinforce class distinctions behind a shroud of apparent 'omnivorousness'. For them, the decline in explicitly snobbish gourmet food consumption has not led to the simplistic democratisation of food and food culture. Instead, what has been deemed 'omnivorous' food is seen to rearticulate the status and meanings associated with a 'poor (wo)mans' dish through the addition of distinctive ingredients. For example, a hamburger is no longer just a hamburger, but a gourmet inspired meal with *elements* resembling the traditional hamburger. Crucially, though, they argue that distinction persists, but remains hidden and misrecognised under the appearances of 'traditional' and 'simple' food. With the reinvention of traditional foods imbued with the distinctive sign of the 'rustic', it is argued that such misrecognition of distinction can be seen in the practices of 'alternative' food consumption today. It is through an analysis of the social relations played out within settings of 'alternative' food consumption that we can come to inform strategies for sustainable consumption that take into account the diverse social and economic landscape within which such initiatives are set. Before commencing an analysis of the social class relations as played out within a setting of 'alternative' food consumption, this chapter will make clear the conceptual lens through which the data is understood.

While it remains imperative to understand income based inequalities in access to food, it seems reasonable to consider that there remain cultural intermediaries that govern food consumption practices. For example, consumers face not only physical consequences from maintaining an 'insufficient' diet, but often moral reprimand for not consuming 'good' food. Whereas accounts of bringing 'good' food to 'bad' neighbourhoods (Guthman, 2008) are interlaced with the politics of race, gender and ethnicity, it seems that social class is considered as but one other means by which spaces for the consumption of 'good' food are created. In this chapter, the accounts of two research participants are explored. While these participants clearly possess the 'sensitive antennae' described by Sayer (2005), these 'antennae' are not particular to these participants but are present across the sample of twenty 'alternative' food consumers. However, it is the 'antennae' of these two interviewees in particular that encapsulate the dynamic of class misrecognition that appears in this case to fuel the social exclusion of working class consumers from spaces of 'alternative' food consumption. Moreover, these respective interviewees represent a dynamic of relational position taking within this field of 'alterative' food consumption.

This concept of relational position taking derives from the sociology of Pierre Bourdieu (1984) and is operationalised for its attention to the interconnections between objective (structural) and subjective (experiential) class. Also, this work takes into account the means by which acquisition of 'capital' – economic, social, symbolic and cultural – amount to the resources of distinction battled for in the social field. Such struggles for distinction are played out through the expression of taste for certain goods over another, certain goods that may be valued by one's social class or indeed 'habitus'. It is the contention of this chapter that 'alternative' food practice has become a terrain that is rich in such 'capital' resources. It follows that spaces of 'alternative' food practice potentially represent sites for achieving distinction from 'others'. To consolidate and maintain a discursive separation from an 'other' it would seem that moral boundaries are formed with the exacting of moral judgement upon such an 'other'. This very process of moral boundary formation through separation from an 'other' within a field of 'alternative' food is explored below. This serves the purpose of demonstrating the need for a nuanced understanding of the (moral) significance of class (Sayer, 2005) in attempting to both imagine and equitably realise a sustainable food future.

12.2 Methods

This chapter is informed by data collected over a two year period of participant observation, although particular attention is paid to data collected via twenty semi-structured interviews. The interview schedule included an invitation to participants to speak of the routines and practices of their every-day lives with particular attention to food. In order to recruit these participants, a survey was completed at both the farmers' market and community food co-operative, which sought to gain information pertaining to their demographic class position. Through this survey, respondents were invited to take part in further research. All participants who took part in the stage of qualitative interviewing were recruited via this method. Questions related to class and to 'alternative' food consumption practices were purposefully absent from the 'aide memoir' in order to illicit accounts of class only *should* they arise of the participant own accord. When speaking of their hobbies and routines of food consumption, participants frequently employed the discursive technique of distinction by 'positioning' themselves in relation to an 'other'. In this way, it seems that participants speak their class 'position'. Of the participants introduced in this chapter, Ken identifies himself as 'working class', a categorisation that is supported by the objective indicators supplied by the survey data. Karen, a farmers' market customer, positions herself as 'middle class'. Each of these participants, when speaking their class position do so by drawing upon cultural references that pertain to their experience of visiting the farmers' market.

12.3 Discursive positioning in the social field

Participants came to be understood as having been 'positioned by and affected through discourse' (Edley and Wetherell, 1997: 205 in Wood and Kroger, 2000: 24). This does not suggest that participants have freely constructed a sense of themselves and their class. Rather, it seems that participants constructed a sense of their class through positioning work that is evident through analysis of discourse (Fairclough, 2003). Such discourses are seen as those that are available to them, and respective to the structures of class to which they are bound by economic limitations or advantages. This analysis builds upon such objective conceptions of class identity by considering the social actor as expressing through discourse the subjective and experiential aspect of class culture as tied to objective and economic structures of class. For example, through discursive means, participants are seen to separate themselves from an 'other'. This discursive separation is seen to involve moral positioning, for to identify an 'other' involved judgement of practices deemed 'good' and as 'right' in comparison to those that were deemed 'bad' or 'wrong'.

Extracts 1-4 demonstrate such acts of 'position taking' as carried out by participants. It is these extracts that carry the strongest message for those seeking to realise sustainable food systems; that of the *moral* significance of class. Here, an example of 'positioning work' is evident, for a community food co-op customer speaks of the one time he visited the farmers' market. When speaking of his visit, Ken constructs a position for himself in relation to the 'poshies'. He begins by speaking of accent, and 'people like us' before identifying that accents, for him, are indicative of class membership. Moreover, the presence of 'other' accents is presented by Ken as making him feel like he 'shouldn't really be there'.

> (Extract 1) *yeah cause there was a lot of people there wearing afghan stuff (big hippy scarves) and 'hello Rodney' and all this you know (laughs) and then there's me with my [local] accent like, there wasn't much evidence of a [local] accent there like you know and you did feel like, what's this doing, this is not for me, you know'* (Ken, Co-op customer).

When asked to say a little more about why he felt this space 'wasn't for him', Ken continues;

> (Extract 2) *yeah, erm yeah I suppose erm a lot of it's down to accent isn't it you recognise people with a local accent and ok we're the same sort of thing, and you recognise people with a posher accent and you're used to working with the people with the local accent and working together, and you're used to the people who are the bosses having the posh accent. So when you're in a room of people and they're talking a certain way, you know they're the 'poshies' and they're the middle classes and you shouldn't really be there sort of thing you know that's my feeling obviously a lot of people don't feel that way but I still do* (Ken, Co-op customer).

Crucially, a sentiment often expressed by interviewees and participants; that working class people do not wish to eat 'good' food, that they are not interested in sustainability, and the judgement that they must therefore be re-educated and retuned into appreciating 'good' food is common amongst the data collected for this project. Such a sentiment is discussed below. Importantly, it is in the dynamic between these two discursive positions being constructed here that good news is brought to those seeking to develop spaces of alternative food consumption. This can be seen as 'good news' for Ken reveals – in opposition to Karen's imaginary of the 'working class' – that he did after all have an interest in alternative food consumption, but was intimidated by the 'other' middle class customers and not by the market or by 'sustainability' itself.

Here, a customer from the farmers' market distances herself culturally from a working class 'other' by suggesting that this 'other' is uncritical of the culture of cheap food. Indeed, she suggests that they do not *want* to engage with an ethical or moral commitment, for they are dedicated to their own vision of the 'good life'. When asked to expand on a previous statement whereby Karen states that 'working class people are different to middle class people', Karen continues;

> (Extract 3) *Well, both economically and socially erm you know the sort of the sort of people I was talking to you about in [place they live]... quite working class parents I know who took their children to ... school with my kids, they've got a hugely different attitude to life than my friends and I have. Erm and it's not just about money, I mean a lot of them are earning more money than I am, probably but they spend their money in different ways, because that's how that kind of culture is very different erm and so it' not just an economic thing it's a kind of cultural thing as well, I think* (Karen, farmers' market customer).

Karen continues;

> (Extract 4) *Well I mean as I say in terms of being very uncritical about that sort of things I'm criticising in terms of the consumerist culture, cheap food, not worrying about where the stuff comes from how far it's come or how it's produced and just saying well something's cheap and that's great and bigger cars and going on foreign holidays and thinking that's you know, taking the attitude that that's the good life* (Karen, farmers' market customer).

In combination with the Extracts 1 and 2, it becomes clear that class positions are in some part articulated through discussion of food consumption and most importantly for this chapter, 'alternative' food consumption practices. Extracts 1 and 2 conveys the discomfort felt by a community food co-operative customer at attending a farmers' market, whereas Extracts 3 and 4 represent a farmers' market customer's assumptions about the lack of presence of working class populations at such a market. These assumptions about alternative attitudes to 'the good life' have been found to be misconceived, for it is the *way* in which the space has been *appropriated* by middle class consumers that a self identified working class consumer does not value.

12.4 Discussion: the moral significance of class for sustainable consumption

In the face of climate change and biodiversity loss, and in recognition of the interdependence between our coupled social, physical and ecological systems, we are tasked with radically reimagining the means by which to maintain human life. One such arena in need of reconsideration is of food provenance (Morgan 2010), a challenge that is framed by discussions of an increasingly degraded natural environment and the pursuit of development (for a full discussion see Baker, 2006). As an offshoot of sustainable development, and within the frame of sustainable consumption, debates have ensued (Jackson, 2006) with regard to the very role of material goods in everyday life, from those goods required in order to provide nutrition and shelter, to those that create and maintain personal and cultural expressions through engagement with an 'identity project'. Both the goods that generate human subsistence and those that create and maintain such personal and cultural expressions have, however, become interwoven with the emerging practices of 'ethical' and/or 'alternative' food consumption, a trend that has incited much criticism as the 'medium (roast) becomes the message' (Low and Davenport, 2005). This mirrors deeper criticism, wherein the radical spirit that was characteristic of environmental movements in past decades is seen to have been overtaken by the mainstreaming of the environmental problematic into market based solutions that dampen the will for transformative environmental politics (Seyfang, 2004). Such colonisation by market forces is made apparent by the case of 'ethical' consumption by virtue of its exclusion of those without the economic capital to take part. Furthermore, the often used label for 'alternative' food consumption known as 'ethical' consumption itself poses a distinct moral challenge, for it could be seen to present a symbolic barrier to developing consumer participation by its very labelling as an 'ethical' and arguably 'elite' practice. The ubiquity of messages left to consumer interpretation (Gabriel and Lang, 2006) evident in the mainstreaming and marketisation of sustainability then appears to create further problems of inclusion and exclusion. It is this dynamic that demands the reinstatement of class analysis in order to understand the implications that class struggle in the social field can have for the future of this very practice. Although the class dimensions of consumer lifestyles have been widely researched with respect to both 'extraordinary' (Devine *et al.*, 2005) as well as 'ordinary' (Gronow and Warde, 2001) consumption, we must explore how class relations affect the possibilities to successfully realise sustainable consumption. If there is a recognition that some can more easily engage with practices of 'ethical' or 'alternative' consumption due to their greater access to cultural, economic, social and symbolic resources of 'capital' (Bourdieu, 1984) then a greater and more nuanced understanding of both class and practical means of achieving sustainability *for all* may ensue.

In further support of this claim, the status of food as a moralised terrain, and therefore as a cultural resource with a long standing reputation for the inflection of the relations of class, one would expect the take-up of 'ethical' and 'environmental' choices to reflect engagement with similar terrains of debate. However, Extracts 2-4, taken from interviews with individuals who engage with ethical food practice illustrate a seemingly endemic perspective; that to make ethical choices in food consumption is simply a matter of rational choice based on education and political awareness. Such an argument appears to misrecognise class (Skeggs, 2004) for Extracts 1-2 clearly demonstrate otherwise, and indeed problematise the notion that environmental choices are the result of rationally calculated decisions. Although consumer choices have throughout history been subject to constraints of cost, time and cultural preferences (Trentmann, 2005) it seems that a further challenge is to be faced by those seeking to realise sustainable consumption. This challenge seems to be located not in seeking to change the attitudes and the behaviours of individual consumers, but one that addresses the need for change to the settings within which practices of alternative food consumption take place. In widening the range of means of alternative consumption, it seems reasonable to suggest that these take into account the moral significance of class (Sayer, 2005).

To suggest that the settings within which alternative consumption take place be modified in order to take into account the relations of social class is to take seriously the embeddedness of class culture. Indeed, given the pervading discourses that suggest that practices of social distinction are at work across almost all practices of consumption, then it is reasonable to conclude that to interfere with the relations of class themselves seems futile. Instead, it is likely that the form of discursive positioning exercised by Karen as she distinguished herself from the imagined working class 'other' as a practice of identity work will continue despite any efforts made to dissipate such a practice of moralising the 'other'. Moreover, it is reasonable to imagine that working class consumers such as Ken will continue to feel uncomfortable in settings such as the farmers' market. It follows, then, to imagine that farmers' market will continue to sell their produce in a manner that middle class consumers have come to enjoy, but that this be *recognised* as a *middle class* consumer institution that is not suitable for all. Crucially, this is not to suggest that those developing settings for alternative consumption 'give up' on the working class consumer. On the contrary, Ken provided some insight as to the absence of working class consumers from such settings. Ken suggests that he recognises this space as middle class setting. Moreover, he suggests that he feels uncomfortable – not due to the price of commodities, or by the sort of consumption that advocates sustainability of the environment and rural livelihoods – but by the class dynamic manifest amongst the regular consumers. To alter such a dynamic in order that working class consumers are comfortably incorporated is a task that may prove impossible. Rather, it seems likely that if faith was given to working class communities by offering 'alternative' food in a context that fosters an ease with which to engage in such practice that this might have the potential to inspire the realisation of sustainable food systems. To move into working class settings and institutions in order to promote sustainable consumption, to foster comfort within the settings on their own terms, and not in a way that is solely reminiscent of middle class practices may aid the transformation necessary to realise an equitable and sustainable food future.

12.5 Conclusion

This chapter argues that the environmental social sciences must take seriously the importance of social class on a subjective as well as objective level. To take seriously the presence of and important affects that social class relations have upon interactions within settings of alternative food consumption may serve the purpose of rendering successful the implementation of sustainable consumption strategies *within* high consumption societies.

Moreover, to understand inequalities at the local level and within Western contexts evidently delivers insight that is of use to those developing sustainable food strategies. As conveyed by Ken; decisions regarding engagement with alterative consumer practice do not solely involve a calculation of price, but of feelings of comfort and discomfort. It seems evident, therefore, that it is how the *spaces* of alternative food consumption have been appropriated, and appropriated for middle class 'distinction' that pose a challenge to equitable sustainable consumption as strived for by those who take on the task of realising the normative goals of the sustainable consumption agenda. It is the contention of this chapter that the challenge here rests not solely on the shoulders of the producer or the consumer, but on those who construct and mediate the *settings* within which alternative consumption takes place.

Acknowledgements

With thanks to the Economic and Social Research Council for funding the doctoral study from which this chapter is drawn, and to my doctoral supervisors who gave support throughout the process of completing the research that informs this chapter.

References

Baker, S., 2006. Sustainable Development. Routledge, Oxon, UK, 245 pp.

Beck, U., 1992. Risk Society: towards a new modernity. Sage, London, UK, 272 pp.

Bourdieu, P., 1984. Distinction: a social critique of the judgement of taste. Routledge, London, UK, 640 pp.

Devine, F., Savage, M. and Crompton, R. (Eds.), 2005. Rethinking class: culture, identities and lifestyle. Palgrave Macmillan, Basingstoke, UK, 248 pp.

Douglas, M. and Isherwood, B., 1979. The world of goods: towards an anthropology of consumption. Basic Books, New York, NY, USA.

Elias, N., 1939,1982. The history of manners: volume 1 of the civilising process. Pantheon Books, New York, NY, USA, 310 pp.

Elias, N., 1969, 2000. The civilising process. Blackwell, Oxford, UK.

Engels, F., 1844, 2009. The condition of the working class in England. Blackwell, Oxford, UK.

Fairclough, N., 2003. Analysing discourse: textual analysis for social research. Routledge, London, UK, 270 pp.

Gabriel, Y. and Lang, T., 2006. The unmanageable consumer. Sage, London, UK, 220 pp.

Giddens, A., 1990. The consequences of modernity. Stanford University Press, Stanford, CA, USA.

Glass, D.V., 1954. Social Mobility in Britain. Routledge and Kegan Paul, London, UK.

Goldthorpe, J., 1980. Social mobility and class structure. Clarendon Press, Oxford, UK, 400 pp.

Gronow, J. and Warde, A., 2001. Ordinary consumption: complexity and emergence in organizations. Routledge, London, UK, 272 pp.

Guthman, J., 2008. Bringing good food to others: Investigating the subjects of alternative food practice. Cultural Geographies 15: 431-447.

Jackson, T. (Ed.), 2006. The Earthscan reader in sustainable consumption. Earthscan, London, UK, 416 pp.

Johnston, J. and Baumann, S., 2007. Democracy versus distinction: A study of omnivorousness in gourmet food writing. American Journal of Sociology 113: 165-204.

Johnston, J. and Baumann, S., 2010. Foodies: democracy and distinction in the gourmet foodscape. Routledge, Oxon, UK, 280 pp.

Kneafsey, M., Cox, R., Holloway, L., Dowler, E., Venn, L. and Tuomainen, H., 2008. Reconnecting consumers, producers and food: exploring alternatives. BERG, Oxford, UK, 224 pp.

Low, W. and Davenport, E., 2005. Has the medium (roast) become the message? The ethics of marketing fair trade in the mainstream. International Marketing Review 22: 494-511.

Lyle, S.A., 2008. (Mis)recognition and the middle-class/bourgeois gaze: A case study of 'wife swap. Critical Discourse Studies 5: 319-330.

Morgan, K., 2010. Local and green, global and fair: The ethical foodscape and the politics of care. Environment and Planning A 42:1852-1867.

Pakulski, J. and Waters, M., 1996. the reshaping and dissolution of social class in advanced society. Theory and Society 25: 667-691.

Peterson, R. and Kern, R., 1996. Changing highbrow taste: from snob to omnivore. American Sociological Review 61: 900-907.

Savage, M, 2000. Class analysis and social transformation. Open University Press, Berkshire, UK, 208p

Sayer, A., 2005. The moral significance of class. Cambridge University Press, Cambridge, UK, 256 pp.

Seyfang, G., 2004. Consuming values and contested cultures: a critical analysis of the UK strategy for sustainable consumption and production. Review of Social Economy 62: 323-338.

Skeggs, B., 1997. Formations of class and gender. Sage, London, UK, 208 pp.

Skeggs, B., 2004. Class, self, culture. Routledge, London, UK, 232 pp.

Trentmann, F., 2005. The making of the consumer: knowledge, power and identity in the modern world. Cultures of Consumption series, Berg Publishers, London, UK, 256 pp.

Veblen, T., 1899, 1994. The theory of the leisure class. Dover Thrift Editions, General Publishing Company Ltd., Toronto, Canada.

Walkerdine, V., 1990. Schoolgirl fictions. Verso, London, UK, 328 pp.

Wood, L. and Kroger, R., 2000. Doing discourse analysis: methods of studying action in talk and text. Sage, London, UK, 256 pp.

Chapter 13

Resilient urban community gardening programmes in the United States and municipal-third sector 'adaptive co-governance'[21]

Henry Barmeier[1] and Xenia K. Morin[2]
[1]Department of Education, University of Oxford, Merton College, Oxford, OX1 4JD, United Kingdom; [2]Princeton Environmental Institute and Woodrow Wilson School for International and Public Affairs, Princeton University, Robertson Hall, Princeton NJ 08544, USA; henry.barmeier@gmail.com

Abstract

Engaging city government in sustainable food system planning may enhance urban health, social justice, and environmental quality. However, little is known about why municipalities engage in food system planning. To gain insight on the genesis and evolution of local government involvement in sustainable food system policy, we examined four long-standing urban community gardening programmes in the United States. At the time of our field work in 2010, municipally-sponsored community gardening programmes in Seattle, Portland and New York exemplified schemes of 'adaptive co-governance' as the mechanism for local-level policy making. Bottom-up policy-making was evident in establishing community gardening programmes in all three case studies. City Council members and mayors who championed the programmes did not lead the charge for community gardening as much as they responded to and facilitated an interest in the community. Over time, however, the relationships among public and third sector partners have cycled between four phases: 'budding,' 'building,' 'bonding,' and 'breaking'– the Four-B framework. The histories of community gardening programmes in Seattle, Portland and New York suggest that resilient relationships in co-governance enable the resilience of the gardens themselves, and hence, of the local food system. This conclusion is further supported by study of a non-municipally-sponsored community gardening programme in Philadelphia, which demonstrates that co-governance can function without municipal government, but is less able to adapt to adverse shocks. In sum, government's presence in the garden is better understood as a process rather than a product; it has not been consistent in the past and likely will not remain consistent in the future. Equally, citizen involvement is inconsistent over time. Resilient co-governance schemes may offer a model for success but persistent commitment from all partners will be needed to build resilient urban food systems.

Keywords: sustainable food system, local food policy, community food security

---

[21] This research was conducted as part of an independent senior thesis research project submitted to and partially funded by the Woodrow Wilson School of International and Public Affairs, Princeton University, 2009-2010. Parts have been published in Barmeier's senior thesis entitled 'Why is government in the garden? case studies of resilient co-governance in urban community gardening programs' available from the Princeton University Library.

Henry Barmeier and Xenia K. Morin

13.1 Introduction: engaging local government in food system planning

Prior to the last decade, food issues were 'low' priorities for most municipal planners. Food system scholars Pothukuchi and Kaufman (1999, 2000) suggest that the lack of consideration of food issues within local government has been driven by the perception that food policy only pertains to agricultural, or rural, areas; and by the opinion that urban issues such as crime, housing and transportation present more pressing problems than those that relate to food.

However, engaging city government in sustainable food system planning and policy may hold the key to enhancing urban health, social justice, and environmental quality. Although US federal policy establishes a powerful framework for how food is grown, distributed and consumed, food policy-making involving local government may offer a better response to localised concerns and interests. According to legal scholar N.D. Hamilton, local food policy 'capitalizes on the ability of people to control their own destiny by using institutions they control.' Federal and state programmes (e.g. SNAP and WIC) can meet some food needs of the community, but not all (Hamilton, 2002).

However, little is known about why municipalities decide to prioritise food planning or decide to maintain this priority over time. To gain insight on the genesis and evolution of local government involvement in sustainable food system planning, we examined four long-standing urban community gardening programmes in the United States.

Community gardens – plots of public or private land cultivated by volunteers – offer an instructive lens for study. These programmes represent a local food system initiative capable of responding to adverse socioeconomic and environmental shocks. In the United States, interest in and support for community gardening has surged at several times of national crisis over the last 120 years. They have empowered citizens to better 'control their own destiny' by providing food, employment, environmental remediation and diversion through two World Wars and periods of economic distress (Lawson, 2005). Because city governments' involvement in community gardening programmes predates their involvement in many other sustainable food system initiatives, these programmes offer a better opportunity to understand how municipal support, in collaboration with the third sector (citizen and non-profit groups), is sustained over time.

In Section 1 of this chapter, we present evidence of adaptive public-third sector co-governance in three municipally-sponsored community gardening programmes: Seattle's P-Patch Community Gardening Program, Portland Community Gardens, and New York City's GreenThumb Community Gardening Program. In Section 2, we contrast the development of these programmes with the Philadelphia Green community gardening program, now operated largely without any municipal support. Table 13.1 presents a basic description of these four programmes.

Table 13.1. Characteristics of case study community gardening programmes.

City	Programme initiation date	Municipal sponsor	Municipal host department	Primary nonprofit sponsor(s)	Gardens (#)	Gardeners (#)
Seattle	1974	P-Patch Community Gardening Program	Department of Neighbourhoods	P-Patch Trust	83[a]	3,800[a]
Portland	1975	Portland Community Gardens	Portland Parks and Recreation	Friends of Portland Community Gardens	35[a]	3,500[a]
New York	1978	GreenThumb Community Gardening Program	Department of Parks and Recreation	Green Guerillas, Open Space Greening, Trust for Public Land	600	8,000
Philadelphia	1974	Not applicable	Not applicable	Philadelphia Green, Neighborhood Gardens Association	226[b]	Not available

[a] Projected for 2010.

[b] Of these, approximately 60 have an 'active' relationship with one of the nonprofit organisations.

31.2 Adaptive co-governance in municipally-sponsored community gardening programmes

Our research finds that the municipally-sponsored community gardening programmes in Seattle, Portland and New York exemplify schemes of 'adaptive co-governance,' as the mechanism for local-level policy making. By 'adaptive co-governance' we mean a system of distributed interdependent responsibility that is capable of change over time. According to one conception, co-governance brings together government and non-governmental actors in a relationship of mutual responsibility and power (Stoker, 1998). In practice, co-governance involves two shifts of authority: a downward 'vertical' shift of authority from central to local levels; and a 'horizontal' shift of authority from public to private entities (Eckerberg and Joas, 2004). By embracing the interdependence of the public, private and third sectors, co-governance aims to be more efficient and responsive than a system of top-down government intervention (Carlsson and Berkes, 2005; Stoker, 1998). Co-governance can even occur without the participation of government as long as partners are able to establish effective mechanisms for regulating a given sphere of activity (Rosenau, 1992; Stoker, 1998). This raises the possibility that community gardening programmes without formal government sponsorship can nonetheless be interpreted as forms of co-governance. Finally, co-governance schemes – with or without government – can be understood as 'adaptive' arrangements to the extent that they are able to facilitate the absorption of external shocks to the governed system, and to respond resiliently to changes in relationships among partners (Holling, 1995; Nkhata *et al.*, 2008).

Although the literature on adaptive co-governance considers separately the resilience of the 'system' being governed and the resilience of the relationships of co-governance, these case studies suggest that these two meanings of resilience are nearly inseparable. To conceptualise this linkage, we propose a new model of co-governance inspired by Nkhata *et al.* (2008). In our 'Four-B framework' (Figure 13.1), the same vocabulary of 'budding,' 'building,' 'bonding,' and 'breaking' describes change both in the system being governed and in the relationships of co-governance. This model is applied to the case studies and its general features are summarised in Table 13.2. More details from specific case studies are given below. Although the Four-B framework over-simplifies co-governance – different public and third sector partners may simultaneously be in different phases at any given time – it nonetheless provides a useful lens for analyzing the development of our municipally-sponsored community gardening programme case studies.

13.2.1 Budding and building: governments engage citizen gardeners in the 1970s

Bottom-up policy-making was evident in establishing community gardening programmes in all three cases studies. In Seattle, Portland and New York, common economic and social

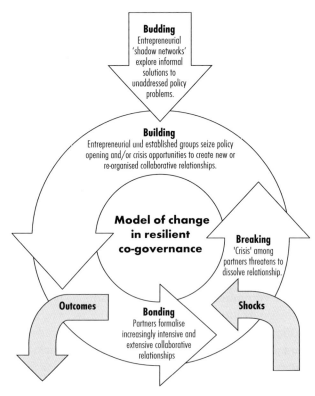

Figure 13.1. The 'Four-B framework': phases of relational change in resilient co-governance (Gunderson, 1999; Nkhata et al., 2008).

Table 13.2. Characteristics of the phases of resilient relational change in co-governance.

Phase	Specific characteristics in urban community gardening
Budding	• Generation of interest in community gardening (in response to perceived problems)
	• Formation of neighbourhood garden groups
	• Creation of purely citizen-driven gardens
Building (1)	• Establishment of formal links between public and third sector actors
	• Process of development of shared interests
	• Emergence of new leaders through advocacy
	• Strategic planning to meet goals
	• Development of community gardens through participatory process involving public partner
Bonding	• Efficient allocation of responsibilities among partners
	• Formalisation of community gardening goals in policies and plans
	• Security of land tenure
	• Maintenance of successful community gardens
Outcomes[1]	• Environmental, social and economic benefits of community gardens
	• Contribution to food system resilience
Shocks[1]	• Changes in municipal or third sector leadership
	• Changes in programme resources
	• Surge in demand for gardening beyond supply of gardens
	• Development and gentrification
	• Gardener turn-over and attrition
	• Vandalism
Breaking	• Threat to/disappearance of individual gardens
	• Threat to capacity of programme to develop and maintain gardens across city
	• Declining interest in gardening
	• Long waiting lists
	• Loss of a partner in co-governance scheme
Building (2)	• Mobilisation of third sector social networks
	• Creation of new supportive institutions (land trusts and advocacy organisations)

[1] See Figure 13.1.

woes fomented the initiation of community gardening efforts. 'Stagflation' from 1973 to 1975 delivered a punishing blow to the country. Inflation topped nine percent while real GDP shrank. In response to the scourges of urban decay, unemployment, and poor access to food, citizens in Seattle, Portland and New York planted vegetables on available plots of land. Embodying an entrepreneurial spirit, the informal 'shadow networks' (Gunderson, 1999) of gardeners in these three cities attempted to privately fix public problems that they believed government had not adequately addressed.

The development of ties between the pioneering gardeners and public partners in each city marked the transition of the previously-independent movements into schemes of co-governance. Through this process, local government officials who championed the programmes did not lead the charge for community gardening as much as they responded to and facilitated a demonstrated interest in the community. In Seattle, the City Council agreed to provide municipal support for incipient community gardening efforts as a means of strengthening ties between local agriculture and the historic Pike Place Market, which had been threatened in the early 1970s by an urban redevelopment plan (B. Chapman, personal communication). In Portland and New York, local government established formal processes for leasing municipal land for gardening in order to regulate citizens' use of vacant and unimproved land (Portland, Oregon 1975, Ordinance No. 139598; New York City Department of Parks and Recreation, 2010).

13.2.2 Bonding: fulfilling public goals efficiently through community gardens

Although public and third-sector actors entered into relationships with very different ideas about the need for a community gardening program, the process of building a stronger sense of shared interests in community gardening has undergirded the development of co-governance by reinforcing partners' commitment to successful collaboration.

Much of the important work of building common goals has involved municipal agencies and policy-makers understanding the ways in which community gardens contribute to their cities' broader social, environmental and economic goals. A crucial step in the co-governance 'bonding' process has been locating the community gardening programme in a specific municipal department whose mission and capabilities most closely align with the local community gardens. Seattle's P-Patch program, for example, blossomed after a move from the Department of Human Services (DHS) to the Department of Neighborhoods (DON) in 1997. In contrast to the DHS, the mission of the DON explicitly values intangible benefits of gardens, including social cohesion and inclusion of low-income and ethnic groups in civic processes (R. Macdonald, personal communication). Portland's community gardens have thrived in the city's Department of Parks and Recreation under the supervision of Commissioner Nick Fish, who recognises that 'within this small piece of dirt we are meeting dozens of core goals and values,' such as community building, promotion of exercise, and protection of the environment (N. Fish, personal communication).

Behind the scenes, individual vision and the cultivation of personal relationships between municipal and third sector leaders has driven the convergence of interests. In Seattle, for example, citizen community organiser Jim Diers learned through 20 years of experience that there was little cooperation between neighbourhood groups and the City; consequently, he endeavored to establish the DON, which absorbed the P-Patch programme during his tenure as department head (Diers, 2004). In both Seattle and Portland, frequent formal and informal communication between the leaders of non-profit community gardening organisations and members of the City Council has helped to keep money flowing into the programmes and to resolve disputes over such issues as land tenure and zoning regulations (R. Schutte, personal communication; L. Pohl-Kasbau, personal communication). And in New York, GreenThumb director Edie Stone has worked to build a 'long track record' of fiscal responsibility and efficiency that has made the programme a favourite among officials who dispense the Community Development Block Grant funds on which the community gardening programme relies (E. Stone, personal communication). The allocation of responsibilities among partners to maximise their comparative advantages marks the second over-arching process in the development of co-governance in the community gardening programmes of Seattle, Portland and New York.

Through their delivery of services, the City staff in Seattle, Portland and New York primarily try to support the autonomy of individual gardens. In all three cities, the primary functions of the municipal programme are garden administration and leadership development. Municipal management of leases and gardener registration helps to reduce bureaucratic transaction costs; ensures that the loss of any individual gardener does not jeopardise the garden's access to land; and allows gardeners to focus on tending their plots rather than handling administrative chores. Municipal departments can also provide more tangible support, including garden materials and some hauling and maintenance services (L. Pohl-Kasbau, personal communication; R. Macdonald, personal communication; E. Stone, personal communication).

Citizen engagement is critical to the establishment and long-term viability of community gardens. All three programmes require a demonstration of commitment from a neighbourhood before the City will contribute resources to building or maintaining a garden. After the gardens' initial construction, citizens are their primary care-takers. Volunteers worked a total of 15,500 hours in Seattle's approximately 80 P-Patch gardens in 2008, and about 9,000 hours in Portland's 32 gardens (City of Seattle Department of Neighborhoods 2009; Portland Parks and Recreation, 2009). The only comparable volunteer data for New York, from 1983, shows that volunteer 'sweat equity' at the time accounted for about 17% of the development cost of community gardens, and 92% of maintenance costs (Lawson, 2005).

Nonprofit organisations in Seattle, Portland and New York also play a critical part in supporting autonomous community gardens. These organisations, such as the P-Patch Trust in Seattle, the Friends of Portland Community Gardens, and the Green Guerillas and Open Space Greening Program in New York, were generally founded to provide services to community gardens that the City was not providing, including money, land security, political legitimacy, and basic garden supplies. Today, these organisations continue to provide various resources to bridge the gap between what the local government agencies are able to contribute and what the gardeners need to establish and sustain their plots (M. Bedard, personal communication; Council on the Environment of New York City, 2009; Green Guerillas, 2010; R. Schutte, personal communication).

The structure of co-governance in these three community gardening programmes has allowed them all to stretch scarce municipal resources. With government outlays ranging from \$250,000 to \$700,000 a year, none of the three community gardening programmes accounts for more than a few hundredths of one percent of its City's respective municipal budget. Municipal directors of all three programmes opined that their frugality contributes to their popularity (R. Macdonald, personal communication; L. Pohl-Kasbau, personal communication; E. Stone, personal communication). With relatively minimal resources, the municipal programmes manage significant quantities of open space in their cities, provide food for thousands, and create positive social externalities. City leaders' support for these programmes in the midst of another economic recession suggests that they view them as wise investments. To address the motivating question of this chapter, it appears as though municipal governments in Seattle, Portland and New York have prioritised sustainable food planning in part because such efforts represent 'good' government: they are fiscally efficient means of empowering citizens to improve their own communities.

13.2.3 Breaking and building anew: adapting resiliently to change

The schemes of co-governance in the community gardening programmes of Seattle, Portland and New York have all demonstrated capacity to adapt to such changes as funding cuts, leadership turnover, development, gentrification and demographic change. The ability of these co-governance schemes to buffer shocks to community gardens has depended on their ability to generate new relationships and re-organise old ones. Resilient schemes of governance transition from a crisis of breaking to a new period of building. Individual champions and groups in public and third sector have responded to, or pre-empted, threats to gardens by creating new supportive institutions and nonprofit organisations, by formalising community garden development in policies and planning goals, and by broadening and strengthening inter- and intra-sector social networks to enhance the capacity of the community gardening constituency to mobilise and resist adverse changes.

A fight to save over 100 community garden properties that former New York Mayor Rudy Giuliani slated for auction in 1999 demonstrates the complicated nature of resilient co-governance. The plan to evict gardeners galvanised neighbourhood groups across the city. Initially disparate, these groups eventually developed a coherent voice through high-profile public demonstrations and communication over the internet (Smith and Hurtz, 2003). Although citizen activism was not, by itself, decisive in resolving the matter, GreenThumb director Stone believes it was instrumental in encouraging then-New York State Attorney General Eliot Spitzer to file a lawsuit that succeeded in blocking the sale of the gardens (E. Stone, personal communication). The non-profit Trust for Public Land supported the citizen and government efforts by paying \$3 million to secure the threatened properties (Elder, 2005). In August 2010, a citizen-led community garden advocacy group and City Council Speaker Christine C. Quinn demonstrated again in public to strengthen the original agreement with the City to protect the gardens. As this case shows, the ability of public and third sector partners to combine efforts in response to change significantly strengthens their adaptive capacity. Responsibility for resilience is not simply divided, but truly shared.

13.3 Adaptive co-governance when federal government support is withdrawn: consequences for resilience

Our case study of the Philadelphia Green community gardening programme supports the proposition of Stoker (1998) and Rosenau (1992) that co-governance is possible without a public partner. However, our analysis shows that co-governance may sacrifice some adaptive capacity by involving only the third sector.

The 'budding' of community gardening in Philadelphia in many ways resembles the budding of community gardening in the three other cities. Initially practiced by ad-hoc groups of citizens, a community gardening programme was formally organised through the non-profit organisation Philadelphia Green in 1974. The programme had no public partner at the outset; however, the United States Department of Agriculture (USDA) established a programme in 1977 that facilitated the creation of the Penn State Urban Gardening Program, which supported the efforts of the existing Philadelphia Green projects. As in the other examples of public-third sector co-governance, the partners achieved an efficient allocation of roles, with Penn State providing horticultural training and Philadelphia Green supplying physical materials (Hynes, 1996). Together, citizens, the Penn State program and Philadelphia Green had developed or helped maintain over 500 gardens throughout the city by 1994 (Vitiello and Nairn, 2009).

Despite having a close relationship with a federally-funded partner, Philadelphia Green never formed strong ties to local government (C. Baker, personal communication). This major difference in the 'building' phases of co-governance schemes in Philadelphia and in Seattle, Portland and New York had significant consequences for the 'breaking' phase. In 1996 the USDA terminated funding for urban agriculture projects. With no capacity to reverse the decision in Washington and with no support from the City to fall back on, community gardens in Philadelphia fell to the management of the third sector alone (Vitiello and Nairn, 2009).

Over the last 15 years, these third sector partners have built a new scheme of co-governance outside of government for the development and management of community gardening. Philadelphia Green conducts trainings and contributes some supplies to groups interested in starting and sustaining community gardens (S. McCabe, personal communication). The non-profit Neighborhood Gardens Association continues to support the long-term viability of gardens by holding land in trust (Neighborhood Gardens Association, 2010). And as they always have in Philadelphia, citizen gardeners continue to carry the largest responsibility for garden development and daily maintenance (C. Baker, personal communication).

While this case study shows that co-governance is clearly possible without a formal public partner, it also illustrates how community gardens become more vulnerable to change under such an arrangement. First, the withdrawal of federal government support has reduced funds available for community garden development, and has made community gardens more dependent on potentially-capricious philanthropic donations (C. Baker, personal communication; B. Bonham, personal communication). Second, the absence of collaboration with a municipal agency has prevented the same kind of convergence of understanding of the multiple benefits of community gardens that took place in Seattle, Portland and New York. Without this understanding, the City continues to view community gardens as a temporary use of land, vulnerable to development at any time, and especially vulnerable in the face of City budget shortfalls (*Ibid.*). Third, lack of a formal public partner undermines gardens' ability to adapt to demographic changes. Unlike in Seattle, Portland and New York, there is no organisation to oversee garden leases and recruit gardeners to replace those who leave. In a system that demands citizen commitment to maintain plots, the attrition of gardeners, and especially garden leaders, without replacement spells a likely end for gardens. The effects of these vulnerabilities are real: whereas Philadelphia boasted about 500 community gardens in 1994, it had just 226 in 2008 (Vitiello and Nairn, 2009).

13.4 Conclusion: understanding why municipal government is in the garden

In this chapter, we have used our 'Four-B framework' to analyse the dynamics of local government's engagement in community gardening. The phases of 'budding,' 'building,' 'bonding' and 'breaking' can describe both the development of community gardens and the development of relationships that public and third sector partners form to govern the gardens. Using a case study from Philadelphia, we have argued that co-governance schemes involving a public partner may have an enhanced adaptive capacity.

Applying the Four-B framework to existing public-third sector co-governance may also yield insights about the involvement of local government in sustainable food planning measures beyond community gardening. The 'budding' and 'building' phases of co-governance in Seattle, Portland and New York provided an avenue to move sustainable food initiatives from the private to the public domain. However, this responsibility was not transferred in order for government to solve problems on its own. Rather, through schemes of public-third sector co-governance, citizens have been empowered to a greater extent than before to participate in the enhancement of their food systems by building and maintaining gardens. Future research could examine how similar processes of budding and building of public-third sector relationships can put other sustainable food initiatives on the municipal agenda. In this regard, it would be productive to examine whether deliberate government actions tend to promote or inhibit the formation of public-third sector co-governance (Folke *et al.*, 2005; Ruitenbeek and Cartier, 2001).

Furthermore, a common element of the 'bonding' phases in Seattle, Portland and New York was the establishment of municipal policies linking community gardening to more ambitious environmental and planning goals. In Seattle, community gardening is embedded in the current Seattle Comprehensive Plan, as well as in the municipally-funded Local Food Action Initiative (City of Seattle, 2005; Conlin, 2010). Portland's Climate Action Plan, adopted in 2009, calls for the creation of 1,300 new garden plots by 2012, provision of more education to citizens regarding home and community gardening, and the more intensive use of public and private land for food production (City of Portland and Multnomah County, 2009). And in New York, a 2009 plan known as FoodWorks, aimed at strengthening the contribution of local food to health and the economy, promises to 'expand urban agriculture through community gardens, green roofs and urban farms' (Quinn, 2009). Because of their collaborative nature, co-governance schemes are natural allies of other, diverse stakeholders working on projects with similar goals. The bridges between these groups of stakeholders (e.g. community gardeners and climate protection advocates) can be mutually beneficial insofar as they share resources and ideas and reinforce each others' causes. Future research can investigate the genesis, development and effects of these cross-issue relationships and policies.

Finally, it is important to note that government's presence in the garden is better understood as a process rather than a product; it has not been consistent in the past and likely will not remain consistent in the future. Equally, citizen involvement is inconsistent over time. Resilient co-governance schemes may offer a model for success but persistent commitment from all partners will be needed to build resilient urban food systems.

Acknowledgements

We are deeply grateful for the patience and insights of everyone who spoke with us in Seattle, Portland, New York, Philadelphia, Trenton, and other cities. We would also like to thank the participants of the 2010 Second European Sustainable Food Planning Conference in Brighton for their invaluable feedback.

References

Carlsson, L. and Berkes, F., 2005. Co-management: concepts and methodological implications. Journal of Environmental Management 75: 65-76.

City of Portland and Multnomah County, 2009. Climate Action Plan 2009. City of Portland Bureau of Planning and Sustainability, Portland, USA. Available at http://www.portlandonline.com/bps/index. cfm?c=49989&.

City of Seattle Department of Neighborhoods. A stroll in the garden: an evaluation of the p-patch program. City of Seattle Department of Neighborhoods, Seattle, USA, pp. 1-82.

City of Seattle, 2010. Seattle's comprehensive plan: toward a sustainable Seattle. city of Seattle department of planning and development, Seattle, USA, 498 pp. Available at http://www.cityofseattle.net/dpd/Planning/ Seattle_s_Comprehensive_Plan/ComprehensivePlan/default.asp.

Conlin, R., 2010. Local food action initiative. Available at http://www.seattle.gov/council/conlin/food_ initiative.htm.

Council on the Environment of New York City, 2009. Community gardens. Available at http://www.cenyc. org/openspace.

Diers, J., 2004. Neighbor power: building community the Seattle way. University of Washington Press, Seattle, USA, 216 pp.

Eckerberg, K. and Joas, M., 2004. Multi-level environmental governance: a concept under stress? Local Environment 9.5: 405-142.

Elder, R.F., 2005. Protecting New York City's community gardens. N.Y.U. Environmental Law Journal 13: 769-800.

Folke, C., Hahn, T., Olsson, P. and Norberg, J., 2005. Adaptive governance of social-ecological systems. Annual Review of Environmental Resources 30: 441-73.

Green Guerillas, 2010. Our programs. Available at http://www.greenguerillas.org/GG_ourprograms.php #helpingcommunity.

Gunderson, L., 1999. Resilience, flexibility and adaptive management: antidotes for spurious certitude? Conservation Ecology 3.1: 7.

Hamilton, N.D., 2002. Putting a face on our food: how state and local food policies can promote the new agriculture. Drake Journal of Agricultural Law 7.2: 416.

Holling, C., 1995. What barriers? what bridges? In: Gunderson, L., Holling, C. and Light, S. (Eds.) Barriers and bridges to the renewal of ecosystems and institutions. Columbia University Press, New York, NY, USA, pp. 1-34.

Hynes, P., 1996. A Patch of Eden: America's Inner-City Gardeners. Chelsea Green Publishing, Inc., White River Junction, VT, USA, 185 pp.

Lawson, L.J., 2005. City bountiful: a century of community gardening in America. University of California Press, Berkeley, CA, USA, 382 pp.

Neighborhood Gardens Association, 2010. About us. Available at http://www.ngalandtrust.org/assist.html.

New York City Department of Parks and Recreation, 2010. The community garden movement. Available at http://www.nycgovparks.org/sub_about/ parks_history/ gardens/community.html.

Nkhata, A.B., Breen, C.M. and Freimund, W.A., 2008. Resilient social relationships and collaboration in the mangement of social-ecological systems. Ecology and Society 13.1: 2. Available at http://www. ecologyandsociety.org/vol13/iss1/art2/.

Portland Parks and Recreation, 2009. Community gardens business plan: fiscal years 2008-09 through 2010-1. Portland Parks and Recreation, Portland, OR, USA, 40 pp.

Pothukuchi, K. and Kaufman, J.L., 1999. Placing the food system on the urban agenda: The role of municipal institutions in food systems planning. Agriculture and Human Values 16.2: 213-24.

Pothukuchi, K. and Kaufman, J.L., 2000. The food system: A stranger to the planning field. Journal of the American Planning Association 66.2: 112-24.

Quinn, C.C., 2009. FoodWorks New York.' Available at http://council.nyc.gov/downloads/pdf/ foodworksny_12_7_09.pdf.

Rosenau, J., 1992. Governance, order and changes in world politics. In: Rosenau, J. and Czempiel, E. (Eds.) Governance without government: order and change in world politics. Cambridge University Press, Cambridge, MA, USA, pp. 1-29.

Ruitenbeek, J. and Cartier, C., 2001. The invisible wand: adaptive co-management as an emergent strategy in complex bio-economic systems. Center for International Forestry Research Occasional Paper No. 34. Available at http://www.cifor.cgiar.org/publications/pdf_files/OccPapers/OP-034.pdf.

Smith, C.M. and Hurtz, H.E., 2003. Community gardens and politics of scale in New York City. Geographical Review 93.2: 193-212.

Stoker, G., 1998. Governance as theory: five propositions. International Social Science Journal 50.1: 17.

Vitiello, D. and Nairn, M., 2009. Community gardening in Philadelphia: 2008 Harvest Report. Penn Planning and Urban Studies, Philadelphia, PN, USA, 68 pp.

Part 2. Integrating health, environment and society

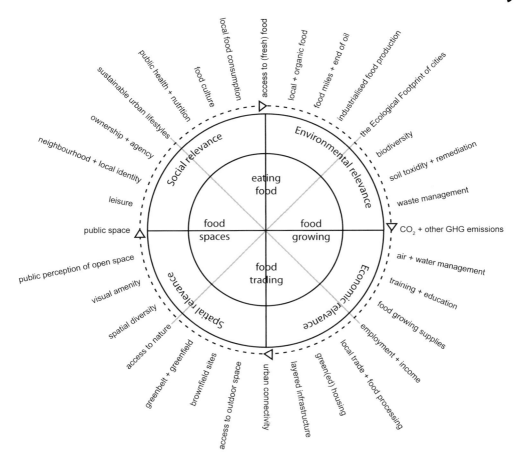

Chapter 14

Integrating health, environment and society – introducing a new arena

Bettina B. Bock[1] and Martin Caraher[2]
[1]*Wageningen University, Department of Social Sciences, Rural Sociology Group, Hollandseweg 1, 6706 KN Wageningen, the Netherlands;* [2]*Centre for Food Policy, Room C 310, School of Health Sciences, Northampton Square, City University, London EC1V OHB, United Kingdom; bettina.bock@wur.nl*

14.1 Background to integrating the issues

Planning for sustainable food is an increasingly important issue for policymakers, activists and scientists alike and includes the manifold problems that are arising around food provision in modern and urbanising societies. One of these problem areas is the relation between food, human health and wellbeing. This relation is becoming 'common knowledge' in modern society and helps to explain why food has become such a topical issue. It is reflected in growing anxiety about food safety but also growing interest in food quality, in growing/sourcing as well as cooking and eating well. Cooks and food activists become celebrities and 'food books' sell and interestingly many of them have built in the areas of health and the environment with cook books and television programmes addressing these issues (Caraher and Seeley, 2010). Good food and healthy nutrition are subjects of public debate and concern but also of public entertainment, as the growing numbers and variations of television cooking programmes tell.

In developing this theme we wanted to see to what extent the apparent public interest in food and human health may offer a promising new route of entrance into the discussion of not only human but environmental health and, hence, the issue of sustainability.

Does the growing awareness of the negative effects of the mainstream 'food industry' on (individual) human health, encourage more ecological awareness and readiness to change behaviour and actively engage in favour of environmental sustainability? Or it is just another fashion, part of the growing list of 'exotic' subjects, to be presented in reality TV programmes?

Some authors are convinced that food will promote ecological awareness and engagement (Lang *et al.*, 2009) and many civil society activities claim to act in defence of both. But there are also those who criticise this approach. Some consider the alternative food movement as an elitist movement, that does no justice to the inaccessibility of the new food markets for lower income groups (Kneafsey *et al.*, 2007). Others refer to the ongoing modernisation of the agro-food industry and their contribution to the sustainability of the planet (Keith, 2009), motivated however primarily by the scarcity and costs of resources in the first place. Some argue that health is only of interest to the industry when it adds value and augments profits;

this is generally realised through modifications, which again is interpreted as 'unhealthy' or 'unnatural' by others (Pollan, 2008, 2009).

The list of concerns may easily be expanded; what they point to is the need to critically address the growing interest in food and health, especially where it feeds into new policymaking and planning. We need to check whether 'new food planning' is split around the global north/ south, class and income differences and may in fact perpetuate and even widen inequalities both within and between countries. Supporting 'cosy' initiatives of alternative movements may, in a similar vain, distract from questioning the problems embedded in the dominant food supply chain. The chapters in this section further elaborate these issues by discussing how the integration of health and environment might add to the development of sustainable food policies in modern societies and how this could be addressed without widening inequalities.

This section presents six chapters which in different ways engage with the concerns presented above. There are two examples of policy initiatives implemented at the City level (Malmö and Ghent) that are designed to contribute to sustainability within the cityscape. The following two chapters report on initiatives run by the third sector, including local charities and NGOs (hereafter referred to as the third sector) that focused on the accessibility of healthy and sustainable foodstuff in two low-income city areas (in Sandwell and Manchester). The two other chapters examine more conceptual issues such as the role of a food co-operative and the attitudes of working class (low income) consumers to healthy and sustainable food.

There are of course issues that run across all the chapters – an awareness of global food problems, a concern with the environment and a focus on locale, whether this be local engagement with communities or the provision of local food.

14.2 The food-arena

In the following we briefly discuss the changing composition of the political arena around food, taking on board food, health and environment issues. We explore the position of local initiatives and the state, as well as their changing interrelations. This, as an attempt to sketch the broader context to be taken into account when studying, evaluating but also supporting and promoting initiatives that are meant to achieve what has proved to be enormously difficult – persuade us to eat healthily and sustainably. In doing so we look into the changing role of the state and the third sector, and the part to be played by food initiatives. We conclude by explaining what the present contributions hope to add.

14.2.1 The state and the third sector

The nation state still plays a key role in food politics, but that it is changing is a likewise obvious yet largely unstated issue although some have begun to research the area (Hawkes, 2005; Public Health Association of Australia, 2008). Many of the initiatives described in these chapters are heavily reliant on the state sector for funding and or support. But equally the state is heavily reliant on the third sector to deliver and respond to needs in quickly changing environments that the state, with its cumbersome processes, would find difficult to realise (Poppendieck 1999, 2010). Third sector actors deliver programmes in ways that are more flexible than the state and or more responsive to local needs (Dowler and Caraher, 2003; Garr, 1995).

Nevertheless it is important to clarify the relative merits of both sectors (Mawson, 2008) – as, without doubt, we need both to make lasting achievement. Unless third sector initiatives are firmly embedded in policy, progress made is subject to the vagaries of funding and changes in policy direction (Lang *et al.*, 2009). With today's prominence of public sectors cuts, initiatives may be endangered, regardless of their successes, as the Sandwell case presented in one of the following chapters shows.

The third sector meets many of the agendas of the philanthropic ideal and they can deliver in a cheaper way and as noted above are more flexible in their delivery (Mawson, 2008; Poppendieck, 1999). Such a movement has been underway across Europe – since the 1940s – following the establishment of the various welfare states.[22] The French system of *solidarité sociale* for social insurance after the Second World War was conceived as a way of healing the ruptures caused by the war (Chamberlayne, 1992) and the British Welfare State based on the same principles of tackling 'want' (Timmins, 1995). These approaches assume a 'common good' where even those who don't benefit themselves, consider it worthwhile to contribute to the overall social benefit. But even within this overall model of state welfare, charity and third sector initiatives developed – often to address gaps or flaws in the system. Indeed the European Union Food Aid Programme for the Most Deprived Persons, which currently has a budget of €500m, is mainly distributed through Catholic Charity agencies. The fund itself is a historical legacy of the surpluses of the Common Agricultural Policy and the feeling that it was best distributed through charities and NGOs than the state (Commission of the European Communities, 2008; Zahrnt, 2008).

14.2.2 Food initiatives: food security versus sustainability?

Following the arguments of Belasco (2007) we may be witnessing a creeping invasion of the philanthropic model into food and sustainability areas in recent years as well as many sustainable projects originally founded on the basis of the common good becoming more commercially oriented. Gibson-Graham (2008) notes alternative movements often use existing economic models and add a transformative element to them without abandoning the dominant economic model of supply and profit. Kneafsey *et al.* (2008) elaborate on this point and make the case for a more elaborate but perhaps more fragile and susceptible capitalism. They argue that the reconnections in the food system represent a form of care. The new

---

[22] In many instances the history of third sector delivery around food has been different in Europe than in the US, where delivery by the charity and private sectors was seen as a better and more efficient way than the state (Poppendieck, 1999). This was certainly true of areas such as food welfare but in the more recent past the model has been taken up and applied to the food and sustainability sectors.

generation of Alternative Food Networks (AFNs) have contributed to the development of a 'cuddlier' and more acceptable form of capitalism so fighting mainstream capitalism with consumer-friendly forms of capitalism. But, the new AFNs have achieved a hegemony of place and product with in most peoples' minds the 'alternative' being associated with locale and local sourcing. Two short examples may suffice at this stage to illustrate some of this.

With the demise of the Soviet Union the Russian people experienced shortages of basic foodstuffs. Rooftop gardening has emerged as one way of addressing urban food shortages. In one district in St. Petersburg 2000+ tonnes of vegetables are grown. This arose out of the need to meet food shortages and food insecurity (WHO, 1999).

On a similar climatic level in Michigan but a few degrees south in latitude, there is group of local food consumer 'activists' – those committed to 'eating locally' in Michigan. The group have adopted the name Edible WOW (WOW takes its name from three densely populated counties in southeast Michigan: Washtenaw, Oakland and Wayne) are part of 'Edible Communities' network of local food publications (see www.ediblecommunities.com). The reasons for these actions in Michigan are very different from those in St Petersburg; the WOW group is focused on eating locally, methods of production and the origins of food.

The two groups are doing similar things but for different reasons, one because they had to, the other because they chose to. The activity in Michigan fits what Winter (2003) has called 'defensive localism' where the development of local and alternative food economies are seen as bulwarks against the dominant system of food supply and delivery. These are essentially different from food initiatives involved in addressing food poverty and hunger, not least because of the issues of voluntarism involved – the poor have little choice. This is not to say that low income communities cannot engage in efforts to be more sustainable, but merely to recognise that the starting points are different.

Another important consideration is that food initiatives while good at tackling issues in the short term, do generally not address the fundamental causes of food insecurity or unsustainability, often because they cannot and also because their energies (and expertise) goes in tackling the immediate problems. Rather than fundamentally changing the global food system, they are often attempting (and succeeding) in creating niches within it. So looking to the third sector to address the issues of food security related to sustainability and health are equally limited/doomed. They clearly have a part to play as providers of service, developing good practice and as advocates for change, but in realising fundamental change the state has an essential role to play.

14.2.3 The present volume

These various elements can be seen in the chapters in this section from the work in Manchester, Sandwell and Malmö through to the possibility of local food co-ops in the UK helping deliver on health and sustainability agendas. The work on food co-ops shows that the wider policy frameworks need to be in place otherwise you end up with policy discordance and a range of initiatives labelled food co-ops, which have little in common and do not make use of their collective power to change the system. However, some might argue that such policy disharmony is not merely a bye-product of the chaos of food initiatives but a deliberate attempt to shift governmentality from the realm of the public to that of the private (Dowler and Caraher, 2003). We will discuss this further below.

The Sandwell work described here has a long and honourable tradition within public health work in the UK but the changes described above, in funding and the changing nature of the public sector, now put the continuation of some of this work in jeopardy. We know from work on food welfare systems that the third sector often steps in where the state is failing. In many instances it props up the gaps in formal welfare provision, but may also have the effect of hiding problems and thus allowing the state to further withdraw from services (Dowler and Caraher, 2003; Poppendieck 1999, 2010, Richies 1997, 2002). It may be in a similar way that issues around food health and sustainability are being addressed by handing them back to the third sector and adopting them as social or consumer norms; sustainability would then become a standard for consumers as does ethics as exemplified by fair trade products. In doing so, the responsibility for the governmentality has shifted from the state to the consumer and retailer interface, which removes government from the task of having to intervene.

None of the chapters in this section deal with the power of the conventional food chain or even of the nature of the urban or cityscape as major contributors to health and sustainability (Lister, 2007; Steel, 2008). This is not a criticism of the chapters but more of an observation on the policy environment and indeed how we phrased our call for chapters. The Malmö, Ghent and Sandwell chapters all mention wider urban policy frameworks. Some chapters describe engagement with the dominant food sector but only on the fringes. The example from Malmö shows how the state sector can influence local growing and provision by way of its procurement standards. Many of the schemes described here are alternatives to the dominant system: alternative in offering an alternative supply chain (the Manchester supply system, the establishment of food co-ops and the Sandwell community gardening developments) or alternative in helping shape a part of an existing system (the Malmö procurement scheme and the Thursday Veggie Day in Ghent). The research by Hawkins shows that the ways in which many consumers equate and operationalise 'healthy and sustainable food', is by creating 'heuristics' so local=sustainable. We should be aware of the need for individuals to develop simple but meaningful messages from the vast morass of competing messages out there.

The chapters highlight the links and connections between the various sectors, those of the state, the citizen as in civic society and those of the supply chain. Where the third sector/civic society stands in this triangle of effect (Lang *et al.*, 2010) is of major import. Traditionally many third sector groups and their delivery of services has arisen from their campaigning and advocacy roles and services delivered such as food growing or food delivery have been 'alternatives' and/or small scale. This gave them a location within the system as voices for change. If now they are delivering services and wishing to become more mainstream, they run the danger of being part of the system and less alternative (Belasco, 2007). Such a move may also compromise their advocacy voice. This may also be a result of the 'state' bringing the third sector into mainstream provision. In doing so they achieve what Foucault (1979) described as the change in governmentality from the state to that of 'idealised communities'.[23] In this instance the 'idealised communities' can be those that promote eating less meat, for example, which does not change the food system (Imhoff, 2010; Lang *et al.*, 2010).

What this may mean in effect is that government can effectively regulate through idealised social norms as opposed to setting laws to regulate. It may also relegate the activities of civic society to niche demonstration projects with limited impact. Our point here is not that these civic society or third sector projects should not deliver services but that they should develop mechanisms to ensure their advocacy role is not diluted. There is an old saying in public health, which relates to action and it goes something like the following: there are three courses of action: (1) large scale social change and regulation- not acceptable; (2) doing nothing, just accept the status quo equally not acceptable and (3) do some small scale intervention to show you are doing something, this is not really effective but shows you are doing something.

What also becomes apparent is the lack of any cohesive or comprehensive policy on food which explicitly links health and the environment, despite the obvious successes of the projects. The modern food system faces a serious dilemma (Flannery, 2005). On the one hand, it has delivered unprecedented quantity and choice of food to hundreds of millions of people. On the other hand, evidence has mounted as to food's impact on health, the environment and social structures. Rightly, the dominant food system claims this as a policy success but the future is less clear, growing global populations along with issues of equity in the global food system raise serious concerns for the future (see the website of Olivier De Schutter the United Nations Special Rapporteur on the Right to Food http://www.srfood.org/).

How to conceptualise, resolve or manage this triple challenge – health, environment and social behaviour – crosses all these chapters. While there are no overall answers, they do present partial solutions and or indicators to how city and urban landscapes might become more sustainable. The challenge we now face is how to encourage – speedily but sensibly – the production, distribution and consumption of a good, health-enhancing and environmentally based diet.

References

Belasco, W. 2007. Appetite for change: how the counterculture took on the food industry and won. 2nd Edition. Cornell University Press, New York, NY, USA, 327 pp.

Caraher, M. and Seeley, A., 2010. Cooking in schools: Lessons from the UK. Journal of the Home Economics Institute of Australia 17: 2-9.

---

[23] see Coveney 2006 for a discourse on food.

Chamberlayne, P., 1992. Income maintenance and institutional forms: a comparison of France, West Germany, Italy and Britain 1945-90. Policy and Politics 20: 299-318.

Commission of the European Communities., 2008. Commission staff working document impact assessment, SEC (2008) 2436/2. Commission of the European Communities, Brussels, Belgium.

Coveney, J., 2006. Food, morals and meaning: The pleasure and anxiety of eating. Routledge, London, UK, 224 pp.

Dowler, E. and Caraher, M., 2003. Local food projects: the new philanthropy? Political Quarterly 74: 57-65.

Flannery, T., 2005. The future eaters: An ecological history of the Australasian lands and people. Reed New Holland, Sydney, Australia, 423 pp.

Foucault, M., 1979. Discipline and punish: The birth of the modern prison. Penguin, London, UK, 352 pp.

Garr, R., 1995. Reinvesting in America the grassroots movements that are feeding the hungry, housing the homeless and putting americans back to work. Addison-Wesley, New York, NY, USA, 271 pp.

Gibson-Graham, J.K., 2008. Diverse economies: performative practices for 'other worlds'. Progress in Human Geography 32: 613-632.

Hawkes, C., 2005. The role of foreign direct investment in the nutrition transition. Public Health Nutrition 8: 357-365.

Imhoff, D., 2010. The CAFO: the tragedy of industrial animal factories. University of California Press, Berkeley, CA, USA, 462 pp.

Keith, L., 2009. The vegetarian myth: food, justice and sustainability. Flashpoint Press, Crescent city, CA, USA, 309 pp.

Kneafsey, M., Holloway, L., Dowler, E., Cox, R., Tuomainen, H., Ricketts-Hein, J. and Venn, L., 2007. Reconnecting consumers, food and producers: exploring 'alternative' networks. Cultures of Consumption, Series. Berg, Oxford, UK, 224 pp.

Lang, T., Barling, D. and Caraher, M., 2009. Food policy: integrating health, environment and society. Oxford University Press, Oxford, UK, 336 pp.

Lang, T., Caraher, M. and Wu, M., 2010. Meat and policy: charting a course through the complexity. In: d'Silva, L. and Webster, J. (Eds.) The meat crisis: developing more sustainable production and consumption. Earthscan, London, UK, pp. 254-274.

Lister, N.M., 2007. Placing food: Toronto's edible landscape. In: J. Knechtel (Ed.) FOOD. MIT Press, Cambridge, MA, USA, pp. 148-185.

Mawson, A., 2008. The social entrepreneur: making communities work. Atlantic Books, London, UK, 192 pp.

Pollan, M., 2008. In defense of food. Penguin, London, UK, 256 pp.

Pollan, M., 2009. Food rules: an eater's manual. Penguin, London, UK, 240 pp.

Poppendieck, J., 1999. Sweet charity? Emergency food and the end of entitlement. Viking, New York, NY, USA, 354 pp.

Poppendieck, J. 2010. Free for all: fixing school food in America. University of California Press, Berkeley, CA, USA, 368 pp.

Public Health Association of Australia, 2008. A Future for food: addressing public health, sustainability and equity from paddock to plate. Public Health Association of Australia, Sydney, Australia.

Riches, G., 2002. Food banks and food security: welfare reform, human rights and social policy. Social Policy and Administration 36: 648-663.

Riches, G., (Ed.), 1997. First world hunger: Food security and welfare politics. Macmillan Press, Basingstoke, UK, 224 pp.

Steel, C., 2008. Hungry city: how food shapes our lives. Chatto and Windus, London, UK, 400 pp.

Timmins, N., 1995. The five giants: A biography of the Welfare State. Fontana, London, UK, 606 pp.

Winter, M., 2003. Embeddedness, the new food economy and defensive localism. Journal of Rural Studies 19: 23-32.

WHO (World Health Organization), 1999. Urban Food and nutrition action plan elements for local community action to promote local production for local consumption. WHO, Copenhagen, Denmark.

Zahrnt, V., 2008. Reforming the EU's common agricultural policy; health check, budget review, Doha Round. ECIPE Policy briefs/ No 06/ 2008. European Centre for International Political Economy, Brussels, Belgium.

Chapter 15

Policy for sustainable development and food for the city of Malmö

Gunilla Andersson and Helen Nilsson
Environment department, Lifestyle and consumer affairs, Bergsgatan 17, 205 80 Malmö, Sweden; gunilla.i.andersson@malmo.se

Abstract

The city of Malmö in Sweden has been striving for many years to be more sustainable, which is reflected in its environmental programmes and policies. The city has taken several measures to improve the food from a sustainable perspective, such as systematically introducing organic food, running a pilot project in one school, training of cooking staff, introducing the 'Eat S.M.A.R.T.' concept and finally launching a policy for sustainable development and food. The policy for the city was approved by the Municipal Executive board in October 2010. The policy will regulate the procurement and quality of food served in public kitchens within the city of Malmö. It will also ensure that food procurement is more sustainable. The implementation of the policy will move steadily forward, step by step. It needs to have clearly formulated targets combined with clear leadership strategies at all levels. Well educated and committed personnel, who have access to continuous training is also a prerequisite for success. There is also a need to have a close dialogue with suppliers. With all these factors fulfilled we are convinced that the aims of the policy will be met in 2020. The outcome of all the measures that are already taken and the measures that will be introduced while implementing the new policy will be diets that are improved from both health and environmental perspectives.

Keywords: climate friendly food, organic food

15.1 Background

The public catering sector is important in Sweden, due to the fact that hot school lunches for children aged from seven to sixteen years old are municipally-funded, as stipulated by Swedish law; which has been the case since the 1960s. Warm lunches are also served in most nurseries.

The Swedish government has established a target which states that 25% (expressed as share of food budget used for purchase of organic food) of the food served by the public sector should be certified as organic by 2010. The figure was 10.2% in 2009, (EkoMatCentrum and KRAV, 2010) so there is still a long way to go on a national level.

In the City of Malmö, Malmö School Restaurants (MSR) are the main supplier of food to Malmö's public schools. They serve 38,000 meals each weekday. (H. Löfven 2010, personal communication) In addition to Malmö School Restaurants, meals are prepared by nurseries as well as service centres for elderly and disabled people.

The city of Malmö has been working for more than a decade to integrate sustainable development in all policies and has an ambition to be a leading eco-city. The city has been awarded both nationally and internationally for its progressive development in sustainability (Globe Award, 2011; Miljöaktuellt, 2010; UN Habitat, 2011). Since 1998, Malmö's environmental programmes have included aims to increase the conversion of agricultural land within the city borders to organic cultivation and increase the procurement of organic food by the city. In the last few years when competition for decreasing public money has become tougher, the need to focus specifically on food from a broader sustainability perspective has arisen. This is also in line with the ambitious aims for sustainable development of the city. Food procurement is an important part of the impact the public sector has on the environment. For example, food consumption contributes to about 25% of the greenhouse gas emissions from Swedish consumption, (Naturvårdsverket, 2008).

The chapter continues with a brief presentation of what Malmö has done so far, highlighting its work in areas such as organic procurement and the Eat S.M.A.R.T. concept. The City of Malmö's plans for the future are then presented before a general discussion on the main points and what barriers still need to be overcome.

15.2 What Malmö has done so far

15.2.1 Organic procurement

Since 1998, the public catering services in Malmö have introduced more and more organic food. In 2009, 43% of the food served by Malmö School Restaurants (MSR) was organic. For the whole city of Malmö the percentage was 24%. (M. Palmkvist 2010, personal communication) The lower share for the entire city is due to the fact that the rest of the city's catering facilities, for example in nurseries and service centres for elderly people, haven't yet started to use as much organic food. As a result of legislation at EU level that regulates the marketing of organic products; organic food can be demanded in procurement tenders, while following public procurement guidelines.

In 2004 Malmö School Restaurants decided to set up a reference school as a model and chose Djupadal, a school of 500 pupils, age six to twelve years in a middleclass suburban area, with a majority of privately owned single family housing. The main scope of the project was to produce 100% organic food, within the same budget as standard food, with the same nutrition and quality and to remain as close as possible to the existing favourite dishes of the pupils. The establishment of a pilot school included careful planning and renovation of the kitchen, additional education for the staff and a firm commitment from the school's headmaster. In 2007 the goal of 100% organic was reached with a cost per portion of food served that was 10% higher than the cost of an average portion served in Malmö. Since 2007, the percentage of organic food at Djupadal has decreased from 100% to 85-90%. This can be explained by the fact that the project has ended and the additional pressure and attention has been removed from the school. The persons in charge, both the project leader and the head chef, have moved to new tasks. Another explanation is that it is hard to achieve the last 15% of organic foodstuffs without causing additional expenditures. (Löfven, H, 2010, personal communication)

15.2.2 Eat S.M.A.R.T. concept

A challenge in composing a good school lunch is to combine food that is healthy, tasty and good for the environment. In this process we have used a concept called 'Eat S.M.A.R.T.' which has been developed by the regional health authorities in the Stockholm region (Centrum för Tillämpad Näringslära, Samhällsmedicin, 2001). It combines health science with Swedish national environmental objectives. We have built both our training courses and our new policy on the 'Eat S.M.A.R.T.' concept. S.M.A.R.T. stands for:

S   : smaller amount of meat;
M : minimise intake of junk food/empty calories;
A   : an increase in organic;
R   : right sort of meat and vegetables;
T   : transport efficient.

This way of thinking has been successful when making menus better for both health and environment without increasing the costs.

15.2.3 The Plate Model

The 'Eat S.M.A.R.T.' concept can also be used together with 'the Plate Model' from the Swedish National Food Administration (Livsmedelsverket, 2010) to encourage a balanced diet. Malmö School Restaurants have printed plastic plates with colour codes on the edge to guide the children when serving themselves, where the plate should consists of 25% of meat, fish or protein rich food such as pulses and the remaining part of the plate equally divided between starch-rich food like rice, pasta or potatoes and non-starchy vegetables. 'The Plate Model' can be compared to the 'Eatwell Plate' from Food Standards Agency, UK (Food Standard Agency, 2011).

15.2.4 Training

Approximately 1000 municipal employees, both catering staff and teachers have received training in the connection between food, health, environment and climate. They have also been taught S.M.A.R.T. cooking with more vegetarian food and more coarse vegetables.

15.2.5 Policy for sustainable development and food

In public operations there is always a scarcity of money and competition between different interests when planning the budget. Food tends to have a low priority, which could in the long term have a negative impact on the health of young people and their school results. To guarantee an acceptable quality for the food served in Malmö, in terms of both health and sustainability, a political decision was made to develop a policy for sustainable development and food.

Based on the above a decision to develop the policy was approved by the Municipal Executive Board in June 2009. The process of developing the policy included several workshops to ensure broad involvement of relevant stakeholders in the development process. The democratic process also enhances the implementation of the policy. The suggested policy was accepted in October 2010. It is valid for all food procurement and preparation within the city's own operations. The policy states: all food procured by Malmö should be organic by 2020 and greenhouse gas (GHG) emissions from food procurement should be decreased by 40% by 2020, compared to the 2002 level, which was 13,360 ton CO_2-eqvivalents (unpublished data). There are six key focus areas for the policy:

- Malmö should serve tasty, safe and healthy food to all customers. The food should, as far as possible, be prepared close to the customer from fresh ingredients with a minimum of additives.
- The kitchen staff should have access to professional support and relevant knowledge development to keep a high skills level.
- There should be a focus on sustainable and healthy products:
  − fair trade products should be the first choice in relevant product groups;
  − all food procurement should be according to the 'Eat S.M.A.R.T.' concept.
- There should be a focus on the economic and environmental value of food procured:
  − purchased food should be within agreed procurement contracts;
  − food wastage should be minimised. Food waste should be used for biogas production;
  − good food should be highly valued and sufficient resources allocated to secure high food quality;
- By following this policy, and offering sustainable meals during events and public dinners, the city should set a good example to other municipalities in Sweden, as well as offering inspiration to all Malmö residents.
- Sustainable agriculture around the city should be supported; as well as supermarkets, markets and restaurants that offer sustainable food.

15.3 The outcome so far and to expect

The outcome of the long term ambitions to improve the public food in Malmö has resulted in several outcomes. The high proportion of organic food is beneficial for the environment. As an example we buy around 1,500,000 litres of milk of which the main part is organic. This has led to 300 kg less pesticide and 24 tonnes less fertiliser being used in Sweden (Konsumentverket,

2007). 53% of the coffee purchased is organic which has led to 700 kg less pesticide being used and thus better health for the coffee growers (Konsumentverket, 2007).

To be able to afford a larger proportion of organic food and to have health and environmental benefits we have used the Eat S.M.A.R.T. concept with more vegetarian dishes offered. Now there is a vegetarian main course once a week and a vegetarian alternative every day in schools. Every day the pupils are also offered a salad buffet which consists more and more of coarse vegetables and different variations of pulses.

The use of the plates coloured according to 'the Plate Model' has changed the pupils eating habits to more healthy ones, at least among the younger children.

It is still to be established if the changed menus have lowered the green house gas (GHG) emissions from the City's food procurement. We have let the Swedish Institute for Food and Biotechnology calculate the GHG emissions caused by the total amount of food purchased by the city of Malmö in 2002 and we will let them do a new calculation of GHG emissions from the food procured in 2010 so that any possible changes can be found.

15.4 Discussion

The level of organic food in schools and other public catering facilities in Malmö is among the highest in Sweden. We also want to be pioneers in taking a more holistic sustainable approach to food and develop menus that are healthier and more climate friendly. As a public catering service we are able to serve healthy food which is also environment friendly by using the 'Eat S.M.A.R.T.' model. The most important measures we can implement, that can have an impact on the environment and the climate, are a reduction in the amount of meat served and an increase in the share of organic products. The most important impact on health is the use of a greater variety of vegetables and a lowering of the intake of empty calories.

While introducing organic food, Malmö School Restaurants have gained a lot of experience, which will be valuable during the implementation of the new food policy. Firstly, personal participation at all levels and a committed leadership working towards the goal is important. To be successful, it is important to move slowly, but steadily towards the goal, and to set up interim targets along the way. It is also important to have committed personnel, which is why education on why and how to introduce the new policy is an important key to the successful implementation of the policy. Finally, in order to make procurement decisions sustainable, close collaboration with the suppliers is necessary. When these factors are fulfilled, we have found that even ambitious goals can be reached (H. Löfven 2010, personal communication).

Implementation of the new policy is the responsibility of the entire city. Every board and unit should break down the policy and adapt their own targets and follow up systems suitable for their operations. A broad education campaign has been launched to reach all employees who are involved in catering or food procurement in the city. It will give both knowledge about food and sustainability as well as practical inspiration on how to change the menus and food procurement so that the food is more sustainable.

If the goal of serving all organic food by 2020 and a reduction in greenhouse gas emissions is to be realised, it is recognised that the current menus have to be radically changed, with less meat, and fewer vegetables that are grown under glass. These will be substituted by more beans and lentils and seasonal vegetables such as cabbage and root vegetables. To get the students to accept these menu changes is a big challenge, and a lot of information is needed. The cooks have to acquire new skills and the pupils have to adopt new eating habits. In Malmö School restaurants the menus are fixed and centrally decided, so some of the decision-making is already done, while in the nurseries there is no central control over the menus and every cook writes his/her own menu. This makes the process longer and increases the need for training for nursery cooks. If we can change the eating habits of young people we have a lot to gain as they are the consumers of the future.

Going from 24% to 100% organic may make the food more expensive. A low share of organic products is easy to reach with products with little price premium such as milk. The larger share, the more you have to buy products with higher price premiums. In these times of financial crisis and economic belt tightening, it is difficult to get the decision-makers to be willing to pay a higher price for school lunches. It is likely that the price premiums for organic products will decrease when the market for organic food has developed further. Through the substitution of meat with vegetable sources of protein, there can be cost savings, as organic meat is very expensive. In addition, cooking more meals from scratch without a lot of processed ingredients can bring costs down as well, although this also requires additional training for catering staff and review of catering equipment in order to ensure that meals can be prepared.

The supply of organic everyday products such as milk, meat and vegetables is now filled fairly satisfactorily while a lot more processed organic food is still lacking from the market. This problem could be solved by increasing the number of meals cooked from scratch; which can also bring down costs, as mentioned above. In addition, as the organic market develops in Sweden, more processed organic food will become available. The city of Malmö is a major buyer of organic products, with an annual procurement budget for Malmö school restaurants alone of 6 M€, it can use its position to negotiate with suppliers and wholesalers to get better prices and find news sources of organic products.

Unfortunately, there is still a lack of knowledge and insufficient research results on the effect on the climate of all the different food products which makes composing climate friendly menus difficult. Research illustrating the climatic impact of food is very important and there are a number of different research institutions that are attempting to calculate food's impact, often using life cycle analysis. This is a research area that is still in its infancy, but it is hoped that within a few years there will be a tool that can be used to calculate the change in greenhouse gas emissions brought about by changing the menus.

To reach our ambitious goals we need to have clearly formulated objectives combined with a dedicated leadership. It is also necessary to realise that we have to move slowly, step by step, but steadily forward. To fulfil these aims it is also necessary to have competent and committed personnel, who have access to continuous education and training. There is also a need to have a close dialogue with suppliers.

Is the City of Malmö going to reach its ambitious goals? We hope so, but it will depend to a large extent upon the will of the politicians to support the policy both with strong leadership and with sufficient resources.

References

Centrum för Tillämpad Näringslära, Samhällsmedicin, 2001. Ät S.M.A.R.T. – ett utbildningsmaterial om maten hälsan och miljön. Stockholms läns landsting, Stockholm, Sweden, 50 pp.

EkoMatCentrum and KRAV, 2010. Ekologiskt i offentliga storhushåll. Available at http://www.ekocentrum. info/files/Rapport%20kommunenkat%202010%20reviderad.pdf.

Food Standards Agency, 2011. Using the eatwell plate. Available at http://www.food.gov.uk/multimedia/pdfs/ publication/eatwellplateguide0310.pdf.

Globe Award, 2011. Nominees – sustainable city award 2010. Available at http://globeaward.org/nominees-sustainable-city-2010.

Konsumentverket, 2007. Handla ekologiskt? Available at http://www.framtidahandel.se/pub/data/doc/a01/ b67.pdf.

Livsmedelsverket, 2010. Tallriksmodellen. Available at http://www.slv.se/sv/grupp1/Mat-och-naring/ Matcirkeln-och-tallriksmoddellen/Tallriksmodellen/.

Miljöaktuellt, 2010. Hela listan – Sveriges alla kommuner miljörankade. Available at http://miljoaktuellt.idg. se/2.1845/1.329754/hela-listan-sveriges-alla-kommuner-miljorankade#.

Naturvårdsverket, 2008. Konsumtionens klimatpåverkan. Rapport 5903. Available at http://www. naturvardsverket.se/Documents/publikationer/978-91-620-5903-3.pdf.

UN Habitat, 2011. The 2009 Scroll of honour award winners. Available at http://www.unhabitat.org/content. asp?typeid=19&catid=588&cid=7291.

Chapter 16

Meat moderation as a challenge for government and civil society: the Thursday Veggie Day campaign in Ghent, Belgium

Tobias Leenaert

Ethical Vegetarian Alternative, St.-Pietersnieuwstraat 130, 9000 Gent, Belgium; tobias@evavzw.be

Abstract

There are few consumer products that touch on so many different issues as the meat we eat. The high consumption of meat and animal products in general in the developed world, has a significant impact on the environment, public health, world food security and animal welfare. In this chapter we state that the issue is so severe and the challenge so great, that in spite of its reluctance to do so, government needs to get involved in order to efficiently tackle it. We discuss the case study of the city of Ghent, which became the first city in the world to officially stimulate its citizens to have a weekly vegetarian day. Probably it was the first time that a government (be it a local one) structurally encouraged meat reduction.

Keywords: vegetarian, meat reduction, sustainability, food, community

16.1 The background and principles

The human population on our planet has a huge cultural appetite for meat and other animal products. Demand is expected to double by 2050, when our numbers are projected to reach nine billion people. The high consumption of meat and animal products is responsible for significant problems in four main domains:

- *Environment*: large scale animal agriculture comes at a considerable environmental cost. According to the Food and Agriculture Organisation (FAO):

  the livestock sector emerges as one of the top two or three most significant contributors to the most serious environmental problems, at every scale from local to global'. According to the report, the livestock sector 'should be a major policy focus when dealing with problems of land degradation, climate change and air pollution, water shortage and water pollution and loss of biodiversity (Steinfeld *et al.*, 2006: xx).

- *Health*: high consumption of animal products is associated with an increased risk of cardiovascular disease, obesity, diabetes and several forms of cancer. As health care costs are reaching untenable levels in many industrialised countries, in the future we may not be able to care in the same way for our sick or elderly population. Much more emphasis on prevention of western chronic diseases is needed.

- *Hunger*: while roughly one billion people in the world are suffering from hunger or malnutrition, we convert about 40% of the world's cereals into meat at a highly inefficient rate. The high demand for meat makes grain less affordable for the poorest, and requires inefficient use of valuable agricultural land.

- *Animal welfare*: the large demand for cheap animal products requires farmers to raise livestock in less than optimal circumstances (so called 'factory farming'), which poses major ethical problems.

All the above reinforce the drive towards lower meat consumption, but opinions on how to secure enough healthy and affordable food for everyone in a sustainable and animal friendly way differ widely. One body of researchers – and obviously the livestock sector itself – believes a solution can only be found in more intensive farming practices, including more intensive animal agriculture and the use of technological innovations like genetic modification. Another camp believes such an evolution would only exacerbate the problems and especially cause even more animal welfare issues, thus basically only substituting one problem for another.

Here we take side with the second position, and suggest that, especially as meat consumption rises in developing countries, the West needs to opt for a significant dietary shift. Meat and other animal products can no longer remain the centre of our everyday meals and food habits. We will assume here that at the very least in the West, lowering consumption of meat and other animal products is not just desirable but also necessary if we want a food production system that can provide in the nutritional needs of this and future generations without further negatively impacting the health of our planet. Other potentially positive consequences, such as public health and animal welfare, will be considered as co-benefits.

Food habits notoriously die hard, and animal products occupy a very prominent place in most Western meals. Ask most Western-Europeans, for instance, if they can name a substantial vegetarian dish from their culinary heritage, and chances are all they will come up with is a blank. The situation is different in most non-Western and often less affluent countries: Indian, Mexican, Moroccan, Thai, Indonesian, Middle-Eastern and other cuisines, for instance, have a large vegetarian repertoire. While the question there will be especially to discourage those cultures from following the West's meat-heavy diet as their standards of living increase, the challenge in Western countries will be to enthuse consumers for vegetarian meals and products through initiatives on many different levels. Breaking food habits and changing consumers' eating behaviour consequently requires a well-developed strategy.

As may be clear from this brief overview, there are consequences both on individual levels and community levels, nationally and internationally. The problems and solutions lie at both the supply side and the demand side, and they concern both preventive and curative approaches. They contain both practical and moral aspects, and cover sociology and psychology, economy, biology, nutrition, international politics, philosophy, agriculture, health economics and many other fields. The complexity of the problem is very big, and finding a solution necessitates an integral approach in which all sectors from production to consumption need to be involved. Government seems to us a necessary partner in any endeavour to tackle the problem of the high production of animal products. Public information campaigns, regulation, taxes, subsidies and other measures could make a significant difference.

16.2 Meat and politics: working with government

Governments have up until now not been very keen to address the issue of high meat production and consumption. Some of the barriers are listed below. They are epistemological, economical and philosophical in nature.

16.2.1 Lack of understanding

The attention given to the relationship between animal products and sustainability is a relatively recent phenomenon.. Whereas environmental questions have been popular, so to speak, for many decades, it is only in the last few years, especially since the publication of the FAO's *Livestock's long shadow* in 2006, that animal agriculture has been publicly linked to problems like climate change, desertification, soil erosion, deforestation etc. Yet still, there are big knowledge gaps. In a poll executed among 1850 inhabitants of Flanders, for instance, a slight majority recognises the detrimental effect of meat production on the environment.[24]

In the Year of Biodiversity (2010), for instance, we have seen no or little attention to the issue in government publications in Belgium. In general, even though more and more is being written and disseminated on the negative impact of the high consumption of animal products (especially meat), these do not yet have a bad rap in the way smoking has, even though, looking at its general impact as a whole, one might argue that the case against (high) consumption of animal products is even stronger than the case against cigarette smoking.

Similarly, while driving to the bakery around the corner and back in an SUV may be considered by many to be an anti-social act these days, high meat consumption does not yet suffer the same perception. Meat's environmental costs are mainly hidden: on the one hand, there is still not enough information out there for the general public to know. On the other hand and more literally, meat's carbon emissions or usage of fossil fuels, are obviously much less visible than in the case of a car. This applies even more to the costs of the feed that was eaten by the animal.

16.2.2 Economy and vested interests

This sector of animal production is a multi-billion euro market that puts many thousands of people to work. A major shift in our diet might have a serious economic impact on this sector, which is something any government prefers to avoid. Needless to say, economic stakeholders in meat production have no interest in seeing an decrease in meat consumption and continue advertising their products with big marketing campaigns, increase their lobbying efforts, and in the case of the USA even pushed for measures (so called 'food disparagement laws') that make it easier for producers to sue their critics for libel.[25]

---

[24] Poll done in March 2011 by IVOX, under assignment of EVA vzw and the cities of Ghent and Brussels.
[25] See for instance http://cspinet.org/foodspeak/laws/existlaw.htm.

16.2.3 Libertarian attitude

The third barrier to government tackling the meat issue is what we may call an *extremely libertarian attitude* towards meat consumption. Government does not want to meddle with what is on the consumer's (or citizen's) plate, and gives him or her the freedom to choose, even though this choice has a profoundly negative impact. Part of this may be explained by the insufficient awareness mentioned above, yet something else seems to be happening here. Strangely, libertarian arguments are also being forwarded in the case of campaigns that merely *suggest* people to decrease their meat consumption and count on an entirely voluntary participation. Moreover, it is not as if 'the plate' is entirely out of bounds for government: we have seen both public campaigns and regulatory measures to combat, for instance, high salt or high sugar intake. It seems that especially meat is a very sensitive subject and is ground where angels – much more so politicians – fear to thread. The attitude boils down to the following: people want to eat meat and despite the fact that it may be damaging, it is what they want.

16.3 Ghent's Veggie Day: local government getting involved

Let's turn to our case study: the Thursday Veggie Day campaign in Ghent, Belgium. What sets this campaign, originally developed by an NGO, apart from other initiatives concerning meat reduction, is the structural government support and involvement it receives.

Much like the Meat free Monday campaign launched by Sir Paul McCartney[26], the Thursday Veggie Day campaign encourages people to have one vegetarian day a week[27]. It was developed by the Belgian non-profit organisation EVA (which stands for Ethical Vegetarian Alternative), with the idea of launching a doable and simple call for action. EVA is a nonprofit organisation, founded in 2000 to raise awareness about the benefits of eating less or no meat. It is the only organisation in Belgium that works actively and exclusively on this topic, and has about 4,500 members and a staff of five people, as well as many volunteers.

The city of Ghent in the North of Belgium has a population of about 240,000, with one out of four a student. It is a very vibrant city and is considered the greenest and most progressive town in Belgium. It is run by a coalition of socialist and liberal parties. Tom Balthazar (SPA, socialist party), councillor for environmental and social affairs, is responsible for the campaign from the city side. He is a member of the executive council of the city and at the time headed over the departments of health, environment, international development and even animal welfare. With 12 vegetarian restaurants, the city has more vegetarian restaurants on a per capita basis than any other Western city we know of.

---

[26] http://www.supportmfm.org.
[27] It should be noted that 'meat free' or 'meatless' do not have the same meaning or effect as 'vegetarian or veggie'. 'Meat free' does not necessarily exclude fish and fish product, while vegetarian does.

On May 13 2009, during a kick-off event, councillor Balthazar officially declared Thursdays henceforth veggie days. The kick-off event was held at the Vegetable Market (Groentenmarkt) and was meant to gather as much publicity for the campaign as possible. About six hundred people came to the square to take the Thursday Veggie Day Pledge, committing themselves to having at least one vegetarian day each week. As a stimulating reward, they got a goodie bag with products, a recipe booklet, an apron, etc. They could taste vegetarian products, follow on stage cooking courses, get more information, and even play veggie games.

At the launch of the campaign, about one hundred thousand vegetarian street maps were distributed via a local paper. The maps indicated about one hundred restaurants with some decent vegetarian options, and offered some background information on the campaign. At the same time, EVA had developed 'Veggie for Chefs' brochures, in which we explain to restaurant and catering staff why and how to put vegetarian dishes on the menu, and which provided a lot of recipe suggestions. About a thousand of these guides were sent to restaurants in the city, announcing that they could also qualify for free vegetarian cooking lessons.

On October 1, the campaign was rolled out in 35 city schools. These schools have about 11,000 pupils, of which about 3,000 eat a hot meal at school every day. A letter about the campaign, signed by the two councillors involved, was sent to all parents, and a Thursday Veggie Day campaign booklet was made available to all parents who wanted to pick it up. EVA and staff from the different city departments sat together with Deliva, the central catering business providing the school lunches to discuss what was probably the biggest order of vegetarian dishes ever in the country. The caterer developed new vegetarian dishes especially for the campaign.

Some of the principles underlying the Veggie Day campaign are:
- *concrete*: the call for action is very clear, and is at the same time the campaign name. Thursday is Veggie Day;
- *positive*: emphasis is on the alternatives and its benefits, not on the lack of meat;
- *feasible*: one day without meat/fish is something that any of us should be able to do (up to fourty years ago, meat was rather a luxury product);
- *challenging and fun*;
- *empowering*: the consumer can do something about the environment and their health by their own action;
- *sticky*: the concept certainly seems to have a high *stickiness factor*, and is easily remembered.

Following an international press release by EVA on the eve of the start of the campaign in May 2009, worldwide press coverage exceeded expectations. The story was brought all over the world, from BBC to CNN and Time Magazine, from Brazil to Japan.

Response among local citizens and local public institutions has generally been very positive. We know that people's awareness of the issues concerning meat (and especially the global warming impact) is rising, and perhaps many of them realise that something truly serious needs to happen, and that it can only be brought about through some political courage. The people of Ghent have been told that a full year of participation by the entire population would have the same beneficial effect in terms of greenhouse gas emissions as leaving 18,000 Ghent cars in the garage for a year. At the public schools, about 94% chooses the (default) vegetarian dish on Thursdays, even though a the meat option is available. The poll of March 2011 shows that awareness of the campaign in Ghent is 70%. About 25% or 60,000 people participate at least a several times a month.

16.4 Lessons learned

As a local government, the Ghent government obviously was not able to install taxes or issue premiums, even if it would have wanted to go that far. Apart from the decision of the councillor of education (who installed the campaign in all city schools), there was no legislation involved, and emphasis was based on awareness raising. While we worked with suppliers (restaurants and caterers), both EVA's and the city's efforts were focused on the consumer: the general audience, city staff and pupils.

Lowering the threshold for participation in vegetarian eating by proposing one vegetarian day a week has opened many doors. Indeed, there are few partners not prepared to at least consider the issue. When opposition occurs, as still happens in many cases where the campaign is presented during a city council meeting, there is often a lobby involved. In the case of the city of Leuven, for instance, the Flemish farmer's union handed out free samples of meat dishes during the council meeting at which the proposal was being discussed. Presently, the campaign is still not backed by the city of Leuven.

Communication is extremely important. When Ghent's councillor of education Rudy Snoeck announced that city schools would participate in the campaign, the city's major local newspaper wrote on its cover page that Ghent wanted to force pupils to eat vegetarian, even though we clearly communicated that there would be no question of any enforcement. People, it seems, are very protective of their right to eat meat, however much they want, whenever they want.

EVA obviously feels that the partnership with the local government has been very important and fruitful. In terms of city marketing, the campaign was a huge success, with pictures of Ghent's wonderful historical centre appearing on websites and in newspapers all over the globe. However, the ease with which the arrangement with the city was set up and the huge press coverage may hide the fact that for any government, involvement with a meat reduction campaign is not an obvious step. Indeed, it requires some courage.

It seems that, in light of changing perceptions and a growing critical mass supporting meat reduction, governments might be more prone to actively promote meat reduction among its citizens. Indeed, very recently, some initiatives have been taken in this direction, by governments in different countries on different levels.

Around the world, local, state and national governments are starting to encourage their residents to fight climate change and improve their health by making one simple choice: eating fewer animal products. From the USA to Sweden, England, and more, governments are increasingly recognising that the fork is one of the most powerful tools to help improve the environment, human health, and animal welfare. The US city of Cincinnati is urging residents to eat less meat to fight global warming as part of the Green Cincinnati Plan. [28] The goal is to replace some meat with fresh fruit and vegetables and for each person to reduce average meat consumption by one day a week by 2012. Chicago health commissioner Dr. Terry Mason encouraged residents to eat vegetarian for a month.[29] A proclamation passed unanimously by the Takoma Park (Maryland) city council encourages residents to participate in Takoma Park Veg Week by 'choosing vegetarian foods as a way to help protect the planet, their health, and animals and to explore the wide variety of vegetarian cuisine offered' in the city.[30] England's Department for Environment, Food and Rural Affairs (DEFRA) includes reducing animal consumption as a good way to reduce the environmental effects of our food choices.[31] Swedish authorities encourage residents to reduce meat consumption to cut greenhouse gas emissions and protect human health.[32] And the example of Ghent has inspired governments of other Belgian cities like Hasselt, Mechelen and Eupen, as well as international cities like Sao Paulo, Cape Town, San Francisco, Washington DC, Bremen and other places.

One principle which may make it easier for governments to play a role in raising awareness around meat consumption is choice architecture. Choice architecture (synonyms are choice editing and nudging) is about trying to influence decisions (for instance in consumer behaviour) by the way choices are presented. American economists Cass Sunstein and Richard Thaler write about choice architecture in their book *Nudge: improving decisions about health, wealth and happiness* (Thaler and Sunstein, 2009). In a cafeteria, for instance, healthy options may literally be pushed to the fore so as to influence more consumers to choose them and not the unhealthy options. Obviously this kind of choice editing has been done for decades, notably for instance by supermarkets, who try to influence consumers to spend as much as possible by the layout of the aisles, placement of the key products, etc. For governments, however, choice architecture may offer an interesting and elegant way out between the extremes of extreme permissiveness on the one hand and enforcement and limiting free choice on the other. While pushing any services, products, habits, or behaviour is obviously territory where government should not go, choice editing may be an acceptable alternative. One could imagine, for instance, changing the 'default option': in our case, instead of offering the dish with meat as the dish of the day (by default) and having other customers special order their vegetarian dish, the vegetarian dish becomes the default option

---

[28] http://ecolocalizer.com/2009/02/03/cincinatti-urges-residents-to-eat-less-meat/.

[29] http://www.healthierchicago.org/index.asp?Type=B_BASIC&SEC=%7B024063B9-C3F2-4F86-A465-9FDDE19DB308%7D.

[30] http://www.reuters.com/article/2009/04/21/idUS186311+21-Apr-2009+PRN20090421.

[31] http://www.reuters.com/article/2007/05/30/us-britain-methane-env-idUSPAR05104120070530.

[32] http://www.euractiv.com/en/cap/sweden-promotes-climate-friendly-food-choices/article-183349.

on Thursdays, as was done in the Ghent public school system. Pupils who still want meat may order it, but they need to make an extra effort.

Thaler and Sunstein framed the oxymoron libertarian paternalism. Libertarian paternalists want to preserve freedom, but find it legitimate 'to try to influence people's behaviour in order to make their lives longer, healthier and better.' (Thaler and Sunstein, 2009: 5).

Choice editing is not the same as limiting choice.[33] If we may believe American psychologist Barry Schwartz, however, there is even reason to think that actually limiting choice may have beneficial effects for society. In his book *The paradox of choice*, subtitled *Why more is less* (2004), Schwartz argues that in our society, freedom of choice has become a typical example of... too much of a good thing. Eliminating some choices would be able to reduce stress, anxiety and the busyness of our lives. His is not a plea to take away freedom of choice, obviously, but rather to investigate from what point on an abundance of options starts to torment us rather free us.

16.5 The way forward

Government institutions already give us suggestions about what to eat and what not to eat: food guides and daily recommendations for minimum and maximum consumption of ingredients and intake of nutrients will be familiar to most. When a government supposedly should not or does not want to discourage high meat consumption, this can therefore not be a matter of not intervening with what consumers eat, because that is being done already. Apparently, the problem really lies with meat and animal products – a very delicate subject. However, a situation in which governments are unable to tell us not to eat unhealthy or environmentally damaging food or the other way around, is in stark contrast with other parties telling us in clear terms to do the exact opposite: producers, manufacturers and supermarkets are spending big money to make us buy food products that are damaging to both our health and the environment. We believe that government can play an active role in regulation, subsidising, taxing products that are bad for our health and for the environment, and raising awareness around these problems as well as the alternatives.

References

Schwartz, B., 2004. The paradox of choice. Why more is less. Harper Collins, New York, NY, USA, 284 pp.

Steinfeld, H., Gerber, P., Wassenaar, T., Castel, V., Rosales, M. and De Haan C., 2006. Livestock's long shadow: Environmental issues and options. FAO, Rome, Italy.

Thaler, R. and Sunstein, C., 2009. Nudge. Improving decisions about health, wealth and happiness. Penguin Books, London, UK, 324 pp.

---

[33] It's hard to find an official definition of choice editing/architecture or nudging, but Thaler and Sunstein (2009) define a nudge as 'any aspect of the choice architecture that alters people's behavior in a predictable way without forbidding any options or significantly changing their economic incentives .... Nudges are not mandates. Putting the fruit at eye level counts as a nudge. Banning junkfood does not.'

Chapter 17

The perilous road from community activism to public policy: fifteen years of community agriculture in Sandwell

Laura Davis[1] and John Middleton[2,3]
[1]*Ideal for All, 100 Oldbury Rd, Smethwick, West Midlands, B66 1JE, United Kingdom;*
[2]*Sandwell Primary Care Trust, 438 High Street West Bromwich, B70 9LD, United Kingdom;*
[3]*Honory Reader in Public Health, University of Birmingham, Edgbaston, Birmingham, B15 2TT, United Kingdom; laura_davis@sandwell.gov.uk*

Abstract

In 2008, Sandwell's Director of Public Health commissioned a community agriculture strategy for the borough, to be developed through a user-led process, which set out the strategic case for an expanded programme to contribute to the development of a sustainable food system, and to the goals of the Sandwell Food Policy. The Policy seeks to make explicit and develop the links between the regeneration of the environment and the regeneration of health, and the links between environmental (in)justice and health inequalities, through food and horticulture in an urban environment. Growing Healthy Communities: A community agriculture strategy for Sandwell 2008-2012 (the Strategy) is grounded in learning through practice from fifteen years of community activism in the regeneration of derelict land for mixed-use food and therapeutic horticulture initiatives.The Strategy represents a pivotal moment in both community activism and public health policy in Sandwell. It recognises the strengths of innovation through community development approaches; the possibilities of developing policy through 'non-rational' pathways; and values people's abilities, and indeed their rights, as well as those of professionals, to make decisions about 'what works' for them. Nevertheless, this unconventional road is as perilous as that from evidence to policy, in terms of the complexities in both framing, designing and delivering effective, accessible, and relevant, public health 'prevention' and 'inequalities' interventions. This chapter tells the story of Sandwell's perilous journey along this road. It offers reflections and insights into the successes and challenges encountered along the way, including the challenges of evaluating and evidencing the complex and wide-ranging health and well-being outcomes of such an approach: a challenge that is shared by both community activists and public sector professionals engaged in promoting health and well-being, and preventing illness, through interventions in community settings.

Keywords: evaluation, public health, local interventions, health inequalities

17.1 Introduction

Those who are familiar with the work Ray Pawson and Annette Boaz will note that the title of this chapter is adapted from their paper *The perilous road from evidence to policy: five journeys compared* (Pawson and Boaz, 2005). In this paper they cite Kitson and colleagues (Kitson *et al.*, 1998), who argue that the debate about evidence based policy and practice focuses on the level and nature of the evidence, at the expense of an understanding of the environment in which a review of evidence is to take place.

This resonated with our experiences in Sandwell of working within the English public health tradition, with an understanding of health inequalities and their relationship to 'place', and with the Sandwell public health team, at a time of a strong drive for evidence based policy in health, and the search for knowledge about 'what works' in tackling health inequalities. This began a long train of thinking about 'what to do' about evaluation for Sandwell's innovative, award winning community agriculture initiative – a third sector, community led programme, with a strong emphasis on promoting health and well-being and tackling health inequalities, and with multifunctional environmental and social dimensions.

This chapter reflects on our sometimes wonderful, sometimes perilous, journey from community food activism in a local setting, to the development of local policy and strategy. It highlights some of the successes and challenges along the way, which both informed and helped to put into practice key goals of the Sandwell Food Policy (2005), and *Growing healthy communities: a community agriculture strategy for Sandwell 2008-2012* (Davis, 2008), the first of its kind in the UK. Here we (the authors, on behalf of the team of community agriculture practitioners) share and reflect on our experiences, over fifteen years, of delivering Sandwell's innovative, award winning, community agriculture initiative, and our thinking about how to develop appropriate and meaningful monitoring and evaluation approaches appropriate to a third sector, community led programme, taking account of its multifunctional dimensions.

17.2 Life and death in Sandwell

Sandwell's origins as a place lie in the 18th century, as the furnace and foundry of the industrial revolution. Coal and metals, and technological innovation in the form of the coal fired steam engine created the economy and the social conditions of the Black Country in the West Midlands, of which the six towns of Sandwell are now a part. It was once dominated by large-scale metal and chemical works including the very first industrial-scale blast furnaces and the bell pits, mines and opencast coal workings that fired them (Middleton, 1989). Today, Sandwell is one of the struggling, almost post-industrial boroughs of the Black Country. It is wholly urban, and has a population of just under 300,000. It is relatively uniformly deprived from end to end. It still suffers from a poor quality environment, although waves of successive regeneration schemes have bought about some improvements. The slag heaps are now gone, along with some of the worst housing, although much still remains. When asked to map out its contaminated lands and landfill, as part of a national mapping programme, the

Council decided to draw a red line around the whole borough. There is an immense legacy of industrial contamination, dereliction, landfill, economic, and social deprivation, resulting from over 200 years of unsustainable development. It has one of the highest unemployment rates in England, especially among young people, about a third of whom are without work. It is tenth in the national scale of average rank of economic and social deprivation (Research Sandwell, 2008). About 17% of the population is made up of people from black and minority ethnic communities.

People here die younger, or experience more years of disability, than in other, wealthier parts of the UK. Again, although there has been improvement over the last few years, this has not closed the gap between Sandwell, the West Midlands and the rest of the UK. Life expectancy, particularly for males, is still relatively low. The killers are the usual suspects, heart and circulatory disease, cancers, diabetes, lung disease, alcohol related deaths. There is some recent evidence that these gains are levelling off, which is documented in Sandwell's annual reports of the DPH (Middleton, 2010).

17.3 The 1990s: a pivotal decade

Despite, or perhaps because of, the challenges faced in Sandwell, key people in the public and voluntary sectors have long held a dream to create a greener, more productive environment for its citizens. As long ago as 1989, the first report of the new director of public health (DPH) for Sandwell contained, towards the end, the idea that the then health promotion unit should work closely with voluntary organisations and community groups to examine the possibility of establishing local food co-operatives, and community gardens/allotments producing vegetables, etc. (Middleton, 1989). In 1996, the year in which a feasibility study on community agriculture (also known as urban agriculture) was commissioned, in the eighth annual public health report, *Regenerating health: a challenge or a lottery?*, the DPH observed, 'Sandwell as the garden of England is a bit far-fetched, but it is not impossible for us to increase the amount of food we supply to ourselves' (Middleton, 1996: 141).

The 1990s proved to be a pivotal decade for the development of initiatives focused on food and healthy eating, both in the mainstream and through alternative approaches (Murcott, 1998). During the decade, Sandwell Primary Care Trust (PCT) employed a food policy advisor, established a food policy board, began to build up teams of food and physical activity practitioners, and to develop, secure funding for, and deliver a raft of healthy eating and activity programmes. The decade also saw the development of 'alternatives' to mainstream provision, including Sandwell's community agriculture and other aspects of its food policy activity, comprising people and projects that locate food and nutrition issues within social welfare, social policy, social justice and sustainable development (aligned to the 'new' public health) (Du Puis and Goodman, 2005; Koc *et al.*, 1999).

By 1998, funding was secured to act on the recommendations of the feasibility study. 12 sites were thoroughly investigated at great cost, searching for landfill and a wide range of contaminants, assessing suitability of access, security and other issues. Only three of these sites (two mainly abandoned, derelict allotment sites and the 'footprint' of a former 20 storey tower block) were suitable for what we hand in mind. In fact, the tower block footprint was a very difficult and contaminated site, and was chosen due to its proximity to the newly constructed Independent Living Centre, where the headquarters of Ideal for All and the community agriculture initiative is now based, rather than for land quality.

Only the smaller tower block site was situated within a prescribed geographical regeneration zone (of which there have been many and various over the years), and was eligible for funding to help construct the project. The two larger sites were not in any such zones, and were not a comfortable 'fit' within the many and various development 'themes' of the statutory and mainstream regeneration funders. Regeneration 'zones' and thematic plans were (and to a significant extent, still are) strongly driven by large-scale projects, by themes designed to tackle, for example, worklessness, a skills mismatch, or environmental degradation, to attract inward investment; or by area. The suitable land available, in places in where communities of people lived their day-to-day lives, and their needs in terms of access to regenerated, safe and accessible green spaces, and opportunities for food growing, gardening, healthy eating and activity, often did not register on the agendas of the mainstream regeneration funders. We were faced with a choice, follow the funding, by area or priority (i.e. become 'funding led'); or, choose those areas in which we felt both communities and available land were optimal for the successful implementation of the shared vision for community agriculture in Sandwell. We chose the latter – to remain grounded in a community based, public health perspective, and to take on the challenge of ensuring the funding was a good fit for the needs of the local populations.

In 1999/2000, the fledgling projects joined together to become part of local disabled people's organisation and charity, Ideal for All, bringing together two prioritised sites for regeneration and development, one the tower block site (now Malthouse Garden), and one a three acre former allotment site, in the heart of a neighbourhood known as Bristnall (now Salop Drive Market Garden). The third, much larger site is in the very early stages of development for a major new scheme. As an established charitable organisation, Ideal for All possessed the capacity provide the initiative with the legal, financial and management systems necessary to bring forward a successful programme. The three-acre derelict site was about two miles away from their flagship Independent Living Centre, in a place in which we felt the conditions were most conducive to the success of a community led approach. Bringing the two initially separate sites together under one banner created the Growing Opportunities programme, for which we had by then secured some modest political support, and some start-up funding.

Critically, Ideal for All is a user led organisation of disabled and disadvantaged people, founded on the social model of disability (Oliver, 1990). The social model proposes that systemic barriers, negative attitudes and exclusion by society are the ultimate factors defining who is disabled and disadvantaged, and who is not, in a particular society. It recognises that while some people have physical, sensory, intellectual, or psychological variations, which may sometimes cause individual functional limitation or impairments, these do not have to lead to disability, unless society fails to take account of and include people regardless of their individual differences. The origins of the approach can be traced to the 1960s and disabled people's civil and human rights movements; the specific term itself emerged from the UK in the 1980s.This social model and user led approach, brought together at an important moment with a strong local public health focus on health inequalities, including commitment to address poor diet and low levels of physical activity, clarified what should be our ways of working, enabling local people and organisations to work together in an extended collaboration, to realise the dream.

17.4 Realising the dream

The turn of the century started a period of frenetic but carefully planned and co-ordinated activity, in which the regeneration and build of the projects went alongside a fully accessible process of community consultation, capacity building and participation. We recognised that it would not be easy for people who were used to a poor environment and poor services, who were poor by anyone's standards, and not in the best of health, to envisage 'better': to look at these derelict, fly tipped or rubble covered parcels of land and to know what they wanted, or what could be achieved, in terms of productive, accessible, welcoming, and safe garden spaces. We knew that it was not reasonable to expect volunteers to set out with spades and forks to clean up these derelict and dangerous land parcels.

The small community agriculture team (at that time two part time job sharers) set about a two year process of community capacity building and participative planning, made fully accessible by providing information in multiple formats, using interpreters and personal assistants, providing transport, and holding meetings and workshops in accessible buildings at times that suited people. This included organised bus tours, taking people out from urban areas to farms, market gardens and visitor centres, to help fire imaginations with possibilities; and community safety workshops, to address people's concerns about safety and security, as many believed that such projects would encounter theft and vandalism.

Throughout this time the team was fundraising, using carefully constructed business plans for the projects, based on the evidence collected during formative stages about the need for and feasibility of community agriculture in Sandwell. Malthouse Garden received a substantial funding award through a local regeneration scheme, and Ideal for All used the small amount of money it had managed to secure for 'start up' for Salop Drive Market Garden to carry out a initial clearance of the site, removing or burying fly tipped rubbish in 'bunds' that formed the basis of what is now a mixed species hedgerow. This helped to mobilise people in the community, as they could see 'something was happening'. It was only then possible to see the real nature of the problems of the derelict land, and make our plans for regeneration and the new build elements of the projects.

Ideal for All supported the formation of a 'garden advisory group', made up of local people, as part if its user led framework (which has now become the Health and Well Being Board, with delegated powers). It ran a series of participative planning and design workshops, in which local people were asked to put forward their ideas for what they wanted to see in the gardens, how they wanted to use the gardens, and what support they would need to do so, in the presence of the appointed architect, who produced a master plan based on the outcomes of this process, and we applied for and were granted full planning permission.

By then, the fundraising tide was turning in our favour, and from 2001, on the basis of plans developed by and for the community in the participative process, Ideal for All procured the groundworks for and construction of a working, professional standard organic market garden and local food project at Salop Drive, in the heart of a deprived community, alongside the substantial regeneration and build of Malthouse Garden, which was to become a dedicated horticulture therapy unit. Starting from land with significant problems with dereliction, demolition and tipping, infrastructure developments included service connections (e.g. power and water, including a pressurised irrigation system capable of irrigating protected and outdoor crops), laying drains, construction of buildings, greenhouses and polytunnels, access improvements, and soil regeneration, using cultivations, green waste composts, and green manures.

Revenue was invested in employing and developing a cohort of community agriculture practitioners, either learning by being apprenticed to an older, experienced market grower, or by more formal training such as gaining qualifications in horticulture therapy and related skills. By 2003, the land was in full production, staff capacity and skills development was well under way, and the sites were open as public amenities to people of all ages, abilities and backgrounds.

During this time, in the statutory sector the main support came from the Health Authority, which later became the PCT. Although we had established good relations with some Council officers, notably in planning and leisure services, the Council on the whole was relatively indifferent to, and even sometimes dismissive of, the scope and role of such projects in the larger scheme of regeneration in the borough. The projects had to stand or fall upon their own abilities and merits, and to this day, although the political and operational support in the Council is now at the highest possible level, no funding has ever been directly received from Sandwell Council. Nearly all of the funding for capital infrastructure has been and continues to be raised by Ideal for All from beyond the boundaries of the borough. In the last four years, however, the PCT has made a major contribution towards revenue costs, in particular salaries, which has enabled the employment of 11 full and part time staff. But this relationship is now threatened by politically driven reorganisation of health services, including the proposed complete abolition of PCTs, in a proposed shift to General Practitioner cluster fund holding and commissioning structures by 2013.

17.5 Informing and contributing to local policy and strategy

In 2005, as members of the Sandwell Food Policy Board, the community agriculture team were able play a part in development of the Sandwell Food Policy (2005). The Policy aims to protect and enhance the health of the population and contribute to tackling health inequalities. It focuses on creating connections between food, health, regeneration of the economy and the environment, demonstrating the goals of sustainable development. The award winning community agriculture initiative is one of the most visible and successful elements of Policy implementation. Goal 8 of the Policy seeks to 'Integrate into mainstream services the social, health, educational and therapeutic benefits of food growing in Sandwell. Share existing good practice from Salop Drive Market Garden and extend opportunities through other community agriculture projects, allotments, school and household gardens.'

In late 2007, Sandwell's DPH commissioned a community agriculture strategy for the borough, to be developed through a user-led process, drawing on the years of experience, creativity and problem solving. People of all ages and backgrounds, from Sandwell's diverse communities, as members of the public, professionals, and organisations we had worked with along the way, were asked in interviews and workshops what had been learned, what was their vision for the future, what did *they* think was important, what were the opportunities, and how we could overcome the challenges that lay ahead of us? The resulting Strategy sets out the case for an expanded programme, to contribute to the development of a sustainable food system, and to the goals of the Sandwell Food Policy. The vision that developed from this process, set out in the Strategy, is:

Community agriculture will make a positive contribution to improving people's well being and quality of life; including physical and mental health, citizen inclusion and participation, encouraging independence and community cohesion; to the development of individual and community capacity and skills; and to the regeneration of the environment by transforming new and underused sites into accessible, well-managed, productive and safe green spaces, providing activities and services that people of all ages and backgrounds can enjoy. Linking the goals of public health, social inclusion, regeneration and sustainable development through food and gardening will help to develop a 'whole life approach' to public health, developing innovative activities and services which contribute to tackling inequalities in health and social disadvantage, through the engagement of the public and professionals in a shared endeavour.

Growing healthy communities: A community agriculture strategy for Sandwell is thus grounded in learning through practice from fifteen years of community activism in the regeneration of derelict land for mixed-use food and therapeutic horticulture initiatives.

17.6 What are the benefits for the environment and people?

For ten years, the gardens have been run by and for local people, and are staffed by qualified horticulture therapists, assistants and 'garden workers' (volunteers). Salop Drive Market

Garden supplies a weekly bag of fresh, seasonal produce for £4 to about 80 households in the vicinity of the site, and to local community cafes and cookery groups. The gardens are run using agro-ecological methods suitable for intensive, small-scale production. As well as the market garden activities, Salop Drive Market Garden is mixed use, providing both full size and smaller plots for individual gardeners, a community garden, a children's area, a plant nursery, packing and grading and classroom facilities.

Both gardens provide the base for the provision of a wide range of fully accessible services and specialist sessions for individuals or groups of people from Sandwell's diverse communities, offering alternatives to mainstream provision. As a model of good practice, Ideal for All works closely with users and commissioners to support disabled and disadvantaged people in the prevention of ill health, maintenance and recovery, promoting improved mental and physical health, independence and quality of life. The sites are multifunctional, in that they provide many ways for people of all ages and abilities to get involved in growing, cooking and eating healthy, fresh food, from structured sessions to informal social events and celebrations, in a safe, welcoming, non-judgemental social environment. The projects bring together activities across a range of domains, spanning physical and psychological health, social relationships, and environment, in a 'whole life' approach. They are fully accessible and fully inclusive, catering for disabled and non-disabled people from ages 0 to 101, including pre and school age children in curriculum linked and menu development activities.

The environmental benefits of community agriculture in poor quality, deprived urban neighbourhoods are clear and unequivocal, and highly visible. There are now high quality, safe, managed, accessible and welcoming green spaces where before there was dereliction, and a notorious 20 storey block of flats. The mixed, agro ecological methods, and specific conservation features (native species, bird boxes, bee 'hotels', and more) contribute to improved biodiversity in an otherwise poor quality urban environment. The regeneration and improvement of derelict spaces has a 'ripple effect' in the entire community, helping to halt and reverse environmental and social decline. The benefits of access to green space, fresh food, and good company, and the impacts on people's health and well-being have been well described in a number of recent studies.

We have, throughout the duration of the community agriculture initiative, been capturing quantitative and qualitative data. For example, we know a lot about how many and what type of people access the projects and the way in which they use them, using in-house databases. We have collected qualitative data on to what extent the projects are supporting people in setting and achieving personal goals, in improving or maintaining their health and well-being, as well as more straightforward satisfaction surveys, customer feedback and evaluations, from people who use the services. This data show that people are highly appreciative of the initiative, and that they report a range of positive health, well-being and 'social capital' type benefits. Overcoming a sense of social isolation, meeting and making new friends, feeling valued, 'putting something back' into the community, feeling fitter, stronger, more confident, happier and healthier, and the benefits of being outdoors, are all regularly recorded in questionnaires and evaluations.

However, 'evidencing' these kinds of well-being, quality of life, and social capital outcomes of the initiatives in robust, verifiable ways, that will show to the commissioners of services that they are contributing to a range of high level targets and outcomes (such as reducing health inequalities, contributing to more cohesive communities, etc.) is problematic. Proving that there is progress, for example if we are succeeding, or failing, in achieving the aims of the Strategy, or moving in the right direction, is reasonably straightforward. But proving that that there is a meaningful contribution to public health and other wider social goals such as, for example, contributing to tackling obesity, reducing preventable deaths or the burden of long-term illness from cardio-vascular disease, cancer, stroke and diabetes, or that people from different backgrounds get on well together in their communities, is immensely difficult in a busy intervention in a community setting.

It is especially difficult when the community agriculture practitioners team are paid to deliver interventions, rather than systematically collect, analyse, and report on a range of data, at the level required to prove the 'worth' of the activities and services, which will satisfy high level service commissioners and decision makers that the benefits represent value for money, and to do this in a shifting governance and policy environment. This challenge is one we are now addressing with our public health colleagues.

17.7 'What to do' about evaluation?

In the past, evaluation approaches and methods have often been driven, or even diverted, by the individual and sometimes random demands for 'evaluation' (often confused with monitoring) by the many and various funders we have worked with, each of which works in total isolation from the other (even different rounds or streams of the same fund have very different requirements), in often short term, 'projectitis' approaches.

Now we are in our twelfth year of delivery, it is time to face up to the challenges of evaluative research in food and nutrition 'interventions' as identified by Beardsworth and Keil (1997), who point out that one of the key challenges for evaluative research in food and nutrition interventions is that, although the linkage between, for example, diet and health (or similarly, access to nature) and well-being, are inescapable facts of life, these linkages can be a subtle and complex; clear causal pathways can be very difficult to establish. This is especially true for interventions that take place in community settings in which the context, or environment, is overwhelmingly important and cannot be discounted.

In terms of policy, practice and places, food and health improvement issues are interventions in what Blackman describes as 'wicked' problems; issues that are hard to treat not because they are complicated but because they are *complex* (Blackman, 2006). This means that the design, delivery and evaluation of intervention programmes needs to embrace the whole system in which the issues are embedded. This calls for a multidisciplinary understanding and approach, and indicates that an on-going re-working of concepts and materials, in an iterative, deductive process of enquiry (which does *not* mean theory is absent), is likely to be most productive.

The experience and thinking in our community agriculture team about the strengths and weaknesses of different approaches to evaluative research has led us to the conclusion that we should focus on 'well-being' and quality of life measures, drawing on existing, validated approaches. But most of the guidance and models we have looked at, such as EQ5 (EuroQol Group, 1990), the Warwick/Edinburgh scale or WEMWBS (Health Scotland, 2010), ASCOT (PPSRU, 2010), and others, are either too medicalised, individualised, or both, with the possible exception of WHOQOL quality of life assessment (2010), which sets out a framework for a broad and comprehensive assessment that has an 'environment' domain. However, none of these is quite the right fit, on its own, for what we aspire to in the scope of our evaluative work.

Recently, the National Institute for Health Research, part of the UK National Health Service, which has an arm that carries out public health research, put out a call for proposals to undertake a commissioned research project to explore the effectiveness and cost effectiveness of community farms, gardens and allotments in contributing to improved health and well-being. This proved to be a good opportunity for community agriculture practitioners to come together with senior research staff in the PCT, and with an academic partner (University of Warwick) to formulate a mixed method programme of research, including cohort design, that will robustly and comprehensively evaluate the impact of community agriculture on the health and well-being of users and residents in Sandwell. The research project, if we are successful in our bid to win the contract, will help us, and others, to understand the value and impact of community agriculture facilities and services on a range of important areas, including maintenance and improvement of physical activity, diet, mental health and use of NHS services. We will also assess impacts on inclusion, social relationships, independence, confidence, learning and skills. Even if we are not successful in this bid, this opportunity has contributed to an understanding that can be developed to strengthen how we do evaluative research in the future, albeit in a more limited way and scale, to apply a more robust framework, and systematise evaluation across activities and services within the community agriculture initiative.

Finding the right ways of evaluating what works, what does not, for who, that is grounded in a whole life approach and the English public health understanding of the relationship between health inequalities and locality, or place, and that is in line with our social model and user led approach, without adequate funding and in the midst of different and often competing demands from funders and other professional interests, is a very significant challenge. Moreover, this challenge of 'what to do' about evaluation is shared by both community activists and public sector professionals engaged in promoting health and well-being, and preventing illness, through interventions in community settings.

Our aim now is to ensure collaborative working between Sandwell's public health team and its very able research experts, community agriculture practitioners, and service users, to develop and apply evaluative methods which will stand the test of time, and which we can make explicable and accessible for the many and diverse people who use our community agriculture programme. It is important we don't get hung up about the level and nature of the evidence, and focus on our chosen whole life, well-being approach, which can both take account of the variations and complexities of both Sandwell as place, and the experiences of the communities and individuals who – as they have been telling us for years – benefit in multiple ways, often in a life transforming experience, from being involved in community agriculture.

Acknowledgement

The authors are members of the Birmingham Black Country Collaboration for Leadership in Applied Health Research and Care (CLAHRC), funded by the National Institute for Health Research, but the views expressed are their own.

References

Beardsworth, A. and Keil, T., 1997. Sociology on the menu: an invitation to the study of food and society. Routledge, London, UK, 288 pp.

Blackman, T., 2006. Placing Health: Neighbourhood renewal, health and complexity. Policy Press, Bristol, UK, 264 pp.

Davis, L., 2008. Growing healthy communities: a community agriculture strategy for Sandwell 2008-12. Sandwell Primary Care Trust and Sandwell Metropolitan Borough Council, Sandwell, West Midlands, UK, 36 pp.

Du Puis, E. and Goodman, D., 2005. Shall we go 'home' to eat? Towards a reflexive politics of localism. Journal of Rural Studies 21: 359-371.

EuroQol Group, 1990. EuroQol – a new facility for the measurement of health-related quality of life. Health Policy 16: 199-208.

Health Scotland. WEMWBS: The Warwick-Edinburgh mental well-being scale. Available at www.healthscotland.com/documents/1467.aspx.

Kitson, A., Harvey, G. and McCormack, B., 1998. Enabling the implementation of evidence based practice: a conceptual framework. Quality in Healthcare 7: 149-158.

Koc, M., McRae. R., Mugeout, L. and Walsh, J., 1999. For hunger proof cities: sustainable urban food systems. International Development Research Centre, Ottowa, Canada, 238 pp.

Middleton, J., 1989. Life and death in Sandwell: being the first annual report of the Director of Public Health in Sandwell of Sandwell Health Authority and an action plan for Health for All by the Year 2000. Sandwell Health Authority, Sandwell, West Midlands, UK.

Middleton, J., 1996. Regenerating health: a challenge or a lottery? The 8th Annual Report of the Director of Public Health. Sandwell Health Authority, Sandwell, West Midlands, UK.

Middleton, J., 2010. 5% for health: the 20th annual public health report for Sandwell. Sandwell Primary Care Trust, Sandwell, West Midlands, UK.

Murcott, A. (Ed.), 1998. The nation's diet: The social science of food choice. Longman. London, UK, 400 pp.

Oliver, M., 1990. The politics of disablement. MacMillan. London, UK, 152 pp.

Pawson, R. and Boaz, A., 2005. The perilous road from evidence to policy: five journeys compared. Journal of Social Policy 34: 175-194.

PPSRU (Personal Social Services Research Unit), 2010. ASCOT: adult social care outcomes toolkit. Personal Social Services Research Unit, University of Kent, UK. Available at www.pssru.ac.uk/ascot/index.php.

Research Sandwell, 2008. Indices of deprivation: Sandwell 2007. Briefing note 3. Available at www.researchsandwell.org.uk/research/briefingnotes/id_2007_briefing_note.pdf.

Sandwell PCT, 2005. Sandwell food policy. Sandwell Primary Care Trust, Sandwell, West Midlands, UK.

United Nations, 1992. Report of the United Nations conference on environment and development. Rio de Janeiro, 3-14 June 1992. Available at http://sedac.ciesin.columbia.edu/entri/texts/a21/a21-06-human-health.html.

WHO Field Centre. WHOQOL (World Health Organisation Quality of Life Assessment). Available at www.bath.ac.uk/whoqol/.

Chapter 18

Making local food sustainable in Manchester

Les Levidow[1] and Katerina Psarikidou[2]
[1]*Open University, Walton Hall, Milton Keynes MK7 6AA, United Kingdom;* [2]*Dept of Sociology, Lancaster University, Bowland North, Lancaster LA1 4YT, United Kingdom; l.levidow@open.ac.uk*

Abstract

In Manchester, environmental and health issues have been integrated into a wider agro-food strategy for making the city 'more sustainable', in several senses of the word. With crucial support from state bodies and charities, several initiatives expand access to fresh, healthy food, especially in 'food deserts'. Through 'community engagement', they mobilise various resources, skills and voluntary labour to create 'community spaces' for social inclusion. Manchester agro-food networks shorten supply chains, e.g. by more directly linking peri-urban agriculture with urban consumers, and by promoting urban agriculture based on local resource mobilisation and personal trust. These networks provide alternatives to conventional agro-food chains. In Manchester, food relocalisation helps to overcome socio-economic inequalities and health problems. In a national policy context advocating food relocalisation but offering little support, Manchester agro-food initiatives cooperate to develop environmentally sustainable, socially just, healthy communities. These also reconstruct local identities and social commitments. Despite that success, the larger food system is still dominated by conventional agro-food chains. Community food initiatives have a marginal role, so practitioners discuss how to overcome the present limitations. From 2010 onwards the UK government's austerity regime undermines state support, so community engagement will become even more important for mobilising resources.

Keywords: local food, food relocalisation, food deserts, community engagement, Manchester, permaculture

18.1 Introduction

'Local food' has been given various meanings about closer linkages between producers and consumers. In this way, the agro-food sector becomes embedded in a complex web of social relations beyond simply market transactions and financial motives (Ilbery and Maye, 2005). In places where they gain commercial success, however, such alternatives may generate socio-economic inequalities. Food relocalisation can simply create a niche market for affluent consumers, rather than challenge the dominant agro-industrial system (Goodman, 2004).

As a conscious challenge to inequalities, especially from conventional agro-food chains, local food initiatives can be 'seeds of social change'. In this perspective, the 'local' acquires multiple connotations:

> ... of the construction of community through the development of links within everyday life, of the incorporation of a moral economy of interaction between neighbors or allies mutually engaged in production and consumption. The local is assumed to enable relationships of aid and trust between producer and consumer, eliding the faceless intermediaries hidden within commodity chains and industrial foods (Allen et al., 2003: 62, 64).

Since the 1990s, food relocalisation has gained prominence in the UK, partly from civil society initiatives (e.g. Sustain, 2002). A government advisory body, the Curry Commission, recommended economic regeneration by reconnecting people with food production: 'Reconnect our farming and food industry; to reconnect farming with its market and the rest of the food chain; to reconnect the food chain with the countryside; and to reconnect consumers with what they eat and how it is produced' (Curry, 2002: 6). It promoted both short-chain local food and 'locality food', e.g. specialty brands which can be valorised in longer-distance supply chains. But supermarket chains and food processors were seen as main conduits: 'Farmers and farmers' groups that work closely with supermarkets and processors, and that are in touch with the consumer, can do good business. They can play their part in helping develop successful brands in home and export markets' (ibid. p. 17).

Partly in response to the Curry report, the relevant Ministry set up a working group on local food initiatives. Its report proposed support measures:

> There are few strategies or frameworks within which local food is explicitly mentioned in public policy at national or regional level. However, projects in the local food sector have successfully gained public funding for their work towards objectives such as neighbourhood renewal, improved diet and promotion of healthy eating, support for the rural economy, urban/rural linkages etc. (DEFRA, 2003a: iii)

The government did fund some support activities:

> Supporting the quality regional food sector through a five year £5 million programme of support (beginning in 2003/04). Activity under this programme is focused on three key areas – trade development, consumer awareness and business competitiveness with the specific objective of creating a flourishing high quality regional food sector (DEFRA, 2006a: 18).

However, its policy left the main responsibility with local authorities and individual consumer choices. It expected supermarket chains to promote locally sourced food. Some have done so by appropriating local quality brands. Yet supermarkets are widely seen as the source of environmental and health problems, which warrant independent alternatives as the remedy.

18.2 Case study methods

This chapter investigates how local food initiatives integrate health and environment in Manchester. For our case study, the initial data sources were internet-based documents of Manchester organisations, policies and practices: the regional development agency (NWDA), local authority funding programmes (Manchester Food Futures), social enterprises (e.g. the Herbie Van and the HeLF partnership), cooperative marketing initiatives (e.g. Unicorn Co-op and Glebelands Market Garden), small-scale businesses (e.g. the Dig box scheme), allotment and community garden initiatives. Those documents provided a basis for 14 semi-structured interviews with practitioners (from the same sources) which were interlinked in networks for regenerating this urban agro-food system. Our interviews were mainly conducted during 2008-2009, with some follow-up in 2010.

We selected initiatives which develop urban short-supply chains and favour cultivation methods with low external inputs. We sought to identify urban food initiatives which feature co-operative economic relations beyond financial motives. Snowballing contacts was an extra method for identifying relevant practitioners, their linkages and wider networks; interviewees mentioned links or analogies with other food initiatives, e.g. via self-description and anecdotes. Some initial interviews with key actors (e.g. Manchester Food Futures and the Permaculture Network) were instrumental for offering useful insights, as well as guiding our further contacts and research. Most interviews were conducted in the working environments of the practitioners, thus giving us a greater familiarity with their everyday practices. These interviews provide the source for all quotes which do not cite a document.

Interview questions were adapted from a larger research project about alternative agro-food networks (see Acknowledgements). Drawing on concepts in the literature on AAFNs, we investigated how practitioners discursively position themselves and their activities in relation to conventional food chains. Interview questions asked how Manchester initiatives differ from conventional food chains – as regards their aims, knowledges, production methods, networks, producer-consumer relations, etc. All interviewees described such differences, with various enthusiasm and emphases.

Our draft analyses were circulated for comment from interviewees and other practitioners. Preliminary results were pre-circulated for a stakeholder workshop on 'Community food networks in Manchester' held in April 2009, where discussions provided data for further analysis and insights for further research. In March 2010 a summary was circulated for comment to ten key individuals.

From the above research methods, we analysed the interviewees' and official documents' language – e.g. especially meanings of local food. A reference point was their practices which integrate health and environment by building community engagement. Although 'local' could simply denote a short geographical distance, many interviewees gave the term broader social, cultural or political meanings. In this way, we could better identify interviewees' various aspirations and strategies.

This chapter first presents self-descriptions of food projects and support bodies from publicly available documents (with sources cited). Then these are juxtaposed with quotes from our interviews as well as observational data.

18.3 Context: policy support for sustainable urban food

Manchester is the UK's third largest city, with approximately a half-million people, located within the surrounding region of Greater Manchester with 2.25 million people. Socio-economic inequalities and social exclusion contribute to rising health problems, including obesity. Some parts of the city, known as 'food deserts', have little access to healthy food. Urban redevelopment favouring supermarket chains has also been blamed for these problems.

In the 1990s many community activists opposed the expansion of supermarket chains, while also proposing alternative pathways for urban development. Emerging from activist networks in 1996, the Manchester Environmental Resource Centre initiative (MERCi) is an independent charity aiming to make Manchester more sustainable. Thanks to funding from the national lottery, MERCi gained a permanent base in a disused cotton mill, Bridge 5 Mill, as part of an urban regeneration development. They developed proposals to improve the urban environment and health through food activities:

> The history of grassroots campaigning in Manchester is a long and successful one. People taking action, separate from established membership groups, have achieved a great deal in the city and beyond. In the 1990's the groups working for a better Manchester were very active, but even amongst the 'activists' single-issue campaigning still seemed to dominate the agenda, and no one seemed to be talking about the links between poverty and energy or food and well-being. There was also little communication between these diverse groups and no real sharing of resources, support, ideas and skills. From these identified 'gaps' grew the idea for Bridge 5 Mill and the Environment Network for Manchester. The rest, as they say, is history! (MERCi, 2009)

Coming from several civil society groups, early initiatives led to policy changes integrating health and environmental issues into local development.

Food Futures is a partnership of Manchester City Council, the National Health Service, the voluntary and private sectors (see logo in Figure 18.1). It has played a central role in supporting sustainable urban agro-food initiatives. According to its strategy (Food Futures, 2007), it aims to create a culture of good food in the city, especially wide access to healthy, sustainably produced food. The Manchester Community Strategy (2006-2015) sets out how public services will be improved, especially a vision for 'making Manchester more sustainable' by 2015.

Figure 18.1. Logo Food Futures.

In those ways, the Food Futures strategy links health, local economy, regeneration, food as a cultural force, its social impact, anti-social behaviour, the environment, childhood diet, vulnerable groups and transport (*ibid.*). By linking these issues, the strategy provides opportunities for residents and local organisations to get involved in food projects, training activities and events around sustainable food. They support sustainable food activities of the voluntary and community sector (VCS) through funding, advice, networking and publicity. Funds are allocated to urban food initiatives, which take many forms – allotments, community gardens, delivery services, etc. In particular, the Growing Manchester programme has provided support to many community growing projects, with an aim to become more environmentally and economically sustainable in the broadest sense.

Extending the tradition of the 19th century cooperative movement in Manchester, many practitioners have promoted cooperative relations in local agro-food initiatives. To overcome various societal problems, practitioners construct networks creating closer relations between producers, retailers and consumers. Food providers include cooperatives, for-profit businesses, grassroots projects and social enterprises; financial support comes from state agencies, charitable organisations, and lottery funds (MACC, 2006). Although most food initiatives in Manchester distribute food grown near the city, some promote urban agriculture. Some distribute conventional food, mainly from peri-urban farms, while others promote permaculture and organic methods for urban agriculture.

In our research interviews, practitioners expressed two types of motivations:
- Concerns around social and economic inequality in Manchester, as grounds to enhance access to healthy food, to improve the immediate environment, and to promote food cultivation as a means to health and community cohesion.
- Broader issues including concerns around environmental protection, climate change, peak oil and food security.

Those motives span all relevant activities – projects that provide food and/or grow food – as described in the following sections.

18.4 Providing healthy food

Illustrating a pervasive cooperative ethos, Unicorn is a workers' co-operative, owned and run by its workforce. Selling wholefoods, it supplies fresh produce on a daily basis. It aims to provide employment for its members and people with learning disabilities. Unicorn promotes fair-traded products, supporting a 'sustainable world environment and economy'. It also encourages co-operation with other local businesses and co-operatives, through a trading model seeking the minimum possible impact on the environment (Unicorn, 2009). According to a founding member: 'It is based on a principle of buying products as directly as possible through a short chain between the producer and the customer – buying produce at a reasonable volume, trying to get the lower price and achieving the shortest time between the fresh produce and the customer'.

Along these lines, alternative food delivery schemes have been set up to address various societal problems. The Herbie Van and Dig box schemes are run by two Unicorn workers. Both initiatives have bought food from peri-urban producers, partly on environmental criteria. Both aim to provide easier access and lower prices for fresh food, thus enhancing health. They also provide social contact for marginalised people.

The Dig box scheme sources food from small-scale, organic-certified growers, thus promoting such cultivation methods. Dig representatives have developed such knowledge: 'they also learned that buying veg and fruit at a fair price from local farmers had a positive and sustainable effect on the local economy and on the environment' (Dig, 2009). On its website, the ordering information distinguishes between local and very local sources: 'We sell good quality food that tastes great that you can get locally'. In these ways, the scheme develops more proximate, solidarity-based relationships between local producers and small-scale businesses, as well as trust-based relationships between producer and consumers.

The Dig scheme seeks to expand its product volume but not its customer base, in order to maintain quality produce. According to a representative of the Dig box-scheme, 'We are at a manageable size for us and would like to sell more produce to our existing customers. We don't want to expand very much'. By contrast to the voluntary sector, it is an independent business: 'We are a self-supporting business; we have no reliance upon state money'.

Funded by Food Futures, Herbie Van operates like a mobile greengrocers. It provides affordable, fresh fruit and vegetables to residents in East Manchester who have poor access to fresh foods. 'Our produce is fresher and more environmentally friendly than the supermarket equivalent because it is usually bought on the day of sale, straight from the suppliers', says a promotional article (Hollinworth, 2006).

The Herbie Van aims to 'reach as many people in their local community as possible and offer customers a good range of affordable fresh produce' (MERCi, 2009). This outreach involves a humanitarian ethos: 'Some of our customers don't see another person for a whole week. When they come on the van and have a chat with the driver, it's more for people's mental well-being that they have someone to talk to. And it's a regular face; it's not just whichever person is on the check-out looking miserable because they don't want to be there. It's a natural interaction.'

This service forms part of a broader local network. Groundwork, focusing on the 'health aspects' of food, as well as Target Wellbeing and some NHS workers, work with the Herbie initiative. ZEST, a 'healthy living network', cooperates with the Herbie initiators to provide joint cookery classes, which are held by the mobile greengrocers, especially for people who live on their own.

Many of these initiatives succeed in implementing their principles for healthy food by getting their supply from Glebelands Market Garden co-operative. The latter has been a novel urban growing project in Sale, producing food certified by the Soil Association for consumption within Greater Manchester. It uses well-established techniques such as on-site composting, crop rotation and green manure crops to maintain soil fertility and plant health. It also uses more novel practices to extend the growing season wherever possible. Their produce is sold to some local restaurants and box schemes (Glebelands, 2009).

18.5 Urban agriculture

Beyond distributing food from peri-urban producers, other initiatives produce food in the urban area. This features producers-as-consumers, who thereby learn about environmentally more sustainable production methods. Citizens' active participation in gardening schemes has been a crucial way to re-establish a community culture.

Since 2004 the Bentley Bulk Local Food Project has found ways to produce more seasonal food and to promote societal integration (see Figure 18.2). An interviewee stressed the aim of social integration:

> We tried to reinvent the cooperative movement and we came up with this project we called the Bentley Bulk. The idea is to make beneficial connections to local producers, retailers and the community. The idea is to turn the people in this community into not just consumers but producers as well, so that they get involved in the production and distribution of food.

In the Bentley Bulk project, seasonal food has been expanded and promoted as environmentally beneficial: 'the growing season is extended through the use of horticultural fleece, mesh covers, a glass house, polytunnels and drip irrigation lines. The environmental cost of synthetic materials is currently considerably less than transport from Spain and other origins' (NWDA, 2006: 2).

Lately the UK has had a rising demand for allotments – mainly urban, municipally owned land divided into small blocks which are rented by the public and used for food production for individuals' private use. According to our interviewee, the Association for Manchester Allotments Societies (AMAS) is a City Council partner, constituted in 1993 with an aim to support the city's local allotments and horticultural societies. They promote allotments as 'a means to enjoy outdoor living and the satisfaction of growing your own food' (AMAS, 2009), as well as 'community spaces' which make people welcome and give them ownership.

Figure 18.2. Bentley Bulk Local Food Project promotion.

Beyond AMAS, community gardens have been set up for food consumers to become collective producers, as a means to enhance the sense of community through growing. Since the launch of Growing Manchester in March 2010, Food Futures has been instrumental in supporting 15 projects in the city and training 40 people in food growing. In many cases, permaculture has been favoured for its distance from commercialisation, its communitarian aspects, and its environmentally friendly methods – even as an alternative to organic methods and its certification requirements. According to the Permaculture Association,

> Permaculture is about creating sustainable human habitats by following nature's patterns. It uses the diversity, stability and resilience of natural ecosystems to provide a framework and guidance for people to develop their own sustainable solutions to the problems facing their world, on a local, national or global scale. It is based on the philosophy of co-operation with nature and caring for the earth and its people.[34]

In the Manchester context, permaculture has been perceived as a main means for creating sustainable local communities. As an interviewee recognises, the Permaculture Network was set up by 'those who fully appreciated that it's about understanding principles found in the natural world and ecosystems'. It links environmental sustainability with alternative

---

[34] http://www.permaculture.org.uk.

distribution systems: 'The permaculture network brings together people with different knowledges. It helps to enhance community cohesion. It includes your growers but also includes people like me who have an interest in local economic systems, community-supported agriculture, box schemes and local exchange trading systems.'

Based on these permaculture principles, as well as the successful results of Bentley Bulk, in 2006 the Health Eating Local Food (HELF) Partnership was established under the Food Futures Strategy. Encompassing the Health Advocacy & Resource Project (HARP), the Mental Health and Social Care Trust ('The Trust') and Zion Therapeutic Garden Project ('Zion Garden'), HELF was established as a community partnership. According to its *Recipe for Success*, the project has aimed 'to develop a city wide social enterprise, to engage people – especially younger people and people with mental health issues, in healthy local food activities, in order to improve skills, confidence and overall health'.

Like some community gardens, the HELF project promotes permaculture. It 'is designed to empower participants to live more sustainable lives, and to assist in the process of developing HELF venues and activities in an ecologically productive and sustainable way'. The project links and builds its members' skills for sustainable production methods: 'Practical skills training in organic gardening, horticulture, forest gardening and maintaining allotments are designed to help service users to get the most out of the plot'. Team-building has been a long-term objective, aiming 'to assist service users in working co-operatively to manage HELF Partnership assets as productively as possible' (HELF Report, 2007: 36).

HELF has been relaunched as Bite, which uses the food grown on its allotments in its numerous cafes.

> BITE works with people who have experience of mental health issues and may have used mental health services, supporting them to gain new skills and experiences through a range of enjoyable and supported volunteering opportunities. Bite aims to make a positive contribution to health and wellbeing through the activities that take place around growing, cooking and eating food (Bite, 2009).

The Bite and Bentley Bulk projects have similarities to the Local Exchange Trading System (LETS). This provides an indirect barter system for an alternative economy. According to a founder of Bite: 'they are basically a social trading network...They are a means for people who define networks to exchange goods and services without using cash... There was a big LETS system in Manchester with about 600 people trading in it.'

An informal exchange system has also been promoted through allotments and community garden schemes. As a representative said: 'You are not allowed to sell anything. So whenever they have too much food, it is given to family and friends. So it's nice from the social aspect.'

18.6 Community engagement linking health and environment

In sum, Manchester agro-food initiatives aim to re-connect producers and consumers, as well as rural and urban communities. These linkages extend social commitments to create sustainable local communities. Agro-food networks have been redesigned to link health and environment as means towards wider societal benefits including community development, cohesion and inclusion. Practitioners see their initiatives as 'innovative', 'verging on mainstream', or aspiring to such a role. This means empowering and expanding their initiatives within the public sphere – rather than creating a niche market for affluent consumers.

As a key term, 'community engagement' expresses active involvement and social inclusion through agro-food activities. According to the Northwest Development Agency strategy, 'Community engagement is more important than cost-efficiency, and procedures are designed to engage the community in the Food System in as many ways as possible' (NWDA, 2006: 4). According to a member of several Manchester food networks, community engagement gets a multifunctional role in achieving health and environmental benefits:

> Not only do we lack a food culture, we just lack a community culture generally. So by setting up local food production, it's a great way of getting people to have exercise and engage with each other. It's a social integration. And they get to grow food and eat healthy food. It's a great way for people who don't have very much money to have access to affordable health organic food.

Under the 'community engagement' pillar, public funds were also helpful generating support for collaborative projects among community groups to develop more allotments sites, some used for training in organic production methods. Towards this 'sustainable communities' model of local food, some practitioners have even advocated an ambitious expansion, which could reinforce a greater societal change: 'People are realising we need to re-localise not just at a communities level but also at a regional level as well. That has environmental and economic benefits, as you are rebuilding the local economy at the same time.'

In response to these aspirations, Food Futures seeks expansion mainly 'to meet needs and expand, extend and replicate models of good practice locally'. Beyond the city boundaries, collaboration with other authorities in Greater Manchester would be a beneficial means to 'share learning or encourage other areas to adopt the MFF approach'. In their view, good practice means increasing income for food producers and distributors, while keeping money flows within the locality.

18.7 Future needs and challenges

Local food initiatives face several difficulties, indicating various needs for sustaining their future viability. Practitioners have expressed views on future needs for revalorising urban agriculture as means to build sustainable, healthy communities. They see the current initiatives as contributing to people's social and mental well-being. Greater land availability to grow food within the city would help to educate people about food and its production, to provide economic independence for producers and to enhance community cohesion. Urban agriculture could also address issues of global food security, though within feasible limits.

From our project workshop, the following proposals emerged: The needs of small local initiatives and businesses, and the societal benefits that they provide, warrant greater recognition; training for employment and business skills would help them. A city-wide hub would be helpful for storing agricultural produce from nearby farms and then distributing it to urban food suppliers and retailers. Public education would help consumers to appreciate growers' work, especially the labour that goes into good-quality food. If consumers are willing to pay more for their food, then local food initiatives will expand.

Despite their successes, Manchester community food initiatives have a marginal role within the larger food system, which remains dominated by conventional agro-food chains. So practitioners discuss how to overcome the present limitations. Through a series of discussions, entitled 'Feeding Manchester', practitioners have elaborated a vision for the city to develop a sustainable food sector by 2020. They link food identity with environmentally more sustainable forms and sources of food production: 'Greater Manchester has a *food identity* and a growing affinity from local residents, who seek out local produce and are aware of seasonal and local recipes.' They discuss several challenges – for example, how to increase local cultivation of fruit and vegetables; how to return more of the overall profits to food producers; and how to reduce food waste (Kindling Trust, 2010).

To build a sustainable local food system in Manchester, a major gap has been public procurement, which has effectively favoured large suppliers with long supply chains rather than local sources. As the relevant agency, Manchester Fayre is seen as seeking the lowest price, following national policy on 'aggregated purchasing' to minimise prices (Gershon, 2004). Government policy has been contradictory – encouraging higher-quality local sources for food procurement (DEFRA, 2003a,b), yet providing unclear criteria to justify a higher price (DEFRA, 2006b). Nevertheless many local authorities and public institutions have found ways to favour better-quality and/or local food (Morgan and Sonnino, 2006).

Along those lines, Food Futures have proposed criteria which would favour local, organic sources in public procurement. The programme has been working with partners on Sustainable Fayre, which is funded by the Carbon Innovation Fund as part of Manchester's Climate Action Plan to reduce the city's CO_2 emissions by 41% by 2020. The project has two parts. First, a feasibility study looks at the viability of supplying fairly traded, locally grown, organic produce to Manchester Fayre school catering service and explores options on how to reduce the environmental impact of school meals. Second, a pilot study at a primary school offers freshly made soup prepared from fairly traded, locally grown, organic produce; parents have been involved (Kindling Trust, 2011). Through these steps towards a policy change, Sustainable Fayre aims to increase public awareness about sustainable production methods, to gain contracts for food supply, and to help local and peri-urban food suppliers.

18.8 Conclusions

Environmental and health issues have been integrated into a food strategy for 'making Manchester more sustainable', in several senses of the word. Several initiatives expand access to fresh, healthy food; delivery services address the problem of 'food deserts' and social isolation. Some initiatives also expand access to land and other resources for urban agriculture, emphasising organic and permaculture cultivation methods. Producers-as-consumers build skills in environmentally more sustainable cultivation methods.

Through a process of 'community engagement', these initiatives mobilise various resources, skills and voluntary labour to create 'community spaces' for social inclusion. Manchester agro-food networks shorten supply chains, e.g. by promoting urban agriculture based on local resource mobilisation and personal trust, and by more directly linking peri-urban agriculture with urban consumers. This cooperative strategy mobilises solidarity networks within and among food initiatives, beyond business or contractual relationships. For Food Futures in particular, the local focus on climate change has been a political driver to reduce environmental impacts.

In all these ways, food relocalisation helps to overcome socio-economic inequalities and health problems from conventional agro-food chains. Going beyond a niche market for affluent consumers, local food becomes a 'seed for social change' through a 'moral economy' (Allen *et al.*, 2003). In a national policy context advocating food relocalisation but offering little support, diverse organisations in Manchester cooperate to develop environmentally sustainable, socially just, healthy communities. Agro-food initiatives reconstruct local identities by linking health, environment and greater access to food as social commitments. These initiatives take responsibility for roles which were formerly externalised in conventional agro-food chains (Brunori, 2007).

All these initiatives illustrate good practices, whose expansion depends on catalysing wider support networks within Manchester. Alongside their achievements, Manchester food initiatives face many challenges that are discussed by practitioners. How to extend the health benefits to more people? How to expand and even mainstream the initiatives, while minimising dependence on state support and voluntary labour? How to return more of the income to food producers (rather than to intermediaries) and thus fund environmentally more sustainable production?

Since the May 2010 general election, which resulted in a government coalition between the Conservative and Liberal-Democrat parties, policy changes have created great difficulty for local authorities. Their budgets depend almost entirely upon government grants, which have undergone enormous cuts, thus attacking public services. Local authorities are being given more responsibilities, such as public health. Food Futures are located in a public health team, based in a Local Authority, so staff foresee themselves 'in a strong position to maintain the current programme of work and level of funding' (MFF, 29.11.10, personal communication). Yet the government's austerity agenda squeezes public-sector resources.

The new government has also reinforced the earlier policy on 'aggregating purchasing'. Its Efficiency and Reform Group aims to use buying power to drive down costs of public procurement. If imposed on local authorities, this policy may further marginalise local food suppliers.

Like other public services under attack, state support for food relocalisation will depend upon political alliances defending current initiatives and finding ways to expand them. Although some are small businesses which have gained financial independence, most Manchester food initiatives depend on support from state agencies, private charities, the national lottery or a combination thereof. If state support declines under the UK's austerity regime, then community engagement will become even more important for mobilising resources, skills and voluntary labour. The question remains: How will Manchester food initiatives be sustained?

Acknowledgements

The research leading to these results received funding from the European Community's Seventh Framework Programme (2007-2013) under grant agreement n° 217280. Entitled 'Facilitating Alternative Agro-Food Networks' (FAAN), the project had five national teams, each linking an academic partner with a civil society organisation partner. The UK team consisted of the Open University and GeneWatch UK. Interview quotes from practitioners date from 2008-2009 unless otherwise noted. More results, especially a Europe-wide summary report (Karner, 2010), can be found at the FAAN project website www.faanweb.eu.

References

Allen, P., FitzSimmons, M., Goodman, M. and Warner, K., 2003. Shifting plates in the agrifood landscape: tectonics of alternative agrifood initiatives in California, Journal of Rural Studies 19: 61-75.

AMAS (Association of Manchester Allotment Societies), 2009. About AMAS, Association of Manchester Allotment Societies. Available at http://www.amas.org.uk/.

Bite, 2009. About us. Available at http://www.harp-project.org/projects/project_bite_about.php.

Brunori, G., 2007. Local food and alternative food networks: a communication perspective. Anthropology of Food. Available at http://aof.revues.org/index430.html.

Curry report, 2002. Farming and food: a sustainable future. policy commission on farming and food. Chaired by Sir Donald Curry. Cabinet Office, London, UK.

DEFRA (Department for Environment Food and Rural Affairs), 2003a. Local food – A snapshot of the sector: Report of the working group on local food. Department for Environment Food and Rural Affairs, London, UK. Available at http://www.defra.gov.uk/foodfarm/food/industry/regional/pdf/local-foods-report.pdf.

DEFRA, 2003b. Unlocking opportunities: lifting the lid on public sector food procurement. DEFRA, London, UK. Available at http://www.southwestfoodanddrink.org/uploads/documents/suppliers1.pdf.

DEFRA, 2006a. Sustainable Farming and food strategy: Forward Look. DEFRA, London, UK. Available at http://www.defra.gov.uk/foodfarm/policy/sustainfarmfood/documents/sffs-fwd-060718.pdf.

DEFRA, 2006b. Procuring the Future – sustainable procurement action plan: recommendations from the Sustainable Procurement Task Force. DEFRA, London, UK. Available at http://purchasing.ulster.ac.uk/procuring_the_future.pdf.

Dig, 2009. Our mission. Available at http://www.digfood.co.uk/.

Food Futures, 2007. A food strategy for Manchester. Available at http://www.foodfutures.info/site/images/stories/food%20futures%20strategy%202007.pdf.

Gershon, P., 2004. Releasing resources to the front line: independent review of public sector efficiency. Crown Copyright, London, UK, 66 pp.

Glebelands Market Garden, 2009. Urban farmers escape. Available at www.glebelandsmarketgarden.co.uk.

HeLF, 2007. Recipe for success. Report by Rob Squires. Avaiable at http://www.harp-project.org/news.php.

Hollinworth, S., 2006. Herbie: the lean green veggie machine. Enterprising: reporting on Manchester's social enterprise sector. Available at http://www.mpen.org.uk/web%20optimised%20pdf.pdf.

Ilbery, B. and Maye, D., 2005. Food supply chains and sustainability: evidence from specialist food producers in the Scottish/English borders. Land Use Policy 22: 331-344.

Karner, S. (Ed.), 2010. Local food systems in Europe: Case studies from five countries and what they imply for policy and practice. Available at www.faanweb.eu.

Kindling Trust, 2010. Visioning Manchester's sustainable food sector in 2020. Available at http://kindling.org.uk.

Kindling Trust, 2011. Seasonal School Soup for Starter. Available at http://kindling.org.uk/seasonal-school-soup-starter.

MACC (Manchester Alliance for Community Care), 2006. Funding for local food projects. Manchester Alliance for Community Care. Available at http://www.macc.org.uk/macc/downloads/health/FundingForLocalFoodProjects.pdf.

MERCi (Manchester Environmental Resource Centre initiative), 2009. About us. Manchester Environmental Resource Centre initiative. Available at http://www.bridge-5.org.

Morgan, K. and Sonnino, R., 2006. Empowering consumers: the creative procurement of school meals in Italy and the UK. International Journal of Consumer Studies 31: 19-25.

NWDA (North West Regional Development Agency), 2006. Strategic development support and business plan for burnley food links: case studies report. Compiled by Robert Soutar Ltd, North West Regional Development Agency. Available at http://www.macc.org.uk/macc/downloads/health/BurnleyFoodLinksCaseStudies.pdf.

Sustain, 2002. Sustainable food chains. Briefing paper 1: local food; benefits, obstacles and opportunities. Available at www.sustainweb.org.

Unicorn Co-Op, 2009. Who we are. Available at http://www.unicorn-grocery.co.uk/.

Chapter 19

Defining food co-ops

Martin Caraher and Georgia Machell
Centre for Food Policy, City University London, Northampton Square, London EC1V 0HB,
United Kingdom; m.caraher@city.ac.uk

Abstract

In the UK, the term food co-op is used to describe a range of food projects and initiatives. This chapter explores the current meaning of the term food co-op and presents original research that draws on data collected for the first phase of an evaluation for the Making Local Food Work Programme. Data for this chapter is based on ranking exercises completed by food co-op stakeholders in the UK as well as semi structured interviews with food co-op volunteers, organisers and customers. The research is part of the first stage of a larger impact evaluation of food co-ops. Different types of food co-op operations will be presented. These include locations in an urban church, a community centre, a primary school, a pub, and a market stall. This is a practical study that aims to analyse the range of benefits food co-ops can have as well as addressing the challenges.

Keywords: food project, alternative food network, evaluation

19.1 Introduction

Food co-ops deliver a range of functions for different demographics. To help define food co-ops it is necessary to explore the role and impacts that food co-ops make within the communities they operate within. The aim of this chapter is to explore how food co-ops respond to the needs of the communities they operate within, to help us come closer to defining the modern British food co-op and to look at the future role of food co-ops as an alternative food network in this age of austerity.

Over a decade ago McGlone *et al.* (1999) established through their research for the Rowntree Foundation, that 'food projects mean different things to different people' (p. 4). McGlone *et al.* scratched the surface of what is a complex and relevant issue. Today, there is still a lot of detail missing from the bigger picture of what food projects are, who uses them and why. It is important to note, that food projects are often aspects of alternative food chains that are responses to the dominant food chain. They do not address the status quo, but rather operates on its periphery (Whatmore *et al.*, 2003). By taking 'food co-ops' as one type of food project that appears to encompass a range of activities we will start to fill in the gaps and investigate the context in which this one type of food project develop and operate. The term 'food co-op' is used to describe a range of food projects including fruit and vegetable box or bag schemes, community owned food retail stores, social enterprise managed markets stalls, urban agriculture projects etc.

19.2 Literature review

Research and evaluation in the area of English food co-ops and buying groups is not extensive. The existing literature and evaluations focus on the potential health benefits of food co-ops (Caraher and Cowburn, 2004; Caraher and Dowler, 2007; Elliot *et al.*, 2006).

There is little research on the impact of food co-ops in the UK. Notably, Elliot *et al.* (2006) present an initial impact evaluation of the Welsh Assembly pilot food co-op project in Wales. Specifically Elliot *et al.* assessed the health impacts on the food co-op stakeholders as well as the impact on the communities they were operating in. The evaluation was limited to one type of co-op, a co-op that worked on a bag scheme from one centralised supplier where customers could order a bag of fruit or vegetables for £1 or £3 and collect it a week later. Findings indicate that food co-ops are most valued in areas of deprivation and emphasised that multi-agency partnership working was necessary to ensure a co-ops sustainability. Elliott *et al.* (2006) emphasise the challenges of conducting real life evaluation of these types of projects. They attribute this to the lack of internal monitoring and data collection by projects themselves. Towers *et al.* (2005) also emphasise the health benefits of food co-ops in their evaluation of the Food Co-operative Groups established by the Rural Regeneration Unit in Cumbria. Both evaluations (Elliot *et al.*, 2006; Towers *et al.*, 2005) evaluate only one model of co-op or community food project and thus cannot necessarily be extrapolated into the wider discussion of food co-ops in general.

The challenge of evaluating community food projects has been recognised (Caraher *et al.*, 2002; Dowler and Caraher, 2003; Freathy and Hare, 2004). Challenges can be attributed to varying food co-op models and inconsistent record keeping. The Caraher *et al.* 2001 study in Hastings, recognised that successful community food projects are often driven by an individual 'champion' who is dedicated and enthused to make the food project work. Their work concludes, that although food co-ops were aligned with the government 5-A-DAY programme, and research demonstrates the impact the Hastings food co-ops had in increasing local residents intake of fruit and vegetables, there was still need for further research into the features of food co-ops that make them sustainable. Freathy and Hare's (2004) work evaluating food co-ops in Scotland led to similar conclusions and began to develop a broad food co-op typology reflecting features of three different phases of food co-op development. The evaluation requirements of food projects are noted as not being as rigorous and relevant enough to have impactful consequences. Evaluations are too general and regarded more as a funding compliance requirement than a valuable asset to inform future project direction (Caraher and Cowburn, 2004).

19.3 Methodology

We undertook this research as part of the scoping phase of a larger impact evaluation of food co-ops supported by Making Local Food Work. The work is commissioned by Sustain: The Alliance for Better Food and Farming. Sustain suggest that social benefits, health benefits, economic benefits and environmental are the key areas where food co-ops make an impact. Using these four categories as a preliminary framework, ranking exercises were developed to gauge the impacts that stakeholders from nine food co-ops in London, the North East and Somerset recognised.

Five different food co-op types are presented as case studies, these include outlets from; a market stall, an urban church, a community centre, a primary school, a pub, to a secondary school. Specifically each outlet will highlight the partnerships necessary to sustain the food co-op and the main impacts as recognised by stakeholders. Methods used for collecting qualitative data involved a number of approaches including: semi-structured interviews, graffiti walls, partnership mapping exercises, ranking exercises and observation.

19.4 Findings from case studies

19.4.1 The Ferrier Estate Food Co-op

The Greenwich Community Food Co-op (GCFC) in South London is a social enterprise that manages 10 food co-ops in Greenwich. For this evaluation, we visited the Ferrier Estate Food Co-op. The Ferrier Estate food co-op is managed by two part-time members of staff and volunteers and is open one morning a week. The food co-ops were set-up in response to food access issues in South Greenwich. The GCFC also supports a number of healthy tuck-shops and is part of the wider Greenwich Community Food Initiative that supports cookery clubs, food growing projects and community cafes. The Ferrier Estate food co-op is a market stall where customers can choose the produce they desire. The Ferrier Estate food co-op sells a range of African produce such as plantain, yams and a range of chilli peppers. There was less emphasis on 'local' produce and more emphasis on central buying from wholesale markets to provide produce of cultural significance. The multi-cultural aspect of this co-op is also reflected in the volunteers – of fifteen volunteers they represent nine different nationalities. The food co-op coordinator reports that four volunteers have gone on to paid employment. From observations it is clear that the food co-op plays a significant cultural role on the estate.

19.4.2 St Andrews Food Co-op

St Andrews Food Co-op is run and set in a church in West London and is open one morning a week. It has a simple structure and is not linked to any other co-ops or community food projects. They buy their fruit and vegetables through a market stall operator at a local street market, at wholesale prices. Customers place an order a week in advance and have the option of a £3 mixed fruit selection and/or a £3 mixed vegetable selection. Volunteers from both the

Church and the local community organise the fruit and vegetable selection into selections that are placed on a table or pew for collection. The food co-op has roughly forty orders per week and customers are split with 50% from the church congregation and 50% from the local community. There were a high percentage of parents with young children, including Polish immigrants.

Semi-structured interviews with the food co-op coordinator highlighted the community outreach mission of the co-op and the use of food as a vehicle to bring people from the community together. The Church also operates a coffee morning that occurs at the same time as the food co-op. An important aspect of this food co-op was the social outlet. The most represented groups were elderly people and mothers with young children or infants. Semi structured interviews with volunteers and co-ordinators illuminated the cross-cutting impact the co-op was making: 'from the fruit and veg we are also then able to produce a very nice soup the following day for the lunch club, and again a lot of the mums and toddlers will make use of that at 12 o'clock on the Wednesday after the stories, nursery rhymes and so on'.

19.4.3 Food Chain North East England

Food Chain North East (FCNE) is in the north east of Engalnd and covers a wide area. It represents a streamlined approach to community food projects. FCNE is a social enterprise fruit and vegetable distribution service that provides bags or boxes of mixed fruit and vegetables to community food initiatives based out of schools and community centres across the North East region. FCNE have also developed their business to provide wholesale fruit and vegetable deliveries to community kitchens and local businesses. From visiting the food co-ops that FCNE support it is clear that FCNE is central to the success and conception of these food co-ops. FCNE are the champions of the food projects in the region. One food co-op coordinator suggested that FCNE were the food co-op and that the project she worked at was a 'community project'.

Many of the areas that FCNE are connecting with are former mining communities and have existing community infrastructures that are poorly resourced and supported. In many of the areas, food access is an issue and the co-ops are seen as offering value for money. FCNE provide strategic support, training and marketing materials to the community co-ops in the NE. Two out of the three co-ops in the NE visited for this study were based in and operated out of community centres. From the semi-structured interviews, ranking exercises and observations it is clear that community centres present opportunities for overlapping activities with the food co-op and has the benefit of full-time employees who can support the tasks associated with the food co-op.

A coordinator from one of the new co-ops indicated that the co-op had enabled them to learn about the community they were operating in and respond to these needs. The first week the co-op was in operation 'Premium Boxes' of vegetables had been delivered by mistake, they should have been Family Boxes. The Premium Boxes are 'premium' as they include more uncommon vegetables such as aubergine and butternut squash. The customers who received the Premium Boxes did not recognise much of produce and did not feel confident cooking it. As a consequence, community cooking classes for parents have been developed.

19.4.4 Froots, Roots and Shoots

This is a co-op located in the south west of the country in a county called Somerset. A dedicated team of nine-to-eleven year-olds at a rural primary school manages Froots, Roots and Shoots. Every Friday during morning break, the group divides into a finance team, customer services team, recipe team, produce checking team and bag packing team. The children collect orders for the following week are placed with the producer/distributer Somerset Organic Link and paid for a week in advance. The children develop skills and an acumen for business at an early age. One pupil in the finance team remarked that she liked her role in the finance team as it was a fun way to help practice maths. The older pupils (age 10-11) are responsible for placing orders, customer service, marketing and the financial aspects of the co-op. As the co-op is well established in the school, the younger children who are responsible for sorting and organising the produce into mixed bags are very enthusiastic about becoming old enough to take on the more managerial responsibilities of the older pupils thus ensuring continuity. The teacher supervising the co-op remarked that the only real responsibility she has, is to sign off on any orders to suppliers as the children are unable to do this as they are under eighteen. Once the mixed bags of fruit and vegetables are collated, customers come and pick them up from the school. Most of the customers are parents of current and former pupils and teachers.

19.4.5 Elderflowers

This co-op is also located in the county of Somerset. Elderflowers developed out of the Transition Town Langport movement and is a new co-op. The co-op operates once a week and is set-up as a shop in a back room of a local pub. Customers can come between 4 and 7 pm on Friday to choose from a range of organic and local where possible produce. The organisers of the co-op emphasised tha they wanted to make the co-op as much like a shop as possible and not a bag or box scheme. The products range from organic fruit and vegetables provide by SOL and dried goods are provided by Essential Trading Co-operative. Local users of the co-op often bring baked goods, jams etc. to sell.

19.5 Perceived benefits of co-ops

Figure 19.1 indicates that health benefits are the primary perceived impact from most co-ops. A striking finding from this chart is the different perceived food co-op impacts from rural and urban stakeholders. Those co-ops located in the largely rural country of Somerset indicated that social and environmental impacts are the biggest impacts that food co-ops make. Whereas, in London and the NE, there is more emphasis on the health and economic impacts.

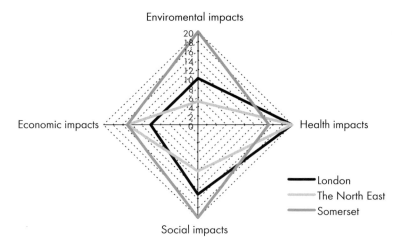

Figure 19.1. Comparison of issues from ranking surveys in three areas in the UK.

The results from the ranking exercise indicate that although co-ops in all three areas ranked health as the primary benefits of food co-ops, other responses varied. Both the North East and London emphasised the value of the social benefits of food co-ops, specifically the impact that food co-ops have on revitalising community facilities and engaging people in community activities. The perceived economic impacts were the most disparate across the three regions. The North Eastern Co-ops ranked 'Help support local producers, growers or other smaller or more ethical suppliers by providing an outlet for their goods'. London ranked 'offer new skills and work experience that could be used in other settings and possibly help them get paid employment' and Somerset ranked highest 'help ensure money spent stays in the local community'.

19.6 Discussion

McGlone *et al.*'s (1999) comment that food projects 'mean different things to different people' is clearly reflected in our first phase of research. The first scoping stage of our study provides valuable insight into the range of both benefits and priorities that food co-ops currently focus on. A common focus that is prevalent across food co-ops are health benefits and in particular increasing access to fresh produce, however beyond this commonality the focus and priorities of food co-ops differ and reflect the varied communities they operate within. Within the ethnically diverse communities – largely in urban areas such as London – co-ops played a crucial role in providing culturally specific produce to a wide range of African immigrants. Whereas in rural food co-ops the co-ops appeared to present an opportunity to connect with local economies and the environmental principles of organically farmed produce.

As McGlone *et al.* (1999) suggest that the term 'food project' is an umbrella term for a range of activities it appears that so is the term 'food co-op'. The first stage of research has led us to recognise the fluid and evolving definitions of the term co-op. Within the co-ops visited, the range of responses to the question 'what makes this operation a co-op?' suggests that the term 'food co-op' is an umbrella term for a range of food projects, while the term 'co-op' is used to describe a range of relationships within each project. Given the wide range of food co-op types, when discussing food co-ops in relation to strategy development, it would be of value to be able to define the type of co-op applicable to the situation and the features that would enable it to be sustainable.

Like the range of food co-op types, there are as many partnership types that keep the wheels turning in all of the food co-ops we visited. Partnerships vary depending on the priorities of each individual food co-op. For example in Somerset where there is a focus on organic produce, maintaining the relationship with the producer networks is fundamental to sustain the goals of the food co-ops. Whereas in the urban settings, low cost fresh food is a primary focus of the food co-ops therefore maintaining a partnership with the wholesale distributors is key. This may indicate a shift in the role of food co-ops in the last decade from largely health focussed responses to primarily low-income food access issues to wider community responses to issues affecting residents.

Funding is an area that demands further examination. The faith based food co-op which has the least financial dependance on outside funding also appears to be the most sustainable. This can in part be attributed to the existing infrastructure of the church in which it operates which is a captive and willing set of people prepared to volunteer provide a range of services linked to the Tuesday morning food co-op.

The sustainability of a co-op is founded in a range of attributes including: volunteers, venues, local completion and individual champions. In discussing sustainability, attention must be paid to strategy and questions need to be asked regarding who is responsible for developing strategies and what information is needed in order for strategies to be developed across the various types of food co-op in operation today. The case studies illustrate both project partners that are key in implementing a co-op model as well as project partners that support models that are developed by communities. Given the varying roles of project partners, there are also different grades of responsibility for project partners and projects. Beyond looking at whether or not the food co-ops themselves are sustainable are questions around whether or not the project partners are sustainable.

It is clear that in order for food co-ops to remain as an alternative food system the dominant mainstream food system would need to remain inaccessible to many or people would have to choose to shop at a food co-op. This is, unless the food co-ops could provide a service that the mainstream food system would be incapable of. The strengths of many of the food co-ops we visited and perhaps the key to their sustainability are the factors that mainstream food systems cannot provide, such as the social aspect of many food co-ops or the volunteer opportunities.

19.7 Conclusion

This phase of research has indicated that food co-ops that have some form of linkage to larger food projects respond to the needs of the communities they operate within in a variety of fashions. There continues to be gaps in the big picture of British food co-ops, however we are more informed of the diversity of food co-op types and food co-op impacts.

Defining food co-ops continues to be a challenge. The food co-ops visited are very different to the original food co-op model based on democratic decision making and equal ownership (Ronco, 1974) and this evolution is still largely unaccounted for. As opposed to defining a British food co-op, at this stage we have concluded it to be of more practical value to define the range of co-op types in operation today and to accept the term 'food co-op' as an umbrella term.

As for the future role of food co-ops, their range of impacts and emphasis on community participation makes them an appealing prospect to the new UK government. One of the partners in the current UK coalition government – the Conservatives – are strong supporters of food co-ops as a concept (Conservative Co-operative Movement, 2010) however the lack of definition in the field makes their support somewhat ambiguous, it is therefore necessary for the big picture of British food co-ops to become clearer before hailing them as a solution to issues that could be more effectively addressed through policy.

References

Caraher, M. and Cowburn, G., 2004. A survey of food projects in the English NHS regions and health action zones in 2001. Health Education Journal 63: 197-219.

Caraher, M. and Dowler, E., 2007. Food projects in London: Lessons for policy and practice--A hidden sector and the need for more unhealthy puddings... sometimes. Health education journal 66: 188.

Caraher, M., Dixon, P., Felton, M., South, L. and Tull, A., 2002. Making fruit and vegetables the east choice – Report of a five-a-day pilot project in Hastings and St Leonards. Hastings and St Leonards PCT, St Leonards-on-Sea, UK.

Conservative Co-operative Movement, 2010. Conservative Co-operative Movement. Available at http://www.conservativecoops.com/

Dowler, E. and Caraher, M. 2003. Local food projects: the new philanthropy? The Political Quarterly 74: 57-65.

Elliott, E., Parry, O. and Ashdown Lambert, J., 2006. Evaluation of community food co-ops pilot in Wales. Cardiff School of Social Sciences, Cardiff, Wales, UK.

Freathy, P. and Hare, C., 2004. Retailing in the voluntary sector: the role of the Scottish food co-operatives. European Journal of Marketing 38: 1562-1576.

McGlone, B., Dobson, B., Dowler, E. and Nelson, M., 1999. Food projects and how they work. Joseph Rowntree Foundation, York, UK.

Ronco, W., 1974. Food co-ops, an alternative to shopping in supermarkets. Beacon Press, Boston, MA, USA, 188 pp.

Towers, A., Nicholson, G. and Judd, P., 2005. Evaluation of the food co-operative groups established by the Rural Regeneration Unit. North West Food and Health Task Force and University of Central Lancashire, Cockermouth, West Cumbria, UK.

Whatmore, S., Stassart, P. and Renting, H., 2003. What's alternative about alternative food networks? Environment and Planning A 35: 389-391.

Chapter 20

Appetite for change: an exploration of attitudes towards dietary change in support of a sustainable food future

Anna Hawkins
Department of Architecture and Planning, Sheffield Hallam University, City Campus, Howard St., Sheffield S11WB, United Kingdom; a.hawkins@shu.ac.uk

Abstract

When considering how to plan for a more sustainable food future we need to decide whether we are willing and able to continue to grow and rear the food that we choose to eat now; or whether we need to plan for a gradual shift in consumer and dietary behaviour to support initiatives aimed at increasing food security and reducing the detrimental impact of food production on the environment and our health. This chapter considers the findings that arose from qualitative data gathered in 2010 through a number of focus groups in the City of Sheffield in Northern England. These groups explored levels of knowledge about environmental issues relating to food and ascertained the level to which people felt willing and able to consider dietary change in order to support a more sustainable food future. The picture that emerges from the focus groups is that there is an overall acceptance of and support for the need to change behaviour in order to reduce the environmental impact of modern lifestyles; but that this generally positive view is severely compromised by the belief that individual contributions are a 'drop in the ocean' and concerns about the agent of this change. Participants felt that they and others would adapt their consumer behaviour in response to changes in product availability and price; but felt less willing or able to make 'good' choices when faced with so many readily available 'bad' ones.

Keywords: behaviour change, food security, environmental diet

20.1 Background

The problematic nature of modern dietary choices is a subject occupying the minds of health and environmental academics and professionals alike (Chatham House Food Supply Project, 2008; Evans, 2009; Lang, 1999). Whether trying to tackle the obesity epidemic or address the UK's food security issues, there is an increasing acknowledgement of the existence of a collective problematic behaviour around food choices and consumption patterns (Evans, 2009; Marlow *et al.*, 2009; Steinfeld *et al.*, 2006). In simple terms, we consume more than we need and we make choices that damage our health; are increasingly acknowledged to have a negative environmental impact; and are reliant upon a supply chain that leaves us vulnerable to economic and climatic changes (Chatham House Food Supply Project, 2008).

Health professionals have, for many years, been developing strategies and interventions to improve dietary choices for health reasons – including a drive to increase consumption of fruit and vegetables to five portions a day. What can we learn from the public health sector about the barriers and motivations to this kind of dietary change upon which so many structural approaches to sustainable food production appear to be based?

The purpose of this chapter is not to engage in debate about the environmental impact of diet or the dietary changes necessary to improve public health; but rather to argue that the proposed 'solutions' to the problems of food security and environmental impact of diet, share many similarities with those of the problems of dietary change and health. Most people now have some level of understanding about what constitutes a healthy diet yet health professionals struggle improve overall levels of public health because there is a gap between what people know is good for them and what they choose to consume (Ashfield-Watt *et al.*, 2004). The same 'value action gap' has been shown to exist in relation to environmental behaviour (Kollmuss and Agyeman, 2002; Redclift and Benton, 1994; Thogersen and Olander, 2002).

We can make structural changes to our food supply system and enable the production of more environmentally and economically sustainable foods but if there is a gap between consumer choice and production, the impact of these interventions will be severely compromised (Chatham House Food Supply Project, 2008; Marlow *et al.*, 2009). In short, can we produce the food that people currently choose to consume in a more environmentally friendly and economically sustainable way but also in the quantities and price-range that consumers currently require and expect, or do we need to consider ways of changing consumer behaviour so that it is possible to develop a more sustainable food future?

This chapter will report on findings from a piece of qualitative research exploring levels of understanding about the environmental impact of dietary choices, and the willingness of participants to consider dietary change for environmental reasons.

20.2 The problematic diet

The dietary choices of UK consumers are in a constant state of flux as travel and immigration diversify our tastes and the availability of products (Lang, 1999). It is too easy to forget, therefore, that there is also a strong continuity of 'traditional' behaviours that underpin these processes of change and diversification (Pryer *et al.*, 2001; Wetherell *et al.*, 2003)

If we look at the dietary choices of poor urban industrial workers in the 19th century we see that their staple diet consisted of a starch base with whatever part of the animal they were able to afford – potatoes flavoured with bacon fat, for example. Without the means to produce their own food they did not waste scarce household funds on vegetables that did not carry a high calorific value and where meat was scarce the majority share would be given to the principal wage earner (Burnett, 1989). Whilst acknowledging the expansion in choice and sophistication of foods available in the UK it is still crucial to acknowledge that those with limited means and hungry mouths to feed will inevitably choose foods that are cheap, filling and tasty (Caraher *et al.*, 2010). Food producers, retailers and marketing professionals know this and supply products that are high in fat, sugar, processed carbohydrates and salt (Monteiro, 2009).

A period oft cited as a 'golden age' in terms of both dietary health and robustness of food supplying the UK, is WWII when popular support for the war effort enabled UK food policy to make sweeping and dramatic changes to dietary choices in support of an impressive increase in domestic food production. We ate very differently during the Second World War – principally the ratio of vegetable to animal foods changed dramatically (Burnett, 1989; Lang, 1999).

Global levels of animal food consumption are increasingly considered problematic. Livestock already dominates agricultural land use and is an intensive consumer of resources including water and a heavy contributor to pollution (Steinfeld *et al.*, 2006). Amongst the plethora of complex recommendations for improving the environmental impact of our diet; the message that what we choose to eat has as much, if not more, impact than simply the location and method of production, has become a little lost (Carlsson-Kanyama and Gonzales, 2009; Powles, 2009).

As a species, we are very protective of our right to eat meat. Meat continues to play an important part in social and cultural rituals and constructions of identity (Helms, 2004; Holm and Mohl, 2000; Lindeman and Sirelius, 2001). Any attempts to discuss changing patterns of animal food consumption must acknowledge this but should not shy away from the discussion for fear of alienating the audience. Animal food consumption is only one of the many factors that make up the big picture of contemporary dietary issues but it is such a significant contributor to the dietary problems both in terms of health and environmental impact, that it deserves higher prominence (Carlsson-Kanyama and Gonzales, 2009; Marlow *et al.*, 2009; Powles, 2009).

Many popular initiatives that form part of the groundswell in interest in food security focus upon the production of more food in the UK. But many of these initiatives such as community gardens, urban farms and allotment projects aim at enabling small scale fruit and vegetable production at a domestic and community level. Alongside this there is a heightened consumer interest in organic and 'locally produced' foods. It is notable, however, that access to these initiatives and products is restricted to those with the financial resources, time, social capital and skills to engage and benefit fully (Buckingham, 2005). In social, health and educational terms; small scale domestic production of fruit and vegetables on an allotment or in garden space is an entirely positive proposal (Buckingham, 2005; Milligan *et al.*, 2004). But arguably production of food on anything approaching the level of domestic consumption is beyond most people living in the UK in high-density urban areas with little or no private growing land, agricultural knowledge and the expectation to go out to work and raise a family.

It is important to recognise the tension that exists between projects aimed at increasing production of plant foods and the ongoing struggle to convince and enable consumers to eat their '5 a day' (Ashfield-Watt *et al.*, 2004) Simply increasing the proportion of fruit and vegetables produced in the UK will only help to improve our health and environmental impact if people feel willing and able to eat the produce.

20.3 Methodology

Four focus groups were recruited from established groups in the Sheffield area. These included a service user group from a homeless charity; an allotment collective and women attending mother and toddler groups. Around 75% of participants were women and participants ages ranged from early 20s to late 70s. Groups were purposively selected for either their perceived interest in the subject or for the value of their particular perspective.

A focus group approach was selected as the most appropriate method for gathering general perceptions, levels of knowledge and emotional responses to the issue of dietary choice and environmental behaviour as it creates an informal and dynamic environment for people to discuss the subject with as little opportunity for interviewer bias as possible (Ruane, 2005).

In order to establish some baseline information about food preferences, participants were asked to identify foods that they would associate with a special occasion, a guilty pleasure, a childhood memory and finally their least favourite food. The groups were then asked about their general attitudes towards environmental issues and whether they support the need to change lifestyles and how they felt about messages from official sources about the need to change behaviours.

The subject of food was then introduced and participants were placed in groups and asked to 'rank' 5 meals in terms of their environmental impact. The purpose of the exercise was not to see how close participants came to a 'correct' ranking but rather to capture their discussions and the rationale for the decisions the groups reached.

The final discussion point was specifically about perceptions and attitudes towards the subject of dietary change in response to environmental issues, drawing upon the results of the food ranking exercise.

20.4 Findings

The following represents a summary of findings from a piece of qualitative research into levels of knowledge about the environmental and economic implications of current dietary patterns and levels of perceived ability and willingness to make dietary changes in support of a more environmentally-friendly, and secure UK food supply.

The majority of participants expressed strong and positive memories about particular foods from their childhood. These tended to be dishes that their mothers had cooked such as shepherd's pie or foods associated with a special occasion – fish and chips on family holidays, for example. Most participants, when asked about food they would choose on a special occasion, selected a savoury dish often with meat as a key ingredient. Examples include several opting for steak or 'roast beef and all the trimmings'.

When asked about general attitudes towards environmental behaviour the vast majority of participants made immediate associations with recycling household waste. A significant number of respondents also mentioned buying environmentally friendly products such as organic food and low-energy consuming products. Many participants expressed a desire to be more environmentally friendly in their consumer choices but felt that cost was a significant barrier. There was also a strong sense from all groups that there was a lack of clear and consistent advice about how to lead a more environmentally friendly lifestyle which was exacerbated by a bombardment of media stories about food safety, climate change and resource depletion. This led to many participants disengaging with the debate or mistrusting official advice and guidelines. Participants also expressed frustration at the lack of infrastructure to support environmental behaviour such as easy access to recycling facilities.

When asked about their understanding of the environmental impact of food there was widespread association with packaging as a problem and a belief that cooking food from raw ingredients rather than relying upon processed meals would have positive environmental impact as well as health benefits. This was the first example of the participants making a positive association between 'healthy' and 'environmentally friendly' lifestyles. There wasn't any sense that these two issues were connected in any way; i.e. a meal wasn't seen as healthier because it was also more environmentally friendly or vice versa; but when participants were asked questions relating to the environmental impact of food their responses often related to the health and benefits of certain food choices.

Participants associated their more 'virtuous' consumer choices with both health and environment benefits; but the majority of participants stated that they made these choices because of the 'quality' of the food and not because of its perceived environmental impact.

An example of this would be organic meat and produce. Regardless of whether they were able or chose to purchase organic food; participants assumed or understood this to be a more environmentally friendly option but most would choose to purchase these products because they were seen as 'better for you'.

Around half of participants had some understanding of the impact of 'food miles' on the carbon footprint of foods and believed that buying local food would be a more environmentally friendly choice. When discussing the ranking of meals, for example, one participant said that they thought the roast beef; vegetables and gravy meal would be the most environmentally friendly because all the ingredients were 'British'. There was less understanding, however, about the difference between the transportation methods or the energy used to keep foods fresh in transit. Distance travelled was the key issue for most participants with a minority having more detailed knowledge of transportation issues. A small group of participants, for example, stated that they thought that a dish containing a large proportion of imported ingredients (Thai green chicken curry and rice) 'wasn't very ethical'. Upon further exploration, the use of the word 'ethical' in this context related to the distance the ingredients had to travel rather than factors such as animal welfare of working conditions/fair trade etc. This sense that transporting food long distances being problematic was reinforced by positive attitudes towards the notion of 'local' food production.

In the meal ranking exercise, for example, one of the groups ranked Thai green curry as being quite 'bad' for the environment because 'we can't produce rice or coconuts in this country' (female participant from Sheffield).

Only a small minority of participants had knowledge about other ways in which food production contributes to environmental degradation, energy consumption and resource depletion. There was very little overall knowledge about the relative impact of animal and plant foods although there was more widespread understanding of the depletion of fish stocks. 'I didn't really realise the impact of meat on the environment; I hadn't really been told much about it' (female participant from Sheffield).

When asked about willingness to change dietary behaviour to, for example, eat less meat or buy seasonally available produce that was locally produced; most participants expressed largely positive views or felt that they had already taken steps to change their dietary choices for other reasons such as an interest in the quality and provenance of food. The actual ability and impetus to change was considered, however, more problematic. Some participants felt that it was not worth them making sacrifices when the vast majority of consumers were not encouraged or forced to do so. Many participants were prepared to, and indeed already did, forsake meat on some days of the week but were unable to afford the more expensive products with associated 'environmental' benefits such as organic food. For a few participants meat was an essential component of at least one meal a day. 'If I made my partner a meat-free meal he'd say "what's happening?"' (female participant from Sheffield).

For others it remained an important status food but they would often cook meals without meat. A participant reflected upon the traditions of meat as a status food in the country he has recently left to settle in the UK. 'If you invite someone – you must bring some meat [to the table], if it is just [your] family it would be ok to have no meat' (male participant from Eritrea).

When asked, many participants expressed the view that consumers would cope with reduced choice (the example of the reducing of the availability of fresh fruits such as strawberries out of season was cited) but felt that making environmental choices within the current range of food available to consumers was problematic and undesirable. They also felt that the price of food would be an important factor in determining food choices and if certain foods such as animal foods or imported fresh foods were to increase significantly in price then overall levels of consumption would fall. 'I think if [chicken] gets much more expensive I'll have to stop buying as much of it, definitely' (female participant from Sheffield). And that, ultimately 'if it wasn't there you wouldn't buy it would you?' (male participant from Sheffield).

20.5 Discussion

The picture that emerges from the focus groups is that there is an overall acceptance of and support for the need to change behaviour in order to reduce the environmental impact of modern lifestyles; but that this generally positive view is severely compromised by the belief that individual contributions are a 'drop in the ocean' and that it is still a minority activity with minimal impact. It is interesting, however, that whilst this view was expressed in reference to food choices; the majority of participants engaged with activities such as recycling household waste. Possibly this is because this activity has moved from one which required pro-active engagement (driving to recycling plants, home composting etc.) to one which is embedded in normal household practices (centralised collection of recycling bins alongside normal household waste) (Diekmann and Preisendorfer, 2003: 441).

Strong associations were made between healthy and environmentally friendly food choices. The majority of participants associated environmentally friendly food choices as 'virtuous' and believed that choosing foods that were healthier or of a higher quality would also equate to more environmentally friendly food choices. This association was made strongly when talking about the relative impact of processed food versus cooking from fresh ingredients, buying fruit and vegetables, and organic meat and produce. This suggests that participants in this study were aware of and, crucially, supportive of the message that processed foods are a significant contributor to dietary problems (Ashfield-Watt *et al.*, 2004; Glanz *et al.*, 1994; Monteiro, 2009).

The need to increase fruit and vegetable consumption relative to animal foods such as red meat, fats and dairy (in high-fat forms such as butter, cheese and cream) and a move away from a reliance upon processed foods in favour of foods prepared from 'raw ingredients' have been clear and consistent messages from the health sector for many years (Ashfield-Watt *et al.*, 2004; Monteiro, 2009). These changes would also, if implemented on a large scale, support the changes that many argue need to take place to our consumption patterns if we are develop a more environmentally and economically just and robust food supply chain.

In health terms, dietary choice is openly acknowledged as a problematic behaviour that needs to be tackled in the same way as other problematic behaviours such as excessive alcohol consumption, substance misuse and smoking. The difference being, crucially, that unlike the other problem behaviours, moderate consumption is not only permissible but essential. Currently the responsibility for improving diets for health reasons rests largely with the individual who is expected to make 'good' dietary choices in terms of type and quantity of food. This is acknowledged to be extremely difficult when the availability of 'unhealthy' food is so dominant and relatively cheap when compared to healthier options (Caraher *et al.*, 2010). This issue of agency also emerged as a problematic issue in relation to making 'environmentally friendly' food choices.

This study has shown that although there was a desire among the groups to access 'better' products such as locally produced organic food, some people reported being unable to afford to make these changes. There was widespread support for the notion of 'local' food – with 'local' being understood as UK produced as well as grown or reared within the region of consumption and this could be significant in the move towards more locally produced foods. In one of the participating groups the participants engaged in small scale food production but for many participants in other groups this was not a practical option because of lack of accessible space, lack of time, caring responsibilities and a lack of interest in the activity.

There was a view that they and other consumers would respond to structural changes such as reduced availability or increased price more positively that simply being asked to make 'good' choices when confronted with a wide availability of 'bad' ones. The only noticeable exception to this was those participants who had already adopted what they considered to be an environmentally sustainable diet by growing some of their own food and only consuming a small amount of organic meat.

It could be argued that the findings of this study show some acceptance of the need for a collective change in behaviour but some reticence about the agent and process of this change. If food policy makers are serious about addressing the issue of food security and the environmental impact of current dietary trends alongside the ongoing and increasing need to improve public health; then evidence would suggest that simply informing people about, or providing them with, 'better' choices will have minimal impact (Diekmann and Preisendoerfer, 1992; Kollmuss and Agyeman, 2002; Sebanz and Knoblich, 2009) and that a more structured approach is needed to bring about meaningful behavioural change (Klesges *et al.*, 2004). It is important to recognise, however, that this is not a case of identifying a 'problem' group who need to change 'their' behaviour but rather a collective change in normal practices such as we, as a nation, have demonstrated that we are able to engage with when the stakes are high enough (Burnett, 1989; Lang, 1999).

What we know is that people struggle to change their dietary choices even in the face of consequences as immediate and tangible as their health and that of their families (Cade *et al.*, 2009). In light of this it is important that we do not repeat the mistakes that those in public health have subsequently learnt from when attempting to grapple with the issue of dietary choice for reasons of food security and environmental impact. Simply telling people what constitutes a 'good' choice is not enough to change behaviour – what is needed are interventions that will enable behaviour change to happen (Glasgow *et al.*, 2004).

If we want to reduce the damaging effects of modern dietary choices on the environment and our health then we need to ask ourselves whether we are willing and able to produce the food that we currently choose to consume at the levels we are accustomed to – or whether we need to consider how to change our collective dietary behaviour in order to support alternative models of food production in the future.

References

Ashfield-Watt, P.A., Welch, A.A., Day, N.E. and Bingham, S.A., 2004. Is 'five-a-day' an effective way of increasing fruit and vegetable. Public Health Nutrition 7: 257-261.

Buckingham, S., 2005, Women (re)construct the plot: the regen(d)eration of urban food growing. Area 37: 171-179.

Burnett, J., 1989. Plenty and want: A social history of food in England from 1815 to the present day (3rd ed.). Routledge, Oxon, UK, 355 pp.

Cade, J.E., Kirk, S.F., Nelson, P., Hollins, L., Deakin, T., Greenwood, D.C. and Harvey, E.L., 2009. Can peer educators influence healthy eating in people with diabetes? Results of a randomized controlled trial. Diabetic Medicine 26: 1048-1054.

Caraher, M., Lloyd, S., Lawton, J., Singh, G., Horsley, K. and Mussa, F., 2010. A tale of two cities: A study of access to food, lessons for public health practice. Health Education Journal 69: 200-210.

Carlsson-Kanyama, A. and Gonzales, A.D., 2009. Potential contributions of food consumption patterns to climate change. American Journal of Clinical Nutrition 89 (suppl): 1704S-1709S.

Chatham House Food Supply Project, 2008. Thinking about the future of food. Royal Institute of International Affairs, London, UK.

Diekmann, A. and Preisendorfer, P., 2003. Green and greenbacks: The behavioural effects of environmental attitudes in low-cost and high-cost situations. Rationality and Society 15: 441-472.

Evans, A., 2009. The feeding of the nine billion: global food security for the 21st century. Royal Institute of International Affairs, Chatham House, London, UK, 59 pp.

Glanz, K., Patterson, R.E., Kristal, A.R., DiClemente, C.C., Heimendinger, J., Linnan, L. and McLerran, D.F., 1994. Stages of change in adopting healthy diets: fat, fiber, and correlates of nutrient intake. Health Education and Behaviour 21: 499-519.

Glasgow, R.E., Klesgas, L.M., Dzewaltowski, D.A., Bull, S.S. and Estabrooks, P., 2004. The future of health behaviour change research: What is needed to improve translation of research into health promotion practice? Annals of Behavioural Medicine 27: 3-12.

Helms, M., 2004, Food sustainability, food security and the environment. British Food Journal 106: 380-387.

Holm, I. and Mohl, M., 2000. The role of meat in everyday food culture: an analysis of an interview study in Copenhagen. Appetite 34: 277-283.

Howe, J., 2002. Planning for urban food: the experience of two UK cities. Planning Practice and Research 17: 125-144.

Klesges, L.M., Glasgow, R.F., Dzewaltowski, D.A., Bull, S.S. and Estabrooks, P., 2004. The future if health behaviour change research: what is needed to improve translation of research into health promotion practice? Annals of Behavioural Medicine 27: 4-12.

Kollmuss, A. and Agyeman, J., 2002. Mind the gap: why do people act environmentally and what are the barriers to pro-environmental behaviour? Environmental Education Research 8: 239-260.

Lang, T., 1999. The complexities of globalization: the UK as a case study of tensions within the food system and the challenge to food policy. Agriculture and Human Values 16: 169-185.

Lindeman, M. and Sirelius, M., 2001. Food choice ideologies: the modern manifestations of normative and humanist views of the world. Appetite 37: 175-184.

Marlow, H.J., Hayes, W.K., Soret, S., Carter, R.L., Schwab, E.R. and Sabate, J., 2009. Diet and the environment: does what you eat matter? American Society for Nutrition 89(suppl): 1699s-1703s.

Milligan, C., Gatrell, A. and Bingley, A., 2004. Cultivating health: therapeutic landscapes and older people in northern England. Social Science and Medicine 58: 1781-1793.

Monteiro, C.A., 2009. Nutrition and health. The issue is not food, nor nutrients, so much as processing. Public Health Nutrition 12: 729-731.

Powles, J., 2009. Commentary: Why diets need to change to avert harm from global warming. International Journal of Epidemiology 38: 1141-1142.

Pryer, J.A., Nichols, R., Elliott, P., Thakrar, B. and Marmot, M., 2001. Dietary patterns among a random sample of British adults. Journal of Epidemiology and Community Health 55: 29-37.

Redclift, M. and Benton, T., 1994. Social theory and the global environment. Routledge, London, UK, 280 pp.

Ruane, J.M., 2005. Essentials of research methods: a guide to social science research. Blackwell, Oxford, UK, 239 pp.

Sebanz, N. and Knoblich, G., 2009. Jumping on the ecological bandwagon? Mind the gap! European Journal of Social Psychology 39: 1230-1233.

Steinfeld, H., Gerber, P., Wassenaar, T., Castel, V., Rosales, M. and De Haan, C., 2006. Livestock's long shadow: environmental issues and opinions. Food and Agriculture Organization of the United Nations, FAO, Rome, Italy.

Thogersen, J. and Olander, F., 2002. Human values and the emergence of a sustainable consumption pattern: A panel study. Journal of Economic Psychology 23: 605-630.

Wetherell, C., Tregear, A. and Allinson, J., 2003. In search of the concerned consumer: UK public perceptions of food, farming and buying local. Journal of Rural Studies 19: 233-244.

Part 3. Urban agriculture

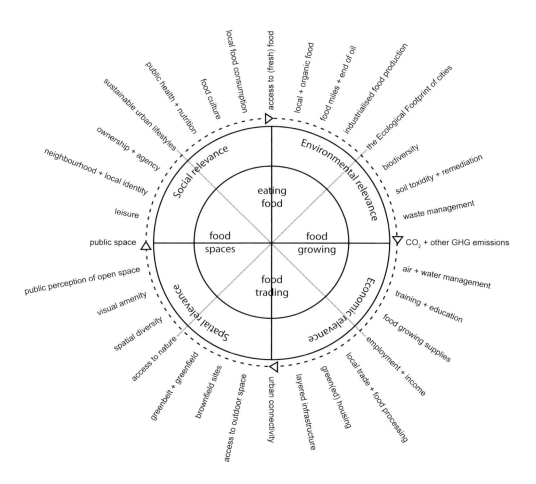

Chapter 21

Urban agriculture in developed economies

Jan Willem van der Schans and Johannes S.C. Wiskerke
Rural Sociology Group, Wageningen University, Hollandseweg 1, 6706 KN Wageningen, the Netherlands; jan-willem.vanderschans@wur.nl

21.1 Introduction

In the first chapter of her book The economy of cities, Jacobs (1969) presents the thought provoking idea that rural economies and agricultural work are directly built upon city economies and city work, rather than the other way round. The more conventional view holds that cities are built on the surplus production created by a rural agricultural base: agriculture first, cities later. Jacobs argues that throughout history rural hinterlands have benefitted from inputs from the city, in fact she suggests that agriculture itself may have originated in cities right from the start: cities first, agriculture later (*ibid.* p. 17). Early forms of agriculture evolved in and around small settlements of hunting gathering people, who used seeds that they gathered and life animals that they hunted for barter trade in natural treasures like glass and stones to make knifes, spearheads and other tools and arms. It is trade rather than agriculture that made these settlements thrive in the first place, agriculture only developed when the stocks of seeds and life animals that were kept for barter trade started to mutate and intermingle and eventually developed into some form of more or less controlled food production. In any case, the history of agriculture and urbanism is intricately intertwined, one cannot exist without the other, a notion that became increasingly obsolete however over the last couple of centuries when food production and city expansion decoupled and each seemed to develop rather independently of the other (Steel, 2008).

21.2 The urban-rural divide

In the preindustrial world, a city's size was limited by the productive capacity of its rural hinterland (*ibid.* p. 70). The exception were cities located on rivers or the seaside, allowing food (in particular grain) to be sailed in from afar. Taking transportation as the limiting factor, Von Thunen analysed that land prices closer to cities rise as costs to bring the food into the city decrease. Highly perishable, voluminous or heavy crops, which are by their nature difficult to transport, will be grown close to the city. A pattern of land use emerged with highly intensive agriculture close to the city and less intensive agriculture further away (in descending proximity to the city one would find market gardens, dairy, forestry, arable land; see Figure 21.1). This pattern of land use existed in the western world before industrialism and it may still exist in some parts of the developing world.

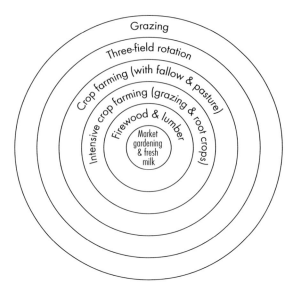

Figure 21.1. The Von Thunen model: land use in the isolated state (Sinclair, 1967).

Gradually agriculture industrialised, technology developed to preserve perishable products, and with the introduction of railways transport over land became much more efficient. As a result, Sinclair (1967) observed, the Von Thunen model did not work anymore, may be even the reverse was true (Figure 21.2).

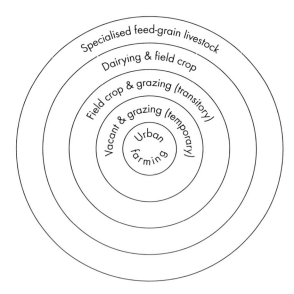

Figure 21.2. Land uses around expanding metropolitan area (Sinclair, 1967).

The further away from the city, the more intensive agricultural land use will be. Expecting the expanding city to take over their land, farmers close to cities seem reluctant to intensify their farm. Further away from the city, farmers are more certain to keep their farm, hence invest more deeply to stay up to date and competitive. Transport costs are almost constant for commodities and relatively small anyway. Closer to a city the structure of farmland is more fragmented, nuisance laws are more restrictive, jobs in the city are more profitable, the agricultural service industry has left, and hence farming tends to become a small scale, extensive, part time occupation. More intensive types of farming leave the peri-urban area and re-locate close to large logistical hubs (auctions) and processing facilities (grain elevators, dairy factories, slaughter houses) which themselves are placed further away from cities (Morgan *et al.*, 2006). A complete segregation of agriculture and urban development emerges, quite often even enshrined in physical planning theory and practice (single use zoning and straight lines of separation in order to reduce mutual interaction).

As it is today, there is rising interest among urban dwellers to re-connect with the origins of their food production, either by growing food themselves or by creating direct links with those who grow their food (Sonnino, 2009; Wiskerke, 2009). Urban planners are increasingly trying to incorporate these concerns in their planning theory and practice (Bohn and Viljoen, 2005). In fact the idea of 'agrarian urbanism', a systematic rethinking of urban form conceived through the spatial, ecological, and infrastructural implications of agricultural production, never completely disappeared from the 20th century agenda of urban planning: from Frank Lloyd Wright's Broadacre City, via Ludwig Hilberseimer's 'New Regional Pattern' to Branzi's Agronica (Waldheim, 2010). And it will return more prominently on the urban planning agenda in the 21st century, as for example new urbanist Andres Duany announced in 2008, when he presented his plan to develop an 'agricultural community' on a 528-acre farm site near Vancouver, British Columbia. Southlands, as the project is called, seeks to integrate agriculture and urbanism at all levels, from high-density units with window boxes to medium-sized farms, it is an as-yet-unbuilt town of 2,000 housing units on one-third of the site's acreage while tripling the value of the land's agricultural production (http://www.imaginesouthlands.ca/).

21.3 Urban agriculture: a definition

Urban agriculture as an entrepreneurial activity receives more and more attention, not only in the developing world (Smit *et al.*, 1996) but quite interestingly also in the developed world (Kaufman and Bailkey, 2000). Urban agriculture can be defined as:

> *an industry that produces, processes and markets food and fuel, largely in response to the daily demand of consumers within a town, city, or metropolis, on land and water dispersed throughout the urban and peri-urban area, applying intensive production methods, using and reusing natural resources and urban wastes, to yield a diversity of crops and livestock* (Smit *et al.*, 1996).

There are several aspects to this definition, which are of relevance when we study urban agriculture in a western context. Urban agriculture is an *industry*, an economic activity professionally carried in response to consumer demand. It can be debated whether hobby type of food growing activities such as allotment gardens, community gardens, window gardens and balcony gardens strictly speaking fall under this definition. The point however is that studying these activities from the perspective of urban 'agriculture' rather than 'gardening' accords them a more prominent place in the urban food system and indicates their potential to produce food complementary to urban and peri-urban food growing activities carried out by professional farmers. To look at urban dwellers as farmers rather than gardeners allows seeing them as co-producers rather than just as consumers of food. This opens up the possibility for a set of new relationships between urban dwellers and rural farmers, such as for example it may re-introduce forms of solidarity based on like-mindedness rather than interdependence, reversing the Durkheimian shift from mechanical to organic solidarity that took place in the process of industrialisation and division of labour. To characterise urban agriculture as an industry rather than a past time also signals the possibility that it is carried out intensively, even if on small plots in urban areas. A typical example is SPIN farming, a market driven system of Small Plot INtensive farming developed by Wally Satzewich and Gail Vandersteen in Saskatoon, Canada (http://spinfarming.com/). They grow food in underused backyards and earn a modest income on multi location sub-acre plots of land.

Urban agriculture is located in *urban and peri-urban* (rather than rural) areas. Exactly where the urban area ends and the rural starts is a matter of definition, and it also depends on the context (in Canada local food is within a 100 miles radius, in the Netherlands 40 km or even 25 km is more often taken as yardstick). The point is that agriculture and urban development are increasingly integrated, whereas the dominant trend has been that they were increasingly segregated (as noted above). Already in 1989 Heimlich showed that agriculture in the US (re-) appeared in metropolitan areas, adapting itself successfully to the urbanising environment through the working of smaller farms, more intensive production, a focus on high-value crops and livestock, and greater off-farm employment (Heimlich, 1989). Olson coined the term Metro Farm for this phenomenon (Olson, 1994). Metro farms are often rather small farms in and close to urban areas which take full advantage of their closeness to urban markets, which exploit weakness of the conventional food system and are focussing on adding value rather than cutting costs as is traditionally the case. On this account, attention is drawn to the positive rather than negative spill-over effects of agricultural activities in urban areas, and vice versa. A good example is peri-urban farm *Hof van Twello*, close to the town of Deventer, the Netherlands, where Gert Jan Jansen focusses on a differentiation strategy growing forgotten vegetables, medieval varieties and vegetables for the ethnic market, all kinds of produce not found in a regular supermarket (http://www.hofvantwello.nl/). He also takes advantage of being located close to the city by engaging city dwellers semi-professionally in production, harvesting and processing activities, compensating them partly in kind and partly through the revenues of on farm sales (Van der Schans, 2010).

Urban agriculture is directed to *local markets,* rather than export oriented. It is to be distinguished in this respect from 'metropolitan agriculture', or 'agriculture in the network society' as defined more recently by Smeets and others (Van Altvorst *et al.*, 2011). 'Agriculture in the network society' is a spatially clustered complex of agrifunctions, including highly sophisticated plant and animal production systems, and using industrial and urban waste flows, in order to compete on the world market (Smeets, 2011). Urban agriculture is by definition about short supply chains. Short supply chains provide a competitive advantage when they involve highly perishable produce (leafy greens, soft fruits, wheat sprouts, etc.), where conventional supply chains typically select for a standard set of products and varieties that resist long distances of transportation. Short supply chains are also particularly suited to transfer services, such as social health care to physically or mentally disabled, or joyful experiences, such as the amenities of an urban farm in an otherwise asphalted environment (Van der Schans, 2011).

The last aspect of the definition of urban agriculture to be taken into account is that it is using urban *waste and underused resources*. The proximity of agriculture to urban populations allows it to benefit from a (re-)closure of material and energy loops (water, nutrients, carbon, etc.). While this aspect of recycling urban wastes is central to urban agriculture in the developing world, it is less prominent in the developed world (Refsgaard *et al.*, 2006). Urban waste management systems in the western world have historically been developed with a view to sanitation standards and with little concern for recycling. As a result nutrients and organic matter are emitted to rivers, precipitated in sludge or, even incinerated. A good example, where urban flows are recycled rather than wasted, is Growing Power in Milwaukee, USA (http://www.growingpower.org/). Here urban farmer Will Allen and his staff collect solid organic urban waste to turn into compost, and re-use nutrients in wastewater from fish farming by cultivating plants in water, a system altogether called aquaponics. Another way in which urban agriculture benefits from the urban environment is by using vacant plots of land and/or other underused urban spaces such as rooftops, or abandoned buildings. In urban agriculture temporary use of space and multifunctional use of space are the norm rather than the exception, in this way urban agriculture ideally avoids competition with other urban activities, a battle it would likely lose anyway.

21.4 Urban food strategies

Cities all around the world are exploring the possibility to reconnect food production (agriculture) and urban life, either by helping to re-establish the link between city dwellers and peri-urban farms or by allowing or creating spaces for food production within the city limits (Sonnino, 2009). These places may involve private allotment gardens, school gardens or community gardens, but there is also growing interest in city farms, run by professional farmers, sometimes in cooperation with dedicated groups of citizens.

To understand the rising interest in urban food production in the western world it is vital to study the social and economic context in which these initiatives take place. The issue of access to good and affordable food is important in certain cities, such as when supermarkets and groceries avoid low income areas (food deserts; Furey *et al.*, 2001), but it cannot be the whole story, as in many western cities fresh food is readily available at every street corner but still initiatives to grow your own food are flourishing (a case in point being the city of Amsterdam). It is clear therefore that urban agriculture also addresses other urban problems, such as redressing a deterioration of public (green) space, overcoming the social exclusion of certain people (with limited access to the regular labour market), creating awareness about healthy lifestyles, or reducing urban environmental problems such as storm water floods or the heat island effect (elevated temperature when too much urban space is covered with buildings and roads).

If urban agriculture can provide so many benefits to urban development the question remains why there still are so few examples in practice of food production properly planned in and around cities as a systematic approach to (re-)built greener, more sustainable metropoles. This is not to say that there are not many inspiring examples, some will be reported in the next chapters, but a systematic approach where food production is quite naturally and quite substantially incorporated in the (peri-)urban landscape, is still hard to come by. One possible answer for this gap between theory and practice may be that when it comes to western cities, it is still not yet clear to what extend urban agriculture can provide a more sustainable answer than conventional agriculture. The current food system where agriculture and city life are comfortably separated may have some cracks but it has not fallen apart yet. And it is re-inventing itself continuously to address societal concerns. Occasionally there may be an issue about food insecurity, such as in London where a strike by truck drivers, a volcano outburst causing dust clouds, or the threat of a terrorist attack may actually derail the logistics of urban food supply ('Nine meals from anarchy', Boycott, 2008). But most western cities continue to depend for a large share on long supply chains, sourcing globally rather than locally. Hence, there is an urgent need for studies that identify the direct and indirect benefits of urban farming (Sonnino, 2009).

One possible way to map out the policy dimensions of urban agriculture is the model developed by RUAF, based primarily on experience gained in the developing world (Figure 21.3) (Dubbeling *et al.*, 2010). This model may also be applicable -with some adaptations- to the developed world (see Figure 21.4).

For example, under the social dimension one could imagine types of urban agriculture such as community gardens, school gardens, food bank production gardens, and social care farms. Social policy goals may include celebrating cultural diversity, providing a network of conviviality to urban dwellers, attracting families with children to certain neighbourhoods, educating a healthy life style to children, providing meaningful activities for disabled people, etc. Under the ecological dimension, one could imagine green roofs to combat the urban heat island effect, storm water containment gardens, urban food waste reduction programmes, and energy crops on urban wastelands. Ecological policy goals may include climate mitigation and adaption, increased biodiversity, and carbon sequestration. Under the economic

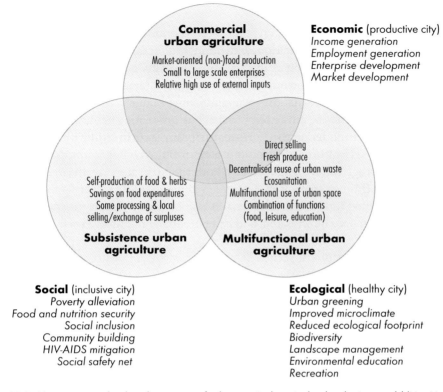

Figure 21.3. Main types and policy dimensions of urban agriculture in the developing world (Van Veenhuizen, 2006).

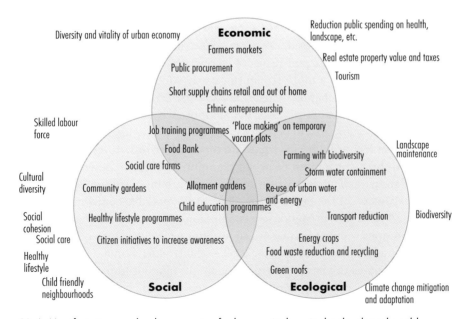

Figure 21.4. Manifestations and policy aspects of urban agriculture in the developed world.

dimension one could imagine urban farming as an economically viable urban profession, landscape maintenance in combination with food production, ethnic entrepreneurship, 'place making' in (re-)development zones, farmers markets, public procurement policies, etc. Economic policy goals may include the reproduction of (artisanal) food production and processing skills, the reduction of landscape maintenance budgets, the increase in real estate prices and tax revenues, and the attraction of a thriving urban food culture for tourists (these adaptations to the RUAF model are partly based on a brainstorm in 2011 with members of the Think tank on urban agriculture in Rotterdam, a platform of civil servants active to explore the potential of urban agriculture in and around Rotterdam). In any case, the model illustrates that looking at cities through the lens of the food system provides an overview of a variety of policy challenges related to modern city life, and also offers a welcome perspective on how various policy measures are interrelated, and can potentially re-enforce each other.

21.5 Outline of urban agriculture section

The series of chapters presented here therefore all deal with the impact of urban agriculture on city life. Some chapters discuss the effect of urban agriculture on the economy of cities; others try to map its effects on urban ecology or social structures. The first chapter by Caputo precisely argues that urban agriculture will only have a bigger impact on urban food supply in the western world if it becomes economically viable. He argues that most initiatives until now mainly focus on social issues, and then offers a few examples in the UK of initiatives taking a more commercial approach. Next, Niwa discusses the prevalence of agriculture in Tokyo, a non-western city in the developed world, as an economically vibrant, urban activity, which is able to survive despite the high competition for space. She discusses the historic roots of urban agriculture in Tokyo but also its ability to re-invent itself in modern urban expressions and shapes. The chapter by Giorda discusses the rise of grass roots urban agriculture projects in Detroit as a response to the city's decay when economic crisis hit the car industry. Giorda assesses the potential of the current projects to rewrite the storyline of the city's future, and signals the emergence of a more straightforward economically motivated initiative in Detroit. The next couple of chapters cluster around the ecological performance of urban agriculture initiatives. Moreau *et al.* describe an approach developed for Canadian local authorities which looks at urban and peri-urban agriculture to reduce and offset greenhouse gas emissions. They identify climate smart urban and peri-urban agricultural practices which will need to be incorporated in urban planning in order to be effective. Jansma *et al.* take as a case study the Dutch city of Almere where urban agriculture will be integrated in the planned expansion of the city, and they look at the possible contribution of the local production of food to reduce greenhouse gas emissions. They find that factors not related to local production per se must be taken into account as well, such as other eating habits (less meat) and other shopping habits (take a bike or introduce home delivery). In another chapter, Denny compares the environmental impact of different tomato varieties and year round sourcing options, including tomatoes produced in semi commercial or commercial urban agriculture. She concludes that local production can be the lowest CO_2 emitting option, especially when people buy tomatoes in season and avoid transport by car to grow or purchase the tomatoes. The last cluster of chapters addresses the social aspects of urban agriculture. Wiltshire and Geoghegan study voluntary urban food production in Britain and argue that a better understanding of its (different modes of) social organisation may indeed be critical to further expansion of urban agriculture. Allotment gardens and community gardens each have their own logic, but both modes will have to be accommodated in order to maximise the supply of unpaid labour. Tornaghi focusses on the emergence of various forms of collective urban agriculture in Leeds, and argues that the ability of local and regional planning organisations to flexibly respond to these new demands for public space is still limited. Only if the control of public urban space is brought back into the realm of democratic decision making, can we move beyond residual use and ad hoc negotiation, and make a structural claim to the commons. Finally Solomon describes her action research introducing structural formats of urban agriculture into under-used green public space in Amsterdam and The Hague. She argues that cultural sector urban agriculture projects bring in innovation and experimentation at a time when the development of practical and resilient formats for urban agriculture are so urgently needed.

21.6 Discussion

If we look at the chapters it is clear that we are only beginning to see systematic evaluations of urban agriculture as compared to conventional agriculture, most notably with respect to ecological performance (Jansma *et al.*, Denny and to some extend also Morreau *et al.*). The evaluations presented here mainly focus on GHG emissions and do not fully take into account other environmental aspects such as land use, biodiversity, water use and water containment, soil fertility, landscape quality, *and* consumption related issues such as product freshness in relation to shelf life, and participation in outdoor activity in relation to a healthy lifestyle. Current evaluations also do not yet take into account other, more innovative forms of agriculture, which may be better suited to urban and peri-urban environments than mainstream agriculture or conventional organic agriculture (which assimilated many of mainstream agricultures production practices and logic, such as a heavy reliance on mechanisation and monocultures). Alternative forms of agriculture include 'closed loop farming', i.e. farming systems in various forms which re-use urban wastes. An example is the Chinese age old system of farming in and close to cities using human excreta and garbage in fertilising the soil in order to balance the heavy drain of a close succession of crops (King, 1911). Other forms of agriculture are being explored as well in a (peri-)urban context, such as 'ecoagriculture' (Scherr *et al.*, 2007), i.e. methods to grow food and support biodiversity and promote ecosystem health at the same time (no till farming, agroforestry, integrated pest management, etc.). There is also need to study the ecological (and social-economic) performance of innovative forms of rooftop farming, indoor farming, and perhaps even vertical farming, especially when these forms of urban agriculture move beyond the drawing table stage. Monitoring programmes, even if they involve experimental set ups, are to be welcomed. The studies presented in this volume more or less take for granted the food choices people make, urban agriculture advocates often hypothesise however that people change their consumption pattern when they grow their own food or have a direct relation with somebody growing their food. Urban farmers explore alternative varieties of crops, to extend the local season, to replace exotic imports, or to substitute meat with legumes. Studies to substantiate the effectiveness of urban agriculture in comparison with other policy instruments to bring about these changes in urban food consumption should be welcomed.

Another issue requiring further study is the business model that (peri-) urban farms use to compete with the conventional agrifood system (which generally has lower production costs), and also to compete with other urban activities (which generally have higher revenues). Many initiatives in the field of urban agriculture are, as Caputo signals, in fact supported for their social, ecological or even artistic qualities, but lack a sound economic underpinning. A study that specifically looked at this aspect in the US also found that only a handful of entrepreneurial urban agriculture projects were beginning to show some profits. More of them were providing a variety of other social, aesthetic, health, and community-building and empowerment benefits (Kaufman and Bailkey, 2000). It should be noted that it is hard enough for conventional farmers already to earn a decent income with growing food, an aspect of the modern food system precisely requiring attention in the first place (Wiskerke, 2009). If we focus on peri-urban farmers, there has been a steady growth of farmers who explore alternative development strategies notably differentiation and diversification (Van der Ploeg *et al.*, 2002), which may inspire city farmers as well (Van der Schans, 2010). The urban farmers of Tokyo, studied by Niwa, offer beautiful examples of differentiation (e.g. capture more value by producing high quality fresh vegetables and selling directly to end consumers) and of diversification (e.g. offer amateur farmers a place to grow food and relax). The level of professionalism may be attributed, as Niwa explains, to the support provided by JA cooperatives (Central Union of Agricultural Co-operatives) operating in urban and peri-urban areas and playing a major role in farm guidance, marketing of farm products, supplies of production inputs, credit and mutual insurance. The JA cooperatives also lobby hard to get the multifunctional values of Japanese agriculture recognised (http://www.zenchu-ja.or.jp/eng/index.html). To strengthen their economic base, urban farmers should actively look for ecological and social services that their farms provide, for which urban parties may be willing to pay, either because the value of their own business operation enhances, or their costs are reduced. More research is needed to capitalise the value of services provided by urban farming (e.g. place making for urban renewal, storm water containment, urban heat island reduction, park management, healthy life style promotion). More research is also needed to identify the institutional mechanisms required to allow the party who invests to capture the benefits spilled over to others (e.g. urban farms on temporary vacant plots play a role in urban district re-development, but they hardly ever benefit from the real estate value increase that they bring about). Urban farming can be a way to make urban green spaces more attractive, as Solomon proves with Urbaniahoeve, but it is still undecided whether this can be marketed as a service to be paid from local government park management budgets or social housing corporations' neighbourhood improvement budgets. Urban farming can be an agent in urban revitalisation, as Giorda illustrates in Detroit. But it is still open for debate which forms of urban agriculture are needed (socially driven or economically driven) and also what effects are desired on the redevelopment of the city as a whole.

Finally, urban farming as a social phenomenon engages city dwellers in food production thereby redefining their social role in the food system. Urban farming as a do it yourself activity has gained in popularity, inspired by social movements like Transition Towns, but also life style magazines increasingly pay attention to backyard, balcony, rooftop or window gardening. Gardening can be a social activity as Wiltshire and Geoghegan show, but also Tornaghi and Solomon. People work together on a common task, they share knowledge and harvests, and they may even actively try to re-establish the link between city dwellers and professional farmers from around the city. For example, Growing Communities started as a box scheme for farmers close to London, but it later developed also in a productive community, largely consisting of volunteers, and growing specific crops on two or three inner city locations (http://www.growingcommunities.org/). Urban gardeners and peri-urban farmers, in this case, are not competing but rather complementing each other. Solomon suggests that artists play a role in experimenting with new forms of urban agriculture which may expand the range of working models available for urban farmers trying to earn an income. It should be noted however that projects positioned as social or artistic often fall under a different regulatory regime than projects classified as economic. Land use conditions, licensing, food safety rules, etc. all change dramatically when a social initiative turns into an economic one. Allotment gardens, for example, often forbid people to sell the crops for an income, in the Netherlands allotment gardens are a recreational activity (not even a productive one, let alone a commercially productive). A powerful metaphor for the social relations between people growing food in public urban spaces, is that of being a commoner, or at least of being a person who reclaims the right to the commons (both Wiltshire and Tornaghi explicitly refer to this social and political concept of the commons). The commons in this line of thought is the land not yet developed, not yet appropriated, which therefore belongs to everybody, or at least to those member of a certain community (respectively the non-exclusive and the exclusive version of the commons; Van der Schans, 2001). More research is needed to study which rules govern the social behaviour of people together growing food in public spaces, which socialisation processes take place, and also what kind of effects these processes have on a person's wider social capabilities. And of course these effects would have to be evaluated against the effects that other social programmes may have.

21.7 Von Thunen revisited

To conclude this introduction to the urban agriculture section of this book, it must be noted that more recently the relation between the urban and the rural may have structurally changed in many western cities, in that the pressure to urbanise agricultural land has decreased. This is so because of the growing popularity of urban design concepts such as 'smart growth' (http://www.smartgrowth.org/) and 'new urbanism' (http://www.newurbanism.org/) which advocate diverse, compact cities and walkable neighbourhoods. These concepts are based on ideas of Jane Jacobs about the vibrancy of mixed use urban districts (Jacobs, 1961) and James Howard Kunstler about the adverse effects of urban sprawl (Kunstler, 1994). Cities adopting these planning strategies should go for infill and redevelopment rather than urban expansion, and they should protect farmland whenever possible (Beasley, 2009). There is a need to explore how urban high density and urban food production can be integrated and urban agricultural lifestyles can be accommodated (Duany, 2011).

The pressure to urbanisation of farmland has also decreased due to the financial crisis and the real estate crisis. Plots remain vacant much longer than expected; temporary use is welcomed in order to prevent further deterioration. Today real estate developers invite city dwellers to grow food on their property, for example in 2008 the development company ONNI turning a vacant site at Seymour and Pacific streets in the heart of Yaletown, Vancouver, into a community garden. Seventy-nine plots were made available on a first-come first-served basis to community groups and residents to grow food on a temporary basis, until the site is developed (Groc, 2008). Compare this to a situation two years earlier, where in order to develop their property real estate developers bulldozed away the South Central Farm, an urban farm and community garden located at East 41st and South Alameda Streets in an industrial area of South Los Angeles, California (http://www.thegardenmovie.com/).[35]

A trend break also seems to take place in urban expansion into peri-urban farmland. According to the National Bureau of Statistics in the Netherlands in 2009 municipalities lost 414 million euro on their land conversion activities, whereas in the year before they still earned more than 600 million euro with converting farmland into development sites (Van Rijn, 2011). Cities like Rotterdam are inviting farmers back on the land which they have bought from them years ago, city expansion will not take place in the near future and perhaps it will never take place anymore, considering the demographic trend of a stable or even decreasing population. But if cities do not expand anymore, this will also reverse the mechanism that farmers close to and inside cities fear to be displaced by urban development, the mechanism identified by Sinclair, as described earlier. Intensifying their farming operation may be rational and profitable, given the proximity to urban markets and urban resources yet unexploited. We are only beginning to see what this means for the development of agriculture in general and urban agriculture in particular. When urbanisation pressure decreases, farmers close to and inside cities can focus again on what they are good at, which is growing fresh, high quality produce, thereby turning urban wastes into food, and offering easily accessible landscape experiences at the same time. This is the Von Thunen model revisited. Welcome to the metropolitan food landscape of the 21st century.

---

[35] It should be noted that real estate companies inviting community gardens on their vacant urban plots can reclassify their property as 'park' or 'garden' in this way reducing their property tax expense. This has generated some debate whether promoting urban agriculture is a strategy to green wash one's image and dodge taxes to weather the real estate crisis, or a genuine effort to redevelop city districts by involving local communities and facilitating organic economic growth (http://www.nowpublic.com/world/those-community-gardens-are-fertile-tax-dodges). More research is needed to evaluate these new alliances between agriculture and urbanism.

References

Beasley, L., 2009. Smart growth in Rotterdam: Considerations from a Vancouver perspective. Report Guest Urban Critic 2009, Van der Leeuwkring Rotterdam, the Netherlands.

Bohn, K. and Viljoen, A., 2005. Continuous productive urban landscapes designing urban agriculture for sustainable cities. Architectural Press, Oxford, UK, 280 pp.

Boycott, R., 2008. Nine meals from anarchy – how Britain is facing a very real food crisis. Daily Mail June 7th 2008.

Duany, A., 2011. Garden cities, theory and practice of agrarian urbanism. The Prince's Foundation for the Built Environment, London, UK, 86 pp.

Dubbeling, M., De Zeeuw, H. and Van Veenhuizen, R., 2010. Cities, poverty and food: multi-stakeholder policy and planning in urban agriculture. RUAF Foundation, Practical Action Publishing, Leusden, the Netherlands, 192 pp.

Furey, S., Strugnell, Ch. and McIlveen, H., 2001. An investigation of the potential existence of "food deserts" in rural and urban areas of Northern Ireland. Agriculture and Human Values 18: 447-457.

Groc, I., 2008. Urban farming grows on Vancouver, Granville. June 03, 2008, available at http://www.granvilleonline.ca/gr/vancouver/2008/06/03/urban-farming-grows-vancouver.

Heimlich, R.E., 1989. Metropolitan agriculture: farming in the city's shadow. Journal of the American Planning Association 55: 457-466.

Jacobs, J., 1961.The death and life of great American cities. Vintage Books, New York, NY, USA, 458 pp.

Jacobs, J., 1969. The economy of cities. Random House, New York, NY, USA, 268 pp.

Kaufman, J. and Bailkey, M., 2000. Farming inside cities: entrepreneurial urban agriculture in the United States. Lincoln Institute of Land Policy, Cambridge, MA, USA, 120 pp.

King F.H., 1911. Farmers of forty centuries, or permanent agriculture in China, Korea, and Japan. Madison, Wisconsin, USA, 441 pp.

Kunstler, J.H., 1994. The geography of nowhere: the rise and decline of America's man-made landscape. Touchstone, New York, NY, USA, 304 pp.

Morgan, K., Marsden, T. and Murdoch, J., 2006. Worlds of food: place, power and provenance in the food chain. Oxford University Press, Oxford, UK, 225 pp.

Olson, M., 1994. Metro farm: the guide to growing for big profit on a small parcel of land. TS Books, Santa Cruz, CA, USA, 576 pp.

Refsgaard, K., Jenssen, P.D. and Magid, J., 2006. Possibilities for closing the urban-rural nutrient cycles. In: Halberg, N., Alrøe, H.F., Knudsen, M.T. and Kristensen, E.S. (Eds.) Global development of organic agriculture: challenges and prospects. CaB International, Oxfordshire, UK, pp. 181-214.

Scherr, S.J. and McNeely, J.A., 2007. Farming with nature: the science and practice of ecoagriculture. Island Press, Washington DC, USA, 445 pp.

Sinclair, R., 1967. Von Thünen and Urban Sprawl. Annals of the Association of American Geographers 57: 72-87.

Smeets, P.J.A.M., 2011. Expedition agroparks: research by design into sustainable development and agriculture in network society. Wageningen Academic Publishers, Wageningen, the Netherlands, 319 pp.

Smit, J., Ratta, A. and Nasr, J., 1996. Urban agriculture: food, jobs and sustainable cities. UNDP, New York, NY, USA, 302 pp.

Sonnino, R., 2009. Feeding the city: towards a new research and planning agenda. International Planning Studies 14: 425-436.

Steel, C., 2008. Hungry city: How food shapes our lives. Random House, London, UK, 383 pp.

Van Altvorst, A.-C., Andeweg, K., Eweg, R., Van Latesteijn, H., Mager, S. and Spaans, L., 2011. Sustainable agricultural entrepreneurship: The six guises of the successful agricultural entrepreneur. Transforum, Zoetermeer, the Netherlands, 180 pp.

Van der Ploeg, J.D., Long, A. and Banks, J., 2002. Living countrysides, rural development processes in Europe, the state of the art. Elsevier Bedrijfsinformatie BV, Doetinchem, the Netherlands, 231 pp.

Van der Schans, J.W., 2001. Governance of marine resources: Conceptual clarifications and two case studies. Eburon, Delft, the Netherlands, 468 pp.

Van der Schans, J.W., 2010. Urban agriculture in the Netherlands. Urban Agriculture Magazine 24: 40-42.

Van der Schans, J.W., 2011. Agrarian urbanism, de nieuwe Utopia? Een bedrijfskundige kijk op stadslandbouw. In: P. de Graaf (Ed.) Ruimte voor stadslandbouw in Rotterdam, Eetbaar Rotterdam. Paul de Graaf Ontwerp en Onderzoek, Rotterdam, the Netherlands, pp. 50-53.

Van Rijn, M.J., 2011. Gemeenten leden in 2009 verlies op bouwgrond. Centraal Bureau voor Statistiek, webmagazine. Available at http://www.cbs.nl/nl-NL/menu/themas/overheid-politiek/publicaties/artikelen/archief/2011/2011-3402-wm.htm.

Van Veenhuizen, R., 2006. Profitability and sustainability of urban and peri-urban agriculture. Food and Agricultural Organization, Rome, Italy, 108 pp.

Waldheim, Ch., 2010. Notes toward a history of agrarian urbanism. Available at http://places.designobserver.com/feature/notes-toward-a-history-of-agrarian-urbanism/15518/.

Wiskerke, J.S.C., 2009. On places lost and places regained: reflections on the alternative food geography and sustainable regional development. International Planning Studies 14: 361-379.

Chapter 22

The purpose of urban food production in developed countries

Silvio Caputo
Department of the Built Environment, Coventry University, Priory Street, Coventry CV1 5FB, United Kingdom; silvio.caputo@coventry.ac.uk

Abstract

Food security is now on the UK government agenda. The perspective of fossil fuel scarcity limiting movements of supply, and recent food price rises constitute a warning sign of the unsustainability of current global food systems. The UK government has produced a series of papers on food production, in which food security, food access, and a more environmentally safe food chain are indicated as important objectives. Urban agriculture is not part of the possible strategies mentioned, which aims at diversifying supply sources, foster healthy and affordable food consumption, and reduce emissions. Given the massive carbon footprint of cities, food production in an urban context ought to be considered as a contributor to a low carbon society, as it can deliver crops with zero food miles, possibly accessible to low income groups. However, for this to happen, urban agriculture needs to become economically viable. Research is needed to establish modalities for highly productive food grown in cities, and in particular, business models based on a network of small and medium sized cultivated spaces, which utilise local urban resources (water, compost, sun). This chapter makes the case for a new role for urban agriculture in developed countries, which is centred not only on environmental and social benefits, but also on economic opportunities. It does so by reviewing the national definition of food security, the role that non-profit organisations play in promoting urban agricultural practices, and the business model on which they are based.

Keywords: urban agriculture, food security, food systems

22.1 Introduction

Urban agriculture is quite difficult to define (Garnett, 1999) since it can take many forms and serve multiple purposes. Mougeot (2006) argues that the production of food in an urban context is not new. Depending on the historical moment and the context, it has been practiced with a vast variety of techniques, connecting with space, ecology, and economy of cities. He gives the following broad definition: 'in very general terms, urban agriculture can be described as the growing, processing, and distribution of food and non food plant and tree crops and the raising of livestock, directly for the urban market, both within and on the fringe of an urban area.' In the last few decades urban agricultural practices, both in developing and developed countries, have been the subject of academic and public interest. Research has been produced which documents experiments and practices undertaken. Some of the related benefits include: health, educational, environmental, social, and economic (Garnett, 1999; Fairholm, 1999; Martin and Marsden, 1999). However, depending on the context, priorities and importance of those benefits vary. In developing countries the economic benefits are generally related to secure food to the poorest share of the urban population (Mougeot, 2006). In 1996 United Nations Food and Agricultural Organisation defined the concept of food security as follows: 'Food security exists when all people, at all times, have access to sufficient, safe and nutritious food to meet their dietary needs and food preferences for an active and healthy life' (FAO, 1996).

The role of programmes such as Agropolis has demonstrated that urban agriculture can help fighting hunger and poverty in cities (Redwood, 2009) by providing low income groups with subsistence food cultivated within the urban precincts. An indicator of the urgency of alimentary issues in developing countries is perhaps given by the percentage of income spent on food by low-income groups in Kinshasa (60%), in Bangkok (60%), or in La Florida, Chile (50%) (Akinbamijo *et al.*, 2002). The urban configuration of the metropolises of the South often offers abundant interstitial, unregulated spaces for crops, as the division between urban and rural is blurred (Mougeot, 2006), and planning enforcement is not efficient.

In developed countries the rationale for urban agriculture seems to be different. In UK the average household expenditure on food is 9%, with the poorest 10% spending as much as 15% (The Strategy Unit, 2008). Poverty is still at high levels (Brandolini and Cipollone, 1992; The Poverty Site, 2010), but to be sure safety nets for the disadvantaged are still in place (Bourque, 2005; Pothukuchi and Kaufman, 1999). The cost of food may be raising but it is still affordable to most, although healthy food (organic, non processed food stuff) is expensive and its importance to human health still not completely acknowledged. Consequently food debate tends to be focussed on quality rather than availability. In this picture the objectives for urban agriculture are primarily environmental and social, namely: preserving biodiversity, tackling waste and reducing energy utilised by food systems (Viljoen *et al.*, 2005). The reconnection of urban dwellers with the food culture it is also important to improve the quality of diets which, as an average, is not balanced (Barling *et al.*, 2008; Food Standards Agency, 2010). Finally urban food can be a catalyst for social amelioration. Growing food can connect minority groups (Mougeot, 2006), or become a tool for community building (Garnett, 1999) or give opportunities to population 'at risk' such as homeless (Bourque, 2005).

It is clearly not possible to generalise and divide urban conditions in two distinct, clear-cut geo-political areas: developed and developing. For example the number of households growing food in Moscow rose from 20% in 1970 to 65% in 1999, suggesting that in some developed countries where 'unemployment is endemic', during hard times, urban agricultural practices could be part of a survival strategy (Deelstra and Girardet, 2005). However, in developing countries, wherever urban agriculture is practiced, it can represent a measure of food security (Mougeot, 2006) and an opportunity to create business ventures and employment (Redwood, 2009). Conversely, in developed countries the contribution to food security, creation of new jobs and economic growth seems to be marginal. Still, the current food crisis is having an impact on food security policies worldwide, and the role of urban agriculture is likely to become increasingly important in mature economies too (Sonnino, 2009). Thus some question arise: can urban agriculture realistically contribute to the national food system? Is there any advantage for this to happen from a perspective of national food security? And if so, are there organisational forms that can better link small, diffused urban farmers with food systems, within market-driven economies? This chapter briefly reviews the current UK food security policy, as well as the organisational forms of groups focusing their activity on alternative food production and/or supply, to then argue that new business models are necessary for urban agriculture to become a consistent contributor to food production. It is not the objective of this chapter to investigate on the technicalities of business models that can spur urban agricultural production, but rather to offer an initial contribution to the understanding of the elements that hinder or can favour the expansion of the role of urban agriculture in developed countries.

22.2 UK food security policy

The environmental impact of food production and consumption, and the risks related to geopolitical uncertainties have informed a series of governmental reports (see DEFRA, 2006, 2008; Foresight, 2011; The Strategy Unit, 2008). The UK food chain accounts for the 22% of all GHG emissions (HM Government, 2010). The unexpected rise of food prices in June 2008, apparently connected with cost of oil and US policies on biofuel, causing a 3% rise in food inflation over the previous three months (Soil Association, 2009), has exposed the fragilities of the global food system. It would seem sensible to consider in any strategy for national food security future uncertainties related to political factors, and fuel and resource scarcity, given the dependence of UK on imported inputs such as oil and gas, fertiliser, feed and machinery (Barling *et al.*, 2008). In official records England produces roughly 60% of the food consumed (Barling *et al.*, 2008). Such figures are somehow misleading as they are based on the commercial value of food rather than either volume or calorific levels (Soil Association, 2010b).

FAO (2010a) estimates that in order to meet food demands, global food production should rise 70% by 2070. DEFRA's consultation paper 'Ensuring the UK's Food Security in a Changing World' draws a picture in which at a global level the increase in production should happen mainly in Africa and Asia. This has to be managed sustainably through technological innovation, increased efficiency in the exploitation of resources, and possibly the use of GM crops. Climate Change is going to have a strong impact on developing countries. It is thus necessary to support them by raising awareness, and harness the market in a way that can stimulate an increased output (DEFRA, 2008). At a national level, the interest towards local food has fostered a business of farmer markets, 'veg boxes' and 'pick-it-yourself' that generates £2 billion a year (DEFRA, 2008). The consultation paper recognises this as an asset for the UK economy, but as the national production already accounts for 60% of the food consumed, the best strategy for food security is to ensure an internal economic stability which would prevent inflation on food prices with a consequent rise in alimentary expenditure for low income groups (DEFRA, 2008).

Although the report acknowledges the risk of instabilities related to energy supply, it seems to over-rely on a positive evolution of the economic and environmental situation of the countries that most contribute to the global food system, apparently ignoring political and structural conditions that can hinder the capacity to overcome exploitation of resources and climate change effects. It seems also to rely excessively on a stable economic situation which would in turn keep food accessible to all, and on innovation which would allow minimisation of energy use in agricultural practices. Still, many factors seem to fuel global instability, and to augment uncertainties related to global supply chains. For example, FAO (2010b) flags up the necessity to increase financial support and knowledge transfer to developing countries in the light of climate change consequences, which can have a disastrous impact on agriculture. Furthermore, the International Energy Agency (2009) suggests that the rising oil demand will impact on economies because of the related environmental damage.

In this perspective, the case for production of food in proximity to places of consumption is not only a way to respond to the increasing demand for local food but also a strategy which can build resilience in the national food chain. Clearly urban agriculture could have an important role in this. Nevertheless, it is not considered as a contributor in national policies (Howe, 2010). Resistance to identify advantages of a diffused food micro-production is understandable for an economy based on traditional industrial logic. Furthermore there is little bottom-up push that can trigger alternative pathways to food security. This last factor may be linked to the difficulty of urban dwellers in developed countries to perceive the food chain as vulnerable in this moment of history (Pothukuchi and Kaufman, 2000).

Pothukuchi and Kaufman (1999) describe the food system in cities as invisible to people. The reasons are: availability taken for granted, invisibility of food chains, production perceived as a rural issue, and lack of interest of policy makers in this matter. Thus the perception of its importance is diminished, reinforced by the cheap availability of some of the foodstuff. It is precisely this altered perception that must be challenged for an alternative and resilient food production, and the important role that urban agriculture can play in it, to become a priority. This is not an easy task as the 'visibility' of food comes with its absence. Furthermore at a governmental level, indicators for food security do not include climate change, soil fertility, water, and other environmental factors (Barling *et al.*, 2008), making it difficult for policies to consider disruption that depletion of resources could bring to national food supply systems.

22.3 The practice of urban agriculture in the UK

Research suggests that in the UK there is research suggests that there is little awareness among planners of benefits associated with urban agriculture (Pothukuchi and Kaufman, 2000; Howe, 2010). This may be a consequence of lack of political will (Martin and Marsden, 1999), demonstrated by the absence of directives regarding spaces and modalities for urban food production. In planning directives the provision of open spaces, including allotments, is indicated as a necessity for the quality of urban space condition (DCLG, 2006), but there is no specific consideration formulated about spaces and modalities for growing food. Nevertheless, concerns about food and the progressive disconnection of urban dwellers from it are leading local authorities to consider modalities that go against this grain. For example the 'London Food Strategy' (LDA, 2006) recognises the importance of locally grown food as an alternative system that can oppose widespread consumption of unhealthy, inexpensive food available through large scale distribution. It also recognises the important role spatial planning can have in making space available for such practice. The Camden Borough in London includes food growing as one of the priorities to make healthy food available in a situation of economic downturn and 'growing concern about food and health' (Borough of Camden, 2009). Conversely in some London Boroughs, Bexley for example, allotment sites that remain underutilised or unused for years are considered for possible commercial alternative use (Mbiba, 2003). Haringey Council in spite of current demand outnumbering the offer, does not believe the option of making space for new allotments to be realistic (Borough of Haringey, 2006), although the 1908 Small Holdings and Allotments Act places 'a duty on local authorities to provide sufficient allotments to meet demand'.

The blurring of the traditional divide between rural and urban, with all that entails, is witnessed in academic research and through a multitude of flourishing individual or community initiatives, from the joyous to the controversial, to the socially engaged. For example Barton *et al.* (2003), in their urban design guide for sustainable neighbourhoods, suggests including in Local Plans a requirement for allocating allotments in brown-field regenerations over a certain net density. Thomas Sieverts (2003) urges policy to integrate agricultural practices in a new urban landscape that merges cultural, social and productive functions. At a different level, movements such as the Transition Network and Guerrilla Gardening explore on the grounds the integration of the natural and the artificial, and test alternative lifestyles which rely on locality and self-reliance.

The organisations that are exclusively focused on food and food production are usually based on a non-profit model or charity. Some of the business models that are currently tested are:
- Community Supported Agriculture (CSA): 'a mutual commitment between a farm and a community' which takes many forms: from customer supported box schemes, to rent or adopt schemes, to urban food growing projects (Soil Association, 2009).
- Food Hubs: 'a hub is an intermediary led by the vision of one or a small number of individuals which by pooling together producers or consumers adds value to the exchange of goods and promotes the development of a local supply chain' (Horrell *et al.*, 2010).
- Cropshare: a scheme through which surplus from allotments is sold (Local Food Link, 2010).

These organisations are essential to spread urban agricultural practices, as they engage people on food matters at a collective level. Rather than dwell on the pleasure of growing food for the family in single allotments, each individual participating in community action is made aware of the role that food plays in society. However, the initial research of the author suggests that these models are very valuable when it comes to creating participation in community issues, offering an alternative to mainstream food distribution, and creating some opportunities for volunteering and part time work. Conversely, with regards to expanding urban agricultural practices and, to an extent, contribute to national food production, they succeed much less.

Groups working in the food sector are mushrooming. The Soil Association has identified more than 100 CSA's with the most diverse organisational forms (Soil Association, 2010b). Most of them rely on direct agreements between a community and a group of farmers. Many of them are not strictly based on urban food production but rather on offering an alternative food chain to local communities by supporting small or medium farmers in the region, and organising bulk purchase from retailers or growers. A few of them, in addition to these activities, grow also crops on urban land. This is the case of Growing Communities, one of the most famous CSA groups based in the London Borough of Hackney, which is growing salad on its patchwork farm (on urban land) as well as collecting produce from local farmers (Growing Communities, 2010). To a diversity of forms of organisation corresponds an overarching shared concern for the same objectives, namely: community building, healthy lifestyle, alternatives to current food systems, and a strong ethical attitude to social life. Urban food production is only one of the components of their programmes. Most of these organisations have a strong focus on social objectives, but it would seem that there is a need for these to be complemented with additional businesses promoting and expanding food production capacity exclusively in an urban context.

This is not easy, considering the difficulties related to urban agricultural practices in developed countries. For example some of factors that can hinder such practices could be:
- availability of non contaminated urban land;
- difficulties in connecting small growers in a network of producers;
- regulation preventing sale of produce coming from allotments;
- complexities related to an efficient organisation of non-profit or profit enterprise.

The list is by no means exhaustive. However, the initial literature research documented in this chapter, suggests an overall tendency of enterprises offering an alternative food production and distribution to shy away from ambitious goals of production, let alone on urban land. Thus, advantages that economy of scale can offer, and attempts to connect small growers under the umbrella of large consortia or cooperatives are still largely unexplored.

22.4 Business models to support urban agriculture expansion

Some of the elements that can hinder or favour the expansion of the role of urban agriculture in developed countries can be: the invisibility of the food systems, the privileging of social (rather than economic) advantages, and, at national level, food-security policies over-reliant on global market forces as opposed to local self-sufficiency, and more. The model of organisation (and business) adopted by the groups working in this area, it is argued herein, could be one of such obstacles. Business models that can spur urban agricultural production may vary according to local conditions, or local demand and offer, or opportunities that existing enterprises can offer. Any of such models, however, need to be able to produce enough to meet demand, represent an attractive alternative to mainstream food offer, and provide economic self-sufficiency to the organisation adopting it.

The Soil Association (2010a) maintains that at present the market for organic food in UK accounts for £1.84 billion. In spite of economic recession and negative growth of the organic food market in 2009, expectations for 2010, and for following years, are positive. However, independent retailers represent only 26% of the market, with 'veg boxes' only 8.4%. More importantly, statistics suggest that 33% of the total amount spent comes from low income group customers. What is encouraging is that independent retailers have a good share of the market, organic food is sought after across the social spectrum, and land used for organic production is increasing (Soil Association, 2010a). This can be taken as an indicator of a persistent and growing public interest in alternatives to mainstream production and distribution. Urban agricultural practices need not to engage with strict organic standards, but they can offer produce that has a significantly lower environmental impact and can meet the growing demand for food with such qualities. In a dynamic market, there is space for growth for those enterprises that can organise their structures effectively, and meet the needs of customers.

Debate about business models that can support urban agriculture is on-going. Some of these models, although not specifically designed for urban food growers but rather at small rural farmers, are nevertheless proposing structures which could be adapted for such purpose. For instance, the Soil Association has developed a toolkit to inform and facilitate the start of small commercial enterprises in agriculture, stressing the importance for businesses to make a profit (Pilley *et al.*, 2005). Although the toolkit is not specifically directed to urban food growers, it could provide guidance since it is addressed to small-scale production. The East Anglia Food Link, an NGO at present focusing on food security, delivered a report in which the case is made for a London food depot that can gather produce from small growers, including urban farmers, recognising the importance of providing easy access to market for surplus production (Saltmarsh, 2009). Sustain, Local Action on Food, and the London Food Link have recently organised an event called 'Getting Down to Business' to disseminate experience in this regard. Ben Reynolds, a speaker from Sustain, maintained that in order to become an alternative to current food system, urban agriculture need to set targets and define its contribution in terms of production (speech at Sustain conference: Getting down to business, London, 25 June 2010). Nevertheless, case studies illustrated by speakers were exclusively related to small scale organisations, which, as one of the organisers mentioned at the close of the event, were engaging mainly with the ethical dimension of food production and supply, not with maximisation. In this vision ethics clashes with maximised production, that needs to be pursued in order to raise the contribution to the national food production.

It is therefore unsurprising if at present alternative urban food enterprises are largely not commercially viable, rely on grants funding or volunteers (Garnett, 2005), and do not prioritise efficient management (Horrell *et al.*, 2010). Organisations struggle with financial aspects and stay small or disappear, as their dependence on funding for survival makes them vulnerable to external and unstable factors such as political will (for the permanence of funding streams) or general economic situation (the permanence of private donations). The risk for this multitude of associations and businesses is to remain a niche phenomenon if an economic contribution to a wider level is not pursued. This would require different models of business organisation, and the recognition that urban agriculture can deliver social, environmental and economic benefits at several scales and at several levels: local and global, at a community level and at an urban or national level.

The non-profit, charity model may not be suitable to such ambitious targets. There is an inherent 'in-built ceiling to the reach and effectiveness' of such organisations (Yunus, 2007) as they constantly rely on grants and donors. Social business can provide a structure that can liaise with existing financial structures and operate to a national level. Yunus, the founder of Grameen Bank, maintains that social business utilises organisational structures similar to profit making businesses although it is finalised 'not to achieve limited personal gain but to pursue specific social goals' (Yunus, 2007). The attainment of social objectives can be effectively accomplished through private companies' organisational logic, provided profits are exclusively reinvested in the organisation and benefits are distributed to the community to which the enterprise belongs. Further advantages related to this model are: commercial viability, independence from institutional or private support, and creation of job opportunities.

At present, there are some initiatives that are going in this direction. For example, Food from the Sky (2010, see http://www.foodfromthesky.org.uk) is an organisation operating in London which is moving towards constitution as a social enterprise. Located on the rooftop of a supermarket, it has established a symbiotic relationship with its host by growing food that is sold below through the commercial enterprise. At present, the organisation uses volunteers to carry out growing and harvesting activities. However, one of its objectives is to becoming financially independent by expanding growing capacity and related revenues. Urban food production, together with a range of educational activities which include workshops for gardening and activities with schools, should produce sufficient income to achieve independence of the enterprise. The particular relationship with the supermarket allows volunteers to exploit managerial skills of the commercial enterprise. Directors of the supermarket, in turn, are aware of the 'marketing tool' that Food from the Sky represents. Through the organisation they connect with the community, sell 'zero food-miles' food, and boost their environmental credentials. This symbiosis has been made possible partly by a managerial vision that attempts to move beyond the traditional role of the supermarket to explore a new model, inverting some of the preconceptions of the retail system. This vision has led to other innovative actions involving staff as well as customers, that further reflect the values being proposed by the roof-top activities: for example, staff members, many of whom are from the Bangladeshi community, are involved in food cultivation and encouraged to bring their own cultural identities to the growing space above; the supermarket also runs a no-plastic bag policy, encouraging customers to bring their own bag and offering a recycled bag to customers should they have forgotten to bring one. Existing commercial models can offer a blueprint for social businesses and cooperatives to develop in larger enterprises. For instance, organic boxes door-to-door distributors rely already on a wide distribution and sometimes solid profits. Synergies could be pursued between cooperatives of micro urban farmers and commercial enterprises.

Garnett suggests that if a reasonable percentage of the green land within the London area had to be used to grow horticultural crops and fruit, the estimated production could contribute 18% of Londoners intake (Garnett, 2005). Using estimates given by the Soil Association (Soil Association, 2010b), food produced on that land by skilled growers could provide a suitable income for more than 14,000 people. Although purely hypothetical such figures give a rough idea of the potential of urban agricultural. Harnessing this potential through commercial structures shaped around the social business model would possibly create incentives for small growers, but also could generate the evidence required to draw the attention of central and local governments to urban food production.

22.5 Conclusions

Urban agriculture is gathering consensus in developed countries. There is acknowledgement of the role that it can play in fostering biodiversity, in reutilising organic waste, and in contributing to reduce energy used in industrial food production. Indeed, there are many other benefits that can be associated with this practice such as improving health of urban dwellers, contributing to social integration, contributing to a positive microclimate and more. Yet, unlike for developing countries, the contribution that urban food growing can deliver to food security is hardly considered, since food security is not widely perceived as a priority. At a governmental level food security is regarded as an important issue, and as such is carefully evaluated. However, this chapter argues that such an evaluation is excessively reliant on a positive evolution of global food production, in the face of warning signs given by increased food prices, and progressive environmental and resource depletion. In this light urban agriculture should be encouraged as a source of production close to consumption (therefore not energy intensive), and as a resource that could contribute to the resilience of current national food systems, excessively depending on unstable global food systems.

In parallel to such a recognition, urban agricultural practices should become economically viable and self-sufficient as well as increasingly productive. However, in order to preserve their important social and environmental contribution to the urban ecology mentioned above, organisations working in this sector need to adopt models capable of conflating social and economic objectives with competitiveness on the market. For this purpose, the social business model seems to offer several advantages by drafting private sector logics into the third sector agenda. For urban agriculture to come to the forefront and provide a potential contribution to UK national food security, two important factors need to be substantiated:

the quantity of food that can be produced and the economic advantages related to such practice. If this happens, economic opportunities can provide incentives to the diffusion of urban food cultivation for commercial purposes. In market centred societies, economic opportunities provide motivation for policy action, which is needed for providing urban land and effective legislation.

An expansion of urban agriculture in developed countries triggered by economic opportunities would inevitably deliver those environmental, social, and educational benefits that today are much sought. It could also create new jobs and new professional roles. Finally it could also contribute to redefine urban models for sustainable cities, as the compact urban model as known today could compete with food production (Howe, 2010).

References

Akinbamijo, O.O., Fall, S.T. and Smith, O.B. (Eds.), 2002. Advances in crop-livestock integration in West African cities. Grafisch Bedrijf Ponsen, Wageningen, the Netherlands, 214 pp.

Barling, D., Sharpe, R. and Lang, T., 2008. Rethinking Britain's food security. Centre for Food Policy, City University London, London, UK, 46 pp.

Barton, H., Grant, M. and Guise, R., 2003. Shaping neighbourhoods: a guide for health, sustainability and vitality. Spon Press, London, UK, 244 pp.

Borough of Camden, 2009. Good food for Camden – The healthy and sustainable food strategy. Borough of Camden, UK, 55 pp.

Borough of Haringey, 2006. Open space strategy – A space for everyone. Borough of Haringey, UK, 31 pp.

Bourque, M., 2005. Thematic paper 5 – policy options for urban agriculture. Available at http://www.ruaf.org/node/56.

Brandolini, A. and Cipollone, P., 1992. Working paper no. 139: Urban poverty in developed countries. Maxwell School of Citizenship and Public Affairs Syracuse University, Syracuse, New York 13244-1020. Available at http://www.heart-intl.net/HEART/HIV/Comp/Urbanpovertyindevelopedcries.pdf.

Deelstra, T. and Girardet, H., 2005. Thematic paper 2 – Urban agriculture and sustainable cities. Resource Centre on Urban Agriculture and Food Security. Available at http://www.ruaf.org/node/56.

Department for Communities and Local Governments, 2006. Planning policy guidance 17: planning for open space, sport, and recreation. Available at http://www.planning-applications.co.uk/ppg17_openspacerecreation.pdf.

DEFRA (Department for Environment, Food and Rural Affairs), 2006. Food security and the UK: An evidence and analysis paper. Department for Environment, Food and Rural Affairs, London, UK, 93p. Available at http://archive.defra.gov.uk/evidence/economics/foodfarm/reports/documents/foodsecurity.pdf.

DEFRA, 2008. Ensuring the UK's Food Security in a Changing World. Department for Environment, Food and Rural Affairs, London, UK, 32 pp.

Fairholm, J., 1999. Urban agriculture and food security initiatives in Canada: A survey of Canadian non-governmental organizations. Cities feeding people series, Report 25. International Development Research Centre, Ottawa, Canada, 75 pp.

FAO (Food and Agriculture Organization), 1996. Rome declaration on world food security and world food summit plan of action. World Food Summit 13-17 November 1996, Rome, Italy. Available at http://www.fao.org/docrep/003/w3613e/w3613e00.HTM.

FAO, 2010a. Growing food for nine billion. Available at http://www.fao.org/docrep/013/am023e/am023e00.pdf.

FAO, 2010b. Harvesting agriculture's multiple benefits: mitigation, adaptation, development and food security. Available at ftp://ftp.fao.org/docrep/fao/012/ak914e/ak914e00.pdf.

Food Standards Agency, 2010. National Diet Nutrition Survey: headline results from year 1 (2008/2009). Food Standards Agency, London, UK.

Foresight, 2011. The future of food and farming: Challenges and choices for global sustainability. Final Project Report. The Government Office for Science, London, UK, 211 pp.

Garnett, T., 1999. CityHarvest – The feasibility of growing more food in London. Sustain, London, UK, 176 pp.

Garnett, T., 2005. Urban agriculture in London: rethinking our food economy – Growing cities, growing food: urban agriculture on the policy agenda. DSE, Feldafing, Germany. Available at http://www.ruaf.org/book/export/html/54.

Growing Communities, 2010. Urban food growing. Available at http://www.growingcommunities.org/food-growing/.

Her Majesty's Government, 2010. Food 2030. Department for Environment, Food and Rural Affairs, London, UK, 84 pp.

Horrell, C., Jones, S.D. and Natelson, S., 2010. An investigation into the workings of small scale food hubs. Sustain, London, UK, 8 pp. Available at http://www.sustainweb.org/pdf/mlfw_hubs_research_summary.pdf.

Howe, J., 2010. Growing food in cities: the implications for land-use policy. Journal of Environmental Policy & Planning 5: 255-268.

International Energy Agency, 2009. World Energy Outlook 2009 Factsheet. IEA. Available at http://www.iea.org/weo/docs/weo2009/fact_sheets_WEO_2009.pdf.

Local Food Link, 2010. Building a Sustainable Community Food Hub: Distribution of surplus from allotments. Sustain, London, UK, 5 pp. Available at http://www.sustainweb.org/pdf/Building_Sustainable_Community_Food_Hub.pdf.

London Development Agency, 2006. Healthy and sustainable food for London – The Mayor's food strategy. London Development Agency, London, UK, 139 pp.

Martin, R. and Marsden, T., 1999. Food for urban spaces: The development of urban food production in England and Wales. International Planning Studies 4: 389-412.

Mbiba B., 2003. Urban agriculture in the London Borough of Bexley. Urban Agriculture Magazine 8 – December 2003.

Mougeot, J.A., 2006. Growing better cities: urban agriculture for sustainable development. International Development Research Centre, Ottawa, Canada, 97 pp.

Pilley, G., Bashford, J., Billyard, P., Langham, C., Button, D. and Barber, H., 2005. Cultivating co-operatives: organisational structures for local food enterprises. The Soil Association, UK, 88 pp.

Pothukuchi, K. and Kaufman, J.L., 1999. Placing the food system on the urban agenda: The role of municipal institutions in food systems planning. Agriculture and Human Values 16: 213-224

Pothukuchi, K. and Kaufman, J. L., 2000. The food system. Journal of the American Planning Association 66: 113-124.

Redwood, M. (Ed.), 2009. Agriculture in urban planning: Generating livelihoods and food security. Earthscan, London, UK, 248 pp.

Saltmarsh, N., 2009. Making local food work supply chain development: a new local and organic food depot for London? East Anglia Food Link and Making Local Food Work. Available at http://www.eafl.org.uk/LondonDepot.asp.

Sieverts, T., 2003. Cities without cities: an interpretation of the Zwischenstad. Spon Press, London, UK, 187 pp.

Soil Association, 2009. A share in the harvest: A feasibility study for community supported agriculture. Soil Association, UK, available at http://www.soilassociation.org/LinkClick.aspx?fileticket=rq9ulg7UAvM%3D&tabid=387.

Soil Association, 2010a. Organic market report 2010. Soil Association, UK, available at http://www.soilassociation.org/LinkClick.aspx?fileticket=bTXno01MTtM=&tabid=116.

Soil Association, 2010b. A share in the harvest: An action manual for community supported agriculture – 2nd edition. Soil Association, UK, available at http://www.soilassociation.org/LinkClick.aspx?fileticket=gi5uOJ9swiI%3D&tabid=204.

Sonnino, R., 2009. Feeding the city: Towards a new research and planning agenda. International Planning Studies 14: 425-435.

The Poverty Site, 2010. Summary – key facts. Available at http://www.poverty.org.uk/summary/key%20facts.shtm.

The Strategy Unit, 2008. Food matters: towards a strategy for the 21st century. Cabinet Office, London, UK, 144 pp.

Viljoen, A., Bloch, K. and Howe, J., 2005. CPULs: designing urban agriculture for sustainable cities. Architectural Press, Oxford, UK, 280 pp.

Yunus, M., 2007. Creating a world without poverty: Social business and the future of capitalism. Public Affairs, New York, NY, USA, 261 pp.

Chapter 23

Farming in Motown: competing narratives for urban development and urban agriculture in Detroit

Erica Giorda

Michigan State University, Department of Sociology, 422 Berkey Hall, East Lansing, MI 48824, USA;.giordaer@msu.edu

Abstract

Detroit used to be the core of the US car industry, the engine of American industrial production. In the 1950s, Detroiters had apparently achieved the American dream, but when the industrial system started to collapse, the city rapidly followed. For many years local administrators have tried to revive the past glory, until the 2008 crisis made clear that that path is closed. Urban gardens started to appear in the last decade as a grassroots response to the urban decay, spurred by the lack of access to fresh food that makes Detroit one of the largest urban food deserts in the US. In the last five years the phenomenon has caught the eyes of the media, and recently the local government has started to recognise it. According to local sources, the number of gardens is steadily growing. Five structures that call themselves farms are now operating in Detroit, along with more than 300 community gardens. Until recently, institutional support for these activities has been limited. Some institutional actors are now investing in the sector. Wayne State University sponsors a farmers market and lets the students grow gardens on campus; the Eastern Market Corporation started an experimental farm on its campus; the City planning department is working on a plan that would permit commercial farming within the city boundaries. Moore's work on sustainable cities could help frame what's happening in Detroit. I argue that urban farmers in Detroit follow a story line that describes the future of Detroit as a city of gardens. As many Detroit gardeners are the heirs of the activists who fought in the past for social justice and equal rights, their quest for locally produced food is informed by the concept of food justice. The arrival of a new, well-capitalised farm on the stage might change this storyline towards a new direction.

Keywords: social imaginaries, urban farming, urban planning

23.1 Introduction

If there is a city that embodies the concept of wicked problem (Rittel and Webber, 1973), that city is Detroit. Contradictions are as thick as the weeds that grow in the abandoned lots, and the interests of the communities, local politicians, and financial groups do not align easily. All the forces at work in Detroit will have to make huge efforts and accept a number of compromises to get the city out of shallow water. Apparently, one of the ways this can be accomplished is by 're-imagining' the city: in 2003 Eisinger first saw the necessity to 're-

'imagine' Detroit in order to create a new perspective for the city to be able to disengage from its automotive past and develop a more diversified economy (Eisinger, 2003). Detroit Press journalist Gallagher titling his 2010 book *Reimagining Detroit* (without quoting Eisinger) hints at how the idea is getting mainstream. In the last five years the debate – in the public sphere as well as in the academic one – has been characterised by the contrast between those who still support the heavy industry perspective that made the fortunes of Detroit in the past (Shaiken, 2009; Steinmetz, 2009; and even the Rev. Jessie Jackson during a visit in 2010), and the ones that suggest time is due for a change in strategies (Eisinger, 2003; Gallagher, 2010; Jacobs, 2009). The urban farming movement as a whole is also pressing for change, although, I will argue, from a radically different political perspective.

In this chapter I try to extrapolate the narratives that synthesise two different positions and highlight the possibility of a third one. According to Taylor (2004), narratives, as expressed by art objects, public discourse, and publicly shared sets of expectations, are one of the ways we can use to access our social imaginaries, the 'common understanding that makes possible common practices and widely shared sense of legitimacy.' (Taylor, 2004: 23). I see these narratives as analytical interpretations: my use of two internationally renowned artworks to represent the two main contrasting positions is intended to provide a symbolic perspective over the struggle that is taking place in the city: I'm sketching the picture of how Detroit represented itself to the world, how that image changed, and how a different one is coming along to compete with the old one as the representation for the Detroit of the future.

In 2011, as the Chrysler's Super Bowl advertisement clearly demonstrated, 'Detroit' is still a shortcut for 'automotive industry', and the fact that citizens might have a different perspective of their city has not been taken into account as much as it could have been. Investments in the city have traditionally favored the financial elite, who in many cases did not repay either the trust or the favorable mortgages, and eventually left the city in breaches (Darden *et al.*, 1987). Time is due to change course. Interestingly enough, even after the 2008 meltdown, the institutional proposals have not been as creative and alternative seeking as they appear. Steinmetz (2009), on the brink of the bankruptcy of GM, still sees a return of the industry as the solution. Not many years ago, Eisinger (2003) envisioned local, car related tourism, middle class housing and investments in the service sector as a possible path. To a certain extent, the current development in the Mid Town neigbourhoods around the Wayne State University campus can be seen as following that line: right now the major employers in Detroit are the medical facilities and Wayne State University. While the redevelopment that revamped the Campus area in the 1970s was criticised from many parts (Darden *et al.*, 1987), this is now one of the most vibrant areas in the city. However, looking at services and higher instruction doesn't necessarily break out from the traditional outlook for development. The other path, almost invisible until very recently, would be to search for new ventures in less explored space. Some people are now betting their dimes on the rehabilitation of the food system and the growth of the urban farming model.

Against this complex background, how can we characterise the growth of the urban farming movement in Detroit? My analysis will draw from Moore (2007) to identify two storylines that represents opposite perspectives on the past and the future of the city, its social and environmental sustainability, and finally, its possible agricultural future.

23.2 The growth and decline of the American automotive capital

Detroit was portrayed in the 1940s as 'the Arsenal of Democracy' and in the 1950s as 'Motor City': for the first half of the 20th century it represented, for the worst and the best, the achievements of the American industry. Sugrue (1996) used Detroit as the paradigmatic example to describe the urban crisis in the US. As of now, that analysis still stands, and Detroit has become the icon of urban decay in the US. There is a direct connection between the fact that, in the 1950s, Detroit was fabled as one of the most modern cities in the US and the way contemporary Detroit is burdened with its deficit, its decaying infrastructure, and its abandoned lots. Already in the 1950s, when it supposedly reached its zenith, the shiny Detroit apple had a hollow core: its glitter and wealth were built on the exploitation of African Americans, segregation, and a sustained ignorance of the effects of the industry on the land. Starting from the early 1960s, the unsustainable basis became evident and decline rapidly followed, setting a path for other industrial cities in the US Midwest: the infamous rust belt has its buckle in Detroit.

After WWII, the federal government distributed low interest loans to veterans, spurring the development of suburbia and the first wave of white flight. Again, blacks did not get those benefits and were only able to move to the areas left behind. At the same time, due to the growing power of the international union of automobile workers UAW and the Keynesian attitude of the central government, industry workers received higher wages and benefits than other sectors in the area and in the rest of the country, leading to the creation of the so-called blue-collar middle class (Jacobs, 2009). Contradictions were already looming. Three factors can be accounted as the main motors of the subsequent decay, and, if they affected the whole US economy, they were particularly strong in Detroit: racial inequality, lack of coordinated economic planning at the regional and state level, and an industry driven infrastructure that favored car transportation over mass transit.

Detroit's black population suffered from the beginning because of segregation and poor housing conditions; anger was already growing in the 1950s when – within the civil rights movement around the US, black citizens were fighting actively to get political representation and equal treatment. The UAW was not responsive: in the plants, racial tension was caused by discrimination, but no action was taken to effectively address the problem. The tension grew constantly for more than 20 years, culminating with the infamous 12th Street riots (Sugrue, 1996; Thompson, 2004).

From an economic perspective, the city of Detroit started to lose its strength at the end of 1950s, well before the 1970s oil crisis, when the auto industry relocated its plants in the suburbs, where taxes were lower and there was more space. The persistent tension between the city and its suburbs dates back to that era, and city built infrastructure played a big role: the construction of the net of highways that literally bonds Detroit favored the suburbs, creating an easy and quick way – for those who could afford to buy a car – to move from the new residential areas to their working places. On the downside, highways were built in areas where the most blighted neigbourhoods used to stand: the only places where blacks were allowed to live. The construction of Interstate 75 literally erased Hastings Street, setting the majority of African American retailers in the city out of business without compensation. Only the owners of the building were compensated, and those were mostly white. Black businessman who rented the commercial spaces and black families living in apartments only received notice to leave and no real option to restart a business or find alternative housing elsewhere. The housing projects that should have hosted the displaced residents were never completed (Darden *et al.*, 1987).

During the 1980s Detroit became increasingly segregated; most of the white population left when the political power went steadily into the hands of the black community. On behalf of the black community, mayor Coleman Young pursued the self-sufficiency of Detroit, withholding almost any form of connection and cooperation with the suburbs, which, on their part, mostly acted as racially segregated enclaves and made no effort to create any relationship with the city (Jacobs, 2009).

23.3 Farming in the city: urban gardens in Detroit

Nowadays Detroit is one of the poorest cities in the US. About 30% of the population receives welfare benefits, one third of the population is food insecure, and one fifth of households do not have access to private transportation (Gallagher, 2007; Pothukuchi, 2011). As of 2009 unemployment reached 29% (Steinmetz, 2009). Lack of trust, resentment, and persistent inequality are the legacy of this historical process, and are key features to understand the urban farming movement in Detroit (White, 2010).

Detroit has a long tradition of urban gardening, from the famous potato patches sponsored by Mayor Hazen Pingree in the late 19th century to the liberty and victory gardens during World War I and II. African Americans who came to Detroit to work in the car industries grew vegetable gardens in their backyards (Treuhaft *et al.*, 2009), and a number of small farms flourished on the outskirts of the city, tended by people belonging to diverse ethnic groups, before the land was all converted to residential areas (Darden *et al.*, 1987). The crucial points to understand are whether (and, if so, how) the contemporary wave differs from the previous one, and why it may represent a sign of change for the city.

In the last decade urban gardens have started to appear again as a grass-roots response to the urban decay, spurred by the lack of access to fresh food that makes Detroit one of the largest urban food deserts in the US. Private and community gardens have sprouted over abandoned lots, with and without authorisation from local authorities or absent owners. In the last five years the phenomenon has caught the eyes of the media, and finally also the local government is recognising it.

A lot of credit should go to Wayne State University professor Kami Pothukuchi, whose research in planning for food access in inner cities and whose indefatigable efforts have brought together local activists, international experts, and university students. The SEED (Summer Educational Experience for Disadvantaged Students) programme she leads developed a farmers market in Mid Town, exclusively served by local farmers, and two community plots for Wayne State students to garden on their own.

Five semi-structured 'farms' are now operating in Detroit, along with more than 200 community gardens. Some of these growers sell their produce at Farmers Markets throughout the city, but the earnings are still low. However, only the plot managed by the Greenings of Detroit on the Eastern Market campus can be legally recognised as a 'farm'. The city has no interest in enforcing regulations everybody deems obsolete, but many cultivated plots are on residential lots, where farming for commercial purposes should not be allowed[36]. The recently passed 'Cottage Food Legislation' at state level eases the production of small batches of processed products and will give those operations a better legal standing, but the city planning department – however committed – is still far away from producing the much wished for rezoning plan. One of the main impediments is the Michigan 'Right to Farm' act, according to which, if a parcel of land is rezoned as agricultural, it cannot be converted back to non-agricultural uses. Moreover, the city would lose its ability to control those parcels and impose health care related restrictions on agricultural practices, and aesthetic restrictions on the appearance and the maintenance of the lots (Figure 23.1).

It is also important to underline that even the most productive urban gardens are tended by volunteers, the level of professionalisation is low, and the actual amount of farmed land is still minuscule compared to the vast expanses of abandoned lots and brown-fields. Colasanti and Hamm (2010) tallied 4,800 acres of vacant, publicly owned land in the city. According to their study, putting 2,000 acres into production would provide at least half of the amount of fresh fruit and vegetables (excluding tropical fruits) that the city currently needs.

While the different groups that manage the larger projects cooperate on a regular basis and organise activities together, they have quite diverse backgrounds. One of the largest farms, D-Town farm, is managed by the Detroit Black Community Food Security Network (DBCFSN) on two acres belonging to the former city nursery and is specifically devoted to empower the black community and improve the health of people of colour. They also aim at involving youth in food and social justice activities (Pothukuchi, 2011; White, 2010). The Catherine Ferguson Academy's farm is mostly an educational tool that enables teen mothers to get a diploma while learning practical skills, and better nutritional habits for themselves

---

[36] Zoning restrictions officially prevent growing food for any purpose other than personal consumption.

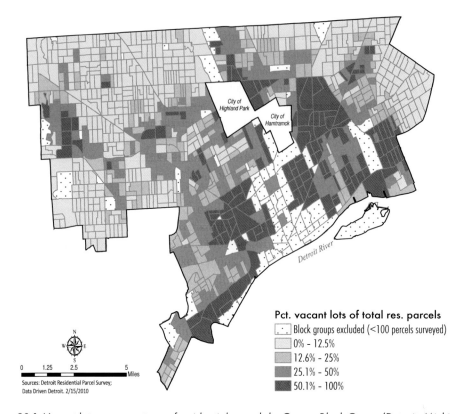

Figure 23.1. Vacant lots as percentage of residential parcels by Census Block Group (Detroit, Michigan). The Detroit Residential Parcel Survey (DRPS) surveyed predominantly residential parcels. The DRPS also includes vacant lots in neigbourhood commercial areas adjacent to residential areas (this map does not include vacant lots in other commercial or industrial areas) (Data Driven Detroit: http://datadrivendetroit.org).

and their babies[37]. Earthworks farm, linked to the Capuchins soup kitchen, is devoted to educational activities and turns over a large proportion of its production to the soup kitchen itself. The Greening of Detroit expanded its beautification and educational activities on the numerous community gardens it supports, and opened a new hoop house in may 2010, intended to be a laboratory for promoting and testing of season extension practices. A new CSA also started its operations on the west side in spring 2010, with about 20 regular clients, and more soup kitchen related gardens are growing on the west side of the city.

Not counting the larger ones and the officially established community gardens, mapping the backyard gardens of Detroit is extremely difficult, as they sprout on abandoned lots, and local residents do not necessarily trust researchers or investigators. Based on the quantity of seedlings distributed by the Garden Resource Program, a collaborative project between some of the actors in the urban gardening scene, the number of gardeners is constantly growing:

---

[37] Due to the ongoing restructuring of the Detroit School District, it is unclear if the school will be permitted to continue its operations.

in 2010 more than 5,000 adults and more than 10,000 youths took part in various gardening activities around the city, and Detroit sported 328 community, 39 market, 63 school and 804 family gardens, producing more than 169 tonnes of vegetables. UrbanFarming, another grassroots organisation, declares to have helped plant more than 1,200 gardens in the larger metropolitan area (Pothukuchi, 2011).

The last actor appearing on the stage is the Hantz Farm project. An entrepreneur with huge financial backings, John Hantz started this new venture in 2009 with the idea of creating a commercial farming activity on about 50 acres of what is now city owned land. The details of the agreement are not defined yet: so far in 2011 only a few acres have been planted with a tree nursery, and the rest of the project is still under development, while the legal issues about zoning and land use wait for clearing. This project is nevertheless creating new tensions within the grassroots organisations that are on the forefront of the urban farming movement. The capitalistic approach Hantz Farm is using seems to many in contradiction with the spirit behind community gardens and the social justice perspective adopted by the majority of Detroit urban farmers insofar. I would describe this reaction as a conflict between the two storylines I outlined: even if Hantz Farm portrays itself as an implementation of the Garden City ideal, it does not fit into the narrative that community activists use to describe their efforts.

23.4 Two competing storylines to design the future of Detroit

The growth of the urban farming movement in Detroit relates with three main themes: social justice, food insecurity, and an extreme environment (see also Giorda, 2010). An extreme environment is the consequence of a technological disaster, has persistent and deleterious effects on residents of the area affected, and has a specific agency that affects the ability of people to react and overcome those effects (Kroll-Smith *et al.*, 1997). I see the efforts of urban farmers in Detroit as a reaction to the alienation caused by extreme environmental conditions and also as a project to rebuild a productive relationship with the land. Along similar lines, White's (2010) interviews with D-Town farm volunteers show 'their efforts to be the agents of their own transformation and the transformation of the city of Detroit by claiming the human right to food.' As it is clear that current urban farming efforts do not produce enough food to sustain gardeners on a daily basis, this expression of agency must be considered for its political and symbolic value, especially as it appears to be only one of the various strategies currently under implementation in Detroit to cope with food insecurity, *and* social and environmental injustice. Urban farming in Detroit satisfies some immediate needs: fresh food is one of them, but gardens also provide community gathering spaces and represent the concrete and visible effort of people who fight urban blight and decay without the ability or the interest to get money and help from outside, be it the government or some philanthropic foundation.

Within this kind of context, the attitude of people, the perception they have of their efforts, and the narrative they use to define them are crucial: Moore's (2007) essay on the relationship between sustainability and urban environments formalises this central aspect with the concept of 'storyline'. Basically, according to Moore, storylines are the expressions of different sectors of the society engaged in the production of alternative plans for the development of a city.

Following Moore's lead I identify two prominent storylines that characterise Detroit's development. I'll call them the 'Motown Mythology' and 'Detroit City of Gardens'. The first one, Motown Mythology, connects Detroit with its industrial origins and its 1950s hall of fame status. It states that today's city is the natural heir of that past, and the only possible lines of development are related to the car companies and their fortunes. I already described how the story unfolded. The GM owned Renaissance Center could be seen – together with the abandoned Packard plant and the still roaring Ford Rouge Factory (which – significantly – is not in Detroit, anyway) as the iconic buildings for that storyline. Future historians might recognise 2008 as the year that definitively ended it, but another powerful representation of this project will remain in the great frescoes that Rivera painted in 1933: the Detroit Industry murals.

Commissioned by Edsel Ford, the frescos celebrate the rise of the automotive industry, the labourers who bore the weight of the production chain, the achievements of human intelligence and the bounty of the earth that nourishes the soil and provides the raw material for the industrial enterprise. It is an incredible work of art, and quite a clear representation of what Detroit wanted to be. It also shows that – with respect to the relationship between humans and earth – the communist and the capitalist shared a common position that places humans as dominators, authorised by the superior power of history to exploit the earth to fulfil their destiny. This is exactly the starting point of Detroit development, and the main reason for its decline. As, apparently, history has not fulfilled the other side of Rivera's representation: the systemic connection of all the parts in a whole. The power of the Detroit Institute of Art frescos comes out of the ensemble of all the elements involved: industry leaders, workers, natural resources and technology. Rivera was keenly aware that technology was a double-edged sword, and the development of industry was heavily based on exploitation of workers.

A few miles away from the cultural district, on an abandoned block on the East Side, another urban masterpiece flourished in more recent years. The Heidelberg project was originally developed in 1986 by Tyree Guyton, as a way to restore and reclaim the neigbourhood he grew up in. Consisting of ruined houses and reclaimed waste, the project aims at improving 'the under-resourced and horribly blighted Detroit community where the project was founded' (Heidelberg project website, 2010: http://www.heidelberg.org/). The symbolic trajectory between the exaltation of the 1930s and the trash bin art of the 1990s is quite explicit. If artists have a special ability to see reality beyond ordinary appearances, then here we have some clues. I see this project as the artistic representation of the second storyline: Detroit City of Gardens.

The idea of Detroit as a Garden City has been around for about five years. It appears as an ideal model in the Eastern Market 2008 development plan, it names a group on Facebook, and even appears in a sermon by an evangelic pastor in Seattle.[38] For sure the urban farming movement has embraced it, as much as the city council planning committee, overseeing the necessity of 'right-sizing' the city, is looking at it with interest. It is quite early to see if this storyline has enough strength to rise as the new paradigm for the city.

It is important to point out that Detroit has always been quite a flat city, and green space, in the form of individual lawns, has always been present. Backyard gardening and community gardening have also a tradition, dating back to the potato gardens in Pole Town during the Depression, to the many victory gardens that flourished during the war. So Motown was already somehow green, but – we might say – a different shade of green. The difference lays in two different transportation and urban growth options. Motown was built for and around cars. The new City of Gardens would aim at dropping cars, as 22% of citizens do not have a car already (Pothukuchi, 2011) and would emphasise public transport. Also, due to resizing, the new idea of green space would comprise more public spaces, such as parks and huge swats of land reconverted to agricultural use. The Eastern Market development project describes this idea as a city made up of villages, surrounded by green areas. However, people do live in some of the semi-abandoned neigbourhoods that would be converted into green spaces, and given the history of the city it is not easy to predict who has the ability to convince these citizens to relocate. The public discourse is just now starting to revolve around the second storyline, as the last blow to the car industry left the city without hope in that direction, and the administration needs to make the idea of 'right sizing'[39] acceptable to citizens.

Finally, as Moore (2007) discovered about Austin, there are also storylines that don't make it to the surface. During a recent series of public workshops and conversations about undoing racism in Detroit, organised by many of the actors that are working on urban agriculture, people of colour of all ages restated how racial barriers are still high, and access to resources jeopardised. The uproar of the urban farming community against the opening of the Hantz Farm is quite telling, as it presents itself as a promoter of City of Gardens storyline. Despite the relatively small size of the project (the projected 40 to 70 acres is very little for a farm), some citizens are uncomfortable because of the attitude of the promoter. Basically, many people within the communities that would be affected by the project fear that Hantz Farm might foster gentrification and jeopardise the efforts of the more community oriented projects. Looking at the past, and how the interests of the citizens have been submitted at the interests of capital (Eisinger, 2003), the risk of misunderstandings and conflicts between those different approaches to urban farming are visible. Even if Hantz Farm aims to be a project by a Detroiter for Detroiters, it might fail to recognise that social justice is a component of the Urban Farming quest as much as the re-localisation of the agricultural production. Allen

---

[38] http://www.seattlefirstbaptist.org/?Locator=Sermons&Header=Sermons&AutoLaunch=12536 retrieved in April 2010.

[39] 'Right sizing' is a contested concept, and in many cases it is used to name what is in fact deep restructuring by downsizing of services.

(2010) underlines that 'localities contain within them wide demographic ranges and social relationships of power and privilege embedded within the place itself.' She recognises that the movement towards the re-localisation of the production of food at the community level does not necessarily imply that all the members of the community will be granted the ability to decide how to implement it. In Detroit this creates a case for a potential conflict.

23.5 Conclusions

The idea of Detroit as a city of gardens is receiving growing media coverage, but it is still to be decided if these gardens will bring about gentrification and will reinforce the existing power structure, or if they will create new opportunities for a different type of growth. Furthermore, despite evidence that growing and processing food within the city boundaries could create economic benefits (Colasanti and Hamm, 2010) and reduce health care and food access problems (Pothukuchi, 2011), the industrialist view to economic recovery is still very strong. It is also unclear how an increased presence of gardens or farms in the city could affect real estate values. The perspective of an increase in property values is seen as a possible danger for some residents, who might not be able to afford higher taxes. In the highly racialised context of Detroit, there is a number of historical precedents (Darden *et al.*, 1987) where revitalisation projects have been used to expel black residents from certain areas, by simply building housing projects that were not affordable by those whose houses were demolished. Memories of these unjust redevelopments are still present, and create suspicion toward aggressive entrepreneurial tactics among long-term residents.

While the traditional storyline that sees Detroit as the 'carmaker' has been polished and reviewed for decades, so that its implications and its weaknesses are well known, the 'Garden City' storyline is still in the making. I described here two partially opposite interpretations of it. In the first one, gardens and farms are tended and developed by existing communities in Detroit, according to their needs and goals. This is the way of grassroots organisations: ordinary people working with limited means to beautify and rescue their neigbourhoods. Its main asset, the strong volunteer support, is also its main weakness, if we look at it from the perspective of economic efficiency, but at the same time these groups do not consider that kind of efficiency their main goal. In the second interpretation, gardens and farms are more explicitly business oriented, and traditional real estate development may mix in together with various tactics to 'right sizing' the city. At this moment, there is no way to tell which one will be successful, but there is indeed a lot of space available in Detroit, and Detroiters might realise that there is enough land to realise both projects.

References

Allen, P., 2010. Realizing justice in local food systems. Cambridge Journal of Regions, Economy and Society 3: 295-308.

Chrysler Corporation. 2011. Super Bowl commercial. Available at http://www.youtube.com/user/chrysler?bid=5079147&adid=233347236&pid=57249858&KWNM=2011+chrysler+super+bowl&KWID=150748009&channel=PS.

Colasanti, K. and Hamm, M.W., 2010. The Local Food Supply Capacity of Detroit, MI. Journal of Agriculture, Food Systems and Community Development 1: 1-18.

Darden, J.T., Hill, R.C., Thomas, J.M. and Thomas, R., 1987. Detroit: Race and uneven development. Temple University Press, Philadelphia, PA, USA, 229 pp.

Eisinger, P., 2003. Reimagining Detroit. City & Community 2: 85-99.

Gallagher, J. 2010. Reimagining Detroit. Opportunities for Redefining an American City. Wayne State University Press, Detroit, MI, USA, 192 pp.

Gallagher, M., 2007. Examining the impact of food deserts on public health in Detroit. Gallagher Research and Consulting Group, Chicago, IL, USA, 16 pp.

Giorda, E., 2010. Fresh veggies in an extreme environment: challenges and opportunities for the urban farming movement in Detroit. Presented at IGU/UGI Local Food Systems in Old Industrial Regions, Toledo, OH, USA.

Jacobs, A.J. , 2009. Embedded contrasts in race, municipal fragmentation, and planning: Divergent outcomes in the Detroit and greater Toronto–Hamilton Regions 1990–2000. Journal Of Urban Affairs 31: 147-172.

Kroll-Smith, S., Couch S.R. and Brent, K.M., 1997. Sociology, extreme environments and social change. Current Sociology 45: 1-18.

Moore, S.A., 2007. Alternative routes to the sustainable city: Austin, Curitiba, and Frankfurt. Lexington Books, Plymouth, UK, 245 pp.

Pothukuchi, K., 2011. The Detroit food system 2009-10 report. Detroit Food Policy Council, Detroit, MI, USA, 76 pp.

Rittel, H.W.J. and Webber M.M., 1973. Dilemmas in a general theory of planning. Policy Sciences 4: 155-169.

Shaiken, H., 2009. Motown blues: What next for Detroit? Dissent 56: 50-56.

Steinmetz, G., 2009. Detroit: A tale of two crises. Environment and Planning D: Society and Space 27: 761-770.

Sugrue, T.J., 1996. The origins of the urban crisis: race and inequality in postwar Detroit. Princeton University Press, Princeton, NJ, USA, 416 pp.

Taylor, C., 2004 Modern social imaginaries. Duke University Press, London, UK, 215 pp.

Thompson, H.A., 2004. Whose Detroit?: politics, labor, and race in a modern American city. Cornell University Press, Ithica, NY, USA, 304 pp.

Treuhaft, S., Hamm, M.J. and Litjiens, C., 2009. Healthy food for all. Building a sustainable Food System in Detroit and Oakland. PolicyLink and Michigan State University, USA, 64 pp.

White, M., 2010. Shouldering responsibility for the delivery of human rights: a case study of the D-town farmers of Detroit. Race/Ethnicity: Multidisciplinary Global Contexts 3: 189-212.

Chapter 24

Why is there agriculture in Tokyo? From the origin of agriculture in the city to the strategies to stay in the city

Nelly Niwa

Institute of Land Use Policies and Human Environment, Lausanne University, Quartier Sorge, Bâtiment Amphipôle, 1015 Lausanne, Switzerland; Ikaga laboratory, Department of System Design Engineering, Faculty of Science and Technology, Keio University, 223-8522 Yokohama, Japan; nelly.niwa@unil.ch

Abstract

Tokyo is one of the densest megalopolises in the world. This leads to extreme pressure on land and intense competition between urban activities. In this context, it is surprising to imagine that there is agriculture in Tokyo. Today, agriculture represents 2% of the city area. If it is less than during the Edo period (1603-1868) – when more than 40% of the land area was devoted to agriculture – it is still more than in many other cities. How can we explain the presence of agriculture in Tokyo? What strategies do the farmers develop to remain in the city? These are the questions we will try to answer in this chapter. First, we will show the historic roots of agriculture in Tokyo and how Tokyo's development was based on agriculture. Secondly, we will give some illustrations of the strategies that have been initiated in Tokyo to maintain agriculture in spite of the high competition for space. We will present how agriculture fits into the urban context and develops strategies to find new spaces in the city and how the farming profession is now being reinvented.

Keywords: intra-urban agriculture, Tokyo, urban planning

24.1 Introduction

For a long time, agriculture was supposed to take place outside of the city. Today, agriculture is winning the right to stand within the city. This phenomenon reveals deep modifications in our society. In a context of environmental and economic crisis, farming seems to be a feasible solution to improve the self-sufficiency of the city, to build a more resilient city and to give its inhabitants a better quality of life. However, if the benefits of urban agriculture are legitimate, it is also true that the future of this agriculture is uncertain. First, because of the proximity of an urban context that generates pressure on farming activities. Secondly, because the intention to build a more sustainable and compact city leads to favour more profitable activities that urbanise the farming land.

In this context, the existence of agriculture in Tokyo, one of the most dense cities in the world, is particularly surprising. Today, the 23 wards of Tokyo include 671 ha of agriculture that represent 2% of the city area (TMG, 2005). A total of 1,916 farms is distributed throughout the whole city, building an agro-urban patchwork.

The goal of this chapter is to briefly describe the state of agriculture in Tokyo. Little research has dealt with this subject. This is particularly surprising as Tokyo presents a very interesting case study of agriculture in a developed megalopolis. First, we will show how Tokyo's development was based on agriculture. We will demonstrate how farming structure, zoning system and cultural factors have favoured agriculture in Tokyo. Secondly, we will give some illustrations of the strategies that have been initiated in Tokyo to maintain agriculture in spite of the high competition for space. We will present how agriculture fits into the urban context by developing strategies to find new spaces in the city and by reinventing the farming profession.

24.2 How can we explain the presence of agriculture in Tokyo?

In Europe and in the west in general, agriculture and the city are clearly separate. In Tokyo, agriculture has always been a component of the urban pattern. Historically, when Tokyo was called Edo, 40% of the city was agricultural (Yokohari and Amati, 2005). Agriculture was supported by the samurai because it allowed them to grow their own food (Naito, 2003). More generally, agriculture was also a way to secure the food supply and manage urban waste (Tajima, 2007).

Today, agriculture still exists in Tokyo. We make the hypothesis that this can be explained by Tokyo's unique urban growth. Like most cities, Tokyo developed following an agricultural pattern. However the originality of Tokyo lies in the fact that agricultural plots were maintained during the expansion of the city and were not urbanised in the same way as in European cities. This kind of development leads to an urban form, which is not clearly defined. Unlike the limits of western cities, which are clearly defined, the limits of Tokyo are forever vague and it is difficult to distinguish the city from the countryside. Figure 24.1 presents a comparison between the limits of a European city (Lausanne, Switzerland) and what can be considered as a border of Tokyo. In Lausanne, there is a clear distinction between urban areas and countryside. In Tokyo, there is a great mix between urban and rural patterns and thus the city limit is difficult to establish.

In practical terms, this hybrid pattern can be explained by a scattered urban growth without a clear logic. The city eats the countryside like the worm eats a leaf (Ashihara, 1989). Rice fields become patterns where urbanisation takes place.

The reasons for this mixed agro-urban form are complex. In this chapter we make the hypothesis that three main factors can explain this development: the structure of farming, the system of zoning and taxes and cultural factors.

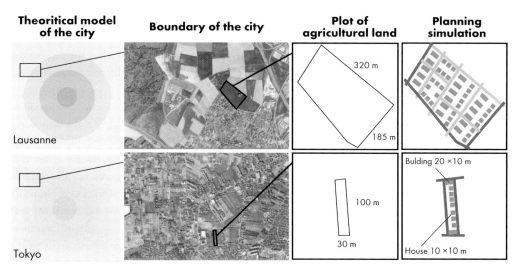

| Theoritical model of the city | Boundary of the city | Plot of agricultural land | Planning simulation |

Figure 24.1. Comparison between the boundary of Lausanne (Suisse) and Tokyo (Japan).

24.2.1 The structure of farming

The emergence of Tokyo's hybrid pattern is favoured by the fact that rice fields are fertile ground for urbanisation. Rice fields are small and flat areas, filled with sunlight and fully equipped in terms of connection to networks (Godo, 2007). In comparison with western agricultural land, they can be easily urbanised.

In Tokyo, 90% of the farms are less than 1 ha and 35% are less than 0.3 ha (Ogino and Ota, 2007). The origin of this repartition can be found in the land reform that was carried out under the occupation. The aim of this reform was to redistribute farmland from landlords to owner cultivators. This new repartition has led to an explosion in the number of small farms and had a long-term impact on later urbanisation (Sorensen, 2002).

As we can see on the two simulations of urbanisation we have made for farm plots in Lausanne and in Tokyo, it is easier to develop a small area. The urbanisation of a rice field does not imply important and complex planning as it may imply for the Lausanne plot. The development will also be easier as rice fields are already connected to water and energy networks, which reduce the cost of turning them into constructible land. Furthermore, the networks of roads used to cultivate the fields can be easily converted for urban infrastructures (Yokohari *et al.*, 2000).

The development of a hybrid agro-urban form in Tokyo is also favoured by the Japanese farming organisation. In Japan, a farm is traditionally made of several different plots. These plots are scattered all around the main farm building but are not necessarily next to one another. This distribution makes it easier for a farmer to sell a part of his land without affecting the other plots. It also means that if a farmer decides to stop his activity, a batch of dispersed land will be put on the real estate market. In the end, the multipolar structure of farms leads to a chaotic urbanisation development and reveals the hidden pattern of the property (Shelton, 1999). This pattern is so complex that it made it difficult for the authorities to plan the urban development of Tokyo. In the end, Tokyo developed from agriculture and agriculture became a part of the urban form.

24.2.2 The system of zoning and taxes

The establishment of a zoning strategy aims to clarify the different allocations of land. In Tokyo, and in Japan in general, the distinction between different zones did not really succeed for many reasons. We make the hypothesis that this non-determination and the flexibility of zones have allowed agriculture to stay and be accepted as an urban affectation in the city.

In the first zoning system in 1919, agriculture was not a specific land use designation. It is only in 1968, with the new planning system that considerations on farming have started in planning. The objective of this new planning system was to control the rampant urban sprawl and more generally the process of conversion of land from agricultural to urban uses. City planning areas (CPA) were divided into 2 zones, the urban promotion area (UPA) where planned urbanisation was to be promoted, and the urbanisation control area where urbanisation was to be restricted. The majority of farmers used strong political lobbying to included their land in UPA and to retain a preferential tax treatment – meaning they commonly paid 1-2% of the tax paid on nearby UPA land (Sorensen, 2002). The farmers' strategy was thus to hold their land for future developments, allowing a precarious stay of agriculture in the city. In 1992, a tax reform was passed which compelled farmers to choose between continuing their activity with lower tax rate (productive green land), or to start paying a higher rate of tax and keep their full rights to development.

The affectation 'productive green land' can finally be considered as an official recognition of the special status of agriculture in the city. The adaptation of the zoning system has allowed agriculture to stay and be accepted as an urban affectation in the city.

As presented in Figure 24.2, there is no clear distinction between the different urban and agricultural zones. The agricultural promotion area can be superposed to the city planning area, which allows the farmer to get subsidies from the government. Furthermore, only one part of the city was defined as a planning area, the other part was left relatively unregulated, allowing many uses of land. The extreme flexibility of zoning in Tokyo creates complexity and allows uses that would be restricted in other western cities. More research should be done to understand the impact of zoning on the persistence of agriculture. However at this stage, it is hard to escape the hypothesis that a more rigid system would have undermined the presence of agriculture in the city.

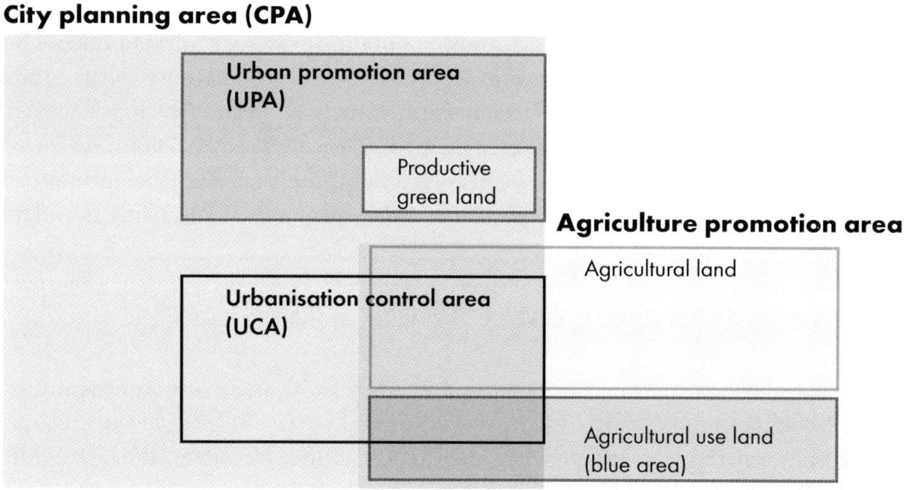

City planning area (CPA)

Figure 24.2. Possible land use zoning for agriculture.

24.2.3 The cultural factors

The presence of agriculture in the city can also be justified by general cultural factors such as the link the Japanese people have with their agriculture. Explanations can also be made regarding specific factors like the dependence of public planning on private interests or the conceptual acceptation of the combination of different activities.

In Japanese society there is still a clear supremacy of economic power over public interests. In Tokyo, this supremacy can be found in the landowners' influence on urban planning. In other words, the liberty of the owner is so strong that it cannot be restrained for the sake of urban form (Ashihara, 1989). As important landowners, farmers are very influential in Japanese society through their cooperatives (Godo, 2006). To the contrary, the national or the city government have a very weak position mostly due to their low economic power. If agriculture is tolerated in Tokyo, it was also because the government did not have the ability to expel it from the city. Even more, they rely on farmers to develop the city as most of the urban land will be developed based on the land readjustment system. Originally, this system was developed in Frankfurt am Main, before being transposed in Tokyo. In essence, the land readjustment system consist in allowing owners to build land in exchange of a portion of their land for public use (usually 30%) or to be sold to contribute for future projects (Sorensen, 2002). Land readjustment is popular in Tokyo because it provides a low-cost means of achieving orderly urban growth. However, it also makes the public infrastructure dependent on private owners (mostly farmers), thus making global planning difficult to manage and allowing the will of the owners an important role in Tokyo urbanisation.

The existence of agriculture in Tokyo can also be explained from the cultural acceptation of the mixture of agriculture and other urban activities. In Europe, there is a clear will to separate the different functions of the city and to contain the city within its limits, whereas in Tokyo, there is a high level of density and mixed uses which contribute to the vitality of the city. If the hybrid urban form also exists in European cities (in suburbs for example), the intention is to separate the empty spaces from the urbanised ones (Niwa, 2009). All European urbanity is based on this distinction between the city and the non-city (the agriculture). In contrast, Japanese cities contain a continuous blurring between the built up areas and nature (Berque, 1993).

24.3 What strategies do the farmers develop to remain in the city?

Nowadays, a lot of research shows how agriculture can bring many environmental, social or economic benefits to the city. In Tokyo, agriculture takes part in feeding the city, making it resistant to natural risks and improving the quality of life. However, there are still few examples, which show how agriculture can really fit into the urban context. The urban context is, to the contrary, presented as unfavourable to agriculture, putting it in competition with some more profitable urban activities and making it incompatible with other ones.

In Tokyo, agriculture demonstrates a huge capacity for flexibility and responsiveness in order to fit into the urban context. First, farming in Tokyo has adapted its activities to the city, developing new types of production, of commercialisation, and strategies to diversify the farmer's income. Secondly, in the context of a shortage of land, farming in Tokyo is being done in unconventional locations in the city. Thirdly, the profile of the Tokyo farmer is being reinvented by his use of urban resources.

24.3.1 In Tokyo, agriculture is really urban

Today, agriculture in Tokyo has become as urban as other activities in the city. This urbanity is based first on the adaptation of agricultural practices to the urban context and secondly on the use of the advantages the urban context provides to agriculture.

In Tokyo, agricultural practices have been adapted to limit their impact on the urban context. The essential part of agricultural labour is done by hand, which restricts noise pollution. A dairy farm situated in the centre of Tokyo has found an original solution to limit olfactory pollution by dispersing coffee powder all around the farm.

Tokyo farmers have also established special links with consumers and have turned their farms into social spaces in the middle of the city. Leisure agriculture for city dwellers (*Taiken noen*) has become more and more popular. It consists of a farmer keeping a part of his farm for the use of city inhabitants to give them the opportunity to grow food. The farmer gives advice on cultivation practices (for example, on the use of fertiliser or pesticides). In exchange, the farmer receives a financial contribution from the users and from the district. More specifically, the ward subsidises two-thirds of maintenance costs and one-third of operating costs of each farm (Shiraishi, 2001).

The benefits of leisure agriculture are also interesting for public authorities as they can provide an effective public programme at low cost and maintain a control of urban development. Frequently, small collective spaces are linked to the farm and professional and amateur farmers can gather together around a table to taste their own products. These encounters strengthen the links within the neighbourhoods, links that will be particularly useful for local emergency help in case of natural disasters (earthquake, flood, etc.).

Tokyo farmers have developed strategies in order to benefit from the proximity of the urban context. They have invented marketing strategies, which allow them to sell their products to urban consumers. Examples may be the establishment of contracts between urban dwellers and farmers (Tekkei) or the direct sale of products on the farm. The last one is really well organised as the union of agricultural cooperatives publishes a map where direct selling points are located. On site, a green flag indicates a selling point. The selling devices can be very simple (a shelf with a moneybox), or complex (coin-operated vegetable lockers), more mobile (vegetable trolley) or more organised (collective farming shop) (some examples in Figure 24.3).

Figure 24.3. The selling device of farmers in Nerima ward (Tokyo): (A) a vegetable shelf, (B) a flag sign, (C) a cooperative shop and (D) some pictures of the farmers.

Tokyo is not only a consumer area but also represents the opportunity to develop a secondary income for farmers. Today, farming has become a part-time activity. In 2000, 81% of Tokyo farms were part-time operations (TMG, 2005). The other strategy to increase the farmer's income is to diversify activities on the farm. For example, in an ad hoc way the production of renewable energy has been established on some farms (mostly through solar panels). More commonly, the farmer is becoming a real estate manager, using his land not only for agriculture. The strategy here is to keep farming, which allows one to receive a state subsidy. At the same time one can also sell a part of the land as constructible areas (the zoning system in Japan is very flexible) or to rent the land for other temporary activities (parking lots, supermarkets) (Niwa, 2009).

Figure 24.4 shows a typical farm in Tokyo. We can see the actual farm and former agricultural land now transformed into a housing complex and parking areas (as indicated in the figure). Without pressure on the urban land, these activities would be less profitable for the farmer, so in the end, the city offers the opportunity to maintain agricultural activity. This should not hide the fact that the city is also a direct threat to agriculture and that it is still more profitable for a farmer to sell his land than to carry on with farming activities. In the end, we can forecast that traditional agriculture taking place on urban land may disappear. However, the tendency in Tokyo actually also shows that agriculture is being relocated in new urban soil.

24.3.2 The new agro-urban soil

The pressure on urban land has led to an efficiency in the usage of land resources. The book *Made in Tokyo* shows some of these strategies such as the multiplication of activities for the same building, the establishment of spatial synergies between activities or the reuse of the 'waste space' of the city (Kaijima *et al.*, 2001).

Figure 24.4. A typical urban farm (Hiyoshi, Kohoku-ku, Yokohama).

In Tokyo, urban agriculture also uses these strategies in order to maintain its activities in the city. Agriculture takes place in the empty spaces of the city. These spaces cannot be devoted to other functions because of their situation or their small size (200 to 1000 m^2). Thus, they are exempt from urban pressure and agriculture can be quite stable in these areas. Agriculture can be found in Tokyo in the proximity of transport infrastructures, under electricity pylons or in flooding areas.

Agriculture also takes place on the roofs of buildings (Figure 24.5), façades and in basements. Buildings have really become a support for farming. In the district of Omotesando, in the hypercentre of Tokyo, Mr Imura has introduced agriculture on the roof of a building. Small agriculture plots (2.50 m × 1 m) are available to rent for 300 euros per month. Although it looks very expensive, this operation was a great success as all the plots are occupied.

Agriculture in Tokyo can also take place inside a building. Pasona O2 farm (Figure 24.6) is situated on the first floor of a building in the centre of Tokyo (Yaesu district). This farm produces 200 types of fruits and vegetables. The main part of the farm is constituted by a rice field and takes place in the lobby of the building. This agricultural space works because it is essentially based on hydroponic culture and artificial light.

For the moment, the spread of agriculture inside, or on top of, buildings is limited. However, it can evolve quickly as the municipality launched a regulation in 2007 to increase the amount of green roof in the city. This regulation compels the owner to plant one quarter of his roof if it is more than 1000 m^2 for a private building and 250 m^2 for a public building.

Figure 24.5. Roof agriculture in Omotesando (Tokyo).

Figure 24.6. Agriculture in Pasona Building (Tokyo).

24.3.3 The new urban farmers

Today, the profile of the farmer in Tokyo is changing. This is particularly interesting in a context where 80% of the farmers are more than 50 years old (TMG, 2005).

Private companies, in their own office buildings, are putting urban agriculture in Tokyo into practice, such as Pasona (human resources) and NTT (telecommunications). In the NTT example, agriculture is taking place on the roof of their main building in Tokyo since 2006. This operation is called 'sweet potato roof'. This roof farm has contributed to the production of sweet potatoes, has created employment and has taken part in reducing the heat island phenomenon.

The profile of the farmer is now being reinvented in Tokyo. People who traditionally have no links with agriculture are developing this activity. The example of Shiho Fujita is particularly striking. Mrs Fujita is a very popular pop singer in Japan. She is affiliated with the 'Shibuya girl' trend, which consists in a questioning of Japanese identity by dying one's r hair blond and by adopting a schoolgirl style. Her background did not really destine her to have an interest in farming and yet, in 2009, she started a rice farm project. Her rice is sold on the internet with packaging decorated with iconography from the fashion district (Shibuya). Mrs Fujita has turned farming into a fashionable trend in Japan, and the word *nogyaru* (agriculture girl) was designated as the word of the year 2009.

These new profiles of the urban farmer are interesting and show a real dynamism in Tokyo. However, we can also question the trivial aspect of these experiences and the benefit these operations may have on urban agriculture in general. The important media coverage they have received may make us fear that they are only undertaken for this goal. Agriculture has a strong image of reliability in Japan. Private companies are thus very interested to be associated with this image. Likewise, if Mrs Fujita started a farm, it is not only for her interest in agriculture or to do business. It may also be to re-gild her image, which has been one of a superficial woman. More generally speaking, in all these new kinds of urban agriculture operations, the productivity and the profit of the activity is not the first goal of the farmer. For example, the productivity of the Pasona farm or the Omotesando roof is very low.

Furthermore, these operations can become a direct threat to the existing traditional farming of Tokyo. As they do not need to be profitable, they create unfair competition with the farmers that depend on their income to survive. Often, the leisure farmer sells his production at a loss, upsetting the local farming market. The leisure farmer is also less personally engaged in his farm than if it was his family heritage. As the leisure farmer is a dilettante, his interest in farming may not last and thus the stability of the farming operation is weakened. That can represent a real danger because, in Japan, a farm, which is abandoned, soon becomes un-exploitable.

So in the end can we still speak of agriculture? Or should we use other expressions like urban gardening or agricultural marketing?

21.4 Conclusion

The goal of this chapter was to understand why there is agriculture in Tokyo. We have shown how agriculture in Tokyo has historic roots and how Tokyo's development was based on agriculture. We have also shown how agriculture fits into the urban context and develops strategies to find new spaces in the city and how the farming profession is now being reinvented.

Another aim of this chapter is to show how agriculture in the city can be dynamic and lead to the implementation of imaginative strategies. Tokyo is very interesting because it shows that agriculture can take place in the city and be viable even in a megalopolis where urban pressure is so extreme. Tokyo's example also shows that an urban form mixed with agriculture can be sustainable. This is particularly interesting in the western context where hybrid urban forms are denigrated and agriculture has difficulties to be considered legitimate in the city.

Acknowledgements

This chapter is based on research that was done during my stay as an invited researcher in Keio University, Yokohama, Japan. I would like to thank particularly Professor Ikaga (Keio University), Professor Godo (Meiji Gakuin University) and students (Tawada Tomomi, Hantani Eriko, Saeko Tsumuraya).

References

Ashihara, Y., 1989. The Hidden Order: Tokyo through the twentieth century. Kodansha International, Tokyo, Japan, 201 pp.

Berque, A., 1993. Du geste à la cite. Gallimard, Paris, France, 244 pp.

Godo, Y., 2007. The puzzle of small farming in Japan. Asia Pacific Economic Paper, No. 365.

Godo, Y., 2006. Financial Liberalization and Japan's Agricultural Cooperatives. Poster paper prepared for presentation at the International Association of Agricultural Economists Conference. Gold Coast, Australia.

Kaijima, M., Kuroda J. and Tsukamoto Y., 2001. Made in Tokyo. Kajima Institute Publishing Co, Tokyo, Japan, 191 pp.

Naito, A., 2003. From Old Edo to Modern Tokyo, NIPPONIA no. 25 June 15 Available at http://web-japan. org/nipponia/nipponia25/en/feature/index.html.

Niwa, N., 2009. La nature en ville peut-elle être agricole? De la Suisse au Japon. Urbia, no. 8. Urbanisme végétal et agriurbanisme. Available at http://www.unil.ch/ouvdd/page74948.html.

Ogino, Y. and Ota S., 2007. The evolution of Japan's rice field drainage and development of technology. Irrigation and Drainage 56: S69-S80.

Shelton, B., 1999. Learning from the Japanese city: West meets East in urban design. Routledge, New York, NY, USA, 210 pp.

Shiraishi, M., 2001. The preservation and use of scarce agricultural land in suburban areas. NLI Research Institute. no. 148. Available at http://www.nli-research.co.jp/english/socioeconomics/2001/li0101a.html.

Sorensen, A., 2002. The making of urban Japan. Nissan Institute/Routledge Japanese Studies Series, London, UK, 386 pp.

Tajima, K., 2007. The marketing of urban waste in the early modern Edo/Tokyo Metropolitan area. Environnement Urbain/Urban Environment 1: 13-30.

TMG (Tokyo Metropolitan Government Bureau Of General Affairs), 2005. Statistic Division Management and Coordination Section, Tokyo statistical yearbook. Available at http://www.toukei.metro.tokyo.jp/homepage/ENGLISH.htm.

Yokohari, M. and Amati, M., 2005. Nature in the city, city in the nature: case study of the restoration of urban nature in Tokyo, Japan and Toronto, Canada. Landscape Ecological Engineering 1: 53-59.

Yokohari, M., Takeuchi, K., Watanabe, T. and Yokota, S., 2000. Beyond greenbelts and zoning: A new planning concept for the environment of Asian Mega-cities. Landscape and Urban Planning 47: 159-171.

Chapter 25

Recommended practices for climate-smart urban and peri-urban agriculture

Tara L. Moreau[1,3], Tegan Adams[2,3], Kent Mullinix[3], Arthur Fallick[3] and Patrick M. Condon[4]
[1]*Pacific Institute for Climate Solutions, University of British Columbia, 2060 Pine Street, Vancouver, BC V6J 4P8, Canada;* [2]*Faculty of Land and Food Systems, University of British Columbia, MCML 248 – 2357 Main Mall, Vancouver, BC V6T 1Z4, Canada;* [3]*Institute for Sustainable Horticulture, Kwantlen Polytechnic University, 12666 – 72 Avenue, Surrey, BC V3W 2M8, Canada;* [4]*Design Centre for Sustainability, University of British Columbia, HR MacMillan Building, Room 394 – 2357 Main Mall, Vancouver, BC V6T 1Z4, Canada; taramoreau@gmail.com*

Abstract

Agriculture's reliance on synthetic fertilisers, pesticides and fossil fuels contributes to greenhouse gas (GHG) emissions and climate change in numerous ways – including land-use changes, machinery operations, chemical manufacture, chemical applications, leaching and runoff. In British Columbia, Canada, provincial policies mandate municipalities to reduce GHG emissions by 33% in 2020 and 80% in 2050. To initiate immediate action, over 170 Municipalities have signed the voluntary Climate Action Charter which commits them to becoming carbon neutral by 2012. For cities, moving towards carbon neutrality requires local governments to quantify, reduce and offset GHG emissions from public operations. Given the rise of agriculture within and around cities, the development and support for climate-smart agriculture is crucial to supporting production systems that can simultaneously address food security, emissions reductions and climate change adaption. A collaborative project between planners, landscape architects, local governments, agronomists and academic researchers is working to design 'Low Carbon Communities' by addressing urban issues related to buildings, transportation, energy, waste and food. Food work is being evaluated through an integrative framework for food systems planning within cities called Municipally Enabled and Supported Agriculture (MESA). With a particular focus on Metro Vancouver, the objective of this study in the context of MESA and Low Carbon Communities was to: categorise GHG emissions from agricultural production in the region, recommend climate-smart urban and peri-urban agricultural practices, explore the potential of carbon sequestration in urban and peri-urban agriculture, and identify measurable indicators for climate-smart practices. Key recommendations for implementing MESA are discussed.

Keywords: agri-food systems, carbon sequestration, greenhouse gas (GHG) emissions, Municipally Enabled and Supported Agriculture (MESA)

25.1 Introduction

Human interest in gardening, food production, lawns, green spaces, forests and parks results in a diversity of plant and animal species cohabiting with people in cities. The ability of urban vegetation (e.g. lawns, green roofs and golf courses) to affect the earth's carbon balance has received some attention in recent years (Byrne *et al.*, 2008; Getter *et al.*, 2009; Niinemets and Penuelas, 2008). However, a lack of ecological understanding of cities was recently identified as a risk to conservation efforts – highlighting a need for additional study of urban ecosystems (Corbyn, 2010). Public reallocation of urban lands into community gardens, farmer's markets, school gardens and urban/peri-urban farms is occurring throughout North America. With over 50% of the global population now residing in cities and given the increase of urban agriculture activities and human population, planning for climate-smart agriculture is essential to addressing food security, emissions reductions and climate change within municipalities. Climate-smart agriculture is production that 'sustainably increases productivity, resilience (adaptation), reduces and removes greenhouse gases (mitigation), and enhances achievement of national food security and development goals' (Food and Agriculture Organization, 2012).

Agriculture, categorised as urban, peri-urban, or rural, broadly refers to the production or transformation of plants and animals into foods, fibres, ornamental, medicines and fuel. The contribution of agricultural production to greenhouse gas (GHG) emissions is estimated at one-quarter of global anthropogenic GHG emissions (Scialabba and Muller-Lindenlauf, 2010). However, agriculture production is merely one facet of the larger food system that encompasses many sectors: farm input manufacturing, food production, transportation, processing, packaging and distribution, retailing, catering and consumption, home preparation and waste management (Ericksen, 2008; Garnett, 2008). It is estimated that if food handling and processing activities were accounted for in global GHG inventories, the agricultural food system would contribute to at least one-third of total anthropogenic emissions (Scialabba and Muller-Lindenlauf, 2010).

Given the contribution of agriculture to global GHG emissions and the eminent rise of local, urban-focused food production and the potential for development of regional agri-food systems, it is fundamental that agri-culture's reintegration to urban-culture include climate-smart production practices in its planning, implementation and management. In British Columbia (BC), Canada, climate change mitigation legislation (Bill 44: Green House Gas Reduction Targets Act 2007 and Bill 27: Local Government Statutes Amendment Act 2008) mandates municipalities to reduce total GHG emissions by 33% in 2020 and by 80% in 2050. To address this challenge, over 170 municipal governments in BC signed the voluntary Climate Action Charter pledging municipalities be carbon neutral by 2012 (BC Ministry of Community and Rural Development, 2010). As part of these initiatives, local governments are required to include targets, policies and actions for GHG mitigation in their Official Community Plans (OCPs).

Carbon sequestration, or, holding of carbon, refers to the transfer of atmospheric carbon dioxide (CO_2) to plants, soils and fauna via photosynthesis and decomposition. Although carbon sequestration can only make modest contributions to mitigating CO_2, it has been identified as a partial solution for short- and medium-term reductions of atmospheric carbon (Hutchinson *et al.*, 2007; Lal, 2009; Morgan *et al.*, 2010). Provincial reports anticipate that BC agricultural soils will become important producers of carbon offsets through soil carbon sequestration (LiveSmartBC, 2009). Identified by Lal and Follett (2009), there is a need to sequester more carbon (C) in urban ecosystem, and management practices can be applied to improve soil C storage in urban soils. However, it is important to recognise that soil C sequestration is non-permanent, difficult to verify and is a not a substitute for, but rather a complement to, GHG emission reduction strategies (Lal, 2009).

The mandate of BC municipalities to become carbon neutral creates an opportunity to transition to resilient and sustainable cities that contribute to climate change mitigation and adaption. Researchers from the Pacific Institute for Climate Solutions (PICS), the Institute for Sustainable Horticulture (ISH, Kwantlen Polytechnic University) and the Design Centre for Sustainability (DCS, University of British Columbia) are working to design and plan better food systems as one dimension of 'Low Carbon Communities'. Low Carbon Communities is an explorative concept that evaluates reducing GHGs from buildings, transportation, energy, waste and food within cities (Condon *et al.*, 2010).

To address the global issues that affect food such as rising food prices, climate change, food security and sovereignty, employment and poor economic returns for farmers, ISH and PICS researchers are advancing Municipally Enabled and Supported Agriculture (MESA) as a framework for having local governments take an active role in reconfiguring agri-food systems. The goal of MESA is to produce a compendium of appropriate and targeted concepts, tools and strategies which can be used at the municipal level to enable and support the design of local-scale, climate-smart, human-intensive sustainable agri-food systems. Diverse models of MESA are currently being implemented in participating municipalities in South-West, BC. As part of the larger MESA project, the goal of this study was to categorise GHG emissions from agricultural production in Metro Vancouver and to recommend practices for climate-smart urban and peri-urban agriculture. An examination of carbon sequestration in cities was conducted and factors affecting its potential are discussed. Recommended climate-smart agriculture practices are linked with progress indicators that could be used by municipalities to measure, report and verify project outcomes.

25.2 Methodology

The scope of this study is limited to the Metro Vancouver region in British Columbia, Canada. Four key aspects of agriculture in the region were studied: (1) agriculture in the Metro Vancouver area; (2) sources of agriculture production GHG emissions; (3) carbon sequestration in cities; and, (4) recommended practices for climate-smart urban and peri-urban agriculture.

25.2.1 Agriculture in Metro Vancouver

Political context

With a population of approximately 2.1 million people, the Metro Vancouver area ($\sim$ 2,900 km^2) comprises 22 municipalities, one electoral area and one treaty First Nation. Fourteen percent of the area ($\sim$ 410 km^2) is covered by agricultural land (Metro Vancouver, 2007). A unique feature within the province of BC is the Agricultural Land Reserve (ALR). Established in 1973, this farmland conservation policy was intended to conserve agricultural land against urban encroachment and enhance agricultural viability. Discussions in BC around the control and management of the ALR are controversial and public opinions tend to be highly divided because urban land economies place a premium on land for housing (Condon *et al.*, 2010).

Regional climate

Agriculture is highly dependent on climatic conditions and is susceptible to projected climate changes resulting from increased GHGs in the atmosphere (e.g. increased temperatures, erratic water availability, extreme weather events, increased sea levels, and amplified pest outbreak) (Ministry of Environment, 2011). BC's Lower Mainland is subject to a favourable yet seasonal food production climate and contains some of the richest soil for agriculture in Canada. The Metro Vancouver region is characterised by an oceanic climate that has rainy winter months (especially between October and March) and dry summer months (drought conditions in July and August). Vancouver's average maximum temperature is 6 °C (43 °F) in January and 22 °C (72 °F) in July, and its average annual rainfall is approximately 117 cm (44 inches) (Hello BC, 2010).

25.2.2 Sources of agriculture production GHG emissions

To understand where sources of emissions from food production in Metro Vancouver occur, this study used methodology similar to a partial life cycle assessment (LCA). In addition to considering emissions from agricultural land use (inventoried in our National and Provincial Emissions Reports), emissions generated from the manufacture and transport of agricultural inputs (e.g. agri-chemicals, agri-machinery, water) and on-farm energy were also considered. Based on information from government documents, the International Panel on Climate Change (IPCC) and peer-reviewed literature, farm practices contributing to emissions were sorted into *Farm Management Categories* (Artemis, 2002; Niggli *et al.*, 2008; Scialabba and Muller-Lindenlauf, 2010; Smith *et al.*, 2007, 2010; Willey and Chameides, 2007) .

25.2.3 Carbon sequestration in cities

One recommended approach to measuring carbon sequestration is to use permanent sampling plots to monitor soil carbon over time (Hamburg, 2000). Accounting for carbon sequestration can involve measuring four pools of carbon in aboveground living biomass, belowground living biomass, necromass and soils. In general, assessments of carbon sequestration in urban soils and land covers are limited.

25.2.4 Recommended practices for climate-smart urban and peri-urban agriculture

Many ambiguities in agriculture-related emissions occur. Based on the Farm Management Categories, practices for climate-smart urban and peri-urban agriculture were identified. As an additional tool for municipalities, recommended practices were linked with progress monitoring indicators as tools to measure initiative baselines, set targets, evaluate project successes and highlight areas for improvement.

25.3 Results

25.3.1 Agriculture in Metro Vancouver

The 2006 Census of Agriculture indicated that 14% (41,035 ha) of total land area in Metro Vancouver is covered by agricultural lands (Metro Vancouver, 2007). Small urban agricultural practices, such as community gardens, urban farms and rooftop gardens were not included in the 2006 census. Most agricultural land within Metro Vancouver is located in three municipalities: Langley (32%), Surrey (23%) and Delta (18%). Cropped land comprises ~59% of the total area while the remaining non-cropped land is pasture or has other uses (such as Christmas tree farms, woodlands ,wetlands and recreational lands). A diversity of crops are grown in this area but commodity production of berries (4,643 ha), vegetables (3,025 ha) or nursery crops (1,192 ha) dominate. Within Metro Vancouver, animal production consists primarily of poultry and eggs (4,075,048 birds), followed by cattle and diary (29,433 cows). Other livestock include horses and ponies, sheep, pigs, goals, llamas, alpacas and bees.

25.3.2 Sources of agriculture production GHG emissions

Factors affecting GHG emissions from agriculture production in BC where divided into nine *Farm Management Categories* including (1) topography, landscape and climate; (2) soils characteristics; (3) farm mechanisation; (4) water: drainage, irrigation monitoring and management; (5) seeding and crop production practices; (6) crop nutrient inputs; (7) manure management practices; (8) animal characteristics; and (9) pest management practices. In general, GHGs emitted from global agriculture include carbon dioxide (CO_2), methane (CH_4) and nitrous oxide (N_2O) (Smith *et al.*, 2007, 2010). CO_2 is released primarily from microbial decay, biomass burning, decomposing soil organic matter and from on-farm fossil fuels used for machinery. N_2O is released through microbial transformation of nitrogen in soils, in manure pits, and from synthetic fertilisers applied to fields. CH_4 is generated through decomposition of organic materials in anaerobic conditions, notably from ruminant livestock digestion, stored manures and from crops grown in flooded conditions. GHG fluxes from agriculture production are complex and highly variable. Most production systems release all three GHGs (CO_2, N_2O, CH_4). For example, livestock systems produce CO_2 from land use and its changes, CH_4 from animal enteric fermentation and N_2O from manure and slurry management (International Food Policy Research Initiative, 2009).

25.3.3 Carbon sequestration in cities

Sequestering carbon in cities presents many possibilities based on a variety of land uses found in urbanised areas. In theory, carbon sequestration occurs in cemeteries, gardens, golf courses, green roofs, hedgerows, lawns, parks, urban forests, sports fields and other sites that contain soils and plants (Getter *et al.*, 2009; Nowak and Crane, 2002). Obtaining estimates of sequestration potential for different land uses is difficult due to the dynamic nature of soil CO_2 fluxes and the multitude of factors that affect carbon cycles. Some factors that influence carbon fluxes and subsequent sequestration include climate, soil types and characteristics, plant species and cultivars, vegetative structure (composition, arrangement, and density), land use histories, land management, and soil fauna. Due to the complexity of quantifying carbon sequestration, this study did not include estimates for urban land sequestration rates. Although it is estimated that carbon sequestration within urban areas can positively influence urban GHG fluxes, a better understanding of the influencing factors and consistent measuring tools are needed to make proper comparisons.

25.3.4 Recommended practices for climate-smart urban and peri-urban agriculture

Recommended practices for climate-smart urban and peri-urban food production geared at sustainably increasing productivity, resilience (adaptation), minimising GHG emissions and maximising carbon sequestration are outlined in Table 25.1. Limited study of urban agro-ecosystems means that this study has very little baseline data from which targets and achievements may be measured. Therefore, recommended practices are linked with potential indicators to provide municipalities a methodology to measure project baselines, set targets, report and verify project outcomes and highlight areas for improvement. This list is not conclusive and will likely evolve with an expanded understanding of urban agriculture production.

25.4 Discussion and conclusions

The need for understanding GHG emissions from global and local food systems has never been more urgent. The PICS-ISH collaborative research team is evaluating the potential of Municipally Enabled and Supported Agriculture (MESA) to enhance sustainable local food production and contribute to atmospheric GHG reduction. Much work is still needed to conceptualise the MESA framework. As part of an effort to develop MESA for Low Carbon Communities, this study has recommended practices with associated indicators for climate-smart urban and peri-urban agricultural systems. Clearly, the adoption of Climate-Smart agriculture will be highly influenced by policy at local, regional and national levels. Examples such as, incentive-based programmes, have been explored by governments to encourage agricultural practices that have a positive impact on their environment and a variety of policy

Table 25.1. Recommended practices for climate-smart urban and peri-urban agriculture production systems and indicators for measuring, reporting and verifying outcomes.

Climate-smart recommended practices	Potential indicators to evaluate impacts of recommended practices
Optimised agro-ecosystem land productivity	• total farm area • % area of land under crop production • % area of land managed with agroforestry techniques • % area of land covered with native vegetation • % area of land covered with buildings • % area of unused or degraded land
Efficient land management and planning	• % area of land managed according to government environmental farm plans • % area of land under organic management • % area of degraded land restored through improved landscape structure
Optimised food production	• number of crop species grown per unit area • yields of crops per unit area • caloric value of crops grown per unit area • nutritional value of crops grown per unit area • ratio of calories produced to land area under production
Maximised farm energy efficiency	• energy audit of farm buildings completed • % machinery fuel consumed as fossil fuels • % energy consumed on-site from fuel sources generated on-site (e.g. biogas tech; on-site incinerator; agri-fuels from crop residues on site)
Maximised soil health and water management efficiency	• % nitrogen inputs applied from organic sources (e.g. manure, compost) • % area of land tilled • % area of land managed with reduced tillage • % area of land with crop residues integrated into soils for nutrients • % area of land with bare soil • % area of land planted with cover and perennial crops • % area of land under mulch • total amount of water applied per specified area (e.g. metres squared, acres, hectares) • % area of land irrigated with water from rainwater collection devices • % area of land receiving water through ultra-efficient irrigation technology and application rates based on crop evapotranspiration rates • % area of watershed land protected or stabilised with vegetative management (or buffer zones).
Maximised carbon sequestration	• estimate of soil organic carbon per specified area (as measured through dry combustion on c:n auto analyser) • estimate of aboveground biomass

(Table continued on next page)

Table 25.1. Continued.

Climate-smart recommended practices	Potential indicators to evaluate impacts of recommended practices
Efficient nutrient management	• % area of total land with soil analysis information available
	• % area of land with synchronised applications of nutrients specific to crop needs
	• mass or weight of nitrogen fertiliser used per unit area
	• % mass or weight of total agri-chemicals displaced or reduced over time (e.g. one year)
	• mass of organic amendments used per unit area
	• % area of crop producing land with legumes (nitrogen-fixing plants)
Productive compost systems	• mass of organics diverted from landfill through composting
	• type of compost system
	• mass of mature compost added to soils per unit area
	• nutrient analysis of mature compost
Closed-loop energy cycling	• % manure managed with methane capture or biogas technologies
	• % manure managed with covered or dry lot storage systems as compared to liquid slurry or open-aired systems
	• % area of land amended with manure from on-site animals
	• % total waste generated removed from site
	• % total waste generated recycled or reused on site
Maximised biodiversity and species conservation	• number of native plants per unit area
	• number of native crops per unit area
	• number of livestock species and breeds (e.g. cattle, pigs, poultry, sheep and goats) per unit area
	• number of regionally adapted plants per unit area
	• number of native pollinators per unit area
	• number of bird, insect, and mammal species per unit area
Diversified cropping systems	• % area of land managed through crop rotation
	• % area of land intercropped/poly-cropped
	• % area of land left as, or mimicking, semi-natural habitats (e.g. hedgerows, buffer strips)
Integrated pest management	• % total land managed with ipm practices
	• % area of land monitored for pests
	• % area of land monitored for beneficial insects
	• % area of land where biological control agents are released
	• mass of pesticide inputs applied per unit area

options are being analysed for climate-smart urban and peri-urban agriculture. However, it is too early in our research to make concrete recommendations. Some concepts we are exploring include carbon offsetting through climate-smart agriculture, payment for environmental services and long-term land tenure agreements.

Key recommendations:
* incorporate climate-smart urban and peri-urban practices into municipal/regional planning;
* increase research and data collection within regions and cities that quantify agriculture on urban and peri-urban land area in order to obtain reliable estimates of agriculture production and carbon sequestration potential;
* create regional networks to support climate-smart education programmes that support farmers and regional research.

MESA aims to identify win-win agronomic practices as 'keystone features' of climate-smart urban and peri-urban agriculture. The term 'keystone feature' is derived from 'keystone species', which is used by ecologists to describe important species within ecosystems. The term is also analogous with the keystone in an arch where, without the keystone, the arch collapses. Soil organic carbon, an indicator for soil carbon sequestration, is an example of a keystone feature. Increasing soil carbon, through application of manure, compost, cover crops and mulches, provides numerous benefits to the soil including increased water holding capacity and infiltration, increased microorganism diversity, buffering capacity, improved salinity management and increased carbon sequestration. Direct sampling of soil carbon within urban landscapes is a feasible technique for farmers and municipalities. Efforts to create local databases of soil carbon are needed in order to accurately assess, model and compare carbon sequestration in urban areas. The research team is currently exploring methods by which climate-smart recommended practices can be prioritised in the Metro Vancouver region. Given the variability and unpredictability of on-farm GHG emissions as influenced by temperature, moisture, etc., regional data is necessary for prioritising actions. However, regional GHG emissions data for BC and Metro Vancouver agriculture is lacking. Nonetheless, recognised practices to reduce on-farm GHG emissions have been identified and could provide guidance on future actions (Eagle *et al.*, 2010; Niggli *et al.*, 2008; Smith *et al.*, 2007; Snyder *et al.*, 2009).

One challenging area of future MESA research is the diverse, wide-reaching and often interconnected nature of agri-food systems. Another challenge involves addressing and anticipating tensions that arise between food security, climate-smart agriculture, social enterprises, food policy and economic vitality. For example, while certain practices, such as the substitution of synthetic fertilisers with organic amendments may lower GHG emissions, apparent and immediate economic efficiencies of conventional products prevents the use of alternatives despite knowledge of environmental unsustainability (Matthews *et al.*, 1993). The costs and benefits of economic incentives or assistance for sustainable practices are being explored within the MESA concept. A further challenge for designing MESA is to create agricultural communities that are resilient to environmental changes expected from climate change. Recent re-framing of climate change policy proposes that climate change be addressed as a persistent condition that must be coped with and managed rather than a discrete problem to be solved (Prins *et al.*, 2010). Some key actions proposed for agriculture in the BC region in 2002 still hold true today (Artemis, 2002) and include: enhancing agriculture's access to capital, increasing technical and scientific research, creating management plans for rural-urban fringes in order to reduce encroachment on farmland and farming, and assisting farmers in adopting sustainable agronomic practices.

Metro Vancouver is well suited to further expansion of urban and peri-agriculture with great potential for increased local food production. In a 2008 provincial survey, 90% of residents agreed that 'it is important that BC produce enough food so we don't have to depend on imports from other places' (Ipsos Reid Public Affairs, 2008). Despite potential for further increasing local production and high consumer interest in local foods, the current agri-food system restricts domestic supply of food in favour of imports. For local producers, obtaining access to mainstream markets is challenging due to buyer demand for a scale and consistency of supply that is difficult to achieve. Changes in the market structure and the use of direct marketing opportunities are necessary to increase regional access to local foods.

In efforts to support BC communities committed to reaching carbon neutrality, this study outlines practices for planning and implementing climate-smart urban and peri-urban agricultural systems. Climate-smart recommended practices combined with potential indicators provides local governments and project planners tools to measure baselines, set targets, report and verify project outcomes and highlight areas for improvement. There is much work to be done to reduce agri-food GHG emissions and dramatic changes in how food is grown, processed, transported, consumed and wasted necessitates collaboration between people, groups, organisations and governments.

References

Artemis, 2002. An economic strategy for agriculture in the lower mainland. Artemis Agri-Strategy Group, Burnaby, BC, Canada, 54 pp.

BC Ministry of Community and Rural Development, 2010. Climate action charter. Available at http://www. cd.gov.bc.ca/ministry/whatsnew/climate_action_charter.htm.

Byrne, L.B., Bruns, M.A. and Kim, K.C., 2008. Ecosystem properties of urban land covers at the aboveground-belowground interface. Ecosystems 11: 1065-1077.

Condon, P.M., Mullinix, K. and Fallick, A., 2010. Agriculture on the edge: Strategies to abate urban encroachment onto agricultural lands by promoting viable human-scale agriculture as an integral element of urbanization. International Journal of Agricultural Sustainability 8:104-115.

Corbyn, Z., 2010. Ecologists shun the urban jungle, Nature News. Available at http://www.nature.com/news/2010/100716/full/news.2010.359.html.

Eagle, A.J., Henry, L.R., Olander, L.P., Haugen-Kozyra, K., Millar, N. and Robertson, G.P., 2010. Greenhouse gas mitigation potential of agricultural land management in the United States: A synthesis of the literature. Nicholas Institute for Environmental Policy Solutions, Report NI R 10-04, Second Edition, 72 pp. Available at http://nicholasinstitute.duke.edu/ecosystem/land/TAGGDLitRev.

Ericksen, P.J., 2008. Conceptualizing food systems for global environmental change research. Global Environmental Change 18: 234-245.

Food and Agriculture Organization, 2012. Climate-Smart Agriculture for Development. Available at http://www.fao.org/climatechange/climatesmart/en/.

Garnett, T., 2008. Cooking up a storm: Food, greenhouse gas emissions and our changing climate. Food Climate Research Network, Centre for Environmental Strategy, University of Surrey, Surrey, UK, 155 pp. Available at http://www.fcrn.org.uk/fcrnPublications/index.php?id=6.

Getter, K.L., Rowe, D.B., Robertson, G.P., Cregg, B.M. and Andresen, J.A., 2009. Carbon sequestration potential of extensive green roofs. Environmental Science and Technology 43: 7564-7570.

Hamburg, S.P., 2000. Simple rules for measuring changes in ecosystem carbon in forestry-offset projects. Mitigation and Adaptation Strategies for Global Change 5: 25-37.

Hello BC, 2010. Climate and weather. Official Tourism site of British Columbia. Available at http://www.hellobc.com/en-CA/AboutBC/ClimateWeather/BritishColumbia.htm.

Hutchinson, J.J., Campbell, C.A. and Desjardins, R.L., 2007. Some perspectives on carbon sequestration in agriculture. Agriculture and Forest Meteorology 142: 288-302.

IFPRI (International Food Policy Research Initiative), 2009. Agriculture and climate change. International Food Policy Research Institute, 29 pp. Available at http://www.ifpri.org/publication/agriculture-and-climate-change.

Ipsos Reid Public Affairs, 2008. Poll of public opinions toward agriculture. Food and Agri-Food Production in BC. Investment Agriculture Foundation of BC. Available at http://www.iafbc.ca/publications_and_resources/other-publications.htm.

Lal, R. and Follett, R.F. (Eds.), 2009. Soil carbon sequestration and the greenhouse effect. Soil Science Society of America. Special Publication 57, Second Edition, Madison WI, USA, 410 pp.

Lal, R., 2009. Sequestering atmospheric carbon dioxide. Critical Reviews in Plant Science 28: 90-96.

LiveSmartBC, 2009. BC climate action plan. British Columbia Provincial Government. Available at http://www.livesmartbc.ca/government/plan.html.

Matthews, S., Pease, S.M., Gordon, A.M. and Williams, P.A., 1993. Landowner perceptions and the adoption of agroforestry practices in southern Ontario, Canada. Agroforestry Systems 21: 159-168.

Metro Vancouver, 2007. 2006 Census Bulletin #2 Census of Agriculture. Metro Vancouver, Burnaby, BC, Canada. Available at http://www.metrovancouver.org/planning/development/agriculture/Pages/default.aspx.

Ministry of Environment, 2011. Climate change: BC provincial impacts. Available at http://www.env.gov.bc.ca/cas/impacts/bc.html.

Morgan, J.A., Follett, R.F., Allen Jr. L.A., Del Grosso, S., Derner, J.D., Dijkstra, F., Franzluebbers, A., Fry, R., Paustian K. and Schoeneberger, M.M., 2010. Carbon sequestration in agricultural lands of the United States. Journal of Soil and Water Conservation 65: 6A-13A.

Niggli, U., Fließbach, A., Hepperly, P. and Scialabba, N., 2008. Low greenhouse gas agriculture: mitigation and adaptation potential of sustainable farming solutions. Food and Agriculture Organization, Rome, Italy, 14 pp.

Niinemets, U. and Penuelas, J., 2008. Gardening and urban landscaping: Significant players in global change. Trends Plant Science 13: 60-65.

Nowak, D. and Crane, D., 2002. Carbon storage and sequestration by urban trees in the USA. Environmental Pollution 116: 381-389.

Prins, G. , Galiana, I., Green, C., Grundmann, R., Korhola, A., Laird, F., Nordhaus, T., Pielke Jnr, R., Rayner, S., Sarewitz, D., Shellenberger, M., Stehr, N. and Hiroyuki, T., 2010. The Hartwell paper: a new direction for climate policy after the crash of 2009. Institute for Science, Innovation & Society, University of Oxford LSE Mackinder Programme, London School of Economics and Political Science, London, UK, 42 pp. Available at http://sciencepolicy.colorado.edu/admin/publication_files/resource-2821-2010.15.pdf.

Scialabba, N. and Muller-Lindenlauf, M., 2010. Organic agriculture and climate change. Renewable Agriculture and Food Systems 25: 158-169.

Smith, P. and Martino, D., 2007. Agriculture. In: Metz, B., Davidson, O.R., Bosch, P.R., Dave, R., Meyer, L.A. (Eds.) Climate Change 2007: Mitigation of Climate Change, Contribution of Working Group III to the Fourth Assessment Report of the Intergovernmental Panel on Climate Change. Cambridge University Press, UK and New York, pp. 497-540.

Smith, W.N., Grant, B.B., Desjardins, R.L., Worth, D., Li, C., Boles, S.H. and Huffman, E.C., 2010. A tool to link agricultural activity data with the DNDC model to estimate GHG emission factors in Canada. Agriculture, Ecosystems and Environment 136: 301-309.

Snyder, C.S., Bruulsema, T.W., Jensen, T.L. and Fixen, P.E., 2009. Review of greenhouse gas emissions from crop production systems and fertilizer management effects. Agriculture, Ecosystems and Environment 133: 247-266.

Willey, Z. and Chameides, B. (Eds.), 2007. Harnessing farms and forests in the low-carbon economy: how to create, measure, and verify greenhouse gas offsets. Duke University Press, Durham, NI, USA, 240 pp.

Chapter 26

The impact of local food production on food miles, fossil energy use and greenhouse gas emission: the case of the Dutch city of Almere

Jan-Eelco Jansma[1], Wijnand Sukkel[1], Eveline S.C. Stilma[1], Alex C. van Oost[2] and Andries J. Visser[1]
[1]Wageningen University & Research Centre, Business Unit of Applied Plant Research, P.O. Box 430, 8200 AK Lelystad, the Netherlands; [2]The Municipality of Almere, P.O. Box 200, 1300 AE Almere, the Netherlands; janeelco.jansma@wur.nl

Abstract

This case study is situated in the Dutch city of Almere (185,000 inhabitants), 30 km east of Amsterdam. Almere has to expand towards 350,000 inhabitants in 2030. As part of this expansion 15,000 new houses are planned in Almere Oosterwold, an area where urban agriculture will be integrated in urban development on approximately 4,000 ha fertile polder land. Which amount of food can be produced locally in Almere Oosterwold and what will be the impact of this local production on food miles (or kilometres), fossil energy use and greenhouse gas (GHG) emissions of the future city of Almere? We calculated that 6,200 ha conventional farming land (or 8,100 ha in case of organic production) is needed to produce approximately 20% of the daily food basket of the future 350,000 citizens. This local produce, mostly fresh food, contains 76% products from plant origins (fruit and vegetables) and 24% from animal origins (milk, eggs and some meat). The local production in the presented case reduces the GHG emissions with an estimated 27,000 tonne/year. This reduction is relative small because most of the products in this food basket are already being produced in the Netherlands. A considerable part of the reduction of GHG emissions is realised through the reduction of the consumers transport needed for shopping, the substitution of fossil fuel by renewable energy sources and the use of organic manure instead of synthetic fertilisers. These measures are more or less independent of local food production. This study underlines that other aspects have to be taken into account regarding the reduction of food miles, fossil energy use and GHG emissions of our food system. Influencing the composition of the food basket (for example less meat), redesigning the local food distribution system, influencing consumers behaviour towards food consumption and reducing the fossil energy use are better instruments to reach that goal.

Keywords: urban agriculture, food strategy, city planning, GHG emissions, local food

26.1 Introduction

After WWII the main Dutch spatial planning policy was to concentrate or cluster urbanisation (Van Remmen and Van der Burg, 2008). The goal of this policy was to keep the landscape open, to limit travel distance and to support amenities. This policy led to extremely sharp fringes between city and countryside in the Netherlands (Figure 26.1) with an increasing distance (both mentally and physically) between city and citizens and the countryside (and agriculture). On the city side consumers have become estranged of food production, nature and the basic values of rural live, like quietness, darkness and the rhythm of seasons. On the rural side, farmers produce food and products for the world market with hardly any connection to their neighbourhood (market).

Urban Agriculture (UA) produces food products within the city, or in the city fringe and simultaneously provides non-food products and services for city dwellers (Mougeot, 2000). UA is as old as our cities, but lost its function in the 19th century mainly due to new means of food conservation and transportation (Steel, 2008). Today, UA is regaining its function in both developing and developed cities worldwide, including the Netherlands (Dekking *et al.*, 2007). In addition to food production, UA has an added value. It can also contribute to the shape and management of the green fringe of the city. Moreover, it can provide green energy, buffer water and process city waste. Urban needs for healthcare, elderly services, childcare and education are already part of some agricultural enterprises in the Netherlands (for examples see Waardewerken, http://www.waardewerken.nl).

Agromere is a virtual suburb with 5,000 inhabitants where city life and UA (production) are re-connected on an area of 250 ha. This virtual suburb is the outcome of a combined

Figure 26.1. Illustration of sharply delineated city-fringes in the Netherlands (Aerodata International Surveys, 2007).

stakeholder and design process (Jansma *et al.*, 2010; Visser *et al.*, 2009). Agromere is situated in the Dutch city of Almere. Almere (2009: 185,000 inhabitants and the 7th largest city in the Netherlands) is a new and fast growing suburban town, 30 km east of Amsterdam. The original design of Almere is unique for Dutch standards. Founded in the seventies, with a town centre surrounded by several satellite towns within and between forests, parks, canals and ponds, the design of Almere at some distance resembles the ideas of Ebenezer Howards garden city. Even forms of UA *avant la lettre* were part of the original design of Almere (Zalm and Oosterhoff, 2010). Today, Almere is still this poly nuclear city with much more green and blue within its borders than average Dutch cities. UA never developed properly in Almere except for one commercial city farm in the city fringe (Dekking *et al.*, 2007).

Almere has to expand towards 350,000 inhabitants (to become the 5th largest city in the Netherlands) in 2030 because of growing housing needs in the Amsterdam area and the absence of locations to build. Some 15,000 new houses are planned northeast of Almere on approximately 4,000 ha of fertile farmland in the so-called Almere Oosterwold area (Figure 26.2). The Agromere design inspired the city developers to include UA as a part of their plans

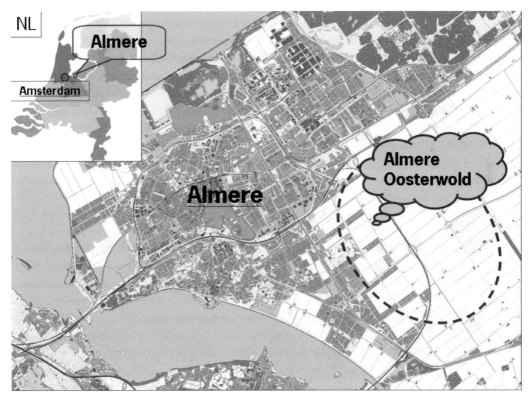

Figure 26.2. The Dutch new town of Almere, 30 km northeast of Amsterdam is situated in the southern part of the Flevopolder. 15,000 houses are planned in Almere Oosterwold, northeast of the city. Arable and dairy farmers producing for the world market predominate in this area.

for the Almere Oosterwold area (Visser *et al.*, 2009). In the draft Strategic Vision Almere 2.0, UA is introduced as one of the potential green and sustainable foundations of the future Almere Oosterwold. The city aims to develop Almere Oosterwold to one of its future food supplying areas (Stuurgroep Almere 2030, 2009).

In order to reach the goals described above, a number of questions have to be answered. What kind of food (products) and which amounts can be produced by future city farmers within the Almere Oosterwold area? What is the impact of local food production in the Almere Oosterwold area on the reduction of food miles (or kilometres), fossil energy use and GHG emissions of the future city of Almere?

26.2 Material and methods

A closer look on the environmental impact of local food production is needed. This underlying study, funded by the city of Almere and the Ministry of Economic affairs, Agriculture and Innovation, explores the impact of local food production in the Almere Oosterwold area on the reduction of food miles, fossil energy use and GHG emissions. In this study, three scenarios were defined for the development of the Almere food system. The core of these scenario's was the ambition to produce locally 20% of the daily food basket of each of the 350,000 city dwellers of future Almere. In this case study locally was defined as within a radius of 20 kilometres of the city centre of Almere. The 20% local production goal is based on two assumptions: (1) the available area under cultivation in the 20 km radius and (2) the potential local production needed for a suburb as calculated within the Agromere study (Jansma *et al.*, 2010).

26.2.1 Scenario 0 (S0): business as usual

Production, processing and distribution of the daily food basket is based on current practices:
A. Production
  – Current Dutch agricultural production methods.
  – Farming systems based on the use of artificial fertilisers and pesticides.
B. Distribution
  – Production partly produced in the Netherlands, partly foreign.
  – Current distribution patterns of food, from primary production through distribution centres and retail to consumers.
  – Current processing and storage of food and food products.
C. Energy use
  – Energy use fully based on fossil fuels.

26.2.2 Scenario 1 (S1): agri business hybride

In this scenario 20% of the daily food basket is locally produced, processed and distributed. Moreover 20% of the energy use, needed for local production, processing and distribution of the produce, is based on renewable energy sources.
A. Production
  – Local production is based on integrated production systems, with re-use of materials and integrated pest management. The production level is based on current Dutch practise.
  – Local production systems are based on both manure and synthetic fertilisers.
B. Distribution
  – Local distribution and processing of the (20%) locally produced food. The remaining 80% is partly produced in the Netherlands and partly foreign (through current production methods).
  – Local food and food products will be delivered or brought to distribution points or shops within walking distance of the consumers. There is no consumers mileage for collecting the local produced food and food products.
C. Energy use
  – Up to 20% of the fossil energy need for local production will be replaced by energy based on renewable sources (for instance wind energy, sun collectors and fermentation of organic waste). Fossil energy use of greenhouse production is reduced with 20% compared to present use.

26.2.3 Scenario 2 (S2): agri business ecology plus

In this scenario 20% of the daily food basket is locally produced, processed and distributed. The local production, processing and distribution in this scenario is based on 100% renewable energy sources, new distribution systems and maximum reuse of waste.
A. Production
  – The local food production is fully organic, with production levels based on the current Dutch production level for organic production.
B. Distribution
  – Local distribution and processing of the (20%) locally produced food. The remaining 80% is partly produced in the Netherlands and partly foreign (through current (non-organic) production methods).
  – Local food and food products will be delivered or brought to distribution points or shops within walking distance of the consumers. There will be no consumer mileage for collecting the local produced food and food products.
C. Energy use
  – The energy needed throughout production, processing and distribution of the local produce is fully based on renewable energy (for instance wind energy, sun collectors and fermentation of organic waste). Greenhouse production uses no fossil energy anymore. Moreover, it produces energy.

Jan-Eelco Jansma, Wijnand Sukkel, Eveline S.C. Stilma, Alex C. van Oost and Andries J. Visser

Replacement (with local production) only encompassed food already on the menu that could possibly be produced locally in Almere Oosterwold. Starting point was the daily food intake of the average Dutch person between 20-30 years (Hulshof *et al.*, 2004). The average age of the inhabitants of Almere is 35 years (Almere, 2010). From this daily food intake we selected the food items that fulfilled the following criteria:

- Possible to produce under the climate and soil conditions for the surroundings of Almere city. Items that can be produced include greenhouse vegetables.
- Only those products are replaced that need little or no processing and can be processed locally.
- Processed foods predominantly consist of regional ingredients.
- Possible to be produced in the area of land available within a 20 km range of the city centre.

Based on these criteria we aimed at a 20% share of local production of the household expenditure on food. The selected food items with comparable production and consumption patterns were aggregated in product categories. Within each product category one or two highly representative or so-called model products were selected to be produced locally. For instance, in the group of greenhouse vegetables tomato was selected as model product. All calculations per product category are based on this model product (Table 26.1). This means that under- or over-estimation of the results depends on the relative share of the model product in the product category.

Table 26.1. Percentage of consumed food categories that are to be produced locally and their share of the total household food expenditure.

Food category	Model product	% product replacement	% of food expenditure
potato and onions	potato & onion	100	1.7
greens	lettuce	60	0.8
greenhouse vegetables	tomato	80	0.5
root vegetables	carrot	100	0.6
cabbage	cauliflower	80	0.6
stem & sprout vegetables	leek	80	2.0
legumes	haricots & peas	80	0.3
fruits	apple	80	2.0
bread	wheat and barley	80	5.2
beer	barley	30	0.9
meat	beef & chicken	4	0.2
eggs	chicken eggs	80	0.7
milk	cow milk	50	1.7
cheese	cow milk	30	1.8
Total replacement of food expenditure			18.9

Based on the daily food intake of these model products, we calculated how much primary production was needed. An average of 30% food losses within the total food chain was taken into account in this figure. The total food chain of each model product, including production, transport, packaging, and storage (conventional or locally produced) was analysed on food miles, fossil energy use and GHG emissions (Figure 26.3). Home preparation and home storage was excluded from the calculations. Also the changes in carbon stocks in soil and vegetation were not included because they are missing in most literature references and the available estimations are highly uncertain. Local production is not expected to influence carbon sequestration in Scenario 1 compared to Scenario 0 because the cultivation methods for both scenarios are assumed the same. In Scenario 2 compared to Scenario 0 there could be an underestimation of the effect on the reduction of GHG emissions because of the change in cultivation method from integrated to organic. Organic agriculture in Western Europe is expected to sequester more carbon in the soil than integrated or conventional production methods (Sukkel *et al.*, 2008). According to the given scenario's the number of food miles, the fossil energy use and the GHG emissions were calculated for the local food production or for the current production. Data for energy use and GHG emissions in primary production were derived from Dutch studies (Bos *et al.*, 2007). Data for the rest of the food chain were derived from various literature sources (Blonk *et al.*, 2008; Dutilh and Kramer, 2000). Data of processed products like milk, beer and bread includes fossil energy use and GHG emissions during processing (and storage). Products like potato are both being consumed as processed (crisps) or non-processed (fresh) products. Data of this type of product only contains the figures of the non-processed variant. Data for food miles were based on estimations of the transport distances in every part of the food chain, including consumers food miles. It is assumed that there will be no differences in the way and the type of packaging of food in the three different scenario's. Consumers food miles (from store till house) in Scenario 0 were estimated to be three kilometres per 12 kg of food. This estimation is based on the typical situation in Almere where food distribution is mainly based on central supermarkets and consumers predominantly shop by car. Food miles were divided into heavy transport, light transport and car. Food miles, fossil energy use and GHG emissions needed to produce the

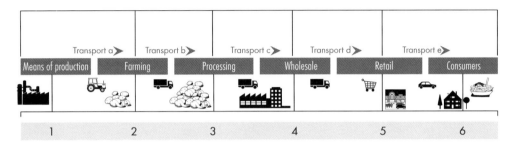

Figure 26.3. The total food chain of each model product, in this figure potato, was analysed on food miles (kilometres), fossil energy use and GHG emissions, from production means (like fertilisers and plant protection) to the consumer's doorstep.

production means like tractors, package machines and the installations to produce renewable energy were not included in the calculations.

26.3 Results

The total household food expenditure based on the daily food intake was € 3,520 per year (2.23 persons per household). Excluded in this figure are out of home food and (bottled) water. We started trying to produce 20% of the consumers expenditure with mainly fruits and vegetables that were in the standard food basket and could be produced locally. This resulted in 8.5% replacement of conventional food with local produce. Adding cereals for bread and beer, which were the other plant based food items, the replacement counted a total of 14.6%. Adding animal products (milk and eggs) was necessary to approximately meet the desired 20% local production. The very limited replacement of meat production consisted only of the by-products of milk and egg production. The food categories that could be produced locally, the percentage of the specific food category that was replaced and the share of this replacement in the total household expenditure can be found in Table 26.1. Except for potato, onion and carrot, none of the model products is replaced more than 80% with local produce, because of:

- local year round delivery is not possible (for example lettuce);
- important products within a product category cannot be produced locally (for example orange is a main fruit that can't be produced in the Netherlands);
- the quality of the local produce does not always meet the standards (Dutch wheat and barley, for example, do not always have bread making quality);
- limited local area (for example milk and meat production).

The needed hectares to produce the 18.9% of local food are presented in Table 26.2. We calculated that 6,200 ha conventional farming land (Scenario 1) or 8,100 ha in case of organic production (Scenario 2) is needed to produce 18.9% of the daily food basket of the future 350,000 citizens of Almere. Locally produced animal products are responsible for 4.4% (Table 26.1.) of the household expenditure, but they need more than half of the necessary local land area. The calculated energy use and GHG emissions for the local food production, processing and distribution in the different scenarios are presented in the Tables 26.3 and 26.4.

Table 26.2. Land area needed for the production of local produce for 18.9% of the household expenditure of the 350,000 future inhabitants of Almere.

	Scenario 1 (ha)	Scenario 2 (ha)
Plant products	2,951	3,968
Animal products (incl. feed production)	3,279	4,109
Total	6,230	8,077

Table 26.3. Fossil energy use in TJ for a 18.9% share of global (S0) or local (S1 and S2) food production for 350,000 inhabitants per scenario and per part of the food chain.

	Fossil energy use in TJ/year					
	Primary production	Transport	Processing	Packaging	Cold storage	Total
Scenario 0	363	99	23	36	30	551
Scenario 1	325	15	19	36	24	418
Scenario 2	152	0	0	36	0	188

Table 26.4. Greenhouse gas emission in Mg CO_2 equivalents for a 18.9% share of global (S0) or local (S1 and S2) food production for 350,000 inhabitants per scenario and per part of the food chain.

	Greenhouse gas emission in Mg CO_2 eq/year					
	Primary production	Transport	Processing	Packaging	Cold storage	Total
Scenario 0	66,261	7,475	3,768	2,339	1,323	81,166
Scenario 1	64,449	872	3,014	2,339	1,058	71,733
Scenario 2	51,725	0	0	2,339	0	54,064

Local production in this case reduces the GHG emissions with an estimated 27,000 tonne per year (Scenario 2 versus Scenario 0) or an equivalent of a yearly carbon uptake of 1,300 ha forest. Primary production has relatively the biggest share in both energy use and GHG emissions. The relatively bigger share of primary production in GHG emissions compared to energy use is caused by the greenhouse gasses methane and nitrous gas (N_2O) mainly occurring in primary production like dairy. Both gasses have a relatively larger contribution to GHG emissions than other greenhouse gasses. The reduction in fossil energy use and GHG emissions in Scenario 1 and Scenario 2 is strongest in the primary production. This reduction is mainly due to the replacement of fossil energy by renewable energy in the greenhouse production. To a smaller extent the reduction in the primary production is caused by the replacement of synthetic nitrogen fertilisers by organic manure.

The reduction of fossil energy use and GHG emissions in transport is mainly caused by the reduction of consumers transport. Reductions in energy use and GHG emissions in Scenario 1 and Scenario 2 in processing and cold storage are caused by the replacement of fossil energy with renewable energy. As a reference, the savings in energy use established with 18.9% local production are comparable to the energy use per year of 3,900 (Scenario 1) or 10,700 (Scenario 2) households.

Scenario 1 and Scenario 2 show a high reduction in food transport kilometres compared to Scenario 0 (Table 26.5). Scenario 1 and Scenario 2 are equal because both use the same distribution system. Because of home delivery or delivery to distribution points, the use of the van is largely increased in Scenario 1 and Scenario 2 compared to Scenario 0. The heavy truck is mainly used for bulk transport from producer to processor. Transport kilometres for the professional transport in Scenario 0 are mainly kilometres on national roads. The persons car transport (Scenario 0) and the van transport are mainly realised locally. The decrease in transport kilometres in Scenario 1 and Scenario 2 compared to Scenario 0 is predominantly caused by the avoidance of persons car use due to another distribution system.

26.4 Discussion and conclusions

The western world strongly relies on long supply chains. For example London, home to 7 million people, imports 80% of its food from overseas. A typical meal travels 3,000 km from farm to fork (Pearce, 2006). Food accounts for about a quarter of London's ecological footprint (Sustainweb, http://www.sustainweb.org/londonfoodlink/about). In the Netherlands food consumption accounts for 33% of national GHG emissions (Vringer *et al.*, 2010). There are hardly any data available on food miles in the Netherlands. Also few studies have been found on the effect of local food systems on energy use and GHG emissions. Hauwermeiren *et al.* (2007) found no relevant differences in energy use and GHG emissions between local and mainstream food systems. A Dutch quick scan case study compared an internet home delivery distribution system with a supermarket distribution system. Both systems showed little differences in GHG emissions (Van der Voort and Luske, 2009). However the internet home delivery in this study was not yet at its optimum scale.

The underlying study explores the impact of local food production in the Almere Oosterwold area on the reduction of food miles, fossil energy use, GHG emissions of the future city of Almere. In this study, three scenarios were defined for the development of the Almere food

Table 26.5. Transport distance in thousands of kilometres for a 18.9% share of global (S0) or local (S1 and S2) food production for 350,000 inhabitants per scenario and per means of transport.

	Distance in 1000 km	
	Scenario 0	Scenario 1 and 2
Heavy truck	1,270	17
Light truck	1,137	425
Van	0	1,628
Person car	15,713	0
Total	18,120	2,070

system. The core of these scenario's was the ambition to produce locally 20% of the daily food basket of each city dweller of future Almere (with 350,000 inhabitants).

Under the constraints given the goal of 20% share of the household expenditure for food that is locally produced, processed and distributed, without changing the current food consumption patterns, was hard to reach. Products that suited the demands for local production predominantly represent the cheaper products in the food basket. This is because these products don't need intensive processing, don't need to be transported for long distances and they consist of a relatively low share of animal products. The share of animal products was restricted because of the limitations in the defined available local area of land in the city fringe. The share of locally produced food could be increased with a prolonged cropping season (greenhouse culture), a bigger share of local produce in restaurants, catering and canteens and with a higher level of seasonal products. Moreover, more products could be produced and processed locally, but the small scale of local production and processing is not always economically viable.

Compared to Scenario 0, the reductions in energy use, GHG emissions and food miles in Scenario 1 and Scenario 2 are not necessarily related to local production. The reductions are mainly realised by replacing fossil energy with renewable energy (possibly to be produced on the farms) and by changing the local distribution system to home delivery.

Local production should normally reduce the (inter)national heavy weight truck transport, but the reduction of heavy transport in this study is relatively small. The main reason is that the products produced locally in this study are already predominantly produced in the Netherlands. Food transport distances within the Netherlands are generally not longer than approximately 150 kilometres. Therefore the heavy truck transport in this study causes much less transport kilometres than the consumers food transport by car.

The main cause of the reduction in food miles, fossil energy use and GHG emission by transport is the change in the distribution system of the local produce. In Scenario 1 and Scenario 2 consumers don't do their food shopping by car but the food is either home delivered or can be obtained in a shop or distribution point at close range. The assumption of completely avoiding person's car kilometres in Scenario 1 and Scenario 2 is probably too extreme, in practice there will be a certain percentage of food transport by person's car. However, the assumption of an average of three kilometre personal transport for an average of 12 kg of products, seems realistic for suburban towns like Almere. It might even be an underestimation (Hubert and Toint, 2002). The results indicate the high potential of decreasing transport kilometres by bringing the food to the consumer instead of the consumer to the food.

The results show that primary production has the largest share in energy use and even stronger in GHG emissions, with transport (being mainly consumers transport by car) as second contributor. The relatively large share of primary production is caused by the type of products that are chosen to be produced locally. Most of them being products already produced in the Netherlands and being only slightly processed and packed. When either more intensively packed, more processed or further transported products would have been replaced, the share of the primary production would have been much lower. On average for the total Dutch food consumption, 29% of the energy use and 39% of the GHG emissions is caused by primary production (Kramer *et al.*, 1999). The energy use and GHG emissions in the Dutch food chain caused by professional transport are respectively 7% and 6% (consumers transport excluded) of the total. Consumers have a 29% share in the energy use in the Dutch food chain of which a substantial part is caused by their food transport by car (Vringer *et al.*, 2010).

Considering the contribution of transport in the Dutch food system, the effects of the different scenarios on the reduction of food miles, energy use and GHG emissions are not surprising. Local food systems are expected to affect mainly professional food transport which however has a relatively small share in the energy use and GHG emissions of the Dutch food system. It should be noted that transport has more negative side effects than energy use and GHG emissions. For example the emission of fine dust, noise, traffic accidents and spreading of pests and diseases. This case study shows that there are other aspects than local food production that have a much larger effect on the reduction of food miles, energy use and GHG emissions of the food system. Influencing the composition of the food basket, redesigning the local food distribution system and influencing consumers behaviour are better instruments to reach that goal. However, local production of food can strengthen these instruments by playing a role in local production of renewable energy, recycling of waste, social and health care and (children) food education.

A change in consumers behaviour related to food consumption (composition pattern of meat and off season products) and food waste could positively influence the impact of local food production and thus the carbon footprint of food consumption. With less meat and more seasonal products on the menu and less waste, more of the daily food basket could be produced locally. The results of this study also suggest to design Almere Oosterwold with a focus on local distribution of food and the use of renewable energy rather than local primary production. The distribution system should be designed in a way that the local produce will be brought to the consumers, rather than that consumers pick it up at central distribution points. The local distribution of renewable energy should be focused on the heavy users like storage and local transport. Moreover, (urban) agriculture could operate as producer of renewable energy through collecting energy out of residual warmth from glass houses or fermentation of waste among others (Jansma *et al.*, 2010).

26.5 Final remarks

The city of Almere finally launched the overall ambition to produce 10% of its food locally by 2030 (Stuurgroep Almere 2030, 2009). We advised to concentrate this ambition on the local production and processing of vegetables and fruit. With approximately 4,000 ha available in Almere Oosterwold, the 10% ambition is realistic. Moreover the soil type (sandy clay soil) is perfectly suitable for fruit and vegetable crops. The local production of food should not only be carried out by professional enterprises. City dwellers themselves could be involved in reconnecting food production with consumption through community gardens and allotments. American research shows that people eat healthier food when they grow this themselves (Bellows *et al.*, 2004). Armstrong (2000) found that gardens generate strong local neighbourhood involvement. Research of Van den Berg *et al.* (2010) highlights the potential contribution of allotment gardens to an active, healthy lifestyle, especially among the elderly.

In the draft Strategic Vision Almere 2.0, UA is highlighted as one of the driving forces in the Almere Oosterwold area (Stuurgroep Almere 2030, 2009). The city's ambition is to develop this area towards a so called continuous productive landscape producing food, energy, resources and water within and for the city. Providing food, energy, products and services, the area will be part of the city's economic, ecologic and social system. Through entrepreneurship and peoples initiatives at local scale this conventional agricultural polder area should transform towards a rural urban area by 2030 (Van Oost and De Nood, 2010). Reconnecting natural, rural-agricultural and urban systems at this size makes Almere Oosterwold unique. The city is now working at a development strategy to realise this transformation. Part of this development strategy will be the design of the infrastructure needed to realise the ambition of local food production, local energy production and distribution and the reuse of waste. Also part of this strategy will be the development of a distribution system of the local produce towards the city (suburbs).

This study concentrated on the effects of local food production (and distribution) through UA on the reduction of city's carbon footprint. But the benefit of UA goes beyond local production of food. The added value of UA could vary from an improved health to cities liveability, from an improved biodiversity to more local employment and from multifunctional land use to social cohesion. Some of these added values are often connected with urban green in general (Vreke *et al.*, 2010). UA could share these positive effects of urban green, but more evidence is needed to persuade planners, designers and developers to incorporate UA in the city system.

References

Stuurgroep Almere 2030, 2009. Concept Structuurvisie Almere 2.0 (Draft Strategic Vision Almere 2.0). Almere kan groeien van 190,000 naar 350,000 inwoners? Wat betekent de schaalsprong voor de stad en de regio? Almere, the Netherlands, 309 pp. (Summary in English).

Almere, 2010. De sociale atlas van Almere 2010: Monitor van wonen, werken en vrije tijd. Gemeente Almere, Almere, the Netherlands, 36 pp.

Armstrong, D., 2000. A survey of community in upstate New York: Implications for health promotion and community development. Health and Place 6: 319-327.

Bellows, A.C., Brown, K. and Smit, J., 2004. Health benefits of urban agriculture. A paper from members of the Community Food Security Coalition's North American Initiative on Urban Agriculture, 27 pp., available at http://foodsecurity.org/pubs.html.

Blonk, H., Kool, A. and Luske, B., 2008. Milieueffecten van Nederlandse consumptie van eiwitrijke producten: Gevolgen van vervanging van dierlijke eiwitten anno 2008. Blonk Milieu Advies BV, Gouda, the Netherlands, 153 pp.

Bos, J., De Haan, J.J. and Sukkel, W., 2007. Energieverbruik, broeikasgasemissies en koolstofopslag: de biologische en gangbare landbouw vergeleken. Wageningen University and Research, Wageningen, the Netherlands, 76 pp.

Dekking, A.J.G., Visser, A.J. and Jansma, J.E., 2007. Urban agricultural guide: urban agriculture in the Netherlands under the magnifying glass. Wageningen University, Applied Plant Research, Lelystad, the Netherlands, 20 pp.

Dutilh, Chr. E. and Kramer, K.J., 2000. Energy consumption in the food chain: comparing alternative options in food production and consumption. Ambio 29: 98-101.

Hubert, J.P. and Toint, P., 2002. La mobilité quotidienne des Belges: Services fédéraux des affaires scientifiques, techniques et culturelles. Enquête nationale sur la mobilité des ménages, Presses universitaires de Namur, Namur, Belgium, 352 pp.

Hulshof, K.F.A.M., Ocké, M.C., Van Rossum, C.T.M., Buurma-Rethans, E.J.M., Brants, H.A.M., Drijvers, J.J.M.M. and Ter Doest, D., 2004. Resultaten van de voedselconsumptie peiling 2003. RIVM rapport 350030002/2004, Biltohoven, the Netherlands, 111 pp.

Jansma, J.E., Dekking, A.J.G., Migchels, G., De Buck, A.J., Ruijs, M.N.A., Galama, P.J. and Visser, A.J., 2010. Agromere, stadslandbouw in Almere, van toekomstbeelden naar het ontwerp. Wageningen University and Research, Applied Plant Research, Lelystad, the Netherland, 98 pp.

Kramer, K.J., Moll, H.C., Nonhebel, S., Wilting, H.C., 1999. Greenhouse gas emission related to Dutch food consumption. Elsevier Energy Policy 27: 203-216.

Mougeot, L.J.A., 2000. Urban agriculture: definition, presence, potentials and risks. In: Bakker, N., Dubbeling, M., Gundel, S., Sabel-Koschella, U. and De Zeeuw, H. (Eds.) Growing cities, growing food: urban agriculture on the policy agenda. German Foundation for International Development (DSE), Bonn, Germany, pp. 1-42.

Pearce, F., 2006. Ecopolis now. New scientist 190: 36-42.

Steel, C., 2008. The hungry city: how food shapes our lives. Chatto & Windus, Londen, UK, 383 pp.

Sukkel, W., Van Geel, W. and De Haan, J.J., 2008. Carbon sequestration in organic and conventional managed soils in the Netherlands. Proceedings 16th IFOAM Organic World Congress, Modena, Italy, June 16-20, 2008, 4 pp.

Van den Berg, A.E., Van Winsum-Westra, M., De Vries, Sj. and Van Dillen, S.M.E., 2010. Allotment gardening and health: a comparative survey among allotment gardeners and their neighbors without an allotment. Environmental Health 9: 74.

Van der Voort, M. and Luske, B., 2009. Energieverbruik en broeikasgasemissies in de keten. Quick scan energie en broeikasgasemissies; Supermarkt versus webwinkel. Praktijkonderzoek Plant en Omgeving, Lelystad, the Netherlands, 21 pp.

Van Hauwermeiren, A., Coene, H., Engelen, G. and Mathijs, E., 2007. Energy lifecycle inputs in food systems: A comparison of local versus mainstream cases. Journal of Environmental Policy & Planning 9: 31-51.

Van Oost, A.C. and De Nood, I., 2010. Almere Oosterwold: toonbeeld van duurzame gebiedsontwikkeling. Groen, vakblad voor ruimte in stad en landschap 11: 41-45.

Van Remmen, Y. and Van der Burg, A.J., 2008. Past and future of Dutch urbanisation policies: growing towards a system in which spatial development and infrastructure contribute to sustainable urbanisation. Urban Growth without Sprawl, 44th ISOCARP International Congress Dalian- China, 19-23 September 2008.

Visser, A.J., Jansma, J.E., Schoorlemmer H. and Slingerland, M.A., 2009. How to deal with competing claims in peri-urban design and development: The DEED framework in the Agromere project. In: Poppe K.J., Termeer, C. and Slingerland M.A. (Eds.) Transitions towards sustainable agriculture and food chains in peri-urban areas. Wageningen Academic Publishers, Wageningen, the Netherlands, 392 pp.

Vreke, J., Salverda, I.E. and Langers, F., 2010. Niet bij rood alleen: buurtgroen en sociale cohesie. Alterra rapport 2070, Wageningen University and Research, Wageningen, the Netherlands, 58 pp.

Vringer, K., Benders, R., Wilting, H., Brink, C., Drisse, E., Nijdam, D. and Hoogervorst, N., 2010. A hybrid multi-region method (HMR) for assessing the environmental impact of private consumption. Ecological Economics 69: 2510-2516.

Zalm, Chr. and Oosterhoff, W., 2010. Het Almere Landschap: drager van polderstad Almere. Groen, vakblad voor ruimte in stad en landschap 11: 8-13.

Chapter 27

Urban agriculture and seasonal food footprints: an LCA study of tomato production and consumption in the UK

Gillean M. Denny

Department of Architecture, University of Cambridge, 1-5 Scroope Terrace, CB2 1PX Cambridge, United Kingdom; gd300@cam.ac.uk

Abstract

Utilizing the Life Cycle Analysis of a tomato as a case study, produce surveys in East Anglia, and household surveys of 183 urban agriculture (UA) practicing households, this chapter explores the ability of UA, in an international food market, to decrease food related greenhouse gas emissions. This study concludes that through UA participation, a decrease in an individual's tomato consumption emissions from 52.4 kg CO_2e/yr to 11.2 kg CO_2e/yr is achievable (equivalent of 2 classic tomatoes/person/week). UA participation, either in the form of growing tomatoes in an allotment, or relying on a semi-commercial or commercial UA shop/market stall, reduces yearly tomato consumption emissions by 44% (from a non-UA, low emission diet) and by 78% (from a supermarket diet). This chapter contributes to UK fresh produce greenhouse gas reduction. First, a direct emission comparison is drawn between UA homegrown tomatoes and purchased commodities, taking into account all identified production burdens and seasonal fluctuations. Next, utilising household survey consumption data of UA practitioners, current tomato production emissions can be calculated for an average UA-participating individual. Finally, a tomato procurement model can be derived for a UK food market with UA inclusion showing the lowest emissions by commodity and average from various procurement locations including commercial, semi-commercial, and allotment forms of UA.

Keywords: commercial UA, semi-commercial UA, g CO_2e/g tomato

27.1 Introduction

Utilizing the Life Cycle Analysis (LCA) of a tomato as a case study, produce surveys in East Anglia, and household surveys of 183 urban agriculture (UA) practicing households, this chapter explores the ability of UA, in an international food market, to decrease food related greenhouse gas (GHG) emissions. UA will encompass any fresh produce both produced (homegrown or commercial) and consumed within the same urban/peri-urban region. Showing a direct relation between food emissions, procurement locations, and UA within an urban plan, emissions reveal a 44% savings beyond non-UA emissions, or a 78% savings beyond supermarket emissions through UA participation. With UA participation, either in the form of growing tomatoes in an allotment, or relying on a semi-commercial or commercial

UA shop/market stall, a decrease of tomato consumption emissions from 52.4 kg CO_2e to 11.2 kg CO_2e are achievable for the year (equivalent of 2 classic tomatoes/person/week).

27.2 Methodology

As a case study, the tomato will illustrate the potential for the GHG emissions of a UK household food to be affected by seasonality and UA involvement. According to the British Tomato Growers'Association (2010), tomatoes are the fourth most popular fruit in the UK, and household surveys conducted for this study revealed the tomato to be one of the most purchased and homegrown fresh produce items for participants, along with potatoes and apples. Likewise, due to wide-spread international standardisation, a comparable tomato production system facilitates a more refined, quantifiable analysis of all tomato LCA metrics.

To determine the contribution of UA tomatoes within the UK food web, comparative emissions for available produce must first be determined. Williams *et al.* (2006, 2009) and Audsley *et al.* (2009) show the need to focus on an LCA of the produce, the product's country of origin, and any specific categorical features which could affect the product's emissions: classic/speciality, loose/vine and packaging, with classic representing beef and classic commodities, and speciality representing all others. A standard percentage of UK tomato emissions are then generated for each procurement location by month: supermarket, local shop, market, UA (commercial), UA (semi-commercial), allotment/veg patch, box delivery. Available produce characteristics (origin, packaging, availability) were based on a monthly survey of three Cambridge local supermarket distributors (Sainsbury's, ASDA, Waitrose) conducted once a month for 14 months. February emissions represent the import season (December-February), April the UK commercial tomato season (March-June and November), and July the outdoor UK tomato season (July-October).

Consumer emissions are then determined and combined with produce emissions to form a total quantifiable product emission and comparative produce emissions across all procurement locations, months, and tomato product subdivisions. Consumer emissions are based on a monthly survey of 183 UA practicing East Anglia and Greater London households, conducted over the course of 14 months. Then, a standard UK tomato's GHG emissions will be determined for different procurement locations based on currently available products. The British Tomato Growers'Association (2010) reports that 5oz (142g) of fresh tomatoes are eaten per person per week (eq. two classic tomatoes), therefore monthly GHG emissions will be shown in 20oz (568g) equivalents. Finally, procurement models for tomato commodity types are drawn based on seasonality, procurement locations available, and UA involvement.

As in Williams *et al.* (2006), all GHG emissions will be aggregated into a single unit for comparative measure and expressed as CO_2e (CO_2 equivalent) incorporating impacts for main agricultural emission sources: nitrous oxide (N_2O), methane (CH_4), and carbon dioxide (CO_2). These are expressed as a Global Warming Potential within 100 years and written as g CO_2e: CO_2 (1 g CO_2e), CH_4 (23 g CO_2e), N_2O (296 g CO_2e), N_2O-N (465 g CO_2e).

27.2.1 Life cycle analysis

Tomato case study findings in Audsley *et al.* (2009) state 3.79 kg CO_2e/kg tomato results from UK production while a reduced 1.30 kg CO_2e/kg tomato is required to grow the same produce for UK consumption elsewhere in Europe. Breaking down the total UK food consumption emissions (234 MTCe) into life cycle processes shows that 34% of emissions are produced during primary production to the Regional Distribution Center (RDC), 26% from the RDC to retail and the consumer (cooking), and 40% attributed to land use changes.

Considerable research by Williams *et al.* (2006, 2009) and Audsley *et al.* (2009) in generating reports to the Department for Environment, Food and Rural Affairs (Defra) and the World Wildlife Fund UK (WWF-UK) clearly outline the life-cycle and emissions of a tomato product, both grown in the UK and imported from Spain. According to Williams *et al.* (2009), a post-farm gate UK tomato is transported, washed and graded, selected, packed, and transported to the RDC. Spanish post-farm gate tomatoes however, after packing are cooled, transported to Spanish consolidation sites, transported to ships, transported as shipping freight to the UK, then transported to the RDC. Or, once reaching the UK, tomatoes could be transported to logistical sites, storage, selection/packaging/re-packing/labelling (including packaging production and transport emissions), and finally transported to RDC.

This complicated system necessitates a streamline LCA with system boundaries, as Pre Farm Gate (pre-FG: building structure, machinery, lighting/heating, fertiliser, pesticide, irrigation, other infrastructure) and Post Farm Gate (post-FG: transport to regional distribution centre, transport to retail, transport by consumer, packaging, storage/refrigeration, chemical ripening, consumer transport, and recycling/disposal). However, both of these reports complete their LCA at the RDC, ignoring the consumer impacts of transport emissions and frequency.

27.2.2 Country of origin

According to the British Tomato Growers'Association (BTGA, 2010), 420,000 t of fresh tomatoes are consumed yearly by UK residents, with approximately 150 ha (370 acres) of glasshouses throughout Britain producing 75,000 t of tomatoes per annum. Nevertheless this accounts for a quarter of the volume of tomatoes consumed every year, requiring over 300,000 t to be imported, ranking the UK as the 6th largest importer of tomatoes according to FAOSTAT (2010). Most imports are from Spain/Canary Islands (190,000 t) and Holland (190,000 t). When in season, the UK produces nearly half the fresh tomatoes consumed in the UK.

UK tomatoes grown in 'protected cropping' glasshouses have extended crop seasons, reduced synthetic pesticides, and increased potential of close-looped systems. Under the open sky, standard UK tomato season runs July to October. However, as seen in Williams *et al.* (2006), commercial glasshouses extend seasons from March into November. International crop trade has similar production methods, since glasshouses are nearly an international standard.

Grown similar to the UK, Belgian tomatoes are available in the UK from late April to December. However, Spain requires 349.65 km² of plastic sheeting for tomato cultivation (Rawstorne, 2005), and the BTGA (2010) reports that Spain's polythene polytunnels only yield one fifth (15 kg/m²) the quantity of tomatoes per unit area as a UK glasshouse.

27.2.3 Tomato commodities

Originally from South America, there are now thousands of different tomato varieties available today, which the BTGA (2010) separates into five general categories: classic (50%) and beef (2%) (which have similar yields), cocktail (14%), cherry (14%), and plum (mini, midi and maxi) (19%). Tomatoes may also be categorised as 'loose' (58%) or 'on-the-vine' (42%). 'Organic' tomatoes amount to only 7% of the UK commercial cultivation area.

Due to the diverse nature of the tomato as a crop, each individual type of tomato has a specific proportion, harvest yield, and weight. There are also different rates of growth (40-60 days) depending on type, with smaller cherry tomatoes ripening faster than the larger varieties (BTGA, 2010). Williams *et al.* (2006) showed that regardless of tomato type, due to crop requirements and unilateral farming methods, all inputs of heat, electricity and fertilisation were equivalent per cultivation area, making the yield the emissions variant. For example, in identical greenhouses, speciality tomatoes (cherry, plum, and beef) had half the yield of classic tomatoes. When examining the CO_2e emissions, producing 1t of speciality on-the-vine tomatoes produced five times more the emissions than 1t of classic tomatoes. The CO_2e burdens of a tomato product are therefore strongly connected to cultivation area required.

27.3 Tomato case study

27.3.1 Pre-FG and post-FG emissions

To generate comparative emissions for available tomato produce in East Anglia and Greater London, a streamline LCA was compiled for tomatoes available in six procurement locations. These emissions include pre-FG and post-FG, but not consumer emissions. Tomato emission totals were subdivided into procurement location, country of origin, packaging, classic/speciality, and loose/vine.

Supermarket emissions were generated for UK produce citing tomato LCA emissions from Williams *et al.* (2006, 2009) and Audsley *et al.* (2009). Based on monthly supermarket surveys (February, April, July), two import countries of origin were selected to represent different import emissions: Spain (Southern Europe), citing Williams *et al.* (2006, 2009) and Audsley *et al.* (2009); and Holland (Northern Europe) citing Biel *et al.* (2006). Typical burdens per type, as expressed by Williams *et al.* (2006), were used to approximate values that were not available for individual categories. Local shops, varying between small grocers and convenience stores, are shown sourcing similar products to supermarkets. Markets represent both local markets and farmers' markets, sourcing wholesale produce, local produce (in season), and small growers. UA (commercial) represents a full-scale commercial tomato growing operation if located within an urban or peri-urban context, comprising UK commercial pre-FG emissions and small grower post-FG emissions. UA (semi-commercial) and UA (allotment) pre-FG comprise small scale allotment farming, with travel to a market stall for semi-commercial. Small scale UA emissions (Carter, 2010) assume GHG burdens of harvested produce (0.93 g CO_2e/annum) and not the total cultivation potential (0.7 g CO_2e/annum). This is to increase comparability between production methods, since produce burdens from other systems account for end product, incorporating emissions from wasted or damaged items in production chain.

UK products generally have greater pre-FG emissions than Spanish/Southern Europe produce, but have reduced post-FG emissions due to their primarily local travel. Packaged, loose, classic UK (2.11 g CO_2e/g tomato) and the Southern Europe (0.27 g CO_2e/g tomato) tomatoes both have lower emissions than Northern Europe products (2.73 g CO_2e/g tomato). Holland's pre-FG process is similar to the UK commercial, but has higher transport needs than the UK, thus exceeding the Spanish total.

Following these assessments, results of the supermarket monthly surveys for February, April, and July 2010 were correlated to provide emissions data of products available in various procurement locations in East Anglia and Greater London. The first three emissions columns of Table 27.1 report the average g CO_2e/g tomato of each available procurement location product by subdivision per month, independent of consumer-related emissions.

Further influences also affect monthly emissions. Supermarket emissions increase in every category from February to July as the lower emission Southern Europe produce fades out and there is greater reliance on UK and Northern Europe sources. Packaged classic vine tomatoes rise from 2.06 to 5.78 g CO_2e/g tomato and non-packaged classic loose tomatoes rise from 0.64 to 2.65 g CO_2e/g tomato, more than doubling emissions.

27.3.2 Consumer emissions

Based on a survey of 183 UA practicing East Anglia and Greater London households, each of these procurement locations will have different consumer transport habits (distance, frequency and mode of transportation) included in the total produce emissions. Figure 27.1 shows the average distance from each survey household to their chosen procurement locations, with a local shop (0.65 miles) and an allotment garden (0.93 miles) being nearest.

The average trips taken per week to each of these locations can be seen in Figure 27.2, with 2.48 trips to the allotment per week and only 0.48 trips/week to a market.

Table 27.1. Tomato monthly emission averages (Audsley et al., 2009; Biel et al., 2006; Carter, 2010; Williams et al., 2006, 2009).

Tomato monthly emission averages (g CO_2/g tomato)				pre-FG + post-FG			Transport			Total		
Procurement location	P/N-P	Produce	Type	Feb	April	July	Feb	April	July	Feb	April	July
Supermarket	packaging	classic	vine	2.07	2.58	5.78	2.41	2.57	2.82	4.48	5.14	8.60
			loose	1.10	1.61	2.49	2.41	2.57	2.82	3.51	4.17	5.31
		speciality	vine	3.11	8.59	9.64	2.41	2.57	2.82	5.52	11.16	12.46
			loose	1.40	1.84	2.50	2.41	2.57	2.82	3.82	4.41	5.32
	N-packaging	classic	vine			6.68	2.41	2.57	2.82			9.51
			loose	0.64	1.60	2.65	2.41	2.57	2.82	3.05	4.17	5.47
		speciality	vine		2.99	13.96	2.41	2.57	2.82		5.56	16.78
			loose		1.29		2.41	2.57	2.82		3.85	2.82
Local shop	packaging	classic	vine	2.07	2.58	5.78	0.32	0.21	0.10	2.38	2.79	5.89
			loose	1.10	1.61	2.49	0.32	0.21	0.10	1.41	1.82	2.59
		speciality	vine	3.11	8.59	9.64	0.32	0.21	0.10	3.43	8.80	9.74
			loose	1.40	1.84	2.50	0.32	0.21	0.10	1.72	2.05	2.60
	N-packaging	classic	vine			6.68	0.32	0.21	0.10			6.79
			loose	0.64	1.60	2.65	0.32	0.21	0.10	0.95	1.82	2.76
		speciality	vine		2.99	13.96	0.32	0.21	0.10		3.21	14.06
			loose		1.29		0.32	0.21	0.10		1.50	0.10
Market	N-packaging	classic	vine			6.68	2.54	2.57	1.71			8.39
			loose	0.64	1.60	2.65	2.54	2.57	1.71	3.18	4.17	4.36
		speciality	vine		2.99	13.96	2.54	2.57	1.71		5.56	15.66
			loose		1.29		2.54	2.57	1.71		3.85	1.71
Market (small grower)	N-packaging	classic	vine			0.92	2.54	2.57	1.71			2.63
			loose			0.92	2.54	2.57	1.71			2.63
		speciality	vine			0.92	2.54	2.57	1.71			2.63
			loose			0.92	2.54	2.57	1.71			2.63
UA (commercial)	N-packaging	classic	vine	5.01	5.01		1.43	1.39	0.90		6.40	5.92
			loose	2.13	2.13		1.43	1.39	0.90		3.52	3.04
		speciality	vine	5.75	5.75		1.43	1.39	0.90		7.14	6.66
			loose	2.39	2.39		1.43	1.39	0.90		3.78	3.29
UA (semi-commercial)	N-Packaging	classic	vine			0.93	1.43	1.39	0.90			1.83
			loose			0.93	1.43	1.39	0.90			1.83
		speciality	vine			0.93	1.43	1.39	0.90			1.83
			loose			0.93	1.43	1.39	0.90			1.83
UA (allotment)	N-Packaging	classic	vine			0.93	0.99	1.20	1.11			2.05
			loose			0.93	0.99	1.20	1.11			2.05
		speciality	vine			0.93	0.99	1.20	1.11			2.05
			loose			0.93	0.99	1.20	1.11			2.05

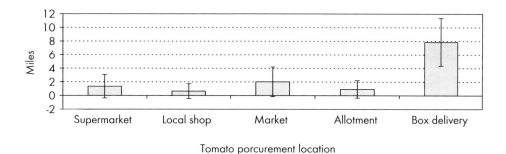

Figure 27.1. Distance to procurement location.

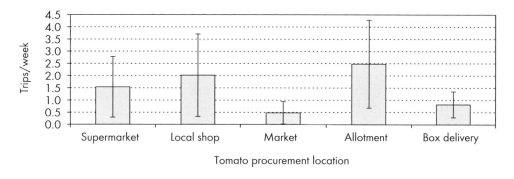

Figure 27.2. Frequency of travel to procurement location.

The mode of transportation to each location also affects emission levels, shifting monthly. Figure 27.3 shows driving a car and walking account for the greatest percentage of transportation, particularly for supermarkets and local shops. Also, while local shops (0.65 miles) and the allotments (0.93 miles) are similar in distance, 25% more allotment holders are taking their cars to the allotment than are driving to local shops, potentially increasing emissions for allotment-based produce. Figure 27.4 highlights shifts in usage within each location by month, potentially influencing the overall transportation modes seen in the previous Figure 27.3.

Combining this survey data with standard UK transport emissions from DEFRA (2008), Figure 27.5 illustrates the calculated consumer transportation emissions based on trip frequency, mode of transportation, distance, and month. Upward shifts in July supermarket transport is largely due to the increased car usage. Market emissions drop from April (2.57 g CO_2e) to July (1.71 g CO_2e) due to increased walking and decreased car usage. The steep drop observed in produce box delivery emissions is potentially a consequence of survey taker error.

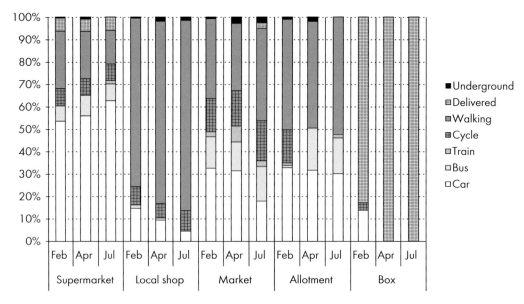

Figure 27.3. Modes of transportation for procurement location by month.

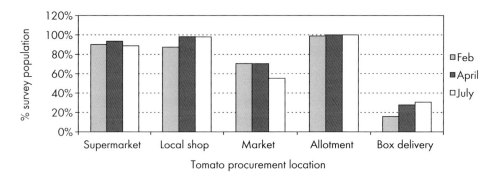

Figure 27.4. Percentage of survey population using procurement location each month.

27.3.3 Total tomato emissions

Total tomato emissions may then be calculated. A shopping trip may comprise multiple items and therefore the single tomato will not incur the total emissions for the trip. Following experimentation, a typical plastic carrier bag can hold around 15 kg and while multi-use carrier bags hold over twice as much, both are infrequently used in this manner. Therefore for the purposes of this study, it will be assumed that a standard shopping trip comprises one full plastic grocery bag, making the typical tomato weekly percentage of shopping weight (5oz or 142 g), around 1% of the total emissions. If desired, multiple persons for a single household could then be calculated, without doubling total trip transport emissions.

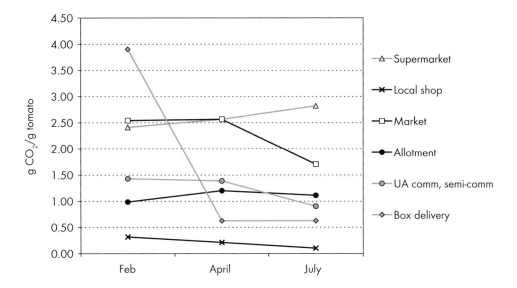

Figure 27.5. Consumer transport emissions.

Final product emissions (g CO_2e/g tomato) (Table 27.1) combine pre-FG, post-FG, and consumer transport. Figure 27.6, illustrates the changes in emissions per month by tomato subdivisions available for month/location. A steady increase for all locations sourcing commercial tomatoes can be seen through April and July as fewer low-emission Southern Europe tomatoes are available. Allotment tomato emissions (2.05 g CO_2e/g tomato) and UA (semi-commercial) (1.83 g CO_2e/g tomato) are seen with very low emissions in July with allotment emissions being higher due to greater use of car transport. The local shop (speciality/loose/no packaging) (0.102 g CO_2e/g tomato) is also seen to be the lowest since these products are still being sourced primarily from Southern European regions and have negligible consumer transport to location for the month of July. Supermarket (speciality/vine/packaging) (16.78 g CO_2e/g tomatoes) emissions however are the highest due to tomato type, use of packaging, consumer dependence on car travel, and product sourcing during July from UK and Northern European countries.

Emissions of commodity (if available) averages can then be drawn for each location/month. Previously, a 'national basket' of tomatoes was derived by Williams *et al.* (2006), which was a percentage of the total annual UK harvest, ignoring imported products or product availability at the moment/location of purchase. Instead, a 'standard' tomato will be determined for each procurement location/month based on the monthly surveyed products in East Anglia grocery stores and by UA household survey results regarding product choices and procurement location. Since 142 g of fresh tomatoes are eaten per person per week, the monthly GHG emissions can be shown in 20 oz (568 g) equivalents. Figure 27.7 shows the standard monthly consumption emissions (g CO_2e/568 g tomatoes) for procurement location/month.

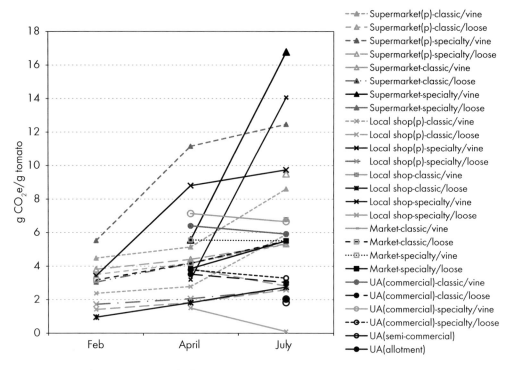

Figure 27.6. Total tomato emissions by location and commodity by month.

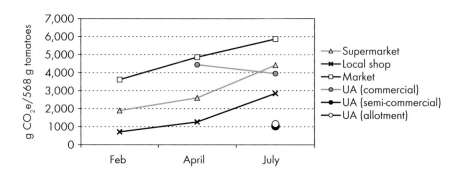

Figure 27.7. Total monthly tomato consumption emissions by month for procurment location (if available).

These figures can then be extrapolated over a year's consumption, with February representing the import season (December-February), April the UK Commercial season (March-June and November), and July the outdoor UK season (July-October). Seen in Table 27.2, an individual consuming the weight equivalent of 2 classic tomatoes per week, generates 52.4 kg CO_2e (from a supermarket diet). Using local shops and markets, without UA participation, this can decrease to 20 kg CO_2e per year. With UA, the lowest emissions achievable are 11.2 kg CO_2e, 44% savings from the lowest non-UA emissions, or a 78.6% savings from a supermarket diet.

Table 27.2. Yearly tomato consumption emissions.

Yearly tomato consumption emissions (568 g tomatoes/month)					Total/year	
Yearly emissions	Dec-Feb	Mar-Jun	Jul-Oct	Nov	g CO_2e	kg CO_2e
Highest emissions	3,612.2	4,858.9	5,866.8	5,866.8	52,381.6	52.4
Lowest emissions (non-UA)[1]	712.2	1,261.4	2,843.7	2,843.7	19,976.6	20.0
Lowest emissions (UA)	712.2	1,261.4	1,037.2	1,261.4	11,168.3	11.2

[1] UA: urban agriculture.

This illustrates that while non-UA products might have lower emissions individually, relying on UA when seasonally available can reduce overall yearly tomato consumption emissions.

A procurement model for tomatoes is then generated, revealing the lowest emissions by location for different subcategories of tomato products by month. Table 27.3 shows the lowest emissions/location for those participating in UA (growing and/or purchasing) and non-UA participants. Please note that transport-related emissions were determined using survey data from UA-participating households, rendering total tomato emission calculation for non-UA households an estimation only and open to further exploration. Likewise, growing seasons shift, making examining monthly boundaries for specific recommendations inexact.

Table 27.3. Lowest tomato emissions by month and commodity.

Tomato commodity	UA-practicing households[1]			Non-UA households		
	Dec-Feb	Mar-Jun	Jul-Nov	Dec-Feb	Mar-Jun	Jul-Nov
Classic/vine(p)	local shop	local shop	local shop	local shop	local shop	local shop
Classic/loose(p)	local shop	local shop	local shop	local shop	local shop	local shop
Speciality/vine(p)	local shop	local shop	local shop	local shop	local shop	local shop
Speciality/loose(p)	local shop	local shop	local shop	local shop	local shop	local shop
Classic/vine		ua (commercial)	ua (semi-commercial)			market
Classic/loose	local shop	local shop	ua (semi-commercial)	local shop	local shop	local shop
Speciality/vine		local shop	ua (semi-commercial)		local shop	market
Speciality/loose		local shop	local shop		local shop	local shop

[1] UA: urban agriculture.

27.3.4 Further consideration

These total emissions will fluctuate based on actual consumption by individuals, e.g. more tomatoes consumed during summer than winter. Outside of strict emissions values, the question of environmental impact of a food product has many avenues of consideration that these numbers have not addressed. These lead to further considerations of nutritional value, waste reduction and embedded water quantities. UA also calls attention to decreased emissions vs. yield/land area, raising questions of specific types of widespread UA suitability.

27.4 Conclusions

When examining procurement locations specifically, the ability to relate food emissions to the urban plan is achieved, showing direct emission connections between where food is purchased or grown, at what time of year, and by varying product specifications. Through the case study of the tomato, inclusion of urban agriculture in the local food market has shown to decrease annual UK tomato consumption embedded GHG emissions. UA participation, either in the form of growing tomatoes in an allotment, or relying on a semi-commercial or commercial UA shop/market stall, reduces yearly tomato consumption emissions by 44% (from a non-UA, low emission diet) and by 78% (from a supermarket diet).

Local shops provide low emission commodities, due to decreased consumer transport emissions and limited packaging. Nevertheless, semi-commercial UA commodities (classic vine, classic loose, speciality vine) from July to October earn the lowest emissions due to low pre-FG, post-FG, packaging, and transport emissions. UA has significant potential to reduce food emissions in the UK, having lower emissions than supermarkets and other sources when in season (July-November). However, semi-commercial UA on average shows the least emissions, since allotments carry heavy consumer transport emissions and commercial UA has the potential to duplicate high UK commercial tomato emissions at the pre-FG level.

Within specific burdens, tomato GHG emissions levels can be reduced by limiting consumer car transport, particularly in allotment and supermarket sectors. Likewise, wasted produce left unharvested currently increases the overall produce GHG emissions of allotments. If allotments were efficiently cultivated, not only as a hobby, this non-commercial production source could achieve the lowest emissions of any UK tomato consumption source.

References

Audsley, E., Brander, M., Chatterton, J., Murphy-Bokern, D., Webster, C. and Williams, A., 2009. How low can we go? An assessment of greenhouse gas emissions from the UK food system and the scope to reduce them by 2050. FCRN and WWF, UK, 85 pp.

Biel, A., Bergström, K., Carlsson-Kanyama, A., Fuentes, C., Grankvist, G., Lagerberg, Fogelberg, C., Shanahan, H. and Solér, C., 2006. Environmental information in the food supply system. Foundation for Strategic Environmental Research. Stockholm, Sweden, 117 pp.

BTGA (British Tomato Growers' Association), 2010. Tomato facts. Available at http://www.britishtomatoes. co.uk/facts/index.html.

Carter, C., 2010. Global warming potential of produce grown on an allotment using a life cycle assessment approach. Case study: Wellesbourne Allotment. MSc Thesis, University of Surrey, Surrey, UK.

DEFRA (Department for Environment, Food and Rural Affairs), 2008. 2008 Guidelines to DEFRA's GHG conversion factors: Methodology paper for transport emission factors. Department for Environment, Food and Rural Affairs, London, UK. Available at http://www.defra.gov.uk.

FAOSTAT (Food and Agriculture Organization Statistics Division), 2010. 2010 agricultural trade statistics. Available at http://faostat.fao.org/site/342/default.aspx.

Williams, A.G., Audsley, E. and Sandars, D.L., 2006. Determining the environmental burdens and resource use in the production of agricultural and horticultural commodities. Main Report. DEFRA Research Project IS0205. Cranfield University and DEFRA, Bedford, UK, 97 pp.

Williams, A.G., Audsley, E., Pell, E., Moorehouse, E. and Webb, J., 2009. Comparative life cycle assessment of food commodities procured for UK consumption through a diversity of supply chains. Main report, DEFRA Research Project FO0103, Cranfield University and DEFRA, Bedford, 18 pp.

Rawstorne, T., 2005. Under vast sheets of white plastic enveloping Spain: Britain's tomatoes are being grown without soil in a soup of chemicals controlled by computers. Daily Mail, London, UK. October 1, 2005.

Chapter 28

Growing alone, growing together, growing apart? Reflections on the social organisation of voluntary urban food production in Britain

Richard Wiltshire and Louise Geoghegan
Department of Geography, King's College London, Strand, London WC2R 2LS, United Kingdom; richard.wiltshire@kcl.ac.uk

Abstract

The organisation of an unpaid supply of labour is an essential but under-researched aspect of contemporary forms of urban agriculture, and a key restraint on its expansion. In Britain there are two dominant modes of practice, collective and individual, the former exemplified by volunteering activity on growing projects, the latter by traditional allotment gardens. This chapter adopts a comparative approach to explore the assumptions about human motivation, individual rights and effective social organisation that underpin these two modes. Both have strengths and weaknesses, and relations between them can be antagonistic, particularly when land is scarce, but areas of convergence have also begun to emerge, and both need to be accommodated if the voluntary supply of labour is to be maximised.

Keywords: volunteers, human motivation, individual rights, collective, allotments, urban agriculture

28.1 Introduction

The successful management of complex relationships between people, land and production is essential to achieving the aggregate sustainability gains associated with the expansion of urban agriculture. This is particularly true when the motivation of individuals to contribute their labour is not solely dependent on material gain, and participation is both voluntary and discretionary.

This chapter critically examines two contrasting forms of social organisation of production in urban food growing projects, the individual and the collective. We contend that the individual and collective forms are aligned with different understandings of the motivations to grow, that each have their own strengths and weaknesses in attracting and retaining volunteers, and that the right mix of these forms is essential to policy to support a sustained expansion of urban agriculture. The distinction between the forms can be interpreted as reflecting conflicting theories of how individuals act, between the 'rational choice' model of calculative self-interest, and the 'social being' model that sees behaviour as driven by culture and social norms (Cleaver, 2001: 39).

The empirical focus of the chapter is on the British experience. The discussion is informed by the principal author's reflections on practical involvement, over two decades, in balancing the contradictions between these two forms of organisation at local and national levels. This included lengthy terms as a volunteer on the national management bodies for two NGOs championing different organisational forms, and as advisor to central government on 'good practice' in running growing projects. While no attempt is made here to generalise empirically beyond the British context, we contend that many of the issues that emerge from a comparison of these two forms are essentially generic. As such, they deserve wider consideration in other contexts, and particularly within attempts to transfer knowledge of 'what works' between nations with contrasting socio-political systems. This observation applies to contemporary enthusiasm for learning from such diverse sources as Cuba (e.g.Viljoen, 2005), and the United States (e.g. Ferris *et al.*, 2001). As regards the latter, Jamison (1985), in his comparison of 'collectivist and bureaucratic cultures' in urban gardening projects, touches on a number of the issues raised in this chapter. His analysis also demonstrates the importance of the socio-political context as a mediating variable, however, contrasting supportive official attitudes towards individualism as an American civic virtue, with the framing of collective 'socialist' projects as a threat to the existing social order (*ibid.* p. 473).

In Britain, the arrangement of growing around individuals is epitomised by the allotment garden, defined in the 1922 Allotments Act as a small plot 'which is wholly or mainly cultivated by the occupier for the production of vegetable of fruit crops for consumption by himself or his family' (OPSI, 2010). The collective form, which has no such statutory underpinnings, is often signified by the use of the adjective 'community', as in 'community gardens'. The use of this adjective is ambiguous, however and, as we shall see, can have both ideological and instrumental roots. In fact, many community gardens in British cities are actually hybrids that make available small plots for individuals to rent – this is true of Culpeper and Cable Street, for example, perhaps the best known community gardens in London – and in other contexts 'community gardening' has very different meanings. In major American cities, for example, the great majority of 'community gardens' are based on individual tenancies, albeit with a restricted right for the general public to access parts of the broader gardening space (Lawson, 2005), a porosity shared by many 'colony gardens' in Scandinavia which are also organised around individual tenancies. Furthermore in Britain, the 'community allotment' occupies a confused middle ground between the two traditions (Wiltshire and Burn, 2008: 4). The contrast being drawn, therefore, is not between 'individual' and 'community', but between 'individual' and 'collective' as the mode for organising the physical practices of horticulture. A distinctive feature of the British case is the power of statutes that favour the individual form, leading to uncertainty over whether urban agriculture should expand through reinforcement and fuller implementation of allotment law, or through collective action outside this legal framework to secure new spaces to meet what are often billed as 'community' needs.

In subsequent sections we will first briefly recount the trajectory of the individual and collective traditions in Britain, using post war historical narratives based on the established literature, that focus on underlying motivations to grow. We then subject these forms of organisation to a comparative analysis aimed at highlighting the strengths and limitations of each. We conclude by noting an emerging convergence between the two forms linked to restrictions on local government finance, which opens up the possibility of more effective use of the existing gardening space in cities, and a more nuanced set of policy options.

28.2 Organisational forms and motivations to grow in historical perspective

Motivations are conditioned by a shifting landscape of needs, choices, social values and supportive institutions. The roots of the mainstream in British allotment gardening lie beyond the city, in the Acts of Enclosure going back to the 17th century, when land was made available for individual dispossessed peasants to grow their own food as a supplement for the meagre wages to be earned as farm labourers (Burchardt, 2002), a practice that migrants subsequently transported to the urban context. The subsequent 'golden age' for allotment gardens, however, stemmed from the existential threat of two world wars. 'Digging for Victory' made sense, both for individuals with no other means to access food, and collectively to ensure national survival. The British state (along with the US, Japan and other combatants) sponsored patriotic campaigns to encourage local food growing, a task which fell mainly to those individuals (such as the old and infirm) who could not make a socially sanctioned contribution to the 'war effort' in other ways. The main consequence on the ground was an expansion of individually tenanted allotment plots, crammed onto every available public space, to a peak of 1.4 million by 1945 (House of Commons, 1998: xiii).

After the war, voluntary urban agriculture declined as the existential threat receded, commercial food supply channels reopened, and economic growth afforded individuals more productive ways of using their working and leisure time (Crouch and Ward, 1997). In rational choice terms, the opportunity costs of growing one's own food increased dramatically for all but a residual hard-core of growers, excluded by income or age from the benefits of renewed national prosperity. The historical association between allotment gardening and poverty was thereby perpetuated, as evidenced by the conclusion of an official Japanese delegation, sent to investigate gardening practices across Europe as the basis for policy reforms, which dismissed the British model as stained by 'the continuing lingering image of allotments as a form of poor relief' (Tsuge, 1987: 186, authors' translation).

In continental Europe, by contrast, many food gardens were transformed after the war into sites for private family-centred leisure. An official inquiry led by Professor Harry Thorpe (Thorpe, 1975) advocated a similar transformation, to establish a contemporary rationale for the provision of allotments in the UK as 'leisure gardens', a moniker quickly adopted by the growers' representative body, the National Society for *Allotment and Leisure Gardeners.* Thorpe's recommendations failed to gain much traction with government or on the ground, but they did reflect more positive social values that had become attached to allotments by the growers themselves, and more complex motivations to grow – or at least to occupy the plot (Way, 2008). The escape from the strains of everyday life in an increasingly complex world, the opportunity to express creativity through the organisation and cultivation of the plot, and the casual (and optional) conviviality of the allotment site (a community in the loosest sense), all added to the pleasure derived from a sense (if not the reality) of individual ownership of a small plot of earth (Crouch and Ward, 1997). That allotments counted for more to individual growers than the value of the food produced was evident from the resistance to site closures documented in the National Society's journal, the *Allotment and Leisure Gardener*, demonstrating a capacity for collective action and social solidarity, but in defence of individual, not collective growing.

Much of the increase in demand for allotments over the past decade, however, may be attributable to the growing importance of environmental values in general and the ethics of food in particular to middle class identities (Buckingham, 2005), as evidenced by the rapid growth in demand for organic food and the popularity of farmers markets (Kirwan, 2004). For those who value 'ethical living', and are repelled by industrialised agriculture and the globalised economy within which it is set, the consumption of own-grown foods yields benefits beyond calories, nutrients and taste: a sense of making a personal contribution to the future of the planet. Direct evidence of this trend is fragmentary, in the absence of any requirement on local authorities to publish statistics on allotment users, but indirect evidence is suggestive. In the first eight years of the current decade, for example, more books were published on allotments than over the preceding five decades (Wiltshire and Burn, 2008: 2), the bulk of them with clear organic and environmental overtones. A strong environmental message has also been embedded in official guidance over the past decade on how best to promote allotments to the public (Crouch *et al.*, 2001).

The vernacular allotment is thus back in fashion, but with new values attached, firmly centred on environmentally friendly production, and shared with many participants in newly emerging forms of collective growing.

As for the roots of contemporary collective gardening projects in Britain, and the values that underpin them, these can be traced back to the 1960s and a post war generation that questioned the material trappings of the new consumer culture and explored alternative values and ways of living. These included philosophies focussed on nature and deep ecology (Pepper, 1993), along with collective living arrangements such as squats and communes. This ideological perspective was reinforced by the march of globalisation, which has been accused of dissolving a wealth of local, economic, social, political and cultural networks, at the cost of environmental and social degradation (Gilchrist, 2000). In contrast to the legislative roots of the allotment, the collective growing project draws its strength from the solidarity of the participants in a shared endeavour, underpinned by a common ideology made manifest through the garden. Living lives more locally, and making better use of the human and natural resources that lie to hand, is seen as an antidote to globalisation processes and their correlates, climate change and food insecurity.

Meanwhile, land rights movements such as *The Land is Ours* rediscovered heroes for the new age, dating back to Gerrard Winstanley and the Diggers (Andrews, 2005), while rejecting institutions based on private property. Against this background, the individual allotment garden could be framed as a petty-bourgeois anachronism, a tool (much like Margaret Thatcher's 'right to buy') for giving ordinary people a misleading sense of a stake in a property-owning system which otherwise oppresses them. Instead, a new style of gardening emerged, led by activists, based on collective principles, often (though ambiguously) bearing the 'community' label as a symbol of moral identity in the face of globalisation (Cohen, 1985), and favouring styles of cultivation in greater harmony with natural systems, including organic growing and permaculture.

The attractions of this radical style remain powerful, particularly amongst younger urban growers. The assumption of collective action is also deeply embedded within contemporary movements supporting a relocalisation of food supplies, including Transition Towns (Hopkins, 2008). The growth of collective cultivation is difficult to quantify in the absence of systematic statistics, but it is evidenced by the growing membership of national supporting organisations that are sympathetic to collective growing. These include Garden Organic, the Permaculture Association, and most notably the Federation of City Farms and Community Gardens. The Federation is a national representative body founded some thirty years ago, which (according to its annual reports) has seen membership growing of late by twenty-five percent a year (www.farmgarden.org.uk), and claims to represent a national movement of more than a thousand groups, the majority of them community gardens with a strong collective element. A current high profile exemplar is Capital Growth (www.capitalgrowth. org), a campaign run by the charity Sustain to create 2012 new food growing spaces in London in time for the Olympics, which specifically excludes the creation of new allotments – notwithstanding the fact that waiting lists for allotments in London are at their longest for decades (Campbell and Campbell, 2010).

Contemporary volunteer-based food growing activities in public spaces in British cities have thus come to be marked by the co-existence of two forms, individual and collective, each with its strengths, but also its limitations, particularly (we would argue) in sustaining the motivation to grow and their capacity to include latent growers, whose mobilisation is essential if volunteer-based urban agriculture is to expand. We develop these points further in the next section, through a critical comparative analysis of the two forms.

28.3 Individual versus collective: a comparative framework

Table 28.1 attempts to conceptualise some of the key attributes of individual and collective gardening in a comparative framework. The list of attributes is not exhaustive, and the choice of descriptors to capture how each attribute is expressed in the two forms of gardening could no doubt be improved by the addition of subtle caveats to cover exceptions. Nevertheless, the table does help to clarify some of the essential differences that must be understood if either form is to be incorporated effectively into an expansion of voluntary urban agriculture.

Table 28.1. Comparative attributes of individual and collective gardening.

	Individual	Collective
Primary motivation	self-interest (rational choice)	common good (social being)
Primary beneficiaries	individuals	participants
Secondary beneficiaries	family	community/environment
Right to participate	tenancy	membership/volunteering
Social participation	discretionary	obligatory
Collective action	latent	axiomatic
Production/distribution	autonomous	negotiated
Accountability	individual	individual/shared
Outcome of transgression	harm to self	harm to group
Administration	bureaucratic	democratic
Primary barrier to participation	land supply	fit to group
Right to access land	statutory	negotiated
Compensation on dispossession	statutory	discretionary
Funding	rents	grants, service agreements, sales
Professional involvement	weak	strong

For the allotment gardener, we suggest that the primary (though not necessarily the sole) motivation for participating in horticulture is self-interest, drawing on the logic of rational choice and the underlying expectation that the main beneficiaries will be the individual grower and her/his family. This expectation is actually hard wired into the definition of an allotment garden in law, and protected by a tenancy agreement that sets explicit and objective conditions, guaranteeing the right to garden on land over which the gardener has exclusive usufruct rights for what is in practice an indefinite period.

In the collective model, however, the primary motivation is more complex: a vision of the common good, to which individuals are inspired to subscribe. Collective gardening is a social act, undertaken to satisfy the ideological preferences of the participants as the primary beneficiaries, but open to interpretation as meeting greater needs, be they of the community or the environment. As noted, many recent converts to allotment gardening are likely to share these preferences, and may indeed be more motivated by them than by more self-interested concerns, but their right to garden is not conditional on a shared ethos. In the collective form, however, rights are conditional either on membership of the group, which brings with it obligations that need not be explicit or confined to horticultural matters, or for the less firmly attached, on a volunteering arrangement.

While social participation is an obligatory feature of the collective model, for allotment gardeners it is discretionary, and may be limited to grudging acceptance of neighbours amongst those who find pleasure in solitude, while others may form spontaneously into vibrant communities of interest. Tenancy conditions relate to the way the land is used: they do not require engagement with others, beyond an injunction against material nuisance. They do not, however, prevent free association or collective action. The latter is always a latent possibility which, when realised, most commonly takes the form of an unincorporated allotment society. When members consider it in their self-interest, a society may assume some responsibility for site management or defence against vandalism or disposal.

In collective gardening, however, collective action is axiomatic, and covers a wide spectrum of decision making: from negotiating what is to be grown, what methods are permitted, and how the resulting produce is to be distributed, to how participation is to be policed and shirkers admonished. In comparison, the allotment gardener enjoys autonomy in production and distribution decisions, within such constraints as the tenancy agreement may impose. The plot holder's autonomy means that production depends little upon the actions of others, except where common property resources (such as water) are shared. The tenant is individually accountable for meeting the terms of the agreement, and if she/he fails to do so the consequences are born personally, in the worst case through termination of the tenancy. Where ambiguity arises, for example over whether cultivation standards have been met, this can be resolved by the application of rules (Wiltshire, 2010), imparting an essentially bureaucratic tone to allotment administration – an attribute which Jamison (1985) also ascribed to American publicly-sponsored community gardening schemes.

For the collective, however, transgressions are more complicated both to define and to manage, because they include contraventions of group beliefs and ideologies (however ambiguous and fluid) which undermine the solidarity that is essential for the group to function. Internal philosophical disagreements, in combination with more prosaic arguments about efforts, rewards and freeloading, can sap the energy and the trust that are essential to any collective endeavour (Falk and Kilpatrick, 2000). Democratic debate to resolve such issues can lead to a conscious and positive effort at renewal or reaffirmation of purpose, but outcomes can also include a reduction of membership to a core of truer believers (and fairer players).

Despite the frequent association with 'community', therefore, the specific solidarities of the collective project can thus be inherently exclusionary of people in the same neighbourhood who harbour other perspectives, and have no entitlement to access the collective gardening space (Littek, 2009). It follows that the capacity of the collective to fuel a sustained expansion in the supply of voluntary labour for urban agriculture depends critically on its ability to tolerate internal diversity, to proselytise its core values, and to expand its client base.

By contrast, the most significant barrier to individuals wishing to participate in allotment gardening is a problem common to both traditions: an inadequate supply of land. On paper, this should be more of a problem for collective growers, who have no statutory rights and must either negotiate for access to land, or take direct action through tactics such as guerilla gardening (Reynolds, 2008). Individual growers, on the other hand have a right under the 1908 Small Holdings and Allotments Act to petition their local authority to provide land. The fact that over 100,000 people are languishing on waiting lists in England, however, demonstrates that supply has lagged well behind the popularity of this form of gardening (Campbell and Campbell, 2010), leading for calls for adaptive responses within the existing allotment estate, such as a reduction in plot sizes (Wiltshire, 2010), as well as support for alternative solutions. The latter include collective cultivation in other public spaces, particularly in inner London, where the scope for Capital Growth to expand has been aided by the removal (under the 1963 London Government Act) of the duty on local authorities to provide allotments.

The reluctance of local authorities to provide more allotments can be attributed in part to the immediate cost, but also to concern over creating legal entitlements to compensation and/or relocation that would make subsequent disposal or conversion of the land to other uses very difficult. Authorities have been accused in the past of undermining the legal protection of allotments, by sapping the motivation of tenants through rumours of impending development, and through poor service to both tenants and inquirers. Indeed, complaints about such behaviour to one parliamentary inquiry (House of Commons, 1998) led the government to tighten rules for site disposal, requiring local authorities to actively promote their allotments before consent for disposal can be earned. Against this background, it is hardly surprising that the National Society has voiced strong misgivings about 'community gardens' and other alternatives that have gained popularity in recent years, as an easy let-out for local authorities seeking to avoid their statutory duty to provide allotments and compensation on disposal, an existential threat that is regularly denounced, for example, in the editorial columns of the *Allotment and Leisure Gardener*.

Funding considerations may also have a bearing on local authority attitudes towards the two forms of gardening in other ways. On paper, again, allotments should have the advantage, as they yield a secure source of rental income, but rents have historically been maintained at a level that fails to meet current expenses for maintenance and administration, let alone depreciation on any capital invested in the site, and attempts to raise rents to a more realistic level face the risk of legal challenge (Clayden, 2008). Allotments, be they council-run or managed by an allotment society, have faced great difficulty in raising alternative funding from charitable sources, due in part to the perception that they *only* benefit private individuals, rather than groups within the local community deemed worthy of support. In fact, the Allotments Regeneration Initiative (www.farmgarden.org.uk/ari) has run the only charitably-funded national funding stream dedicated specifically to allotments in the past decade. Collective projects, however, with their long-standing claims to community benefits and inclusivity and the flexibility that comes from the absence of a statutory definition of their functions, have been much better placed to secure funding outside of local authority sources, including grants, revenues from service level agreements (e.g. in providing horticulture-based therapy for disadvantaged groups), and commercial sales of produce.

Dependence on external funding sources places demands on collective projects, however, which are qualitatively different from those facing allotment managers. Hinchliffe *et al.* (2007), in an insightful case study, highlight the importance of framing applications to meet the priorities of funders, which typically favour innovations rather than the unrestricted core funding needed to sustain projects with a proven track record. As they succinctly note: 'Gardens are matters of process and ongoing achievements. Once created they require continuing care. In this they are something of a challenge to the one-stop temporality of current funding models.' (*ibid.* p. 275). In addition, the need to commit to measurable outputs to justify funding decisions means that what gets funded is what can be measured, which may not be what matters from the project's perspective. Indeed, when what matters is ideologically inspired and thus politically contentious, the exigencies of funding may lead to these concerns being excluded from applications for fear of prejudicing outcomes (*ibid.* p. 270).

To secure funding over the longer term, therefore, projects face an ongoing challenge of reinventing themselves to simulate demonstrable innovation. The *Concrete to Coriander* project in Birmingham, which Hinchliffe *et al.* (2007) use as a case study, is an interesting case in point. What lies undocumented in their paper is the fact that this project actually originated on a disused allotment, but faced the usual difficulties of securing renewed funding. It resolved the problem by moving off the allotment onto nearby derelict land, thus simulating innovation by incorporating the element of reclamation highlighted in the project's attractive and freshly minted title, although in functional terms there was nothing to be gained by the move (Rozanski, 2011, personal communication).

New deliverables, moreover, can transform the relationship between individual growers and the collective body. Some deliverables require a recruitment process to create a population of client growers defined by personal attributes of interest to funders, who must then be convinced to participate in sufficient numbers, as Hinchliffe *et al.* (2007) found in their case study. And deliverables involving specific outputs, such as vegetables for community cafés and box schemes, necessitate the recruitment of volunteers with sufficient motivation to meet the deadlines and other disciplines required for a successful social enterprise.

The need to ensure continuity of funding also fuels the professionalisation of collective garden projects, as people with specific skills in fundraising, reporting and volunteer management are brought on board to augment the original enthusiasts. Professionalisation introduces a further dynamic, as professionals are inevitably motivated to ensure their own incomes and career prospects, based on skills which are marketable elsewhere, creating pressure for fundraising that is independent of the perceived need that gave rise to the project in the first place. Under these circumstances, the 'local voices' of volunteers are at risk of being overridden by the authoritative voices of the 'hired hands', as Goodwin (1998) has demonstrated in the context of environmental projects more broadly.

By comparison, the local authority officers responsible for running urban allotments typically have very different skill sets, such as in horticulture or grounds maintenance. They operate within set budgets handed down to them, and are largely preoccupied with comparatively mundane tasks, such as resolving disputes and ensuring rents and bills are properly administered. The level of professional development is low: the last body to offer formal (but unaccredited) courses in allotment management, the Institute of Leisure and Amenity Management, closed its doors in 2007.

28.4 Growing apart – or space for convergence?

In a future expansion of urban food growing, it is likely that both the individualistic and collective models will continue to attract adherents, and that for reasons relating to underlying philosophies of the relationship between the individual, society and the environment, common ground will be hard to find between strong adherents to either tradition, particularly when land suitable for horticulture in the city is a scarce commodity. Both modes will have to be accommodated, however, if the supply of voluntary labour is to be maximised. Nevertheless, in the British case there are also grounds to anticipate areas of convergence, long advocated as good practice in the management of allotment portfolios (Crouch *et al.*, 2001, Wiltshire and Burn, 2008), which may inject greater enthusiasm for collective action amongst those who might otherwise have gardened alone, while building on the strengths of the allotment model (particularly its accountability) to accommodate more collective growing.

On the allotments side, deficit reduction makes it likely that the quality of local government support will decline, resulting in pressure for growers to accept devolved management agreements as the collective basis for action to protect and improve sites. The London Borough of Barnet, for example, has decided that all its sites will be placed under devolved management, as an exemplar of the coalition government's 'Big Society' policy in action (*Hendon and Finchley Times* 1 December 2009). Simultaneously, the demand for new sites is unlikely to be met by appeals to cash-strapped local authorities to deliver on rather limited statutory requirements. Instead, it will require a more politically nuanced appeal (Wiltshire, 2010) emphasising the benefits of food growing in meeting a wide range of public goals. Some of these, including adaptation to climate change, are also claimed by the collective approach, and sometimes best delivered collectively, particularly to enable inclusion of latent growers who really do need support. Complementarities exist therefore, and should be further supported through public policy, with adaptations to the interpretation of allotment law to encourage tenancies by formally constituted collective projects where appropriate.

Diversity of approaches and participants within urban agriculture need not be a problem, but a means to foster resilience by drawing on the widest possible range of talents and capacities amongst local citizens (Newman and Dale, 2004). In terms of the two approaches examined in this chapter, a lesson drawn from community gardens (American style) in New York captures well the process that is required: 'The struggle on behalf of community gardens has been the context for on-going renegotiations of concepts of community and individuality, linked to a self-conscious examination of how community gardens are perceived and can represent themselves in the public eye' (Von Hassell, 2005: 93).

References

Andrews, A., 2005. The allotment handbook. Eco-logic Books, Bristol, UK, 124 pp.

Buckingham, S., 2005. Women (re)construct the plot: the regen(d)eration of urban food growing. Area 37: 171-179.

Burchardt, D., 2002.The allotment movement in England, 1793-1873. The Boydell Press, Woodbridge, UK, 287 pp.

Campbell, M. and Campbell, I., 2010. Allotment waiting lists in England 2010. Transition Town West Kirby in conjunction with the National Society of Allotment and Leisure Gardeners, UK. Available at http://www.transitiontownwestkirby.org.uk/files/ttwk_nsalg_survey_2010.pdf.

Clayden, P., 2008. The law of allotments. Shaw and Sons, London, UK, 104 pp.

Cleaver, F., 2001. Institutions, agency and the limitations of participatory approaches to development. In: Cooke, B. and Kothari, U. (Eds.) Participation: the new tyranny. Zed Books, London, UK, pp. 36-55.

Cohen, A.P., 1985. The symbolic construction of community. Routledge, London, UK, 128 pp.

Crouch, D. and Ward, C., 1997. The allotment: its landscape and culture. Five Leaves Publications, Nottingham, UK, 320 pp.

Crouch, D., Sempik, J. and Wiltshire, R., 2001. Growing in the community. First edition. Local Government Association, London, UK, 88 pp.

Falk, I. and Kilpatrick, S., 2000. What is social capital? A study of interactions in a rural community. Sociologia Ruralis 40: 87-110.

Ferris, J., Morris, M., Norman, C. and Sempik, J. (Eds.), 2001. People, land and sustainability: a global view of community gardening. PLS Publications, Nottingham, UK, 64 pp.

Gilchrist, A., 2000. Design for living: the challenges of sustainable communities'. In: Barton, E. (Ed.) Sustainable communities: the potential for eco-neighbourhoods. Earthscan Publications, London, UK, pp.147-159.

Goodwin, P., 1998. 'Hired hands' or 'local voice': understandings and experience of local participation in conservation. Transactions of the Institute of British Geographers 23: 481-499.

Hinchliffe, S., Kearnes, M, Degan, M. and Whatmore, S., 2007. Ecologies and economies of action – sustainability, calculations, and other things. Environment and Planning A 39: 260-282.

Hopkins, R., 2008. The transition handbook: from oil dependence to local resilience. Green Books, Totnes, UK, 240 pp.

House of Commons, 1998. The future for allotments. Fifth report of the House of Commons Environment, Transport and Regional Affairs Committee, Her Majesty's Stationary Office, London, UK, 44 pp.

Jamison, M., 1985. The joys of gardening: collectivist and bureaucratic cultures in conflict. Sociological Quarterly 26: 473-490.

Kirwan, J., 2004. Alternative strategies in the UK agro-food system: interrogating the alterity of farmers' markets. Sociologia Ruralis 44: 395-415.

Lawson, L., 2005. City bountiful: a century of community gardening in America. University of California Press, Berkeley, USA, 363 pp.

Littek, M., 2009. 'This ain't Ethiopia you know!' Barriers to participation in urban agriculture on social housing estates in London, with particular reference to the ABUNDANCE project, Brixton. MA Dissertation, Department of Geography, King's College London, London, UK, 52 pp.

Newman, L. and Dale, A., 2004. Network structure, diversity and proactive resilience building: a response to Tompkins and Adger. Ecology and Society 10: 1-4.

OPSI (Office of Public Sector Information), 2010. Allotments Act 1922. Office of Public Sector Information, London, UK, available at http://www.opsi.gov.uk/RevisedStatutes/Acts/ukpga/1922/cukpga_19220051_en_1.

Pepper, D., 1993. Ecosocialism: from deep ecology to social justice. Routledge, London, UK, 288 pp.

Reynolds, R., 2008. On guerrilla gardening. Bloomsbury Publishing, London, UK, 256 pp.

Thorpe, H., 1975. The homely allotment: from rural dole to urban amenity: a neglected aspect of urban land use. Geography 268: 169-183.

Tsuge, N.,1987. Igirisu toshi ameniti-gata e no seijuku no nayami (England – pain on the road to recognition as an urban amenity). In: Egaitsu, F. and Tsubata, S. (Eds.) Shimin noen: kuraingaruten no teisho (Allotment gardens: an advocacy of the kleingarten). Ie No Hikari Publications, Tokyo, Japan.

Viljoen, A. (Ed.), 2005. Continuous productive urban landscapes: designing urban agriculture for sustainable cities. The Architectural Press, London, UK, 304 pp.

Von Hassell, M., 2005. Community gardens in New York City: place, community, and individuality. In: Barlett, P.F. (Ed.) Urban place: reconnecting with the natural world. The MIT Press, Cambridge, MA, USA, pp. 91-116.

Way, T., 2008. Allotments. Shire Publications, Oxford UK, 64 pp.

Wiltshire, R. and Burn, D., 2008. Growing in the community. Second edition. Local Government Association, London, UK, 61 pp.

Wiltshire, R., 2010. A place to grow. Local Government Association, London, UK, 15 pp.

Chapter 29

Public space, urban agriculture and the grassroots creation of new commons: lessons and challenges for policy makers

Chiara Tornaghi
School of Geography, University of Leeds, Woodhouse Lane, Leeds LS2 9JT, United Kingdom;
chiara.tornaghi@gmail.com

Abstract

This chapter is based on academic-activist research into grassroots urban agricultural initiatives which are experimenting – with different degree of legality – with new and convivial ways of sharing urban spaces while producing food. Some of these initiatives start and evolve within marginal/liminal urban spaces while others are seeking a more systematic dialogue with local institutions. Nonetheless, it seems that the ability of local and regional institutions to respond to these new demands are limited and constrained by planning traditions that have not been permeable to emerging urban cultures and their needs, failing to create flexible or more adaptable public spaces which reflect the fluidity of society. As a result, short term satisfaction to these needs is found in the possibilities left open by 'loose spaces' or ad-hoc negotiations between grassroots groups and local councils, but none of these go beyond the status of emergency or residual practices. Looking in particular at four Leeds based projects – Landshare, Urban Harvest, Headingley Community Orchard and Edible Public Space – this chapter aims to present and discuss *the challenges that these practices pose to the political and planning agenda of urban public space managers and designers*, and the potential lessons we can draw from them to inform responsive policy-making in times of energy and financial scarcity.

Keywords: community activism, food growing, urban harvest, relational approach, policy making

29.1 Introduction

In recent years, a new emerging environmental culture has posed new challenges to public space management. Urban social movements in the Global North are claiming public spaces in a way that reflects stronger concerns for sustainability, climate change and environmental quality. From Landshare to Urban Harvest, from Guerrilla Gardeners to the diverse practices of collective urban agriculture, a wide range of initiatives are experimenting and enacting – with different degrees of legality – new ways of sharing spaces while producing food and experiencing conviviality in public spaces.

In some cities these initiatives start and evolve within marginal/liminal spaces (illegal allotments, guerrilla gardening, etc.) while other groups and organisations are seeking a more systematic dialogue with – or support from – local institutions (Transition Towns, Permaculture associations, etc.). Nonetheless, it seems that the ability of local and regional institutions to respond to these new demands are limited and constrained by planning traditions that have not been permeable to emerging urban cultures and their needs, failing to create flexible or more adaptable public spaces which reflect the fluidity of society. As a result, short term satisfaction of these needs is found in the possibilities left open by 'loose spaces' (Franck and Stevens, 2006) or ad-hoc negotiations (such as for the concession of land) between grassroots groups and various local institutions (councils, local businesses or organisations) but none of these go beyond the status of 'emergency' or residual practices.

Drawing on preliminary academic-activist research into grassroots urban agricultural initiatives that have emerged recently within the environmental movement in Leeds (UK) this chapter aims to present and discuss *the challenges that these practices pose to the political and planning agenda of urban public space managers and designers,* and the potential lessons we can draw from them to inform responsive policy-making in times of energy and financial scarcity.

In order to do so it builds on a theoretical perspective which considers socio-spatial relations as a social construct, where daily practices and conflicts around the use, meaning and regulation of urban public spaces are analysed by adopting a relational perspective.

A *relational perspective* to the analysis of the socio-spatial relations which produce and reproduce urban space – among them, urban agricultural projects – draws attention to the agency-structure configurations, the power dynamics and the exclusionary mechanisms that enable or constrain these projects, reproducing or challenging forms of socio-environmental injustice. Such a relational perspective is not specific to a particular disciplinary field, yet it is a quite recent outcome of an interdisciplinary attempt to combine studies in governance and institutionalism, in particular the strategic relational approach of Bob Jessop developed in the understanding of agency-structure dynamics (Jessop, 2001; Jessop *et al.*, 2008), with the urban political ecology approach (Brand and Thomas, 2005; Heynen *et al.*, 2005; Heynen and Swyngedouw, 2003;). Both are embedded in an approach to analysing the social production of space and the built environment, initiated by Henri Lefebvre (Lefebvre, 1974) and further developed by urban and political geographers (Brenner, 2004; Harvey, 2005; Merrifield, 2002; Soja, 2000).

A relational perspective on the analysis of urban public space therefore does not have a unified body of theory, but it is an attempt to bridge disciplines focused on the mutual interconnectedness of territory, place, scale and networks, highlighting how each category's empirical object cannot be analysed in isolation from the social actions and socio-spatial organisation that create them. It therefore addresses the socio-cultural-economic mechanisms of the production of space and nature in an urban environment, and their politics.

While relational perspectives have *de facto* been formulated since the development of urbanism/planning as a discipline by a small number of progressive scholars aware of the by-directional and recursive influence between humans and space (Geddes, 1915; Jacobs, 1961; Riesman, 1950), the marginal role of this perspective in the science of the built environment, until the present, has been largely determined by the dominant pragmatic and normative nature of architecture and planning, which was based on criteria such as functionalism, aesthetics and efficiency, and the primacy of professional knowledge.

In the last two decades this traditional, top down and technocratic approach has been widely challenged by the development of the participatory planning theory (Healey 1990), which is not a fully relational approach but has produced the effect of directing planning practitioners towards various forms of participation of civil society being embedded into their working practice. Nonetheless, attempts to translate the contingent and contextual life of urban space, its fluidity and innovations, into guidelines for responsive forms of construction and regulation of public space are still very marginal and relational approaches do not inform place making.

With the aim of addressing this gap, this chapter argues that policy lessons in urban management and design can be drawn from an understanding of the innovative grassroots practices which represent forms of self-fulfilment of emerging social needs, new organisational experiments, and forms of community resilience.

29.2 Case studies: grassroots initiatives recreating the commons

The last two years have seen an unprecedented proliferation of urban agricultural practices in the Global North. From rooftop gardens and beehives to street containers, more or less institutionalised groups have been experimenting with different forms of urban food growing. These represent a varying mix of growing concerns for environmental conservation, climate change, financial crunch, and the increasingly appealing grow-your-own culture. While some of these initiatives have been promoted and supported by local government bodies and embedded into health, educational or community building policies and programmes, a high number of them remain in the 'grey' field of the grassroots, often self-funded, marginal, or even illegal.

This chapter looks in particular at some of these less institutionalised practices focussing on the different ways in which they contribute to recreating 'the commons'.

This term is intended to cover a wide range of resources that are collectively owned and/or shared and of benefit to a community, and which includes environmental resources, such as land, water and air, and cultural resources such as specific forms of knowledge and technology. Within this chapter I refer mainly to the debate regarding 'land' and its management as a common resource. I propose some preliminary reflections on the contribution that the debate on the commons can bring to the analysis of public space design and management from a planning perspective.

Initiated by Hardin in 1968 with his famous article 'The tragedy of the commons', and more recently developed around the analysis of primitive capital accumulation which enclosed common land over recent centuries, and its current variants (De Angelis, 2001; Glassman, 2006), and widely known in the work of Nobel prize winner Elinor Ostrom (Ostrom, 1990), the almost complete absence of debate in architectural and planning circles over public space is surprising.

As van Laerhoven and Ostrom point out in a recent contribution (2007), an interdisciplinary perspective is fundamental in enhancing the understanding of the evolving and interrelated conditions that make possible the successful maintenance and management of the commons.

Although several critical analysts have pointed out the severe conditions of social exclusions and dispossession of people, environmental collapse and the mechanism of capitalist re-adjustment which make these processes consistent over time (Harvey, 2005), I prefer to follow the perspective highlighted by Federici (Federici, 2009) and Moulaert *et al.* (2010) in looking at the grassroots, the forms of resilience, resistance and empowerment which are enacted around local projects, and the new food sovereignty and food consciousness which is built around them.

The commons – or the potential for their full re-creation – within the case studies highlighted here are not a simple collection of recommunised spaces, nor a bunch of fully 'commonising' re-production sites as De Angelis would have put it (2010: 955), but rather, as Reid and Taylor suggest (2010: 25) a complex web of public space, civic and ecological commons, that is to say the specific contextual cultural and political assets of social and ecological reproduction and their interdependences.

The chapter builds on empirical material collected by the author during 2009 and 2010 through direct involvement in three projects currently running in Leeds (West Yorkshire, UK): Tinwolf/Landshare, Leeds Urban Harvest, and Headingley Community Orchard. For each of them the chapter will provide a description of aims and objectives, a provisional evaluation of their achievements, an analysis of their contribution to a re-creation of 'the commons' and a final assessment of the lessons we can learn in order to inform policy making in the field of public space management and design.

29.2.1 Case 1: TINWOLF Landshare (Transition Inner North West of Leeds Forum)

TINWOLF is a Transition Initiative started about two and a half years ago when the Transition City Leeds group decided to implement a new outreach strategy by setting up neighbourhood groups. Covering a wide area of Leeds, inhabited mainly by students (Headingley, Hyde Park, Burley, Woodhouse), Tinwolf has been the most active of the local groups: promoting film shows, seminars and activities aimed at raising awareness about the cheap oil economy and climate change, renewable energies, food growing, and a wide range of initiatives aimed at increasing community resilience and sustainable modes of urban living. In February 2009, at the peak of its participation, Tinwolf started to promote a specific food project inspired by the LandShare initiative, made known nation-wide by Hugh Fearnley-Whittingstall (River Cottage). For Tinwolf, however, Landshare was more than a matching agency for land givers and growers. With its broader view of promoting a philosophy of reducing food miles, increasing community and individual self-confidence and self-reliance, Landshare is more than a bottom up answer to increasingly long allotment waiting lists (Figure 29.1).

Seen from a relational perspective, Tinwolf's engagement with land share and food emerge as an attempt to build a culture of community solidarity and mutual learning around food production, which has been wiped out by an extreme culture of dwelling isolation, passive citizenship and the consumption based use of non-working time. The traditional nuclear family use of the garden, inscribed in the planned private space of the garden is challenged and reverted. Tinwolf's food group worked as a support team which ran 'activity days' at different gardens, to help with start-of-the-season weeding and soil preparation, new garden start-up or greenhouse and compost heap construction. Skills building for all the people signing up the scheme was promoted not only through gardening coaching during the activity days, but also through linking up with workshops promoted by the Leeds Permaculture Network (LPN) and the Leeds University Union gardening project which runs regular workshops at the Bardon Grange site, north of Leeds.

Figure 29.1. Tinwolf Landshare: a convivial event in a shared garden, august 2009.

However, the fluidity of the group members, the tenancy turnover and the particularly mobile population involved in the Tinwolf activities (students/young citizens) sometimes created a lack of continuity in tending gardens, and therefore required a much greater group involvement/support than expected and longer periods of time spent gardening collectively, which led to tensions regarding alternative ways to manage the initiative which ultimately ended up in a reconsideration of the sustainability of the project itself.

The project's achievements, therefore, seem to be more on the social level: it provided socialisation opportunities for lone dwellers and new urban populations through collective gardening and moments of conviviality, for example, the exchange of recipes and shared meals offered by the garden owners. Looking at the outcomes in terms of the enhancement of gardening skills and confidence in food growing in the group members, however, the project was definitely successful in re-creating commons in the form of shared time, skills, tools and produce.

Among the cases presented in this chapter Tinwolf is one of the most controversial in terms of re-collectivising privatised land and in actually returning it to the realm of the commons. The peculiar design of enclosed gardens, the difficulty of building trust within a community-in-the-making (garden owners and willing-to-grow people did not know each other at the beginning of the project), the occasional lack of proximity between the grower and the garden and the rhythm of working life, did not always make it easy for new and unskilled gardeners to properly tend the new plots and provide produce for self-sufficiency.

Nonetheless, a few elements remain remarkable. Several of the Landshare participants were not on the allotment waiting lists, and there were definitely a higher number of available gardens compared to the number of people wishing to grow.

This points to a prevailing interest in *shared* spaces to grow food. While allotments provide space for mainly individual/family growing, and usually lack facilities such as toilets and cooking areas, shared gardens have the advantage of providing an ideal space for various forms of exchange between the landowner and the gardener: shared produce, shared meals, cooking lessons, gardening tips and coaching, company and friendship.

Existing educational growing sites are mainly promoted by educational institutions and tend to target younger students, rather than adults. A *relational approach* to the planning of housing estates and residential areas, could therefore take into account the emerging practice of food growing in collective spaces around community centres and tower blocks, as well as in private gardens, as long as water, storage and cooking facilities are available and easily accessible. A first lesson we can therefore draw from this case is on the changing meaning of the boundary between the space of the house and the surrounding outdoor space, which becomes an experimental field for community building and alternative models of land share. The similar projects mushrooming around the country, particularly within housing estates, pose new challenges and raise specific spatial arrangements which are relevant to planners, urban designers and housing companies who deal with the design and the management of these in-between spaces.

29.2.2 Case 2: Leeds Urban Harvest

Leeds Urban Harvest is a project which aims to harvest unwanted fruit from public spaces and private gardens for collective food processing (jam, juice, cider making) and free distribution. Inspired by the Abundance project (Sheffield), Leeds Urban Harvest was initiated in the summer of 2009 with a small grant which allowed the group to buy the basic tools required for picking, processing and preserving.

The project is run by a core group of 8-10 people and a variable number of volunteers (200 people have been involved, with a peak of 12 for some sessions) who come along with varying degrees of continuity for the fortnightly picking or preserving sessions, as well as for distributing food and promoting the initiative at local festivals (see Box 29.1).

The group's self-evaluation at the end of the first season has shown that the project has been successful particularly in increasing the awareness of available local healthy food among people in Leeds, has reduced wastage, has developed the skills and knowledge of participants around food harvesting and preserving and has been an opportunity for different generations and communities to network and work together. The group is also building a map of Leeds' unpicked fruit trees which is a valuable tool for current and future food distribution projects.

Urban Harvest initiatives are an exemplar of the emerging forms of practice using private gardens and urban public space in ways which resemble the traditional use of commons for food provision and animal husbandry, and which was the main source of food in rural England until the process of enclosures of the seventeen century which dispossessed the population and supported the rise of industrial capitalism. Fruit picked from private gardens on Saturday is usually shared with the owners (Figure 29.2), if they are interested in having part of the harvest, and then brought to a processing site, usually a kitchen of a community

---

Box 29.1. Leeds Urban Harvest objectives (Leeds Urban Harvest organisers' wikispace).

- Harvest, process and distribute for free unwanted fruit, made possible by sharing equipment and transport.
- Distribute the harvest to those who currently have limited access to fresh fruit.
- Raise awareness of the great abundance of local tasty and healthy food that is available for everyone and for free.
- Empower groups to harvest fruit in their local area.
- Increase the urban harvest by identifying, maintaining, and planting more fruit trees and bushes.
- Do all of this in an ecologically sound way, non-hierarchically, and accessible to everyone.
- Enable groups to harvest, process and distribute fruit that is currently left unpicked on trees and bushes.
- Encourage people to use their skills and talents and get involved.
- Provide opportunities for people to develop new skills in harvesting and preserving fruit.
- Use renewable energy and reuse equipment wherever possible.

Figure 29.2. Apple picking, autumn 2009 (Leeds Urban Harvest Organisers).

organisation to be sorted (the blemish-free fruit is distributed to charitable organisations) and preserved on Sunday afternoons. Part of the preserved food, most commonly fruit juice or chutney, is then usually distributed among the participants, who are allowed to take a small amount of it, or sold to community organisations or citizens at stalls in local festivals.

Leeds Urban Harvest not only promotes a more collective use of trees available in private gardens, enacting a form of temporary collectivisation of the fruit harvest which aims to tackle the dominant culture of food which has emerged in times of cheap energy and abundance, but also helps to challenge the idea of 'public' space simply for its customary aesthetic value. Instead, by raising awareness of existing food available for free in an urban environment, they build a culture of foraging for food in the natural environment, and open up a reflection on the potential of urban spaces to become sites for food allocation in times of recession (and beyond).

The preserving training sessions for new volunteers, run during September 2010, have acted as skills building and empowering devices by having new community members coordinate and run food preserving working days or by providing the necessary training to other local organisations in order that they can safely borrow and use the processing machinery bought by the urban harvesters and stored in an accessible community centre. Urban Harvest therefore contributes to the reconstruction of commons in the sense of the know-how within community groups.

Looking at the mushrooming of urban harvesting and urban wild food walks initiatives around the UK, there are some observations and lessons we can draw, in a relational perspective, for responsive urban design and management. These self-organising and grassroots initiatives represent the immediate fulfilling of emerging needs and interests which go beyond neo-communitarian movements and individualistic needs. Along with re-education on food choices and plant growing, public spaces, parks and urban greens can therefore be designed in order to enhance the benefits of a wider urban community, targeting public enjoyment firstly in terms of its aesthetic and ludic properties but also through the possibilities of using it as a source for healthy and locally grown food and as a ground for food production re-skilling and community building.

It is not only public parks (like in Glasgow or Middlesbrough), but also streets, squares and council greens that can be planted with fruit trees and bushes: something that cities such as Portland (Oregon), or Davenport (Iowa) are already experimenting with (Nordahl, 2009). The emerging need to reduce expenditure on food can be met by appropriate urban design devices.

In terms of commons, these forms of food provision would easily avoid the concerns about over-exploitation of natural resources and environmental sustainability implied by the sustainers of the 'tragedy of the commons' view. Instead, evidence from Urban Harvest demonstrates that the quantity of wasted fruit in a relatively small city like Leeds is enormous, and that with 6 hours of work carried out by a few volunteers it is possible to harvest half a ton of apples, produce 150 litres of juice and still have two full crates of unprocessed fruit to distribute. In the current mainstream culture of individualism and monetised relations, I believe the way to a more sustainable and continuously productive urban landscape has first to challenge the general scepticism towards the commons – i.e. vegetables freely available in public space – rather than their overuse and depletion.

29.2.3 Case 3: Headingley Community Orchard

The Headingley Community Orchard is an urban agricultural initiative promoted by an informal group of residents and workers from the Headingley neighbourhood of Leeds. After a failed attempt to collectively purchase a large piece of land in the nearby Meanwood neighbourhood, the group decided to create a collective dispersed orchard by seeking use-rights on available underused land owned by others, and possibly buying smaller pieces of land, in order to create community fruit patches and orchards.

Having noticed the scarcity of green land in the neighbourhood, in spring 2010 the group started to identify school and church land, and spaces owned by the council. This strategy has the advantage of bypassing the lack of capital needed to kick start such a large scale project; it can positively impact upon local community activity, engaging school students and other existing groups and can enhance local green spaces around the neighbourhood.

The Headingley Community Orchard project is supported by another community initiative, the Headingley Development Trust, which provides administrative support and guidance for further funding applications. Membership of the group is widely accessible through a £1 annual subscription.

There are three plots currently under development: the back of a community owned shop (Natural Food Store), the courtyard of the Shire Oak Primary School, and the green of the St. Chad Parish Church.

The overall project aims are to improve the neighbourhood environment by increasing its biodiversity and the amount of green spaces, promoting learning opportunities through the involvement of local groups in grafting workshops, promote a healthy eating culture and eventually provide locally grown food to be sold at a reasonable price through the Headingley Farmers Market (http://www.headingleycommunityorchard.org.uk/).

Looking into this experience we can explicitly see that the acquisition of the right to use is a second best choice to forms of collective ownership, which has a necessarily temporarily nature. This element of unpredictability is the object of a current strong debate launched by Sharon Zukin on a well-known online urban sociology forum (Comurb_21) on the New York City community gardens and the risk of increasing gentrification. While the previous British government had announced its intention to redistribute unused brown fields and land for growing purposes (Communities and Local Government, 2010), the criteria for such a distribution have not been set, nor has the new coalition government so far expressed interest in ratifying the decision.

In the speculative manoeuvres around wheat prices in the recent economic crisis, concerns for the affordability of fresh vegetables and fruit are rapidly emerging and self-organising strategies are being put into practice by individuals and local organisations. This highlights the almost total eviction of collective agricultural practices in cities since the time of the enclosures (except the example of the Dig for Victory campaign during the Second World War).

The recent debate over world population size and climate change has already pointed to the difficulties of feeding an exponentially growing population when areas of cultivable land are expected to shrink and without further depletion of environmental resources.

While 'rurbanisation' does not appear to be the easier and most immediately viable arrangement – nor is it necessarily the ideal solution, considering the 'friction between ecological scales and temporal dynamics' of institutional change (Van Laerhoven and Ostrom, 2007), the mode of self fulfilment expressed by these forms of self organisation can be a learning opportunity for governing institutions. These projects not only could be mapped into an updated urban food system, but also used to provide suggestions for a responsive form of public space design and urban planning which might meet the most pressing financial constraints as well as ethical concerns about food miles.

A responsive adjustment of urban assets would for example redesign green spaces in and around institutional buildings, providing land which can be contracted out to local organisations for a common food scheme. As many critical thinkers have already pointed out, (Feindt, 2010) the current financial crisis and the collapse of many speculative developers means the time is right for the exploration of alternative forms to capitalist organisation, property rights and urban planning, and to enact public policies which buy back undeveloped land for an ordinary integration of collective urban agriculture into urban planning.

29.3 Conclusions

The brief overview presented here has shown emerging and informal experiments in urban food production which can be read as spontaneous forms of recreating the commons: more or less organised forms of community use of private, public and institutional land are established.

While a sociological analysis of the success or failure of the consolidation of these projects is not the scope of this chapter, a few points can be made on the controversies that they raise, which might point in the direction of a research agenda in itself.

Whether it relates to the actual access or to the share of produce, the 'governance of use' has still to be developed from within these initiatives. The control over the share of produce and processed food, the accessibility of land and the management of the project/funding/revenues has still to be assessed as long as the projects evolve and consolidate, and there are surely grey areas around the definition of the core group in each of these projects, which might make us question the definition of 'community' shared among the group members, and the consequent entitlements. Studies in the field of sociology of organisation are in this sense a valuable tool for a critical reflection on the group dynamics at play in these experiments. The reader might also wonder to what extent Elinor Ostrom's book *Governing the Commons* (1990) might be of relevance when dealing with the uncertainty these cases raise, and for a possible institutional answer to their claims. Considering that the current trends in public space design and management are dominated by the tendency to over-regulate or privatise common spaces, Ostrom's book represents a useful advocacy for self management of common property resources (CPR) as opposed to the need for state coercion or privatisation, which could be a useful starting point by which to remind governmental institutions of existing alternative forms of self-management of common resources which do not result in the 'tragedy of the commons'. However, her work is of less relevance in respect of an analysis willing to take into account ethical and value based agency occurring in a highly unstable situation characterised by economic crisis, neglected needs, long term community disempowering and the necessity to act in residual, marginal spaces with radical, if not illegal, practices.

Considering this situation, my aim is rather to point out how a relational approach to the construction of places can be a useful way of learning from these practices, demands and challenges and work towards a more responsive and socially just form of urban planning and design.

As we have seen in the cases described above, all the three projects mobilise resources existing in the local community (e.g. land owned by local institutions), challenge traditional space management and zoning (i.e. fenced schoolyards, parish greens, public recreational parks), by creating collective resources such as food growing or learning opportunities in unexpected places, and creating new community projects and networks which enhance social cohesion and can act as new generative devices (Lanzara, 2003).

In a conjunctural time in which planning and governmental institutions are faced with the failures of the mainstream economy and its impact on the local communities and the environment, these projects represent important and very accessible learning opportunities. Their location on public land, or through the opening up of private land to the public, creates an *exposure* to the view, which is the first step of a learning trajectory for a wider public.

But how do we learn from them? How can we go beyond these local projects? How can we take into account the contextual specificities of the local community and reflect on transferability and upscale? What tools do we use to learn from the grassroots and to inform socially innovative policy making in the field of urban design and management?

I suggest a threefold methodology for socially engaged individuals (scholars, practitioners, activists and citizens in general) willing to shape a possible way forward: contextual analysis, learning devices and structured claims.

29.3.1 Structural and agency elements and their relational dynamics: contextual analysis

The first element is a contextual analysis which takes into account the relational dynamics between structural and agency elements (Jessop, 2001), between interest groups and regulatory institutions, community cultures and bureaucratic practices, including the contextualised cultural and historical framework in which projects give rise to meanings, where meanings take a specific shape, and in which existing resources become actual resources. In this specific field it is important to consider, for example, the overall availability of spaces for food growing and the risks of incurring food deserts; people's residential mobility; sensitivity to environmental issues/environmental culture; safety/security issues and public behaviour in public spaces which might have an impact with regards to vandalism on exposed growing sites or which require more appropriate selection of sites and design solutions for food growing in urban spaces. These elements, which are best collected with a participatory methodology, should clarify what is the current window of opportunities for new organisational arrangements, and inform the specific type of experimental project which can be proposed as an initial way of applying a relational perspective to planning and urban design.

29.3.2 Learning devices for policy innovators: action research and social platforms as training grounds

The second element is the use of action-research as a training ground for policy makers and governing institutions. Recently gaining attention beyond the educational field in which it has been developed, for its promising impact beyond the academic community, this methodology combines inquiry and dissemination goals, and can act as a tool to enhance the working practices of policy makers and civil servants. The recently developed 'social platform' tool (for example the coordinated action Social Polis, funded within an EU Framework Programme, and directed by Frank Moulaert) is a form of action research which exposes a heterogeneous group of participants to each other's discourse, reasons, and practices over an extended period of time (http://www.socialpolis.eu/).

Developed within studies in social innovation (Moulaert *et al.*, 2010), the social platform is a transdisciplinary tool which values different forms of knowledge (academic and non-academic, practice based and theory based) and which has the principal aims of providing the space for bridging languages, unveiling and confronting interests and analytical perspectives, and providing knowledge for innovative forms of research/policy design. This tool, I believe, can act both as an activator of new multilayered and complex food production projects and policies, as well as a learning device to strengthen existing grassroots initiatives.

The experimental construction of a specific social platform on urban agriculture is currently under development by the author in Yorkshire thanks to a research grant awarded by the Economic and Social Research Council (ESRC), UK. The platform will be co-led with different actors variously involved in urban agriculture, land management and food systems, and will provide a training ground for activists, social entrepreneurs, civil society groups and governmental bodies in negotiating their needs, claims and interests and design a region-wide policy for supporting urban agriculture as a key player in the local food system.

29.3.3 From residual practices to structured claims: re-politicise the commons

The third element I am proposing as a way forward is the re-politicisation of the construction of the commons, and in particular of the land as a common (resource). While many of the practices I have highlighted in this chapter have found their way in the interstices of the city, in ad-hoc negotiations, and in voluntarism, their innovative potential runs the risk of remaining residual and invisible. Their progressive and critical attitude towards the mainstream capitalist land assets and their contribution towards overcoming social isolation and environmental injustice, qualify them as socially innovative in principle (Vicari and Tornaghi, forthcoming), but they fail to fully articulate a political statement which would be an important part in their trajectory of consolidation. One way – although surely not the only way – to escape their residuality, and to upscale, is the possibility of self-reflexivity within these practices, and a political reformulation, which could bring them into the public sphere for democratic discussion.

Among the many that have taken this route (see for example the groups networked around the pan-European movement 'Reclaim the fields'; www.reclaimthefields.org/) is the example of PSPP (Public Space Public Produce), an informal group based in Leeds which is currently running the Edible Public Space project (www.ediblepublicspace.org). Edible Public Space (EPS) is a project which aims to bring back public spaces to the core of the public life of local communities, and which aims to create community-led, convivial and productive public spaces which reconnect individuals to their natural, and thus far alienated, ecological commons (Tornaghi, 2011a,b). EPS, like many others, has not gone down the route of re-appropriating and regaining control over public resources through residual practices (temporary appropriation, squatting of liminal spaces, random interventions), but has deliberately chosen to work within the local community while, at the same time, articulating claims over public resources with governing institutions. While negotiating the right to use a common resource, such as a piece of public land, which is alienated on a regular basis through bureaucratic regulation, paternalistic practices or market driven interests is annoying when not even unethical, the squatting route or the direct action approach (for example guerrilla gardening or land appropriations) has often been ineffective and short lasting, and has not led to long term, widely and diversely participated community projects.

The re-articulation of these claims in discourses over the meaning of 'public', 'community', and 'justice', as well as the commitment to activate a wide public debate in the multiple public spheres is the path towards a re-politicisation of the commons and a challenge to urban planners and managers which can no longer be ignored.

References

Brand, P. and Thomas, M., 2005. Urban environmentalism, global change and the mediation of local conflict. Routledge, London, UK, 237 pp.

Brenner, N., 2004, New state spaces, urban governance and the rescaling of statehood. Oxford University Press, Oxford, UK, 351 pp.

Communties and Local Government, 2010. Grow your own revolution gets major land boost. City Farmer News, available at http://www.cityfarmer.info/2010/03/08/allotment-boost-from-under-used-land-planned/.

De Angelis M., 2001. Marx and primitive accumulation: 'The continuous character of capital's "enclosures"'. The Commoner. Available at http://www.commoner.org.uk/02deangelis.pdf.

De Angelis, M., 2010. The production of commons and the 'explosion' of the middle class. Antipode 42: 954-977.

Federici, S., 2009. Silvia Federici: on capitalism, colonialism, women and food politics. interview by Max Haiven, Politics and Culture, Nr. 2, November 2009. Available at http://www.politicsandculture.org/2009/11/03/silvia-federici-on-capitalism-colonialism-women-and-food-politics/.

Feindt, P.H., 2010. The Great Recession – A transformative moment for planning. International Planning Studies 15: 169-174.

Franck, K. and Stevens, Q. (Eds.), 2006. Loose space. Possibility and diversity in urban life. Routledge, London, UK, 296 pp.

Geddes, P., 1915. Cities in evolution: an introduction to the town planning movement and to the study of civics. Williams and Norgate, London, UK, 409 pp.

Glassman, J., 2006. Primitive accumulation, accumulation by dispossession, accumulation by "extra-economic" means. Progress in Human Geography 30: 608-625.

Healey, P., 1990. Policy Processes in Planning? Policy and Politics 18: 91-103.

Hardin, G., 1968. The tragedy of the commons. Science 162:1243-1248.

Harvey, D., 2005. The new imperialism. Oxford University Press, Oxford, UK, 275 pp.

Heynen, N., Kaika, M. and Swyngedouw, E. (Eds.), 2005. In the nature of cities: urban political ecology and the politics of urban metabolism. Routledge, London, UK, 272 pp.

Heynen, N. and Swyngedouw, E., 2003. Urban political ecology, justice and the politics of scale. Antipode 35: 898-918.

Jacobs, J. 1961. The death and life of great American cities. Random House, New York, NY, USA, 458 pp.

Jessop, B., 2001. Institutional re(turns) and the strategic-relational approach. Environment and Planning A 33: 1213-1235.

Jessop, B., Brenner, N. and Jones, M., 2008. Theorizing sociospatial relations. Environment and Planning D: Society and Space 26: 389-401.

Lanzara, G.F., 2003. Nuove forme di azione organizzata nelle città: comunità difensive o laboratori per l'innovazione? Urbanistica 1: 29-34.

Lefebvre, H., 1974. La production de l'espace. Anthropos, Paris, France, 485 pp.

Merrifield, A., 2002. Metromarxism: a Marxist tale of the city. Routledge, New York, NY, USA, 212 pp.

Moulaert F., Swyngedouw, E., Martinelli, F. and Gonzalez S., (Eds.), 2010. Can Neighbourhoods Save the City?: Community Development and Social Innovation. Routledge, London, UK, 264 pp.

Nordahl, D., 2009. Public produce: the new urban agriculture. Island Press, Washington DC, USA, 177 pp.

Ostrom, E., 1990. Governing the commons: the evolution of institutions for collective action (political economy of institutions and decisions). Cambridge University Press, Cambridge, UK, 280 pp.

Riesman, D., 1950. The lonely crowd: a study of the changing American character. Yale University Press, New Haven, CT, USA, 315 pp.

Reid, H. and Taylor, B., 2010. Recovering the commons: democracy, place, and global justice. University of Illinois Press, Champaign, IL, USA, 288 pp.

Soja, E.W., 2000. Postmetropolis: critical studies of cities and regions. Basil Blackwell, Oxford, UK, 440 pp.

Tornaghi, C., 2011a. Steal this veg. The Red Pepper, London, UK, July 2011. Available at http://www.redpepper.org.uk/steal-this-veg/.

Tornaghi, C., 2011b. Edible public space and urban agriculture. Interview for New Left Project/Red Pepper, edited by Sean Gittins. Available at http://www.archive.org/details/RedPepperAudio-Issue1stJune2011.

Van Laerhoven, F. and Ostrom E., 2007. Traditions and trends in the study of the commons. International Journal of the Commons 1: 3-28.

Vicari Haddock, S. and Tornaghi, C. (forthcoming). A transversal reading of social innovation: Questions to the theorists. In: Moulaert, F., MacCallum, D., Mehmood, A. and Hamdouch, A. (Eds.) International handbook on social innovation. Social innovation: Collective action, Social Learning and Transdisciplinary Research. Edward Elgar, Northampton, MA, USA.

Chapter 30

Urbaniahoeve: expanded urban agriculture

Debra Solomon
Social Design Lab for Urban Agriculture, Barentszplein 3 III, 1013 NJ Amsterdam, the Netherlands; debra@urbaniahoeve.nl

Abstract

Since September 2009 Urbaniahoeve, Social Design Lab for Urban Agriculture, has been conducting action research in the Netherlands towards developing a signature, suitable, and resilient form of urban agriculture (UA) that could be widely adopted in the Netherlands and other Northern European cities. The physical and conceptual starting points for these UA typologies presuppose the use and integration of existing green and social infrastructure in addition to offering a clear benefit to the immediate surroundings. Urbaniahoeve has begun initiating structural formats of UA in the Dutch cities the Hague and Amsterdam by developing landscape architectural plantings for integrating urban food production into the public space. Together with local participants, Urbaniahoeve is creating spatially contiguous *foodscapes*, in underused, under programmed and over-paved urban typologies, infusing them with edible landscaping and socially-driven food-system activities. This chapter charts Urbaniahoeve's specific areas of research and the resulting public greens and social typologies that range from planting designs, to food facilities such as (playground) cooking equipment, and from collaboration protocols and community involvement to the strategic re-appropriation of under-used public space. The chapter concludes with a statement about the characteristics that cultural sector UA project initiators like Urbaniahoeve typically bring to the field, and how these characteristics aid experimentation and innovation at a time when the development of resilient and immediately deployable formats for urban agriculture are so urgently needed.

Keywords: permaculture, foodscape, edible landscape architecture, food growing, public space, urban regeneration, cultural sector

30.1 Urbaniahoeve in context

Urbaniahoeve develops action research models that are examples of publicly maintained urban agriculture, strategically located in formerly underused public space. In 2010 Urbaniahoeve launched three projects comprising a (1) contiguous, productive *foodscape*; (2) public (playground) cooking infrastructure; and (3) a range of (educational) programming at both locations to provide skilling-up and engagement with the physical project locations. The choice to locate projects in the Hague/Schilderswijk and Amsterdam Nieuw West, were opportunity-based, motivated by considerations related to the practice of art and design.

Though the 'public green spaces' of the Schilderswijk and those of Amsterdam Nieuw West differ as widely as their municipal agencies and institutions, Urbaniahoeve is working on the phased development of an urban agriculture platform within an urban neighbourhood context, one that can be accessed and maintained by multiple local stakeholders.

The next section describes three Urbaniahoeve urban agriculture project modules in development; *Foodscape Schilderswijk*, *DIY Mmmmuseum of Oven Typologies*, and *All that Rot!*

30.1.1 Foodscape Schilderswijk (2010, ongoing; Solomon, 2010a)

Foodscape Schilderswijk is comprised of community supported planting interventions that will eventually sustain a physically contiguous, food-producing biotope in the Hague's Schilderswijk. The project originated as an 'Art in the Public Space' commission from Stroom (Hague Centre for Art and Architecture) as part of their multi-year manifestation, *Foodprint. Food for the city*. Both the project format and the term 'foodscape' are inspired by Katrin Bohn and André Viljoen's 2005 book *CPULs: Continuous Productive Urban Landscapes* and refers to a physical landscape in which horticultural knowledge, cultural practice, food-system infrastructure, and a real food-producing biotope are co-located (Viljoen, 2005). In 2009, project *Foodscape Schilderswijk* began mapping the existing green and social infrastructure

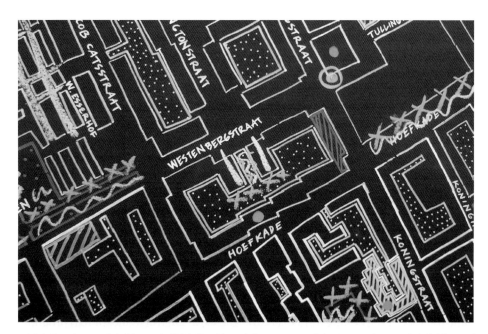

Figure 30.1. Foodscape Schilderswijk (2009), of hand drawn map of the Hague borough Schilderswijk. The map depicts existing green and social infrastructure, and suggests locations and typologies for implementing borough-wide urban agriculture (Solomon and Abelman).

of the Hague borough the 'Schilderswijk' (Figure 30.1), a lower-income, ethnically diverse neighbourhood, to reveal potential action research models to form the *foodscape*. In May 2010, the Urbaniahoeve/*Foodscape Schilderswijk* team and local project participants from the Westenberg and Wellington streets began planting the borough's first community orchard in the Westenberg Hof (Figure 30.2). As of this writing (June 2011), project participants have established three orchard locations in formerly inaccessible and undesirable 'visual green spaces' with nearly 50 fruit trees, hundreds of berry bushes, artichokes, herbs and nectar-rich ground covers, and two tree planter mini-gardens.

In the Schilderswijk, like other boroughs subject to 1980s and '90s-style urban regeneration, most 'green (public) space' is landscaped to be solely 'visual'. Consistently lacking adequate (spatial) programming, one can easily observe that these green spaces were not designed for (physical) use (Alexander *et al.*, 1977). This provokes a perception by a vociferous minority that the public space is being misused (Jacobs, 1993). Bearing this in mind, *Foodscape Schilderswijk* roots into neighbourhood infrastructures, both green and social, to generating edible forms of landscape with local stakeholders that re-activate public spaces like these. The planting and social design typologies under development are listed below.

But first a word about permaculture and on being involved with multiple partners. Permaculture is a design strategy for human settlements and agricultural systems inspired by the relationships found in natural ecologies. It aims to create stable, productive systems that provide for human needs, harmoniously integrating the land with its inhabitants.

Figure 30.2. Planting the Westenberg community orchard. Neighbours from the Schilderswijk nurture their fruit trees planted in a formerly inaccessible piece of 'public space'.

There is a permaculture principle stating that every function (in a given system) must be supported by many elements. Should any element fail, the remaining elements can still support the function. An adjacent principle states that every element should serve many functions; its usefulness should be plural such that should any function falter, the element remains valuable within the larger system (Mollison, 1988). These are principles about building resilience into systems, both environmental and social, and are embodied in the *Foodscape Schilderswijk* project design.

Foodscape Schilderswijk is not a community kitchen garden, but landscape architectural food system infrastructure, primarily *espalier* orchard with perennials such as berries, herbs and nectar-carrying, self-seeding (annual) ground covers, grown in the public space. As of this writing the project engages five different groups of participants, each with a different role and level of engagement towards the project as a whole:

- Neighbours are involved as instigators, whose enthusiasm triggers the development of a new location (within e.g. the enclosed 'yards' of the Westenberg and Wellington Hofs), and the general nature of the planting typology. They participate in 'fun' community-style activities like Sunday afternoon tree-planting, but to date show no interest in the finer aspects of horticulture, *espalier* pruning, landscape architecture, urban agriculture or the politics of civic agency and public space. Recently the one year old *espalier* orchard beds located on the 'library side' of the Westenberg Hof have sparked interest in the neighbours on the opposite side to request collaborating on a new orchard location there, specifically because the existing beds are perceived by these neighbours to have a positive effect on eradicating litter and vandalism.
- A group of mothers and children from 't Palet Elementary School have been involved since March 2011 in planting a classic *espalier* fruit wall on a formerly inaccessible strip of land adjacent to the school and a large local playground (*Hanneman Hoek*). Though this project location was intended for this specific group of mothers and children, its situation lures children from the adjacent playground to drop by and participate in planting activities. A regular group of 'drop-bys' has formed, some of whom are knowledgeable and appear to have a real interest in gardening.
- The (*Hanneman Hoek*) location's high level of visibility (at a pedestrian crossroad) has also brought new adult participants to the *Foodscape Schilderswijk* project as a whole. Though unrelated to the group of mothers and children, they participate with the *Foodscape Schilderswijk* core team at this location (and others) twice a week. One of these participants is independently involved in the development of a permaculture community garden one block away *(Permacultuur moestuin de Groene Mus).*
- Nineteen high school students (and their professor) from the nearby Nova College, (a programme geared towards integrating high school aged newcomers to the Netherlands into the Dutch educational system) participate in *Foodscape Schilderswijk* within the context of their biology class. This group is quite adept, able to carry out the tasks of orchard bed setup and maintenance independently after instruction. Whether this is due to the formal nature of their engagement or it is a coincidence of their individual intelligence and history remains unclear. The Nova College students prove that it is

possible to develop educational modules to teach permaculture and to address the subject of (public space) urban agriculture in a trans-disciplinary manner.

- A group of gardeners from the municipality's Department of Public Greens collaborates with *Foodscape Schilderswijk* on a professional level and regularly provides the project with technical assistance.

The *Foodscape Schilderswijk* Spring 2011 programme included a series of *(espalier)* pruning workshops in order to test project assumptions on the level of engagement of the project participants. In hindsight, it is only natural that of the 5 groups, only the high school students, the professionals from the Department of Public Greens and a group of permaculture enthusiasts participated in the workshops.

With regard to plantings intended for community use, it is currently in vogue to say that all project initiative should 'come from the neighbours'. One of the lessons of the first year of *Foodscape Schilderswijk* activities proves otherwise. Urban agriculture as defined by this project, *productive landscape architecture in the public space,* is food system infrastructure. As with any other system, its resilience is dependent on the engagement of multiple partners, including but not exclusively neighbours participating in their free time. 'Low-dynamic plantings' like fruit trees and perennial herbs that do not require the continual nurturing of say, a vegetable garden, have the drawback that participants that are not formally engaged (e.g. high school biology class) can lose interest. Nevertheless, the bits of public space that constitute the *Foodscape Schilderswijk* project locations have never received as much traffic or attention by so many disparate groups as they do now. The current strategy of hyper-programming (stacking) (Mollison, 1988; Whitefield, 2004) participants with different levels of engagement appears to be effective in developing a base to ensure project continuity and broad community engagement.

30.1.2 DIY Mmmmuseum of Oven Typologies (DIYMOT) (piloted June 2010; Solomon, 2010b)

According to the Urbaniahoeve vision, a successfully networked urban agriculture depends on, and is co-located with thoughtful public space programming stocked with (playful) food-system infrastructure. The *DIY Mmmmuseum of Oven Typologies* (DIYMOT) is a platform for public space cooking under the guise of community oven building. The DIYMOT was successfully piloted at the *Sloterplas Festival* in Amsterdam, a large-scale community event on and around the banks of the Sloterplas (lake) in June 2010. The pilot demonstrated that at the very least, neighbourhood children would show an interest and auto-didactically develop cooking skills by using the ovens. On the island where the DIYMOT was installed, participating children became so absorbed in their newly assumed roles as assistant chefs that they refused to return home with their parents until they were done preparing the on-site dinner (for 64 guests). The lesson of this project anecdote may be that domestic disobedience is a small price to pay for adequate food-system infrastructure.

Ovens made of tamped earth, underground ovens, solar ovens made from wasted umbrellas and/or pizza boxes, the DIYMOT is a collection of manuals and materials with which people can build actual working ovens (Figure 30.3; Alexander *et al.*, 1977). As of this writing, the DIYMOT is negotiating placement in three locations in Amsterdam Nieuw West that will share the project development costs and subsequent programming; a nature park with a focus on energy use, a local build-your-own fort playground, and (abandoned) building sites that, due to the current financial transformation may remain unbuilt for the foreseeable future. In its 2011 planning, Urbaniahoeve has designed a full year's programme that uses the various oven installations and kitchen kit in monthly activities with a strategic selection of local groups. Classes from nearby schools, women's groups, religious organisations, caterers, cooking clubs, illegal restaurants, and foragers unions will each be invited to co-develop the oven platform in the interest of spreading the word that the local public space now includes accessible self-built cooking infrastructure.

30.1.3 All that Rot!: festive fermentation & community food preservation skills (October 1, 2010, ongoing; Solomon, 2010c)

Community food production makes fresh fruit and vegetables available to urban populations. But without planning and preservation the gluts of local food it can produce will go uneaten, becoming a source of food waste not unlike we find in our supermarkets. Resilient urban agriculture values enhanced (cooking) skills to preserve summer's harvest throughout the winter and beyond the Hungry Gap (Astyk, 2008; Katz, 2006). Urbaniahoeve's *All that*

Figure 30.3. DIY Mmmmuseum of Oven Typologies *participants play with the tamped earth oven during the* Sloterplas Festival Art at the Pool Manifestation. *The kids independently prepared tamago, a Japanese omelette as one of the courses of the diner ambulant.*

Rot! celebrates the culture of community food preservation by tapping the technology and traditions of lacto-fermentation (Figure 30.4).

Reigning in the transformative tendencies of naturally occurring lactobacilli, fermentation turns cabbage into powerful sauerkraut, a pro-biotic (living) food that gives the local gut biotope and immune system, an uplifting kick. Eaten as part of a seasonal diet, the lacto-bacilli give a nutritional boost just when the body needs it most, at the point when fresh vegetables become absent from the Winter table. The fermentation process requires neither gas nor electricity when salt and anaerobic conditions bring forth a nutritive force by which sauerkraut's raw brother pales anaemic in comparison (Katz, 2003).

In 2010 Urbaniahoeve organised a series of fermentation workshops with *Vrouw & Vaart*, an Amsterdam Nieuw West women's development centre (that cultivates an on-site permaculture garden). *All that rot!* workshops educated the participants about the techniques of sauerkraut and kimchi (Yoon, 2005), as well as brewing kombucha and kefir (Katz, 2003). The workshops proved that this urban agriculture skill set can be easily and enthusiastically adopted.

Figure 30.4. Making sauerkraut as part of the 2010 Makers Festival. Workshop held with gardeners from de Groene Vaart, a women's gardening group. Urbaniahoeve initiated the All that Rot! workshop to celebrate the traditions of garden-related food preservation.

30.2 Urbaniahoeve planting typologies – now in development

30.2.1 Collaboration protocol and methodology

Because Urbaniahoeve project locations (i.e. *Foodscape Schilderswijk*) access public space, a good working relationship with the municipality's Department of Public Greens is essential to project success. In order to facilitate common ground between the two organisations and to make the project expectations lucid, Urbaniahoeve forged a protocol. Listed below are the most cogent points.

Foodscape Schilderswijk develops community-maintained, visually resplendent edible landscaping and chooses and develops project locations on the basis of the following principles:
- work with existing green and social infrastructure, no tabula rasa situations;
- focus on activating under programmed 'visual greens' as locations for productive landscapes;
- connect project locations to form a contiguous biotope;
- stack programming on single locations;
- work in phases;
- work with non-dynamic (low maintenance, resilient, and perennial) plantings;
- incorporate the techniques and design principles of permaculture;
- develop programming that supports community project maintenance;
- strive for project continuity;
- keep the ground covered with planting as much as is seasonally possible;
- prioritise urban soil fertility;
- generate socially thriving and resilient food producing urban biotopes.

The protocol goes on to further describe workflows and responsibilities, timelines and moments of project evaluation and delivery.

One of Urbaniahoeve's goals for *Foodscape Schilderswijk* is that all future green space plantings in the borough become both edible and biotope enhancing (Alexander *et al.*, 1977). Urbaniahoeve's partner at the Department of Public Greens agreed that however aspirational, budgetary constraints make this goal unachievable. Projects like *Foodscape Schilderswijk* have the potential to connect community-supported edible landscaping with works carried out by the Department of Public Greens to achieve project goals. *Foodscape Schilderswijk* is working on an indexation of its planting modules that corresponds to the technical requirements of the Department of Public Greens such that these modules become replicable throughout the Schilderswijk and later, by other departments in other boroughs.

In order to develop working models for low-maintenance community food producing biotopes, Urbaniahoeve's foodscapes focus on using perennial and non-dynamic plants that require minimal care, that prioritise soil fertility and are planted systematically according to the plant guilds.

30.2.2 Freestanding *espalier* fruit beds

Urbaniahoeve is developing *espalier* fruit beds as food producing, biotope enriching, living screens and to optimise the use of vertical space. *Espalier* is the agricultural practice of training, pruning, and tying branches to grow in relatively flat planes, frequently in formal patterns. Aside from its aesthetic impact the technique is good for fruit production and is part of the Dutch (horti)cultural heritage (Kuitert and Freriks, 1994). *Foodscape Schilderswijk* fruit beds intercrop apples, pears, plums and cherries with herbs, flowers, and bulbs in order to create multi-layered, productive planting beds that will in 4-5 years be easily maintainable and largely self-sustaining (Figure 30.5). A series of pruning workshops, intended to increase project engagement were successful in identifying specific local groups that embrace learning advanced orchard skills.

30.2.3 Asteraceae areas

Experimental beds of artichokes alongside other asteraceae family perennials, the varieties that include echinacea, stevia, and self-seeding annuals like sunflowers and chamomile, have been chosen for their qualities as a food source, as cut flowers, and for their use in traditional medicine. The aesthetic of the asteraceae areas, is inspired by the visually attractive 'natural style' beds of Dutch Wave gardeners Henk Gerritsen and Anton Schlepers masters of bio-diversity (Gerritsen, 2008; Oudolf *et al.*, 2009).

Figure 30.5. Community Fruit Orchard located in the Westenberg Hof in the Hague borough Schilderswijk planted with freestanding espalier fruit trees.

30.2.4 Wildly Attractive Edges

Lacking spatial programming and enclosed by 3-storey apartment blocks, the interior gardens of the Westenberg and Wellington Hofs are monocultures of grass and large, rarely pruned trees. As of April 2011, the frequency of finding garbage thrown in the enclosed gardens had significantly decreased and the downstairs neighbours perceive the problem to be solved by the orchard beds.

Urbaniahoeve is experimenting with a planting typology that will further fill the edges of the enclosed spaces. Bushy, blossoming plants will at the very least, provide an inexpensive experiment in social control through landscape architecture (Alexander *et al.*, 1977). In terms of the garden biotope, the *Buddleia davidii*, *Sambuca nigra* (elder), self-seeding borage, will certainly provide voluminous cover and a source of nectar for the pollinators of the nearby fruit trees, and an invasive patch of comfrey can aside from its qualities as a bee attractor offer 'live mulch' to the beds. Suffering drought in 2011, planting the Wildly Attractive Edges typology has been rescheduled until autumn so that seasonal rains can at least help the young bushes to establish themselves.

30.2.5 Instead of Grass (autumn 2011)

The vast lawns of the interior gardens as well as grassy strips along the edges of the streets require expensive maintenance eight months per year, offer precious little to the biotope, both above and below the ground. Additionally, of those affected by hay fever, 90% are allergic to this form of ground cover (Ogren, 2008). In autumn 2011, the edges of the *espalier* fruit beds, will be seeded with winter-hardy clovers that in the spring require less frequent mowing, provide a source of nectar whilst slowly ameliorating the soil and developing a fragrant, tread-worthy matt in the process.

30.2.6 Herb Carpet (autumn 2011)

In a highly paved area at the corner of Westenberg and Wellington streets a border around the tree planter will provide an experimental location for a workshop with local children and professional gardeners to make an herb carpet (Alexander *et al.*, 1977). Semi-open paving and open-work tiles will be planted with low-lying varieties of creeping herbs like thyme, oregano, and chamomile (Figure 30.6). One of the many aims is to discover the sturdiness and rate of water absorption of the 'herb carpet', to assess whether this typology is suitable for higher foot traffic and whether it can replace traditional paving. Although herbs grown under such conditions may not have value as food, they remain aromatic when tread upon, set an aesthetic tone, and contribute to the contiguous biotope both above and below the soil's surface.

Figure 30.6. Schilderswijk Chamomile Paving, wild chamomile bulging out of paving proving possibility for natural forms of paving reduction as with variations on Urbaniahoeve planting typologies, Instead Of Grass and Herb Carpet.

30.2.7 Foraging Forest

In a shady alcove of the Wellington Hof, *Foodscape Schilderswijk* child participants planted berries, currants, wild garlic (ramps), and chives to develop a natural-style foraging forest (Figure 30.7). Having recently discovered that the Department of Public Greens uses the location to store branches pruned from the flanking sycamores, a portion of the area will be converted to host a mushroom growing workshop in the shadiest corner. Neighbourhood children will enjoy inoculating the logs with oyster mushroom and shitake spores and learning about the role of mycelia in an inner-city forest garden (Kellogg and Pettigrew, 2008).

30.2.8 Tree planter nurseries

In the Schilderswijk the tree planters that line the streets have the potential to link the gardens into a continuous productive space, as informal nurseries for the larger planting spaces elsewhere. Planted with self-seeding and perennial nectar-rich flowers and herbs, tree planters mini-gardens have the power to seduce the sceptical, set the tone for the public space (Alexander *et al.*, 1977), de-fragment the biotope, reduce litter, and initiate a phased reduction of non-absorbent paving in the neighbourhood.

Figure 30.7. Forager Forest, Foodscape Schilderswijk participants plant a foraging forest with berries, currants, and ramps.

30.2.9 Tree planter mini-gardens

Initially an experiment in testing the neighbourhood's tolerance of self-maintained edible landscaping, local children planted two tree planters on Wellington street, one with a selection of Mediterranean kitchen herbs and the other with rhubarb and strawberry (Figure 30.8; Alexander *et al.*, 1977). One year after planting, both planters remain well maintained, vandalism-free, watered, free of litter, and the locals harvest the public space produce for home use.

Figure 30.8. Tree planter mini-gardens.

30.3 Are artists and designers expanding urban agriculture?

Considering the potential of projects initiated by cultural sector producers like Urbaniahoeve, is at the very least, worthy of specific research with regard to the diversity of its output formats; community produced edible landscape architecture, hyper-programming/multiple use of one location, and community up-skilling programmes. Urban agriculture implemented by artists and designers as part of a social design or art in the public space praxis, provides a platform for rapid experimentation and action research that may be better at expanding the range of UA working models than projects initiated by the agricultural or other sectors. This statement is based upon this author's observation and experience from her own practice and involvement in UA projects of various scale since 2006, from extensive contact with UA project initiators in other sectors, as well as contact with local farmers bordering Amsterdam Nieuw West.

30.3.1 Personal/professional autonomy and flexibility of practice

Art and design education is centred on developing a signature approach to research, methodology, project building, and developing a fingerprint with regard to material output. In the context of expanding urban agriculture, this means a playful approach towards project production, agile prototyping, and unique formats (Jencks and Silver, 1972).

It is valuable to mention that art and design 'products' are rarely subjected to the rules, conventions, and regulations of agricultural or industrial food production. Artists and designers produce experiments and prototypes within a very free context and cultural tradition. As an anecdote, the author has designed and implemented concept restaurants as art installations in musea, developed 'kitchen playgrounds' and run 'free kitchens' from market surplus that produced food, but that were never subject to prohibitive hygiene laws and conventions. Officially, the product was not 'food', but a 'art' or 'design'. The fact that guests could put this product in their mouths, chew, swallow and digest, was of no concern to anyone. Because of this, the project focus remained strictly on the array of food one could make from market surplus (a project in collaboration with van Heeswijk and Kaspori, called *Lucky Mi Fortune Cooking*; Van Heeswijk *et al.*, 2007), or if it is possible to run a restaurant that serves a menu of food grown without land or light (Figure 30.9; Solomon, 2006).

30.3.2 Unusual skill sets

Artists and designers routinely at work in the public domain maintain broad skill sets and can engage a wide network of expertise. They are comfortable working in a trans-disciplinary setting with a multi-disciplinary team. The knowledge that practitioners of projects such as *Foodscape Schilderswijk*, *The Cook, The Farmer, His Wife and Their Neighbour* (Wilde Westen and Potrc, 2009), or the Freehouse Collective's *Market of Tomorrow* (Figure 30.10; Van Heeswijk and Kaspori, 2008) encompass, would likely yield an astonishing Venn diagram. In the case of urban agriculture, aside from cultural and historical knowledge of the location and

Figure 30.9. Sprout Restaurant installation at Mediamatic Amsterdam: food production without land or water.

Figure 30.10. Soep Pop, urban intervention Soup Doll, during an iteration of the Freehouse Collective's Market of Tomorrow, selling vegetables in doll-form in collaboration with Afrikaander-market grocer.

its socio-political context, the contours of their expertise may extend from culinary history, high-level cooking and food preservation skills, material and physical infrastructure design, to significant experience with landscape architecture, organic gardening, and permaculture.

30.3.3 Tradition of working in the public space

Since the 'de-materialisation of art' of the 1960s and '70s, the public space is considered an acceptable artistic subject and/or platform (Lippard, 1973). Politically engaged artists produce works of urban and social critique, addressing topics such as land/property use and (social) agency. This cultural heritage informs artists' work in the public domain today. Additionally, a focus on public engagement or 'agency' can result in projects that ultimately provide models for municipal infrastructure or that actually produce urban regeneration (Alexander, 1979), such as artist collective Freehouse in Rotterdam, who have been implementing a regeneration project in the Afrikaanderwijk since 2006.

30.3.4 Opportunities in public space

The urban regeneration of locations like the Hague's Schilderswijk occurred in part as a reaction to perceived problems of the use of public space, (though strangely not of its design). Whether the public spaces of the newly regenerated locations are considered to be well designed or not, some pitfalls remain (Jacobs, 1993):
* visual greens are still implemented in their re-design;
* the visual greens are held and maintained by municipal organisations and are rarely programmed with activity;
* the visual greens are planted with ornamental plants, in deference to the architecture, or in a defensive style (providing barriers).

Even the most non-critical of analysis yields an assertion that such spaces are not examples of a resilient system design for the public space (Thackara, 2005).

Urbaniahoeve's alternative approach to public space through projects such as *Foodscape Schilderswijk* and DIYMOT focus on 'site repair' (Alexander *et al.*, 1977), agency, reactivation, and re-appropriation through (hyper) programming. Although it is too early to determine if a layered approach to i.e. maintaining the orchards or contiguous biotope will become and remain a success, the activities and interventions have at the very least centred attention on these locations by a wider group of stakeholders than has ever occurred in the past. Time will tell if the combination of low-dynamic productive landscape architecture and stacked programming is the way to build a *foodscape* intended for urban agriculture, but already the orchards offer a visual improvement. Time will tell if parks with playful public cooking infrastructure, programming and multi-stakeholder involvement is the way to positively activate the public space (while supporting local urban agriculture).

Continued engagement by municipal agencies such as the Department of Public Greens, and involvement by local housing corporations, in addition to neighbourhood and local institutional participation, all point to a positive result. Urbaniahoeve sees shared responsibility for public space maintenance, particularly if it accesses 'unconventional' groups for this role, as a strategic intervention. Urbaniahoeve sees all physical space that is not specifically private as having potential for integration into the *foodscape*, and as a platform for activation limited only by imagination and passion.

30.4 Conclusions

Although the recent rise in popularity of community gardens and the number of small-scale practitioners is increasing, to actually thrive, a viable culture for urban agriculture will require a critical mass before it can achieve a palpable positive impact. This chapter addresses a range of possible urban agriculture (UA) formats and also the skills and methodologies of its cultural sector practitioners. However, support and at the very least, devoted cooperation from all civic levels will be a requirement for coherent implementation.

The term 'urban agriculture' may well be comprised of the words 'urban' and 'agriculture', but the heritage of typically centralist urban planning and agriculture, as many practitioners will agree, is proving insufficient in dealing with the issues surrounding its development, innovation, and ultimately its deployment. In expanding UA beyond the traditional context of farming it would be wise to include the heritage of conceptual art practice of the 1960s and '70s in which land use critique (Jencks and Silver, 1972), the notion of public space re-appropriation, and citizen agency are understood in terms of a social practice. Benefits of expanding UA by cultural sector initiators include the effects of their professional autonomy, their flexibility with regard to project implementation, and the fact that they are frequently unencumbered by restrictions that negatively impact the production of working models and prototypes. The broad skill sets and range of expertise that artists and designers typically bring to the table supports a culture of plural solutions.

Expanding models of urban agriculture will necessarily address the many types of existing urban spatial typologies, e.g. visual greens, public space locations of denied access, restricted locations already in use and misused by the surrounding public. An additional boost in knowledge is necessary to grow and nurture local actors that will ultimately champion the resulting food-related activities in their communities after project delivery. Flexible civic organisations and schools that can easily see the value of an edible landscape architecture as an adaptable platform, can add an additional layer of programming to locations already in use by neighbourhood gardeners. An expanded vision of urban agriculture can engender a sense of agency in locals accessing the public space for community food (production), and can spawn cooperation between civic agencies and local practitioners. Socially engaged neighbourhoods, in the process of developing consensus about the use of their public space by working in it together, will inadvertently increase food autonomy in their area.

An expanded urban agriculture will include practitioners with a heritage of working within the realm of the social, to reconfigure the city for optimal public (food) production and access. Like any resilient (eco)system, the evolution of urban agriculture will rely on a plurality of locations, of programmes, and of practice.

References

Alexander, C., Ishikawa, S., Silverstein, M., Jacobson, M., Fiksdahl-King, I. and Angel, S., 1977. A pattern language. Oxford University Press, New York, NY, USA.

Alexander, C., 1979. The timeless way of building. Oxford University Press, New York, NY, USA, 552 pp.

Astyk, S., 2008. Depletion and abundance. New Society Publishers, Gabriola Island, British Columbia, Canada, 288 pp.

Gerritsen, H., 2008. Essay on gardening. Architectura & Natura Press, Amsterdam, the Netherlands, 396 pp.

Jacobs, J., 1993. The death and life of great American cities. The Modern Library, New York, NY, USA, 458 pp.

Jencks, C. and Silver, N., 1972. Adhocism: The case for improvisation. Doubleday & Company, Garden City, New York, NY, USA, 216 pp.

Katz, S.E., 2003. Wild fermentation. Chelsea Green Publishing Company, White River Junction, VT, USA, 178 pp.

Katz, S.E., 2006. The revolution will not be microwaved: Inside America's underground food movements. Chelsea Green Publishing, White River Junction, VT, USA, 378 pp.

Kellogg, S. and Pettigrew, S., 2008. Toolbox for sustainable city living. South End Press, Cambridge, MA, USA, 242 pp.

Kuitert, W. and Freriks, J., 1994. Hovenierskunst in palmet en pauwstaart. Uitgeverij de Hef, Rotterdam, the Netherlands, 178 pp.

Lippard, L., 1973. Six years: The dematerialization of the art object from 1966 to 1972: a cross-reference book of information on some esthetic boundaries. University of California Press, Berkeley, CA, USA, 294 pp.

Mollison, B., 1988. Permaculture: A designers manual. Tagari Publications, Tyalgum New South Wales, Australia, 579 pp.

Ogren, T.L., 2008. City trees and urban health. In: Borasi, G. and Zardini, M. (Eds.) Actions: what you can do with the city. Canadian Centre for Architecture SUN Publishers, Amsterdam, the Netherlands, pp. 198-203.

Oudolf, P., Kingsbury, N. and Cooper, S. (Eds.), 2009. Designing with plants. Conran Octopus Limited, London, UK, 160 pp.

Solomon, D., 2006. Grow yer own dang food. Sprout restaurant. Available at http://culiblog.org/?s=grow+yer+own+dang+food+sprout+restaurant.

Solomon, D., 2010a. Foodscape Schilderswijk. Available at http://www.urbaniahoeve.nl/?page_id=54.

Solomon, D.A., 2010b. DIY Mmmmusem of Oven Typologies. Available at http://www.urbaniahoeve.nl/?page_id=107.

Solomon, D., 2010c. All that rot! Available at http://www.urbaniahoeve.nl/?page_id=95.

Thackara, J., 2005. In the bubble: designing in a complex world. MIT Press, Cambridge, MA, USA, 331 pp.

Viljoen, A. (Ed.) CPULs – continuous productive urban landscapes. Designing urban agriculture for sustainable cities. Architectural Press, London, UK, 304 pp.

Van Heeswijk, J., Kaspori, D. and Solomon, D., 2007- ongoing. Lucky mi fortune cooking. Available at http://www.freehouse.nl/freehouse/index.php?option=com_content&view=article&id=53&Itemid=58 (Dutch) and http://culiblog.org/category/lucky-mi-fortune-cooking/.

Van Heeswijk J. and Kaspori, D., 2008 – ongoing. Market of tomorrow. Available at http://www.freehouse.nl/freehouse/index.php?option=com_content&view=article&id=21&Itemid=53.

Wilde Westen and Potrc, M., 2009. The Cook, the Farmer, his Wife and their Neighbour. Available at http://www.kkvb-cfwn.blogspot.com.

Whitefield, P., 2004. The earth care manual: a permaculture handbook for Britain & other temperate climates. Permanent Publications, Hyden House Limited/The Sustainability Centre. East Meon, Hampshire, UK, 469 pp.

Yoon, S., translated by Han, Y., 2005. Good Morning, Kimchi! Hollym Corporation Publishers, Elizabeth, New Jersey/Seoul, 128 pp.

Part 4. Planning and design

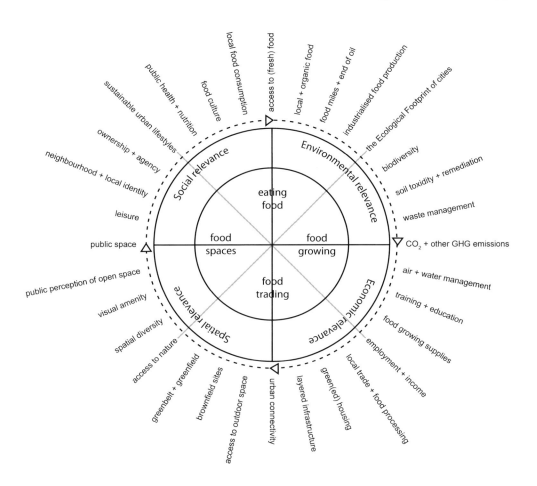

local food consumption

access to (fresh) food

local + organic food

food miles + end of oil

industrialised food production

food culture

public health + nutrition

the Ecological Footprint of cities

sustainable urban lifestyles

biodiversity

ownership + agency

soil toxicity + remediation

neighbourhood + local identity

waste management

leisure

public space

CO_2 + other GHG emissions

air + water management

public perception of open space

training + education

visual amenity

food growing supplies

spatial diversity

employment + income

access to nature

local trade + food processing

greenbelt + greenfield

green(ed) housing

brownfield sites

layered infrastructure

access to outdoor space

urban connectivity

Social relevance

Environmental relevance

Spatial relevance

Economic relevance

eating food

food spaces

food growing

food trading

Chapter 31

Planning and designing food systems, moving to the physical

André Viljoen and Katrin Bohn
University of Brighton, School of Architecture and Design, Grand Parade, Brighton BN2 0JY, United Kingdom; a.viljoen@brighton.ac.uk

31.1 Food and the built environment

What are the physical consequences of a more sustainable food system? How would cities and their hinterlands change? What types of agriculture and distribution systems are envisaged by 'food planners'? These questions frame the discussion and positions presented by practitioners and academics in the chapters that follow.

The infrastructure for food systems in towns and cities has various manifestations including places for food trading, processing and growing. In most countries that have embraced the globalised agri-business model, local food infrastructures have been severely weakened or no longer exist. Many believe that they have got to be reintroduced, and architects and landscape architects are fascinated by the possibilities of, in particular, urban agriculture set within a productive multifunctional landscape, and the spatial, social and environmental opportunities this suggests. The reintroduction of these spaces into European and North American cities, as opposed to their retention (as is the case in many Asian cities), presents many theoretical and practical challenges. Authors in this section of the book focus on urban and suburban areas describing programmes that range in scale from the personal to the metropolitan, describing bottom-up and top-down initiatives. Theory, policy, live practice and pedagogy are covered.

31.2 Planners and designers

Broekhof and Van de Valk's chapter sets the context by comparing papers in academic planning journals that argue for a continuation of the dominant agri-business model and those proposing the so called alternative food systems model. Their findings suggest a highly polarised environment, and they argue for a 'third way' taking elements from both approaches. The chapters that follow generally come from the position of the alternative food movement, but they are far from proposing a totally localised food system. Our impression is that there may be less polarisation amongst the design community than exists between planners and we suggest that the apparently polarised views identified by Broekhof and Van de Valk arise in part from the different approaches that planners and designers take to solving problems and to a general lack of dialogue between the two disciplines. If the views of all can be discussed more openly, then the ambiguity of meaning and assumptions that lead to polarisation may at least be reduced, if not eliminated. Finding areas of common ground and areas for collaboration is essential if we are to achieve a more equitable and less resource intensive food system. At the risk of a gross generalisation, planners, even physical planners, deal with issues at a very large scale, so terms like 're-territorialisation' associated with the reintroduction of more localised food systems, remain somewhat abstract. But once a specific site or project is dealt with the issues become very real and specific, and this necessitates a focus on the particular. It is our impression that this focus on the particular can be misinterpreted as advocating the particular as a general solution.

For example Chapter 33 by Oldroyd and Clavin, 'Nested scales and design activism', documents a programme called 'Back to Front' that encourages the use of front gardens for food growing in a relatively deprived part of England; the actual yield from this, in terms of crops will not by itself obviate the need for large scale food production elsewhere, but as Oldroyd and Clavin put it: 'Back to Front aims to work on the interface between both the human and natural environment, between differing spatial scales and between private and municipal management structures to promote behavioral change on a personal level, which, in turn, may impact on the wider geographical area and the way it is managed'.

Projects like this represent a move from theory to practice and should be understood as one component and perhaps one stage within the evolution of a more sustainable food system.

Tomkins' chapter continues to explore the motivations for individuals engaged in community-based food growing in London, at a scale similar to the Back to Front initiative. He concludes that participants in these programmes are motivated by a wide range of issues that go beyond food, including a desire to exercise control over their personal environments. As with Oldroye and Clavin's work the consequences of small-scale actions, that might be dismissed if measured only by the output of food, may have a much wider and long-term impact through the building of social capacity and the ability for people to co-author their environments. Will this be enough to build the political will necessary to make larger changes to the current food system? Community gardeners in New York provide an answer, having demonstrated the resilience of local initiatives when they are grounded in a genuine desire for something better. Mees and Stone's chapter traces the transitions within New York's community gardening movement from an illicit activity to one that has gained recognition from the city authorities. Their chapter documents the benefits, compromises and constraints that come hand in hand with legitimisation. Food growing has always occurred within these gardens, and this is now receiving more attention and analysis as its role in food security and sovereignty is more fully recognised. Mees and Stone include key facts and figures about the controls on use, construction and food sales from community gardens and these provide useful comparisons for other cities.

31.3 Food and space

Lee responds directly to Broekhof and Van de Valk's observations about the gulf between alternative (local) food production systems and global industrial agriculture, by articulating the need for a rigorous assessment of the yields achievable by urban agriculture. Although historically cities have included urban agriculture in the form of market gardens, dairies and livestock, relearning old cultivation techniques and testing new ones makes it difficult to predict potential yields without field trials. This observation reinforces the need for practical prototypes such as those being attempted by Giseke and her partners working on the Urban Agriculture Casablanca project (see chapter by Kasper, Giseke and Martin Han). Significantly Lee's background in agronomy provides the complimentary knowledge that planners and designers seek, when trying to assess the viability of their proposals and when briefing clients.

Whereas Lee discusses the space required for urban agriculture, Everett proposes a strategy for identifying space to sell fresh produce. Working in New Orleans, she makes a case for locating markets in zones between affluent and deprived residential areas, thereby simultaneously addressing some of the food access issues that bedevil many cities in the USA while also providing a level of purchasing power from wealthier areas that help to make markets viable for traders. Drawing on the work of Jane Jacobs and Kevin Lynch, this strategy has the potential to reduce urban segregation. How the different purchasing power of different users would be accommodated, accessibility, food diversity and the programming of ancillary activities within market space are all identified as issues critical to the success of urban markets.

31.4 Food and place

Everett's proposition and case studies can usefully be compared to another two successful and innovative markets in the USA, the relatively modestly sized Fondy Market in Milwaukee (http://www.fondymarket.org/) and Detroit's large Eastern Market (http://detroiteasternmarket.com/index.php). Both of these long established markets are leading innovative practice to ensure greater food security for local populations, while also providing social space and facilities for citizens, and significantly, they are partnering in new ventures to pilot economically viable urban and peri-urban agriculture projects. The degree to which fresh fruit and vegetables are unobtainable within the less affluent parts of cities in the USA is extreme, but a similar trend, of small local stores selling only processed food and alcohol, is becoming more and more evident within parts of Europe, in particular the UK. Working, accessible and affordable markets represent another component of a more sustainable food system, but not all markets are successful or financially viable, and so finding out what does work is a priority. Dan Carmody, who oversees Detroit's Eastern Market (and has a background in planning, urban regeneration and food retail) puts it like this:

> *You can't have a food system where 20 percent of the people at the lower end of the economic totem pole eat only one-seventh of the amount of fresh fruits and vegetables that the people at the top eat. Society can't afford the health care costs of treating diabetes and hypertension and coronary disease after they're contracted because of their poor diet. We as a society have to figure out a way that people of all incomes can afford and can access healthy eating* (Broder, 2011).

Carmody goes on to suggest that the social environment created by the market is as important as the access it provides to healthy food;

> The Eastern Market is much loved in this region because it's where food and place come together. So the conviviality is every bit as important in that engagement as the good food is here is [sic] to changing that mix as it affects our diet, as it affects the health of the region, and as it affects the country in terms of what people eat (Broder, 2011).

The contributors to this section would recognise this notion that food and place can come together to create better urban spaces. Bohn and Viljoen have for some time made the case that spaces for food production and distribution can beneficially enhance cities as part of a wider landscape strategy, and believe that enough knowledge and experience exists to be able to sketch out the multiple actions and interactions between individuals, organisations, communities and disciplines that together can achieve the infrastructure required to support a more sustainable food system. In our contribution to this section based the concept of Continuous Productive Urban Landscapes (CPUL) we outline a proposal for the CPUL City Toolkit, a four stage action plan for planning and implementing urban agriculture as part of a coherent productive urban landscape. Ideas proposed in the CPUL City Toolkit are being tested in extensive multi-disciplinary, multi stakeholder research and planning programmes in many regions and cities, and examples follow from Greater Casablanca, Switzerland, Rotterdam and Chicago.

31.5 Context and strategies

Casablanca, in marked contrast to Detroit, is projected to increase in population size to the extent that it can be defined as a mega city. Working in this dynamic situation, the ambitious Urban Agriculture Casablanca project, described by Kasper and colleagues, aims to bridge theory and practice using an action research strategy to explore the potential of productive green infrastructure, with urban agriculture at the centre of a proposal for climate optimised multifunctional urban landscapes. Elements of the Casablanca project have been tested elsewhere, and the practical field work projects described have resonances with the exemplary work of the Dutch organisation RUAF (www.ruaf.org), but as far as we are aware this is the first and most comprehensive designer led attempt to research, at a range of scales, a more sustainable food system, taking full account of its spatial, social and administrative impact. As with other propositions for large comprehensive urban strategies incorporating productive urban landscapes, such as those that have been proposed for Detroit, implementation is unlikely to be straightforward (Gallagher, 2010). A major obstacle in the case of Detroit and Casablanca appears to be access to land. In the case of Detroit, within its vastly depopulated suburbs residents who do remain, and highly articulate not for profit organisations working from within communities, are fearful of large scale top-down approaches, especially from financially motivated corporations, that disregard the will and established initiatives of residents. In Casablanca, the context is completely different, as it appears that agricultural land is being 'banked' on the assumption that its value will increase once the city expands (Stollmann, 2011).

Detroit and Casablanca are examples at opposite ends of the spectrum of shrinking and expanding cities. In other relatively more stable situations planners and enterprising individuals are working together in an attempt to develop more environmentally sustainable and equitable proposals for food production and food security. Economic uncertainty, recognition of the impact of peak oil and a changing global power balance are all driving this reappraisal. Rocky Marcoux, commissioner for city development in Milwaukee, with strong support from the city's mayor, is clear that visions for a new urban model including productive urban landscapes are necessary if cities are to become resilient and 'prosper'. He believes that city officials and planning policy should support a range of well thought-out prototypes that individuals and organisations wish to establish (personal communication, 2011). In Milwaukee's case Will Allen's highly publicised food and social justice organisation 'Growing Power' and other ecologically driven urban agriculture initiatives inspired by Growing Power, such as Sweet Water Organics (http://sweetwater-organic.com/), provide evidence of a city's attempt to remove barriers and positively support urban agriculture initiatives by, for example, providing access on reasonable terms to land controlled by the city. Elsewhere, where land is scarce, in for example New York City or Berlin, enterprising individuals are setting up experimental roof top farms that aim to be financially viable, as Brooklyn Grange Farm (http://www.brooklyngrangefarm.com/), Gotham Greens (http://gothamgreens.com/) and Urban Aquaponics Berlin (Reuters, 2011) demonstrate.

White and Natelson, in their chapter about the English planning system and how it can support more sustainable food and farming provide more details about the types of policy and control mechanisms that may be used in this field. Their chapter makes clear that planning legislation can provide a framework to support future actions, but has limited impact on the status quo without a directed vision, reinforcing the same point made by Rocky Marcoux in Milwaukee. White and Natalson also believe that the planning process can become 'a mechanism through which a more equal balance can be established by enabling less powerful sectors of society to have an influence', especially important in a sector like food, where a relatively few powerful 'big players' have the resources to exercise considerable influence.

Working in a very different context, but with a similar scope to Urban Agriculture Casablanca, Verzone's chapter on the Swiss Food Urbanism research programme presents further evidence of the rapidly increasing demand for knowledge and evidence about how a more sustainable food system would impact on the physical form of cities. Of particular note is the Swiss National Science Foundation's title for the funding programme under which Food Urbanism falls: 'New Urban Qualities'. The evolving paradigms for integrating productive landscapes in cities will certainly bring about new qualities, and many have been arguing for, and trying to articulate, these through a variety of means including installation, exhibition and event. Verzone's chapter, like the Urban Agriculture Casablanca chapter, helpfully articulates the cross- and interdisciplinary collaboration required for projects attempting to develop a comprehensive regional model. While the Swiss context is particular, especially when compared to less wealthy nations, the attempt to articulate environmental benefits accruing from urban agriculture, which in the Swiss case may release payments for urban farmers, will in all cases provide a useful evidence base for other nations.

De Graaf and Peemoeller present targeted studies each in different ways aiming to facilitate the development of more resilient local food systems in Rotterdam and Chicago. De Graaf presents what he calls a 'top down' review and assessment of opportunities for introducing commercially viable urban agriculture into Rotterdam; this results in a number of opportunity maps that provide guidance for 'bottom up' initiatives. He describes this work as an illustration of how planners can contribute to the co-creation of a more sustainable city. This process is highly replicable and suggests a number of useful parameters within which such a study can operate and provide useful guidance without getting into a site-by-site analysis of current opportunities. The result is a resource-efficient guide for potential urban farmers in their search for sites and the selection of appropriate types of urban agriculture. As with other contributors, the Rotterdam study introduces the idea that urban agriculture should beneficially tap into resource and waste streams, thus 'closing loops' and reducing the costs and requirements of other urban infrastructures.

Peemoeller's chapter critically assesses the methodology used to produce the first food systems report, which she led, for the Chicago metropolitan region. In common with the Casablanca, Swiss and Rotterdam studies the Chicago report which was commissioned as part of an official planning study looking forward to 2040, recognised the need for top-down and bottom-up approaches. It worked directly with community organisations in a participatory community planning process. Peemoeller articulates the difficulties in convening such a large group of professional and community contributors and the difficulty of arriving at a consensus. The outcome of the nine-month-long study did not result in an action plan, but it did get a diverse range of people talking, and it did produce a shared vision for a desired future food system while embedding food planning within the regional planning process.

31.6 Back to planning and design

It might be concluded that this approach, of working with an extremely wide and diverse group of stakeholders is effective at generating a vision, or goal, but not the best way to generate particular propositions for achieving that vision. The number of participants engaged in trying to reach a consensus would of itself be challenging if such a large and diverse group where trying to arrive at consensus about about particular projects. This brings us back to an earlier point about the difference between generating general 'planning' solutions and particular 'design' solutions. An action plan will require a number of specific propositions, related to particular places and contexts. Trying to achieve a consensus about specific proposals is probably more realistically done with a smaller number of stakeholders, be they a community or individuals directly associated or affected by the proposal. A two-stage approach, in which a shared vision is arrived at collectively and then as a separate task, particular proposals are developed by smaller groups of stakeholders, might provide a better model. As Peemoeller notes; 'It would be a valuable contribution to the professional or academic field to survey different food systems planning methodologies and compare outcomes. This would help to further establish guidelines for the practice of food systems planning and expectations from the process.'

Despite Peemoeller's disappointment at not achieving an action plan, the city is developing planning policy in support of urban agriculture. Within two years of the report's instigation the city's zoning code has been amended to remove certain barriers to urban agriculture (The City of Chicago, 2011). Erica Allen, the noted urban agriculture activist based in Chicago, considers these to be significant changes,

> It legitimises urban agriculture as an enterprise or a business that hasn't been on the books before. Chicago always had farms within the city limits, but the new ordinance creates a space where we can begin to create economic opportunities within our communities, especially in areas where food deserts are a direct result of unemployment and little economic opportunity (Nemes, 2011).

31.7 Pedagogy: design education and food systems

Elsewhere in this book Nasr and Komisar describe the process by which teaching within schools of architecture is resulting in a new generation of practitioners committed to integrating productive landscapes into their practice. The final two chapters in this section address pedagogy with respect to undergraduate students. Crotch describes a programme run for third year Bachelor of Architecture students at the Macintosh School of Architecture in Glasgow, while Moya Pellitero and da Silva Eliziário describe an ambitious European cross-disciplinary and cross-institutional travelling summer school that they have established.

The Glasgow programme is underpinned by a strong polemic, drawing from the slow food and slow city movements, and one of its graduates achieved the accolade winning the Royal Institute of British Architects Bronze Medal for the best undergraduate design project in 2010.

Moya Pellitero and da Silva Eliziário's summer schools set intensive and highly structured weeklong workshops using principals derived from concepts of Landscape Urbanism and Continuous Productive Urban Landscapes. These two chapters demonstrate the challenge of introduction multidisciplinary topics like food systems planning to undergraduate students who are still mastering their own disciplines. Undoubtedly students benefit from being exposed to the complexities of designing for more sustainable food systems, but judging the right balance of the general, to the particular, so that students are not overwhelmed remains a challenge. Nasr and Komisar's chapter makes clear that integrating this relatively new and compelling challenge into design and planning education is appreciated and does result in real change.

To repeat a common theme in this book, the chapters that follow do not yet provide definitive answers to the question of how alternative food systems will impact on future cities, but they do provide models for approaches and experiments that may be replicated, critiqued, modified and tested elsewhere as planners and designers contribute to the definition of a more sustainable and equitable food system.

Refrences

Broder, J., 2011, Eastern Market heats up. In The Metro Times. June 15, 2011. Available at http://www. detroiteasternmarket.com/news_page.php?id=131&p=7&s=. Accessed 17th September 2011.

Gallagher, J., 2010, Reimagining Detroit, opportunities for redefining an American city. Wayne State University Press, Detroit, USA.

Reuters, 2011, Urban farming brings the country to the city. Available at http://www.reuters.com/video/2011/07/25/urban-farming-brings-the-country-to-the?videoId=217512288. Accessed 3rd October 2011.

Stollmann, J., 2011, Douar Ouled Ahmed – A Future Urban Landscape as a Framework for Traditions, Aspirations and Business Opportunities. Urban Agriculture Casablanca, Design as an Integrative Factor of Research. Ed. Giseke, G. Technische Universität Berlin (TUB), Germany, pp. 14-15.

The City of Chicago, 2011. Urban Agriculture Zoning Code Amendment Approved. Available at http://www.cityofchicago.org/city/en/depts/dcd/provdrs/sustain/news/2011/sep/urban_agriculturezoningcodeamendmentapproved.html. Accessed 2nd October 2011.

Nemes, J., 2011, Green Scene: New Chicago ordinance encourages urban farmers to start planting. Available at http://www.cityfarmer.info/2011/09/22/green-scene-new-chicago-ordinance-encourages-urban-farmers-to-start-planting/#more-14560. Accessed 2nd October 2011.

Chapter 32

Planning and the quest for sustainable food systems: explorations of unknown territory in planning research

Sanne M. Broekhof and Arnold J.J. van der Valk
Wageningen University, Land Use Planning, P.O. Box 47, 6700 AA Wageningen, the Netherlands; sanne.broekhof@wur.nl

Abstract

Planning scholars in the United States and in Europe explore the contours of an emerging field of sustainable food systems. Tiny fragments of the immensely complex food cycle have been covered so far. Planning literature is currently focusing on production processing and retail, leaving consumption of food and treatment of waste still largely unexplored. This chapter sets out to explore the state of the art. The research builds on data collected in a survey of international scientific planning journals over a period of a decade (1999-2010). The outcomes of the analysis are put in perspective by contrasting them with the overall picture sketched in topical books on food planning. Preliminary conclusions were reviewed by and discussed with a panel of international food planning experts. The bulk of the journal papers under scrutiny starts from the premise of an unbridgeable gap between two conflicting worlds of food, i.e. a dominant conventional global-industrial system and an emerging alternative complex of local-organic systems. The selected planning papers bring out a strong preference for the alternative food planning discourse. Global and local food systems are construed as thesis and anti-thesis. This chapter explores arguments in favour of a synthesis, i.e. a 'third way' discourse, bridging the contested middle-ground between conventional de-localised and alternative re-localised agri-food systems.

Keywords: spatial planning research, food systems planning, discourse

32.1 Introduction

Food is no longer a stranger in the land of planning (Pothukuchi and Kaufman, 2000). Planners take up the challenge of food systems planning because they understand that food shapes our lives. It relates to a myriad of issues which are recognised as genuine planning objects and vehicles on the road to sustainability. Food related issues play a major role in climate change, they impact on obesity and hunger, they are a major cause of waste problems and finally they keep intact inequalities between developed and developing countries (Despommier, 2010; Pollan, 2006; Ponting, 2007; Salle and Holland, 2010; Steel, 2008). Food relates to transportation planning, planning green spaces, commercial strips and retail markets, conservation of fertile land, mixed land uses, localising urban farms, and much more. The production of food takes up a major part of land used by humans, adding up to seventy percent in densely populated countries such as the UK and the Netherlands (Viljoen, 2005).

The most pressing issue from the planning perspective, is the tension between diverging forces of de-territorialisation and re-territorialisation of food systems (Morgan *et al.*, 2006). De-territorialisation is a major characteristic of the agri-business system making reference to methods of food production and processing which are no longer attached to traditional farming methods, soil-types or regions. The aim of re-territorialisation is at the heart of the alternative local-ecological food systems. Re-territorialisation refers to attempts to restore the link between farming and the land. The contrasting worlds of food, i.e. agri-business and local-ecological farming will be pictured in detail in the following paragraphs.

The concept of de-territorialisation brings out territorially relevant consequences of the global-industrial food system – otherwise agri-business complex – such as monoculture, spatial homogeneity, uniformity in the landscape, decline of mixed farming systems, growing dependence on bulk production and a declining rural labour force. The agri-business complex is dynamic, being the key common denominator and a constant pressure to de-valorisation and depreciation of rural space, through scale enlargement and increased production. The negative consequences of cost-price squeeze and the 'technological treadmill' spurred by super-concentration of retail corporations have generated counter-forces of contest and protest among independent farmers. Attempts by struggling farmers to opt out of the industrial system include experiments in diversification of farming styles, emphasising quality ecological impacts, restoring face-to-face contact between producer and consumer and creating local and regional producer networks.

Advocates of an alternative system aim – among other things – for re-territorialisation of food production and processing. This effort is associated with a re-valuation of regional artisanal and ecological quality standards and the shortening of chains for local and regional producers. The distinctive goal of this process is value-capture for the producers. The resulting dynamics play out differently in different regions.

Specification of a territorial dimension has convinced planners and their professional organisations in North-America and Europe that food and agriculture is a planning issue. In 2007, the American Planning Association issued a white paper on food, thus putting food on a par with established sectors in planning such as housing, energy, transportation and green spaces. The Association of European Schools of Planning established a Food Planning Working Party in 2009.

This chapter explores two diverging planning discourses associated with an emerging field of (sustainable) food systems planning. Discourse is perceived as a set of concepts that structure participant's contributions to a discussion – otherwise – a frame of understanding, an ensemble of ideas, concepts and categorisations through which meaning is transferred to social and physical phenomena. (Torfing, 2005) One aim is to find answers to the question: is food planning a fad or a promise? This aim is inextricably mixed with questions about the potential contribution of spatial planning to the relaxation of the tension between two competing worlds of agri-food, i.e. the agri-business complex and the local-ecological movement. (Morgan *et al.*, 2006) The underlying question is: will the twain ever meet?

The text draws primarily on an exploration of planning literature. Twelve scientific planning journals over a period of a decade (1999-2010) were selected for further reading about the intricacies of food and spatial planning. The literature scan in planning journals is based upon three criteria that all papers have to meet. First, papers to be labelled 'food system' have been selected for the database. Moreover, the paper takes a planning perspective and according to our criteria at least one of the authors has a background in planning. The data were compared and complemented with the outcomes of a scan of books about the state of the art in food policy and food geography. Intermediate results of the literature search have been presented to and discussed with an international panel of experts being dr. Roberta Sonnino (Department of City and Regional Planning of Cardiff University), dr. Kami Pothukuchi (College of Liberal Arts and Sciences of Wayne State University), dr. Joe Nasr (Centre for Studies in Food Security of Ryerson University), Craig Verzone (Verzone Woods Architects) and drs. Pim Vermeulen (Amsterdam municipality).

The results have been obtained with the help of interpretative, qualitative and explorative research methods (Creswell, 2003; Rowlands, 2005; Yanow, 2006).

32.2 Worlds of food: contrasting discourses in planning

Agri-business nowadays encompasses a large industry of research, transport, product development, processing, coordinating and packaging of food. This conventional agri-food system is perceived by a growing group of experts, politicians and opinion leaders as unsustainable, unhealthy, unjust and a danger to landscape heritage. The system enforces landscape uniformity, low prices for producers, bulk production, global injustice and alienation between producers and consumers. Scholars have introduced the term alternative food system (AFS) to cover many different forms of food production that started as alternatives to agri-business, from the idea that the conventional food system is unsustainable.

The agri-business complex supports high-productivity farming, hyper-efficient logistics, complex webs of bankers retailers scientific research and food processing industry. Its major merit is that it has created conditions for the provision of large quantities of cheap processed food, which have eradicated seasonal food scarcities in developed countries. Stability in the procurement of food is perceived to be a major contribution to post-war social stability in most Western countries. Availability of cheap food is also a major driver in the process of urbanisation world-wide which has caused an increase of population living in cities from thirteen percent in 1900 to fifty percent in 2007 (United Nations, 2006). The industrial model is firmly entrenched in national and international food politics in the North-America and Europe. Consequently, industrial food products are easily accessible to most people in the western world (Steel, 2008). But, on the downside it is noted that powerful coalitions fuelled by agri-business, manage to keep a hold on USA agricultural politics and EU common agricultural policy. The global General Agreement on Tariffs and Trade (GATT) and the North American Free Trade Agreement (NAFTA) are supportive for the proliferation of the agri-business model (Morgan *et al.*, 2006; Paarlberg, 2010).

32.2.1 The agri-business complex

Advocates of agri-business are a shrinking and defensive minority in the community of planning scholars. Agri-business-oriented planners accuse their opponents of spreading unscientific speculations and reactionary myths. (Smeets, 2011; c.f. McWilliams, 2009) Political scientist Robert Paarlberg (2010), an advocate of the application of Western technology, expresses the feelings of the proponents of the silent majority in agri-business:

> *Most of the academically trained specialist I have worked with – nutritionists, agricultural economists, toxicologists, biologists, soil scientists, and irrigation engineers – have serious reservations about these most recent food fashions...These assertions that the best food will be organic, local, and slow require critical scrutiny by someone other than a journalist or a 'food writer'* (Paarlberg, 2010: 112).

Supporters of agri-business believe that the environment will benefit from highly capitalised and specialised high-yield farming systems using the latest precision technology. Still, advocates of the agri-business model recognise the need for a transition towards sustainable practices. They put faith in the development of agro-parks and vertical farms (Despommier, 2010; Smeets, 2011). An agro park is an integrated cluster of agri-food facilities. Production of animals, fresh produce, flowers, fodder and plant nutrition is combined with packaging, transportation, waste management, labour management, market research and much more. A vertical farm is an agro-park in a multi-story building. Efficiency and productivity are core concerns in the design of innovative production, marketing and logistics in agro-parks. Table 32.1 presents a schematic overview of food issues from the perspective of advocates of the agri-business model as expressed in 35 papers published in scientific planning journals (for an overview see Appendix 32.1).

32.2.2 The alternative food movement

Critics put their faith in an emerging alternative food movement (Hewitt, 2009; Pollan, 2006, 2008; Salle and Holland, 2010; Steel, 2008). This approach to agri-food unites advocates of local food and ecologically sound production methods. The alternative sector is first and foremost a multi-vocal choir of experts and pressure groups blaming the global-industrial food system for a list of vices such as its contribution to global warming, the decrease of nutritional content and proteins in pre-packaged food products, livestock pandemics, lack

Table 32.1. Schematic overview of food issues perceived by advocates of agri-business (derived from 35 selected papers from scientific planning journals between 1999 and 2010, see Appendix 32.1).

Perceived problems	Proposed solutions
Decreasing quantity agricultural land vs. growing urban population	Mechanised intensive food production in agro- parks
Ineffective (planning) regulations	-
Producer-consumer gap	Provide (school) visits to agro-parks or Develop food labels from agro-parks
License to produce endangered	Agro-parks providing sustainable, local and animal friendly products
Unsustainable waste flows	Closed waste cycles in agro-parks
Food miles	Cluster agri-business activities in agro-parks or Greenport
Landscape destruction	Limit to intensive agro parks in order to protect the rest of the landscape
Animal well-being	Agro-park offering a healthy, save and spacious environment for animals

of stewardship of soils and the neglect of cultural landscapes. The alternative discourse starts from 'quality' instead of cheap bulk products. Only sustainable production methods are accepted. Operational definitions of sustainability are open for discussion between groups of believers though. Some emphasise optimisation of impacts on soil, others prefer economical use of water and then again others prefer those practices which are most beneficial for farmer's income. The advocates of alternative systems promote value capture for the producers and reduction of the distance (literally and metaphorically) between producer and consumer. Consumer knowledge of place, methods, products and spatial conditions is vital. The chain is kept as short and transparent as possible. Regional brands and health certificates are used to build up trust with consumers. Waste is to be reduced and preferably recycled. Table 32.2 summarise the results of a scan of the previously mentioned 35 papers taken from scientific planning journals.

32.2.3 A third way?

Since sustainability is at the heart of shared values underlying both discourses as reflected in planning journals, it is tempting to hypothesise common ground for a synthesis, a 'third way'. But, papers taking an intermediate course are extremely rare. In Table 32.3, the results of Table 32.1 (the agri-business complex) and Table 32.2 (the alternative food movement) are summarised and compared. It is hard to conceive of greys in between the blacks and whites of the opposite discourses of globalists and environmentalists (McWilliams, 2009; Paarlberg, 2010). One rare exemplar of a fully-fledged third way solution is Carolyn Steele's concept of 'sitopia', utopian food-scapes grounded in architectural realism (Steel, 2008). Third-way strategies are rare and mostly imaginative – which need not surprise us taking into consideration the complexity of the global food system. One conclusion stands out from the analysis of the sampled papers: the obstacles for a merger of the separate worlds of food are strong and manifold economically, technically, logistically, ethically, and politically. Ideology

Table 32.2. Overview of food issues as perceived by advocates of the alternative food movement (derived from 35 selected papers from scientific planning journals between 1999 and 2010, see Appendix 32.1).

Perceived problems	Towards solutions
Consumers disconnecting from the food chain, broken linkages between food products and their environment, food chain is split into an increasing amount of activities.	Local agriculture (on-farm shops, farmers markets, community supported agriculture CSA, box schemes, region specific produce), farm-to-school programmes, community gardens, allotment gardens, continuous productive urban landscapes (CPULs)
Food deserts	Farmers markets, connections to shops offering fresh produce, internet sales, community food security
Obesity and malnutrition	Farm-to-school programmes, allotment gardens, community gardens
Economic problems for small farmers	Local agriculture, multifunctional agriculture, organic production
Globalisation on the retail market	
Environmental pollutions	Organic production, perma-culture, crop rotation
Landscape destruction	Landscape protection policy, buffer strips, Small-scale farms, organic production, crop rotation
Food miles	Local agriculture, urban agriculture
Ghost acres	Local agriculture, urban agriculture
Unsustainable food system in cities	Urban agriculture, CPULs, Urban Waste cycles

may be the worst obstacle. This conclusion raises fundamental questions such as: can we reframe the transition of global food systems as a planning issue? Does the alternative food discourse go against the grain of contemporary technologically advanced solutions? Can planners conceive a viable compromise between global and alternative food systems?

Table 32.4 highlights nuances in the valuation of food problems and solutions by advocates of the two major conflicting approaches.

Proponents of the alternative food discourse – which constitute the vast majority of food writers in the planning community – start from the premise that the industrial food system will collapse in the long term because it is not sustainable. The ultimate solution to this problem cannot be deduced from scientific literature. The range of preferred solutions takes inspiration from educated speculation, wishful thinking and casuistry. Advocates of the alternative system do not expect the immediate cure to all evils of the global system as soon as the bulk of their proposed measures have been implemented on the local scale. Rather, the alternatives want to bring back food production to the city and increase community commitment with the food system. The alternative discourse covers a broad spectre of small scale initiatives and projects which count up to a web of ecological and local food systems (Jarosz, 2008). The basic melody in the tune is: it's still small and fragile but it's growing. For

Table 32.3. Overview of problem definitions and solutions.

Problems	Agri-business solutions	Alternative solutions
Decreasing agricultural area vs. growing urban population	High intensive food production in Agro- parks	Continuous Productive Urban Landscapes (CPULs)
Obesity and malnutrition	The creation of healthy food stuffs by the food processing industry.	Farm-to-school programmes, allotment gardens, community gardens
AgB: Producer-consumer gap	Provide (school) visits to Agro- parks, Food labels	Local Agriculture (On-farm shops, farmers markets, CSA, box schemes, region specific produce), farm-to-school programmes, community gardens, allotment gardens
Alt: Consumers disconnecting from the food chain, broken linkages between food products and their environment, food chain is split into an increasing amount of activities.	-	
Critics from society	Agro-park providing sustainable, local and animal friendly products	Local Agriculture (On-farm shops, farmers markets, CSA, box schemes, region specific produce), farm-to-school programmes, community gardens, allotment gardens
AgB: Unsustainable waste flows	Closed waste cycles in Agro-park	Urban agriculture, CPULs, urban food strategies
Alt: Unsustainable food system in cities	-	
Food miles	Cluster agri-business activities in Agro-park	Local agriculture, urban agriculture
Landscape destruction	Limit to intensive Agro-park in order to protect the rest of the landscape	Landscape protection policy, buffer strips, Small-scale farms, organic production, crop rotation
Animal well-being	Agro- parks offering a healthy, safe, and wide environment for animals	Organic produce
Food Deserts	-	Farmers markets, connections to shops offering fresh produce, internet sales
Environmental Pollutions	Innovation of artificial manure, technical innovations for cleaner production techniques	Organic production, perma-culture, crop rotation
Economic problems for small farmers	-	Local agriculture, multifunctional agriculture
Globalisation on the retail market	-	-
Ineffective (planning) regulations	-	-
Ghost Acres	Agro-parks	Local agriculture, urban agriculture

AgB: problem as perceived by authors in the Agri-business discourse
Alt: problem as perceived by authors in the Alternative discourse

example, according to statistics published by the Dutch Ministry of Food in 2007 4.4% of the stock of potatoes, vegetables and fruits and 3.8% of the fresh dairy products is ecological.

Table 32.4. Evaluation of problem definitions and proposed solutions (derived from 35 selected papers from scientific planning journals between 1999 and 2010, see Appendix 32.1).

Perceived problems	problem AgB	problem Alt	solution AgB	solution Alt
Decreasing agricultural area vs. growing urban population	+ +	?	+	-
Obesity and malnutrition	-	+ +	+ -	-
Gap between producer and consumer	+	+ +	+	+
Consumers disconnecting from food chain, broken linkages between food products and their environment, food chain split into increasing nr. of activities.	+ -	+	+ -	+
Dissatisfaction in society with impacts	+ +	+	+	+
(Un)sustainable waste management	+	+	+	+
(Un)sustainable food system in cities	-	+ +	-	+
Food miles	+ +	+ +	+	+
Landscape destruction/conservation	+	+ +	+	+
Animal well-being	+	+ -	+	+
Food deserts	-	+	-	+
Environmental pollution	+ +	+ +	+	+
Economic problems for small farmers	-	+	-	+
Globalisation retail market	-	+ - / +	-	-
Ineffective regulations	+	+ -	-	+
Land needed to produce food, outside the region of those consuming (Ghost acres)	+ -	+	+	+

Problem AgB: problem as perceived by authors in the Agri-business discourse;

Problem Alt: problem as perceived by authors in the Alternative discourse;

Solution AgB: solution as proposed by authors in the Agri-business discourse;

Solution Alt: solution as proposed by authors in the Alternative discourse.

+ +: issue mentioned frequently

+: issue mentioned more than once

+ -: issue mentioned once

-: issue missing from selected papers

32.2.5 Expert opinion

Binary thinking pervades agri-food literature and scholarly planning papers. This conceptual tradition keeps intact two highly segmented food production discourses. (Morgan *et al.*, 2006)

So, will the diverging systems of food ever meet? Few authors bother about answering the question. Those that do bring up many restrictions or flee in wishful thinking. Morgan *et al.* (2006) have identified openings in the borders between the worlds of food. Arguments in favour of a third way are rooted in an emerging new food morality taking into consideration matters of sustainability, public health and fair trade. So far planners have largely ignored the option of a synthesis.

The authors have shared their findings with five food planning experts in Europe and the US. The interviews corroborate the hypothesis of the predominance of two contrasting discourses and the existence of an un-abridgeable gap. Both systems may be construed as competitive spaces separated by a 'battlefield' of changing competitive spatial boundaries. However, linkages do exist, for example, seventy percent of organic sales in the UK are dominated by corporate retailer sales. Sonnino (2010, personal communication) stated that she is developing 'the idea that a truly sustainable food system is actually a mix of all of them, so it has attributes and features of all the different food systems' Nasr (2010, personal communication) stresses the importance of bearing in mind the existence of the conventional agri-business system: 'We should acknowledge the existence, the role of the conventional food distribution system at least and then developing alternatives at the same time. That's why I think it is important to not just read the industrial food system as a single block, but as a complex, as pieces.' Other characteristics of alternative food sectors such as increased interaction with consumers, transparency, and consumer-goodwill are easily discredited in the conventional supermarket. Yet, not all interviewed experts wholly agree on the way we should deal with conventional supermarkets. 'If you can cooperate with Albert-Heijn, then you work with a serious large party. That's one of the reasons why you should deal with Supermarkets. We do need to create some critical mass' (Vermeulen, 2010, personal communication). Pothukuchi (2010, personal communication) does not fully reject the idea to work with supermarkets and more mainstream entities. Yet, she is concerned about the power and scale of large players in the conventional food system. 'Wal-Mart's scale itself is problematic... A hundred farmers markets cannot compete with Wal-Mart's power in the local market.' Moreover, Pothukuchi emphasises the need to think both critically and pragmatically in our search towards more sustainable food systems.

32.3 Conclusion

Planning scholars have embarked upon the field of sustainable food systems planning. Fragments of cycles of production, processing and retail have been covered. Papers and books take the premise of co-existence of two conflicting worlds of food for a starting point. Papers written by planning scholars bring out a strong preference for the alternative food discourse. Global and local food systems are construed as thesis and anti-thesis. A third-way discourse bridging the contested middle-ground is currently perceived as a remote ideal.

Planners focus on local best practices hoping to replicate them in various contexts. Scaling up is necessary since the bulk of local initiatives is fragile and vulnerable. But so far up-scaling meets with un-surmountable problems. These pertain to global trade mechanisms, logistics, distribution, knowledge, deep rooted cultural habits, food provenance and corporate power. So the proponents of the alternative discourse settle for an upward battle – starting from local food initiatives – because they cannot see any substantial changes in the performance of the market system and global trade mechanisms. Agricultural subsidies, land policy, transportation systems, food logistics and environmental regulations have privileged the conventional agricultural system. The local and alternative food scene is still in the phase of a take-off. The foundation under the alternative discourse is 'life-style activism', marrying worlds of glossy health-food magazines and back-to-the-land hippy activism. (Hewitt, 2010) Consecutive prescriptive theories of food planning are rich in ideology and poor in empirical grounding. The quest has just begun.

References

American Planning Association, 2007. Policy guide on community and regional food planning. Available at http://www.planning.org/policyguides.

Creswell, J.W., 2003. Research design. qualitative, quantitative and mixed methods approaches. Sage Publications, London, UK, 245 pp.

De la Salle, J. and Holland, M., 2010. Agricultural urbanism. Handbook for building sustainable food & agriculture systems in 21st century Systems. Green Frigate Books, Winnipeg, MB, Canada, 200 pp.

Despommier, D., 2010. The vertical farm. feeding the world in the 21st century. Thomas Dunne Books, New York, NU, USA, 305 pp.

Hewitt, B., 2009. The town that food saved. How one community found vitality in local food. Rodale, New York, NY, USA, 234 pp.

Jarosz, L., 2000. Understanding agri-food networks as social relations. Agriculture and Human Values 17: 279-283.

Jarosz, L., 2008. The city in the country: growing alternative food networks in Metropolitan areas. Journal of Rural Studies 24: 231-244.

McWilliams, J.E., 2009. Just food. Where locavores get it wrong and how we can truly eat responsibly. Back Bay Books, New York, NY, USA, 272 pp.

Morgan, K., Marsden, T. and Murdoch, J., 2006. Worlds of food. Place, power and provenance in the food chain. Oxford University Press, Oxford, UK, 225 pp.

Paarlberg, R., 2010. Food politics: What everyone needs to know. Oxford University Press, Oxford, UK, 240 pp.

Pollan, M., 2006. The omnivore's dilemma. The search for a perfect meal in a fast-food world. Bloomsbury, London, UK, 450 pp.

Pollan, M., 2008. In defense of food. An eater's manifesto. Penguin Books, London, UK, 244 pp.

Ponting, C., 2007. A green history of the world: The environment and the collapse of great civilizations. Penguin Books, London, UK, 430 pp.

Pothukuchi, K. and Kaufman, J.L., 2000. The food system, a stranger to the planning field. Journal of the American Planning Association 66: 113-124.

Rowlands, B.H., 2005. Grounded in practice: Using interpretative research to build theory. The electronic journal of Business Research Methodology 3: 81-92.

Smeets, J.A.M., 2011., Expedition Agroparks. Research by design into sustainable development and agriculture in the network society. Wageningen Academic Publishers: Wageningen.

Sonnino, R., 2009. Feeding the city: Towards a new research and planning agenda. International Planning Studies 14: 425-435.

Steel, C., 2008. Hungry city. How food shapes our lives. Vintage Books, London, UK, 400 pp.

Torfing, J., 2005. Discourse theory: Achievements, arguments, and challenges. In: Howarth, D. and Torfing, J. (Eds.) Discourse theory in European politics. Identity, policy and governance. Palgrave, Houndmills Basingstoke, UK, pp. 1-31.

United Nations, 2006. World urbanization prospects: The 2005 Revision. United Nations, New York, NY, USA.

Viljoen, A., Howe, J. and Bohn, K. (Eds.), 2005. CPULs. Continuous productive urban landscapes. Designing urban agriculture for sustainable cities. Elsevier Architectural Press, Oxford, UK, 295 pp.

Yanow, D., 2006. Thinking interpretatively: philosophical presuppositions and the human sciences. In: Yanow, D. and Schwartz-Shea, P. (Eds.) Interpretation and method: Empirical research method and the interpretive turn. ME Sharpe inc. New York, NY, USA, pp. 5-26.

Appendix 32.1

The 35 selected papers from scientific planning journals between 1999 and 2010

Banks, J. and Bristow, G., 1999. Developing quality in agro-food supply chains: A Welsh perspective. International Planning Studies 4: 317-331.

Blay-Palmer, A., 2009. The Canadian pioneer: The genesis of urban food policy in Toronto. International Planning Studies 14: 401-416.

Born, B. and Purcell, M., 2006. Avoiding the local trap, scale and food systems in planning research. Journal of Planning Education and Research 26: 195-207.

Campbell, M.C., 2004. Building a common table: The role for planning in community food systems. Journal of Planning Education and Research 23: 341-355.

DeSilvey, C., 2003. Cultivated histories in a Scottish allotment garden. Cultural Geographies 10: 442-468.

Dunkley, B., Helling, A. and Sawicki, D.S., 2004. Accessibility versus scale. Examining the tradeoffs in grocery stores. Journal of Planning Education and Research 23: 387-401.

Gallent, N. and Shaw, D., 2007. Spatial planning, area action plans and the rural-urban fringe. Environmental Planning and Management 50: 617-638.

Hammer, J., 2004. Community food systems and planning curricula. Journal of Planning Education and Research 23: 424-434.

Hinrichs, C.C., 2003. The practice and politics of food system localisation. Journal of Rural Studies 19: 33-45.

Howe, J., 2002. Planning for urban food: The experience of two UK cities. Planning Practice and Research 17: 125-144.

Illbery, B. and Maye, D., 2005. Alternative (shorter) food supply chains and specialist livestock products in the Scottish – English borders. Environment and Planning A 37: 823-844.

Irvine, S., Johnson, L. and Peters, K., 1999. Community gardens and sustainable land use planning: A case-study of the Alex Wilson community garden. Local Environment 4: 33-46.

Jarosz, L., 2008. The city in the country: growing alternative food networks in Metropolitan areas. Journal of Rural Studies 24: 231-244.

Kaufman, J., Pothukuchi, K. and Glosser, D., 2007. Community and regional food planning: a policy guide of the American Planning Association. APA Legislative and Policy Committee and chapter delegates, at the 2007 APA national conference, Philadelphia.

Lang, T., 1999. The complexities of globalization: The UK as a case study of tensions within the food system and the challenge to food policy. Agriculture and Human Values 16: 169-185.

La Trobe, H.L. and Acott, T.G., 2000. Localising the global food system. International Journal of Sustainable Development & World Ecology 7: 309-320.

Lawson, L., 2004. The planner in the garden: A historical view into the relationship between planning and community gardens. Journal of Planning History 3: 151-176.

Marsden, T., 2000. Food matters and the matter of food: Towards a new food governance. Sociologica Ruralis 40: 20-29.

Marsden, T., Murdoch, J. and Morgan, K., 1999. Sustainable agriculture, food-supply chains and regional development: Editorial introduction. International Planning Studies 4: 295-301.

Martin, R. and Marsden, T., 1999. Food for urban spaces: The development of urban food production in England and Wales. International Planning Studies 4: 389-412.

Morgan, K., 2009. Feeding the city: The challenge of urban food planning. International Planning Studies 14: 341-348.

Nichol, L., 2003. Local food production: Some implications for planning. Planning Theory and Practice 4: 409-427.

Pothukuchi, K., 2004. Community food assessment. A first step in planning for community food security. Journal of Planning Education and Research 23: 356-377.

Pothukuchi, K., 2009. Community and regional food planning: Building institutional support in the United States. International Planning Studies 14: 349-367.

Pothukuchi, K. and Kaufman, J.L., 1999. Placing the food system on the community agenda: The role of municipal institutions in food systems planning. Agriculture and Human Values 16: 213-224.

Pothukuchi, K. and Kaufman, J.L., 2000. The food system, a stranger to the planning field. Journal of the American Planning Association 66: 113-124.

Renting, H., Marsden, T. and Banks, J., 2003. Understanding alternative food networks: exploring the role of short food supply chains in rural development, Environment and Planning A 35: 393-411.

Smith, A., 2006. Green niches in sustainable development: the case of organic food in the United Kingdom. Environment and Planning C: Government and Policy 24: 439-458.

Sonnino, R. and Marsden, T. 2006. Beyond the divide: rethinking relationships between alternative and conventional food networks in Europe. Journal of Economic Geography 6: 181-199.

Sonnino, R., 2009. Feeding the city: Towards a new research and planning agenda. International Planning Studies 14: 425-435.

Sonnino, R., 2010. Escaping the local trap: Insights on re-localization from school food reform. Journal of Environmental Policy and Planning 12: 23-40.

Vallianatos, M., Gottlieb, R. and Ann Haase, M., 2004. Farm-to-school. Strategies for urban health, combating sprawl, and establishing a community food systems approach. Journal of Planning Education and Research 23: 414-423.

Van den Brink, A., Van der Valk, A. and Van Dijk, T., 2006. Planning and the challenges of the metropolitan landscape: Innovation in the Netherlands. International Planning Studies 11: 147-165.

Van der Ploeg, J.D. and Frouws, J., 1999. On power and weakness, capacity and impotence: Rigidity and flexibility in food chains. International Planning Studies 4: 333-347.

Wiskerke, J.S.C., 2009. On places lost and places regained: Reflections on the alternative food geography and sustainable regional development. International Planning Studies 14: 369-387.

Chapter 33

Nested scales and design activism: an integrated approach to food growing in inner city Leeds

Emma Oldroyd and Alma Anne Clavin
Leeds School of Architecture, Landscape and Design, Leeds Metropolitan University, Arts Building, Broadcasting Place, Woodhouse Lane, Leeds, LS2 9EN, United Kingdom;
e.oldroyd@leedsmet.ac.uk

Abstract

Back to Front is a cross-disciplinary urban food growing project based in Leeds. The staff and students of Leeds School of Architecture, Landscape and Design developed a partnership with the British Trust for Conservation Volunteers (BTCV), the NHS and Leeds Permaculture Network to form a nested approach to growing food that connects multiple scales of design and delivery to enhance pro-growing behaviors in a number of deprived inner city neigbourhoods. The study area is the East and North East wedge of Leeds. It is a highly deprived area with some parts falling within the 3% of the most deprived nationally. However, the area is vibrant and has a rich cultural mix with Chapeltown having twice the city's average of black and ethnic minority communities and there is strong local support for growing food. Research undertaken by BTCV and the NHS examined the interest in and requirements for food growing in the area. This empirical research was adopted by staff and students to inform multi-scale designs for cultivating edible crops in a number of different housing types and neigbourhood spaces in the study area. Resulting designs ranged from movable modular systems in front gardens, to rooftop banana plantations. Project delivery is taking place using a bottom-up approach working with the city council and developing local links in the neigbourhood, which will create the foundations for an appropriate neigbourhood scale network and resource. The chapter examines the rationale for the approach and then describes its delivery and outputs. In addition, an explanation of how such a live interdisciplinary design project can provide multiple outputs for both teaching and action research will be provided.

Keywords: interdisciplinarity, partnership approach, live project, urban food planning

33.1 Nested scales and design activism: an integrated approach to food growing in inner city Leeds

In the UK, the rise in local food growing projects (Iles, 2010) has been, in part, a reaction to the globalisation of our food resources and issues of food security. Using food growing as an activity for re-designing an inner city area of Leeds (UK), a partnership approach has developed to progress a project called Back to Front, which aims to address urban sustainability at a local level. Using the front garden as a locus of activity, the project explores nesting scales of food growing activity. The aim is to enhance ecological sustainability and to rethink how local resources are utilised. From a pedagogical perspective, the project has provided a trans-disciplinary vehicle to resolve and develop a research-teaching-design nexus and method in Landscape Architecture and related disciplines, which provides an excellent student experience with research outputs.

33.2 Ecological sustainability and the city

There is a growing comprehension that local sustainability cannot be discussed in isolation from global issues. Barton *et al* (1995) argue that a home, an estate or a town should be considered as whole systems in the sense that they provide the essential local habitat for humans and should provide as far as possible for their comfort and sustenance at all of these scales. Furthermore, Massey (2007) suggests that actions that impact the global scale are just as grounded in everyday behaviors as actions that impact local scales.

There is no one correct perspective, rather a diversity of perspectives and scales are required for understanding the complexity of local systems for sustainability. Such an understanding has its origins in Von Bertalanffy's (1969) general systems theory. As natural systems evolve or mature they develop more complex structures and processes with greater diversity, more cycling and recycling of resources allowing for the emergence and mutual support of structures, enhancing ecological integrity.

Ecological sustainability is about maintaining and enhancing the integrity of combined ecological-societal communities. All human based organisational processes, such as municipalities, rest on local ecological landscapes as an enabling resource base and stage for the playing out of human behaviors (Kibert *et al.*, 2001). However, these local landscapes are made up of both public and private spaces, which are often managed very differently and without a sense of working towards a common goal (Barton *et al.*, 2000). Back to Front aims to work on the interface between both the human and natural environment, between differing spatial scales and between private and municipal management structures to promote behavioral change on a personal level, which, in turn, may impact on the wider geographical area and the way it is managed.

33.3 Ecological sustainability and design

The interface between man-made systems and natural ecosystems must reflect the limited ability of natural ecosystems to provide energy and absorb waste before their survival potential is significantly altered. Therefore the impact of a design intervention on natural processes and in turn ecological sustainability must be measured using a non-linear mindset that works upon the assumption that resources are not limitless (Kibert *et al.*, 2001). Design solutions must be both adaptable and applicable to a range of variables. This approach views humans as part of rather than apart from the ecological system. Urban dwellers may gain from learning and from the opportunities that are embedded within the structure of natural and organisational systems that they have access to.

Taking lessons from natural systems may be termed ecoliteracy (Orr, 1992) or sustainability literacy (Stibbe, 2009) and has its intellectual grounding in systems thinking. Being ecologically literate requires the capacity to observe nature with insight, 'a merger of landscape and mindscape' (Orr, 1992: 85). Ecological design therefore requires not just a change in our conceptual capabilities but also a change in our language capabilities. Benyus (2000) proposes that participation in design promotes learning and awareness, a concept mirrored in design activism thinking (Fuad-Luke, 2009). Similarly, ecological designers Van der Ryn and Cowen (1996: 162) state that 'design transforms awareness' and ecological design purports that in designing our built and natural environment we can learn valuable lessons from ecosystems and that each level, cell, organism, ecosystem, bioregion and biosphere – presents a series of critical design opportunities and constraints.

The limits to knowledge implied by such complex systems suggest that we cannot scientifically 'manage' systems beyond a certain scale. Without a sufficient 'eyes to acres' ratio (Van der Ryn and Cowen, 1996: 68), we will be overwhelmed by complexity and so design at a local scale promotes action within the limits of ecosystems and of local understanding. Wendell Berry (1987) describes two different kinds of limits: those on our ability to coordinate and comprehend things beyond some scale, and those inherent in our nature as creatures with a limited sense of the good and willingness to do it (*ibid*. p. 67). At some larger scale it becomes difficult to detect diversity and subtle difference in local environments. Orr (1992) argues, for example, that the ecological knowledge and level of attention necessary for good land stewardship is limited by variations in topography, microclimate and time. As size increases, the individual becomes a manager who must simplify complexity and homogenise differences in order to control. Beyond some threshold Orr argues that it is then not possible to see the outcomes of your actions and where your waste outputs go (*ibid.*).

In order to achieve ecological sustainability, there is a need for a reflexive response in urban design but how can this be inclusive, democratic and enabling? This leads us to consider the appropriate scale that we should begin to plan for ecological sustainability. The Back to Front project uses the front garden as a locus for local food growing activity i.e. all gardening outside the front door of the dwelling. Such spaces were found to be frequently overlooked by residents and are in close proximity to the home where food is prepared and cooked. Such an idea of placing items of highest intensity of use nearest to the focal point of activity (in this case the dwelling) is central to the permaculture concept of zoning. One of the founders of the ecological design system permaculture argues that 'starting at the back door' (and in this case the front door) will prevent 'overreach' (Holmgren. 2002: 138) when developing a site (*ibid.*). If we extend our activities too far and too fast while our immediate territory is not organised and working well, we find our human and resource energy dissipated.

The challenge with Back to Front is in the application of this zoning concept to (relatively) high density urban environments such as Harehills and Chapeltown in Inner city North East Leeds. The Permaculture zoning concept was initially developed in the 1970s in Australia (Mollison and Holmgren, 1978). In developing the permaculture concept in urban areas, Anderson (2006) suggests that a city's zones could be informed by the fossil fuel used by a range of modes of transport, retaining the home as the central locus of activity (see Figure 33.1). Anderson's model enables us to think on a larger scale about permaculture, while still focusing on the dwelling. The model begins to consider public open space as a resource but more comprehensive detail can be added by considering public open space hierarchy's and food zones.

Leeds City Council has not published its own accessibility standards for its open space typology listed within its Parks and Green Space Strategy (n.d.). In its absence, the London's Public Open Space Hierarchy (GLA, 2008) provides a useful link between public open spaces and distance, and can be applied to Anderson's model (see Figure 33.1). When we consider Brown's Food Zones for Hackney developed for local food organisation Growing Communities based in London (cited in Pinkerton and Hopkins, 2009), we see that the urban

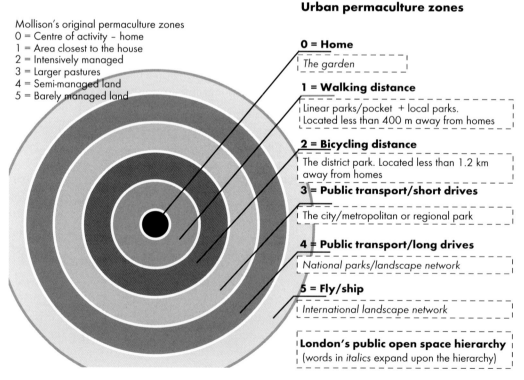

Urban permaculture zones

Mollison's original permaculture zones
0 = Centre of activity – home
1 = Area closest to the house
2 = Intensively managed
3 = Larger pastures
4 = Semi-managed land
5 = Barely managed land

0 = Home
The garden

1 = Walking distance
Linear parks/pocket + local parks. Located less than 400 m away from homes

2 = Bicycling distance
The district park. Located less than 1.2 km away from homes

3 = Public transport/short drives
The city/metropolitan or regional park

4 = Public transport/long drives
National parks/landscape network

5 = Fly/ship
International landscape network

London's public open space hierarchy
(words in *italics* expand upon the hierarchy)

Figure 33.1. Linking Anderson's Urban Permaculture zones to London's Open Space Hierarchy (adapted from Anderson's model of fossil fuel usage zones (2006)).

environment could potentially provide for 7.5% of the Borough's food needs. 2.5% of which come from domestic spaces. This highlights the importance of gardens in responding to the global food security issue and how albeit on a very small scale, Back to Front may contribute to a more sustainable Leeds.

The front garden provides a contained space of appropriate scale for a household to grow edible crops and hence engage in ecologically sustainable activity – linking the home occupier to the wider built and natural environment. In developing a partnership approach with the National Health Service (NHS), Leeds City Council and local interest groups, project engagement and longevity may be realised for local people through the widening of social networks and contacts. Such interventions may also have a positive impact upon the visual quality of the neigbourhood and the wider productive landscape and ecosystem.

33.4 Back to Front

The aim of Back to Front is simply to promote growing food crops in front gardens as a socially acceptable norm in North East Leeds. To develop long term design solutions to promote such activity, a food growing manual for the area which will contain 'off the shelf' designs for people wishing to convert their front gardens into productive spaces will be produced.

Back to Front had its inception in 2009 with a three-stage project plan:
• Stage one: preliminary research undertaken to provide evidence of local 'needs' and interest in urban food growing in the area to inform a series of trans-disciplinary design solutions;
• Stage two: the construction of three gardens (undertaken in July 2010);
• Stage three: the production of the Back to Front Manual.

33.4.1 A partnership approach

Project partners include Leeds Metropolitan University (Leeds Met), Leeds City Council, Leeds Permaculture Network, Groundwork and NHS Leeds along with a number of other local food associated organisations to form a project think tank. The project is classified within the environment category of the 'Social Determinants of Health Development Fund'. The funding stream followed the publication of a review on inequalities in health (Marmot Review) established to support the development of up to only four national projects that demonstrated creative and innovative practice in addressing health inequalities.

The project has been singled out by the Local Government Improvement and Development Agency (LGIDA) as an example of innovative good practice. As such, the project is benefiting from funding to develop the Back to Front manual.

33.4.2 Rationale, scope and design response

In a bid to tackle unequal access to opportunities, Back to Front is limited to the most deprived Super Output Areas in East and North East Leeds with some parts falling within the 3% of the most deprived nationally. Harehills and Chapeltown are culturally rich areas of Leeds with twice the city's average of black and ethnic minority communities (LCC, 2009). First generation immigrants to the UK often arrive with a strong food growing culture and in Leeds, have attempted to grow at least some of their food from home, often at the front of the house. This project aims to embrace this activity and promote it to a wider audience.

Stage one of this project saw Leeds British Trust for Conservation and Volunteering (BTCV) and NHS Leeds develop a bid for the Special Grants Panel to undertake preliminary research into 'local need' for urban food growing. A case worker was appointed whose team of 15 coordinated 400 door to door surveys and 10 focus groups with local people in the Chapeltown and Harehills area. With these results (Preston, 2009), NHS Leeds approached Leeds Metropolitan University to discuss whether students could use the empirical data to inform designs for food growing in the area.

The initial research (Preston, 2009) concluded that just under a fifth (19.4% or 45 people) of participants would consider reconverting their front garden for food production. Those unwilling to do so gave some particular justifications namely: problematic animals and pests, lack of time and lack of interest. These three categories accounted for 40% of the replies given. Almost half of the reasons provided by those unwilling to use the front garden for food production (47.4%) could be tackled with appropriate design solutions. Further detail is provided in Table 33.1.

Table 33.1. Enabling design solutions to be addressed in the Back to Front manual.

Prohibiting factors	Design solutions
Space: concerns about the lack of space	Employ responsive design ideas such as vertical gardens and supporting structures (for example an entrance porch used for supporting beans, peas, trailing pumpkins).
Security: increased likelihood of theft	To be reduced when greater numbers join the scheme. None of the residents involved in the demonstration gardens reported theft as a problematic outcome of the scheme.
Costs	Design solutions to promote the re-use and re-cycling of materials however where this is not possible, savings can be made by residents clubbing together for particular high volume materials.
Site aspect	The manual should provide guidance about which plants are appropriate for each aspect.
Cultural aspects: 'not customary to grow stuff in the front garden' or 'not a habit'	Concern to be reduced when greater numbers join the scheme supporting Back to Front's aim to make front garden growing a cultural norm.
Perceived need: there needs to be investment in other facilities for the area (such as children's play areas)	Links to public space designs can accommodate a wide range of facilities as well as promoting productive landscapes.

The above prohibiting factors will be transformed into enabling features through design solutions. These are firmly placed within the local neigbourhood through links made to local resources. They cater for individual circumstances in matching one's capability to the growing task e.g. beginning with 'container' gardening of simple herbs and salads and moving on to more complex schemes when and if the required physical capabilities and personal circumstances present themselves.

Further thinking resulted in a series of design responses that linked to three scales of intervention; the neigbourhood or district, the street or local park and the home. Students from Landscape Architecture, Garden Design, Architecture and 3D design developed the project in different ways. The 3D design student's work was mostly associated with the locus of activity around the home and garden (see example in Figure 33.2), while architecture students fully embraced the neigbourhood scale (see Figure 33.3). Landscape architects tended to consider links between the street scale and the gardens (see Figure 33.4).

All the student work was displayed as part of a public event in March 2010. The work informed the development of the Back to Front manual, which although is focused on front garden spaces, is supported by interventions within the public realm. The exact nature of these design interventions must respond to local needs and housing type and therefore should be place specific. For example, a flat with a balcony presents different opportunities

Figure 33.2. Mariam Aomar Perez's growing ladder.

Figure 33.3. Christopher Hartshorne's banana world.

Figure 33.4. Liam Clark's design for Rashid's garden.

and constraints for growing and recycling materials to a semi-detached 1930's property. The public realm has the potential to become a highly useful space to accommodate productive food growing space especially when space outside the home is restricted. Links between public and private spaces require negotiation with land owners and local authorities. The project has inspired more integrated activity in a local neigbourhood regeneration project and its ethos has been embraced by a local housing association. However, the negotiation of productive space in the public realm is beyond the scope of the manual.

Stage two of Back to Front involved building three gardens in the area to act as demonstration gardens to promote and test ideas. In order to achieve this, £1000 was obtained from Leeds City Council's area management fund. Youth members of Groundwork and staff and students of Leeds Met designed and constructed three gardens in the area in summer 2010. Each served different purposes: 'A Garden for Time' (see Figure 33.5) is a permanent garden for an owner occupier, 'A Garden to Share' (see Figure 33.6) is a community garden for a housing co-operative in the area and 'A Garden Above Ground' (see Figure 33.7) is a movable modular garden that avoids digging below the ground or changing the structure of the garden. It is appropriate for tenants in a rented property. This typology emerged as a result of the preliminary research stage of the project and occupancy information available from LCC (2009).

Building the demonstration gardens in summer 2010 instigated a shift in the design process from a more traditional way that is initiated with design development (Motloch, 2001) to one which is heavily informed by the availability of local resources and materials. Back to Front

Figure 33.5. A Garden for Time.

Figure 33.6. A Garden to Share.

Figure 33.7. A Garden Above Ground.

has interpreted 'local resources' in two ways; the first of which relates to materials and plants to build gardens, and has been closely linked to business 'waste' products in the immediate local area. The second interpretation relates to resources available from local organisations, service areas and knowledge centres. Back to Front has been working closely with the local authority to explore how individuals involved in the project can be supported by council service areas on a long term basis. This outcome is significant because it links the front garden to the local community, local place and wider scale systems and spaces. This in turn enhances ecological integrity and in turn, ecological sustainability.

Stage three of the project will result in the publication of a growing manual in winter 2011/12. The 2011 growing season will be used to gather feedback on the demonstration gardens and 'off the shelf' growing ideas such as Rachel Forbes' modular planter (see Figure 33.8). As agreed by the project think tank, consultation, and led by designers at Leeds Met, the manual is being driven by the following three objectives;
- to make information available on practical designs for food production especially designed for Northern English inner city dwellers;
- to develop training guidelines for community use to form part the above;
- to raise awareness about the project and a future scheme through available channels.

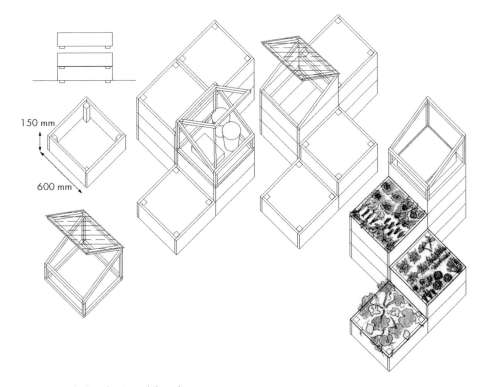

Figure 33.8. Rachel Forbes' modular planter.

The progress of the project has been documented on its website which is available from www.
backtofront.org.uk. Development of further design solutions and testing of the impact of the
work completed to date is currently being undertaken. In addition to being the impetus for
a hive of food growing activity in the Harehills area, the project has also been successful as
a live student project. It has forged links with community, interest groups and stakeholder
organisations whilst producing both novel teaching and research outputs.

33.4.1 Back to Front and design activism

The exploration of the relationship between research and teaching in higher education has
been much debated in academia (e.g. Henkel, 2004; Robertson, 2007; Rolfe, 2003), with a
significant body of evidence in support of the benefits that exposure to both activities bring
to the student and staff experience (Griffiths, 2004; Healey, 2005). In addition and in response
to professional and personal obligations, design academics look to the integration of design
action (which describes all design and related activities) into their work. Building upon the
work of Visser-Wijnveen *et al.* (2010) who have proposed a series of profiles that describe the
research-teaching nexus, Back to Front has allowed the exploration of a three way research-
teaching-design action nexus where design action can be a research method and outcome
as long as the extrapolated outcomes are original, lead to new insights and are applicable to
multiple locations. It is the socially driven project ethos that turns design action into design
activism.

33.5 Conclusions

Using the front garden as a locus for their research and designs, the student work formed an
integrated approach to designing for food growing, linking the nested scales of the residential
front gardens to the neigbourhood and to the city of Leeds. The focus on local resources
informs the modular designs and the content of the manual. This is emerging as one of the
main outcomes of the project and leads its future direction. The iterative process of trialing
the manual will ensure ecologically sustainable design solutions which are place specific but
are based on a series of replicable, adaptable processes that can be applied to other areas of
the city and beyond.

References

Anderson, B., 2006. Adapting zones and sectors for the city. 13 January 2006, available at http://www.cwo.
com/~bart/essays/z_s_urban.htm.
Barton, H. (Ed.), 2000. Sustainable communities: The potential for eco-neighbourhoods. Earthscan, London,
UK, 288 pp.
Barton, H., Davies, G. and Guise, R., 1995. Sustainable Settlements – A guide for planners, designers and
developers, University of the West of England and Local Government Management Board, Luton, 247 pp.
Barton, H., Grant, M. and Guise, R., 2003. Shaping neighbourhoods: A guide for health, sustainability and
vitality. Spon Press, London, UK, 256 pp.

Benyus, J., 2000. Biomimicry. In: Ausubel, K. and Harpignies, J.P. (Eds.) The Bioneers Series – Nature's Operating Instructions. University of California Press, Berkely, CA, USA, 256 pp.

Berry, W., 1987. Home economics. North Point Press, San Francisco, CA, USA, 192 pp.

Fuad-Luke, A., 2009. Design activism: Beautiful strangeness for a sustainable world. Earthscan, London, UK, 192 pp.

Greater London Authority, 2008. The London plan. Greater London Authority, London, UK. Available at http://www.london.gov.uk.

Griffiths, R., 2004. Knowledge production and the research-teaching nexus: The case of the built environment disciplines. Studies in Higher Education 29: 710-725.

Healey, M., 2005. Linking research and teaching to benefit student learning. Journal of Geography in Higher Education 29: 183-201.

Henkel, M., 2004. Teaching and research: The idea of a nexus. Higher Education Management and Policy 16: 19-30.

Holmgren, D., 2002. Permaculture principle and pathways beyond sustainability. Holmgren Design Services, Victoria, Australia, 286 pp.

Iles, J., 2010. Community land bank consultation. Executive summary, conclusion and next steps. Federation of City Farms and Community Gardens, Bristol, UK.

Kibert, C., Sendzimer, J. and Bradley, G., 2001. Construction ecology: Nature as the basis for green buildings. Spon Press, London, UK, 336 pp.

Leeds City Council, n.d. A parks and green space strategy for Leeds. Leeds City Council, Leeds, UK. Available at http://www.leedsinitiative.org.

Leeds City Council., 2009. Neighbourhood index. 1 June 2009, Leeds City Council, Leeds, UK. Available at http://www.leeds.gov.uk.

Massey, D., 2007. World city. Polity, Cambridge, UK, 272 pp.

Mollison, B and Holmgren, D., 1978. Permaculture one: A perennial agriculture for human settlements. Tagari Publications, Tasmania, Australia, 127 pp.

Motlock, J., 2001. Introduction to landscape design. John Wiley and Sons, Hoboken, NJ, USA, 352 pp.

Orr, D., 1992. Ecological literacy: Education and the transition to a postmodern world. State University of New York Press, Albany, NY, USA, 210 pp.

Pinkerton, T. and Hopkins, R., 2009, Local food: How to make it happen in your community. Transition Books, Devon, UK, 216 pp.

Preston, J., 2009. Garden to eat: Encouraging greater domestic food growing. Public presentation. 1 October 2009 BTVC and NHS Leeds, Leeds, UK.

Robertson, J., 2007. Beyond the research / teaching nexus: Exploring the complexity of academic experience. Studies in Higher Education 32: 541-556.

Rolfe, H., 2003. University strategy in an age of uncertainty: The effect of higher education funding in old and new universities. Higher Education Quarterly 57: 24-47.

Stibbe, A., 2009. Handbook of sustainability literacy. Green Books, Totnes, UK, 224 pp.

Von Bertalanffy, L., 2008 [1969]. General systems theory. Brazillier, NewYork, NY, USA, 296 pp.

Van der Ryn, S. and Cowan, S., 1996. Ecological design. Island Press, Washington DC, USA, 246 pp.

Visser-Wijnveen, G., Van Driel, J., Roeland, M., Van der Rijst, Verloop, N. and Visser, A., 2010. The ideal research-teaching nexus in the eyes of academics: Building profiles. Higher Education Research and Development 29: 195-210.

Chapter 34

Architecture *et al.*: food gardening as spatial co-authorship on London housing estates

Mikey Tomkins
University of Brighton, 13 City Pavilion, 59 Chilton Street, London E2 6EA, United Kingdom;
mikeytomkins@gmail.com

Abstract

The French poet and artist Jean Cocteau (1983) once stated that all art is anonymous, in that authors are rarely present when their art is viewed. This is also true of the built environment, which is largely silent of architects' voices, filled instead by the hubbub of residents daily intervention. This chapter explores one intervention – community food-gardening – as it is presently emerging within six inner London housing estates. The estate residents explicitly expressed frustration at the 'blank', 'bleak', 'disused', 'neglected', 'barren', 'grey' and 'derelict' 'blankscapes' surrounding their homes, voicing a desire to re-use them 'productively' through food-gardening. However, this chapter argues that while food is enunciated as the principle agenda, it is a set of related practices, such as the construction of the self-built landscape, the creation of shared social narratives, and the interaction with natural resources that dominate. Using multi-site participant observation, semi-structured interviews and photography, the research throws new light into this overshadowed everyday food-gardening activity that often falls within the penumbra of the productive, economic and environmental 'feeding cities' UA discourse. Rather, what needs to be explored is the re-linking of residents to architecture, landscape, and the planning of cities via food gardening. As one resident put it, gardeners are 'amateur architects', augmenting the pre-planned architecture with a bricolage of seasonal, quotidian, and playful performances. This challenges formal architectural discourse through soil-based interventions, user modification of space, and social harvesting. In this sense, the built environment is seen as a space of multiple authors whose numerous interactions create an architecture *et al*.

Keywords: urban agriculture, participant observation, ethnography, the everyday

34.1 Methods

Six inner London estates and their community-garden projects were selected for participant observation (PO), requiring a participatory role while observing (DeWalt and DeWalt, 2002, Spradley, 1980). Participation or 'hanging out' requires researchers to combine the use of 'their social selves as their primary research tool' (Hume and Mulcock, 2004) with the academic requirements of turning experience into analysis. The six community gardens in this study were selected because they clearly operate within the boundaries of the residents' estates, linking the landscape, the architecture and the home, and they have all chosen to create community food gardens. Community gardens in general were chosen because they represent a shared action on a shared piece of land, responding to the argument that environmental concerns are collective and not necessarily individual (Gaynor, 2006).

The research is approached via grounded-theory method (GTM) and phenomenology (Annells, 2006; Wimpenny and Gass, 2000). GTM (Bryant and Charmaz, 2007; Strauss and Corbin, 1997) advocates a heuristic approach, emphasising repetitive field study and collect data without a principle theory, which allows relevant data to emerge from the exploration of phenomena, which is subject to 'constant comparison and reduction' (Wimpenny and Gass, 2000) from which theory will be generated. Clarke argues that 'the meaning of phenomena, including social worlds, are to be found in their embeddedness in relationships – in society ... because people would not concern themselves with a given phenomenon if they were not concerned to *do* something with it or about it' (Clarke, 1997). The practice of participating in food gardening was continuous from October 2009 to August 2010. Twenty-two semi-structured interviews were recorded alongside field notes and photographs.

34.2 Introduction

The emphasis within literature is to explore consumption discourses, indubitably linked to agricultural (Freidberg, 2001; Morris and Kirwan, 2010), which subsuming the practice of food gardening as an adjunct to agro-industry. For example Carolan (2007) writes, 'short of growing our own food ... can an ecological citizen exist in today's global world that is full of asymmetrical flows?'. Corolan discusses the way in which tactile practices such as food gardening can collapse the 'epistemologically distant' and 'asymmetrical flows' that commodity systems create, concealing environmental relations. Such literature argues for a shift towards 'knowing food, growing food' (Fonte, 2008; Goodman and DuPuis, 2002), that connects producers and consumers. However, I would argue that what the current research project demonstrates is that the 'asymmetrical flows' and 'epistemological distances' that need to be attended to are those segregating the user and designer of space (Hill, 1998), between multifunctional collaboration and technological unity (Lefebvre, 1991).

While much research has been done on the relationship of self-grown food to health (Wakefield *et al.*, 2007), sustainability (Holland, 2004; Koc *et al.*, 1999), education, (Pudup, 2008) consumption and production (Morris and Kirwan, 2010), little research has been done which examines the way in which food gardeners explicitly turn disused architecture into food productivity, exposing the asymmetry of design intentions to user desire. Thus we are not attempting to follow the food (Cook, 2006; Fonte, 2008), but instead follow the narrative from architect to assumed passive user (Hill, 2003) to self-builder (Hardy and Ward, 1984; Szczelkun, 1993). Following this narrative leads not upwards towards a strategy of advocating food growing spaces via policy (Howe and Wheeler, 1999), a new design paradigm (Lim and Liu, 2010; Steel, 2008) or the non-plan of Hall and Banham (Hughes and Sadler, 1999) but stays grounded by recording the food-related, everyday practices, and creative aspirations of residents, within the emic or 'local' language of gardeners. A language rarely observed, recorded, or interpreted. It should be noted that while this research investigates those that are choosing to garden, not all residents are in agreement with the growing of food on estates, with some demonstrating vigorously against the sometimes messy landscape of the food gardeners. This attitude, when backed by landlords, may create barriers to the wide implementation of community gardens, which also needs to be fully research.

34.3 Who makes landscape, describes landscape

While some argue for coherence within UA (Bakker *et al.*, 2000; Mougeot, 2000), there is also a wealth of literature that attempts to capture how individuals practise self-grown food in their own language. For example, the supporting literature documents 'community vegetable gardening' (Domene, 2006), 'backyard' (Blaylock and Gallo, 1983), 'dooryard' (Wilhelm, 1975), home, kitchen, or food gardens (Niñez, 1984), specifically 'urban cultivations' (Sanyal, 1985), or simply 'food gardening' (Vasey, 1985). As Premat (2005) writes, ignoring 'diversity in the actual use of space is to be trapped in a particularly restrictive and impoverished notion of UA that does not take account of reality on the ground'. Within the research sites, the narrative of augmentation often begins with language, as Samuels writes, 'who makes landscape, describes landscape' (Samuels, 1979). In the six community gardens investigated, several are referred to as 'gardens' combined with their geographic location, while others are described as 'kitchen gardens', or 'growing', as well as 'edible'. These food gardens are sequestered into the 'rational, passionless landscape' (Cullen and Knox, 1982): designed open spaces left unnamed and anonymous, 'a barrier to their recognition and hence utilisation as part of the estate environment' (Ravetz, 2001).

Commenced in 1965, Erno Goldfinger designed a large estate in east London called Brownfield. It is overshadowed by two tower blocks, whose appellation 'Balfron', and 'Carradale' are places in Scotland; some 35 years later the 'biography of landscape' (Samuels, 1979) continues when residents conceive of the moniker 'greening' Brownfield. Subsequently, they turn the disused, or according to some residents, never used tennis court into a food garden. I participate in pacing out – on top of the faded white lines of the tennis court – the new lines for the raised food growing beds. Disused underground car parks lie underfoot and due to structural issues, we use a copy of the architect's plans to locate supporting walls. The layers build up, a palimpsest of architect, disused sport arena, returning wildlife, and now food-gardener. The latter a self-built landscape of old tyres, bags of soil, and salvaged wooden raised beds (Figure 34.1).

In some ways, Goldfinger would have agreed with the project, in others, not. He is quoted as saying, 'the whole point of building high is to free the ground for child and grown-ups to enjoy Mother Earth'. But this is a gardening project and Goldfinger wanted London to be a city of parks, 'not, and I emphasis not, a garden city' (Warburton, 2005). However, Goldfinger is not here and the residents are, engaged in acquiring 'direct knowledge of their environment in the course of their practical activities through engagement' (Okely, 2001), an engagement which was so evidently never present within the original design.

Figure 34.1. 'Greening' Brownfields estate.

34.4 The amateur architect

Such small, daily, unplanned changes are termed 'incremental communitarian development' by Szczelkun (1993), while Allen (2007) describes these incremental changes as a 'restless landscape': under the assumed control of planning and architecture, yet under the constant persuasion of residents daily desires. We see this exemplified as residents in Plaistow, east London, take over a disused basketball pitch, or on the Haberdasher estate in Hackney, north east London, where residents remove 'municipal' shrubs to expose land for gardening, or when residents of St Johns estate transmogrify a piece of 'empty' grass into a 'community kitchen garden'. We see more self-building as we go for a walk with Neil around the Haberdasher estate in Hackney. Neil has converted the demolished pram sheds on the upper walkways into a 'tomato factory', the disused underground bike shed into a wormery, and vanity landscapes that surround the estate have been replaced with grow bags.

The idea that the food product is merely the stone that ripples the pond is clear in Neil's description regarding his interaction with potential fellow gardeners. Neil says 'if a new person came and wanted to get involved, then no problem for me, what I'd do is I'll give up one of my bags y'know...' (05/2010).[40] These 'bags', common on most community gardens, are one metre square and usually placed in lines sympathetic to the spaces they inhabit. They gradually sink or fall to one side, evolving over the year as gravity pulls the soil and water down. Neil is willing to sacrifice an already small harvest for greater community involvement.

---

[40] This date refers to the month and year of the audio recording.

The community involvement is evident during my afternoon with Neil as gardening is interspersed with vertical conversations with residents in the flats that overlook the garden. Here, broad beans grow up two storeys high, wrapping themselves around the architecture and the satellite dishes on the second floor. Above us, Lee and John come out onto their balconies to converse. While the bags belong to one particular resident, that waters and tends the plants, the crops are shared by all whose balconies the plants cross (Figure 34.2).

This is not planned, but negotiated, as the plants grow *in situ*. Here cultivation, architecture, and everyday life melt together. As Kimber (2004) comments, 'the needs of human dwelling are achieved when they are allowed to arise spontaneously out of the requirements and concerns of particular people and landscape'. One requirement is the basic need to connect with other residents within what was otherwise an empty landscape. As Samuels (1979) comments, 'there is something unreasonable about a human landscape lacking in inhabitants'. Gardening for food, and constructing its requisite structures, seems to have emerged as a legitimate tool for residents to modify, participate in, and augment planned estates.

Over on De Beauvoir Estate in Hackney, northeast London Cindy rails against the failures of the professional classes who have designed and maintained the estate environment over the last 30 years, to the exclusion of residents. Remarking 'we are all amateurs, I know, amateur architects, amateur landscapers, but if they would just let us get on with it, we would get

Figure 34.2. Lee on the 2nd floor balcony at Haberdasher estate, surrounded by runner beans.

better at it, wouldn't we?' (06/2010). This notion of the 'amateur architect' gets repeated by other gardeners, but never with such lucidity as when Cindy talks. It is often implicit in actions on other estates, such as the planning of raised beds that frame the estate buildings (St Johns estate), or the creation of a shelter, tables and chairs from reclaimed wood (Dirty Hands Project).

34.5 Production and productivity

Neil's memory of the estate landscape he has known all his life is an intimate part of the process of transformation. Perhaps the last time Neil played on the estate may have been as a child, now two decades later he is returning to the planned landscape to play and transform it as an adult. The sense that the gardening project is playful and creative comes across again and again. Simon on St Johns estate melds food growing with other creative practices:

> there is something really satisfying about a radish that you pluck out the ground and just eat, it's somehow different ... it's partly a creative thing isn't it? It's like music or art, you're making something, you're watching it sort of evolve ... and pride in making something... like I'm making the sign [for the garden] and I'm quite pleased with it, it will be quite nice to see it hanging up there ... it's claiming something back [from the city]... (08/2010).

The urban landscape is no longer simply form and function, a machine, but it is also an emotion, a taste, and a story. This 'claiming back' is simultaneously the food and the built landscape. We later wash a tiny tomato, plucked from one of a dozen raised beds. We wash it and eat it surrounded by the estate, briefly escaping into what Gardiner describes as 'non-commodified social relations' (Gardiner, 2000) before Simon lights another cigarette and suggests returning to the local, aptly named Beehive pub, to continue the interview.

Head *et al.* (2004) differentiate between self-grown food production, and discussions of productivity and creativity. Head goes on to describe productivity as encompassing 'three important, distinctly physical engagements between the gardener and the soil, smell and taste, and the temporal rhythms of life'. Once the subject of food gardening is enunciated, all these phenomenological experiences are brought into focus, as food becomes a sharp tool to excavate the impotence of the landscape, the redundancy of the designed grass areas on estates, and its thin underbelly of topsoil. Sitting on the Haberdasher estate with Neil, he takes a deep breath, 'smell that, good isn't it' (07/2010), referring to the odour emanating from the 15 tonnes of compost newly deposited in the centre of the estate (Figure 34.3). It is a scene that is only unveiled by spending time with residents participating with the many environmental tasks necessary to grow food.

The olfaction of local food production is discussed by Malcolm Thick (1998) who describes the everyday market gardens that once spread across a fifth of London prior to the 1830s (Glanville, 1972), as having the constant smell of manure. This he calls the 'ugly sister' of gardening history that now lies 'buried beneath the towns and cities they once served' (Thick, 1998). Angela, from the St Johns estate remarks, 'you dig down six inches and you get brick

Figure 34.3. Fifteen tonnes of soil, Haberdaser estate, London.

and rubble, what did they ever expect us to do with these grassed areas?' (07/2010). All the sites studied had to import vast quantities of soil, with residents being conscious of the soil's quality, because it would be used to grow food they would consume. Soil is the vital addition, it turns inert architecture and space into food. Graeme, Angela's neighbour, is more succinct, 'London has no soil' (08/2010).

34.6 Culture and agriculture

Lee, on the Haberdasher estate, explains that for him he wants to feed the soil that feeds him – this cycle creates a narrative that starts with the soil and ends with the gift of the edible, and the death of the plant. Lee says, 'just to watch how it grows … to do things like benefit my family, the people I love and care for sorta gets ya talking, just opens me up…' (05/2010). Bonsdorff (2005), in her discussion of the way that nature and culture interact with buildings repeats this point, reminding us that the cultural part of 'the agricultural context also makes us more aware of the organic conditions of existence: growth, nutrition, birth, and death'. For Natasha on De Beauvoir estate, 'food is concrete, you have something to show' (05/2010), linking architecture directly with the narrative of food growing and its produce.

However the final 'show' is often small, almost votive in nature. In mid-July two Lansbury gardeners, Francis and June, are harvesting the garlic. Two bulbs are pulled from the ground. The women are evidently excited. June remembers planting the bulbs in autumn, attending the gardening club nearly every week since. For her, the cultivation of the disused land on the estate goes hand in hand with the cultivation of friendships amongst the gardeners. Without the food, there would be no stories. Every session, the group spends nearly 40 minutes walking amongst the growing bags, discussing what has changed. Multiple conversations creep across the site, framed by the pumpkins that creep around the garden walls. The one small cucumber that is grown by the group is sliced, placed on a plate and handed around (Figure 34.4). It is a ritual, a preparation, a reminder that, as Barthes (1975) states, 'No doubt, food is, anthropologically speaking ... the first need.' Currently, within a burgeoning community garden 'movement', which can be witnessed across London, this 'first need' is being met ... or is it? Undertaken for perhaps pleasure, memory, taste, these spaces are rehearsals of our 'first need', producing edibles, but certainly not something of agricultural significance giving researchers a weak link to the agro-food discourse.

Moreover their function and form remain rudimentary, often forming a 'bed and bag' system, giving them a weak link to the formal licensure of planning and architectural design. And yet we chose to call this urban agriculture? There is a danger that UA definitions and initiatives that do not attempt to understand the rich phenomenological life that remains hidden within the daily routines these social spaces manufacture will simply replicate the most dominant and repressive aspects embedded in 'urban' and 'agriculture', inevitably leading to the diversity and ambiguity present in the everyday life of plots being ignored by those patriarchal institutions that currently dominate land use and food production.

Figure 34.4. Self-grown cucumber offered to fellow gardeners, Lansbury estate, east London.

34.6. Summary

Part of the drive within this participatory research project is to record the everyday narrative of food gardening within the built environment, week by week *in vivo*. While our current environmental dystopia can be viewed as a product of incremental diurnal actions, these actions are often recorded as abstract quantitative tables of data. This chapter seeks to refocus food gardening research towards actual 'everyday interactions of people and environment' (Gaynor, 2006) within the local scale of the immediate urban landscape as a narrative.

As Goffman (1956) reminds us by quoting Thomas, 'we do not ... lead our lives, make our decisions, and reach our goals in everyday life either statistically or scientifically.' In this sense, we are not able to do real-time analysis on the effects our lives have on the environment and data we receive is retrospective. Once these environmental actions have been translated to the international stage, and fed back to the user as 'climate change', then the scale has shifted too dramatically for individual households to feel they can impact on their relationship to the environment locally.

Central to this is the realisation that the design, construction and maintenance of the public realm are not left to accident, they are a result of tightly controlled, deliberated design decisions of professional classes and government legislation (Ravetz, 1986). This research seeks to capture the ephemeral, oral, and bodily performance of gardeners informally constructing their environment. Perhaps, as Cindy reminds us, left to its own devices it might move beyond the 'amateur' to develop its own involved vocabulary, aesthetic, and purpose.

This question turns UA research around from the indubitable deference of grow-your-own linked to food consumption, and therefore agriculture; to investigating how growing connects to dwelling (Kimber, 2004; Sharr, 2007), both as a reaction to architecture but also a desire to be part of a multifunctional architectural narrative (Boudon and Onn, 1972; Turner and Fichter, 1972). Specifically, connecting 'growing' and 'knowing' to architecture and the dynamic narrative of the 'dweller landscapers' (Lassus, 1993), busy modifying, designing, and building on the naturally planned landscape for subsistence production (Bonsdorff, 2005).

As Pugh writes, 'natural' is the cultural meaning read into nature ... by those with power and money to use nature instrumentally, as a disguise, as a subterfuge ... that things were always thus, unchangeable and inevitable' (Pugh, 1988). Therefore, food gardeners challenge the 'suspension of disbelief' required by residents to accept this 'subterfuge', projected into the theatre of the built environment, which is clearly not 'unchangeable and inevitable' but, as the research demonstrates, a continuously co-authored space, an architecture *et al.*

References

Allen, S., 2007. Pamphlet architecture 28: Augmented landscapes. Princeton Architectural Press, New York, NY, USA, 80 pp.

Annells, M., 2006. Triangulation of qualitative approaches: hermeneutical phenomenology and grounded theory. Journal of Advanced Nursing 56: 55-61.

Bakker, N., Dubbeling, M., Guendel, S., Sabel-Koschella, U. and De Zeeuw, H.O., 2000. Growing cities, growing food: urban agriculture on the policy agenda. A reader on urban agriculture Deutsche Stiftung fuer Internationale Entwicklung. DSE, Feldafing, Germany.

Barthes, R., 1975. Towards a psychology of contemporary food consumption. European diet from pre-industrial to modern times. [S.l.]: Harper, New York, NY, USA.

Blaylock, J.R. and Gallo, A.E., 1983. Modeling the decision to produce vegetables at home. American Journal of Agricultural Economics 65: 722-729.

Bonsdorff, P.V., 2005. Building and the naturally unplanned. In: Light, A. and Smith, J.M. (Eds.) The aesthetics of everyday life. Columbia University Press, New York, NY, USA, pp. 73-91.

Boudon, P. and Onn, G., 1972. Lived-in architecture: Le Corbusier's Pessac revisited. Lund Humphries, London, UK, 200 pp.

Bryant, A. and Charmaz, K., 2007. The SAGE handbook of grounded theory. SAGE, London, UK, 656 pp.

Carolan, M.S., 2007. Introducing the concept of tactile space: Creating lasting social and environmental commitments. Geoforum 38: 1264-1275.

C.J. Lim and Liu, E., 2010. Smartcities and eco-warriors. Routledge, New York, NY, USA, 256 pp.

Clarke, A.E., 1997. A social worlds research adventure: The case of reproductive science. In: Strauss, A.L. and Corbin, J.M. (Eds.) Grounded theory in practice. SAGE, London, UK, pp. 63-94.

Cocteau, J., 1983. The White paper (anonymous). Brilliant Books, Bedfordshire, UK.

Cook, I.E.A., 2006. Geographies of food: following. Progress in Human Geography 30: 655-666.

Cullen, J. and Knox, P., 1982. The city, the self and urban society. Transactions of the Institute of British Geographers 7: 276-291.

Dewalt, K.M. and Dewalt, B.R., 2002. Participant observation: a guide for fieldworkers. Altimira Press, Walnut Creek, CA, USA, 285 pp.

Domene, E. and Sauri. D., 2006. New urban lifestyles and welfare: water consumption in the suburbs of Barcelona. Sixth European Urban & Regional Studies Conference. Barcelona, Spain.

Fonte, M., 2008. Knowledge, food and place. A way of producing, a way of knowing. Sociologia Ruralis 48: 200-222.

Freidberg, S., 2001. On the trail of the global green bean: methodological considerations in multi-site ethnography. Global Networks, 1: 353-368.

Gardiner, M., 2000. Critiques of everyday life: an introduction. Routledge, New York; London, 224 pp.

Gaynor, A., 2006. Harvest of the suburbs: an environmental history of growing food in Australian cities. Crawley, W.A., University of Western Australia Press, Perth, Australia, 264 pp.

Glanville, P.J., 1972. London in maps. The Connoisseur, London, UK, 212 pp.

Goffman, E., 1956. The presentation of self in everyday life. Anchor, New York, USA, 161 pp.

Goodman, D. and Dupuis, E.M., 2002. Knowing food and growing food: Beyond the production – consumption debate in the sociology of agriculture. Sociologia Ruralis 42: 5-22.

Hardy, D. and Ward, C., 1984. Arcadia for all: the legacy of a makeshift landscape. Mansell, London, UK, 320 pp.

Head, L., Muir, P. and Hampel, E., 2004. Australian Backyard Gardens and the Journey of Migration. Geographical Review 94: 326-347.

Hill, J., 1998. Occupying architecture: between the architect and the user. Routledge, London, UK, 253 pp.

Hill, J., 2003. Actions of architecture: architects and creative users. Routledge, London, UK, 222 pp.

Holland, L., 2004. Diversity and connections in community gardens: a contribution to local sustainability. Local Environment 9: 285-305.

Howe, J. and Wheeler, P., 1999. Urban food growing: the experience of two uk cities. Sustainable Development 24: 13-24.

Hughes, J. and Sadler, S., 1999. Non-plan: essays on freedom and change in modern architecture and urbanism. Architectural, Oxford, 243 pp.

Hume, L. and Mulcock, J. 2004. Anthropologists in the field: cases in participant observation. Columbia University Press, New York, Chichester, 265 pp.

Kimber, C.T., 2004. Gardens and Dwelling: People in vernacular gardens. Geographical Review 94: 263-283.

Koc, M., Macrae, R., Mougeot, L.J.A. and Welsh, J. (Eds.), 1999. For Hunger-proof Cities: Sustainable Urban Food Systems. IDRC, Ottowa, OR, Canada, 239 pp.

Lassus, B., 1993. The garden landscape: a popular aesthetic. In: Hunt, J.D.E. and Wolschke-Bulmahn, J. (Ed.) The vernacular garden: 14th Dumbarton Oaks colloquium on the history of landscape architecture: Revised papers. Dumbarton Oaks Research Library and Collection, Washington, DC, USA.

Lefebvre, H., 1991. The production of space. Basil Blackwell, Oxford, UK, 454 pp.

Morris, C. and Kirwan, J., 2010. Food commodities, geographical knowledges and the reconnection of production and consumption: The case of naturally embedded food products. Geoforum 41: 131-143.

Mougeot, L.J.A., 2000. Urban agriculture: concept and definitions. The Urban Agriculture Magazine 1.

Niñez, V.K., 1984. Household gardens: theoretical considerations on an old survival strategy. International Potato Center, Lima, Peru, 43 pp.

Okely, J., 2001. Visualism and landscape: Looking and seeing in Normandy. Ethnos: Journal of Anthropology 66: 99-120.

Premat, A., 2005. Moving between the plan and the ground: Shifting perspectives on urban agriculture in Havana, Cuba. AGROPOLIS, The Social, Political and Environmental Dimensions of Urban Agriculture, Earthscan, London, UK.

Pudup, M.B., 2008. It takes a garden: Cultivating citizen-subjects in organised garden projects. Geoforum 39: 1228-1240.

Pugh, S., 1988. Garden _ nature _ language. Manchester University Press, Manchester, UK, 148 pp.

Ravetz, A., 1986. The government of space: town planning in modern society. Faber and Faber, London, UK, 154 pp.

Ravetz, A., 2001. Council housing and culture: the history of a social experiment. Routledge, London, UK, 262 pp.

Samuels, M.S., 1979. The Biography of landscape. In: Meinig, D.W. and Jackson, J.B. (Eds.) The Interpretation of ordinary landscapes: geographical essays. Oxford University Press, New York, NY, USA, pp. 51-88.

Sanyal, B., 1985. Urban agriculture: Who cultivates and why? a case-study of Lusaka, Zambia. Food and Nutrition Bulletin 7: 15-24.

Sharr, A., 2007. Heidegger for architects. Routledge, London, UK, 144 pp.

Spradley, J.P., 1980. Participant observation. Holt, Rinehart and Winston, New York, London, 285 pp.

Steel, C., 2008. Hungry city: How food shapes our lives. Chatto and Windus, London, Uk, 400 pp.

Strauss, A.L. and Corbin, J.M., 1997. Grounded theory in practice. SAGE, London, UK, 288 pp.

Szczelkun, S., 1993. The conspiracy of good taste: William Morris, Cecil Sharp, Clough Williams_Ellis and the repression of working class culture in the 20th century. Working Press, London, UK, 128 pp.

Thick, M., 1998. The neat house gardens: early market gardening around London. Prospect, London, UK, 175 pp.

Turner, J.F.C. and Fichter, R. (Eds.), 1972. Freedom to build; dweller control of the housing process. Macmillan, New York, NY, USA, 301 pp.

Vasey, D.E., 1985. Household gardens and their niche in Port Moresby, Papua New Guinea. Food and Nutrition Bulletin 7: 15-24.

Wakefield, S., Yeudall, F., Taron, C., Reynolds, J. and Skinner, A., 2007. Growing urban health: Community gardening in South-East Toronto. Health Promotion International 22: 92-101.

Warburton, N., 2005. Ernö Goldfinger: the life of an architect. Routledge, London, UK, 197 pp.

Wilhelm, G., Jr., 1975. Dooryard gardens and gardening in the black community in Brushy, Texas. Geographical Review 65: 73-92.

Wimpenny, P. and Gass, J., 2000. Interviewing in phenomenology and grounded theory: is there a difference? Journal of Advanced Nursing 31: 1485-1492.

Appendix 34.1

London sites studied in this chapter

'De Beauvoir gardening club'. De Beauvoir council estate, Downham Road, Hackney. Site description: several sites around the estate.

'Dirty Hands Project'. Valetta Grove, Plaistow. Site description: disused backetball pitch.

'Greening Brownfields'. Brownfields estate, Andrew Street, Tower Hamlets. Site description: disused tennis court designed into the estate.

'Haberdasher Tenants and Residents Association Gardening Club project'. Haberdasher estate, Haberdasher Street, Hackney. Site description: several sites around the estate.

'Lansbury Gardeners'. Hind Grove Community Hall, Hind Grove, Tower Hamlets. Site description: unused open space connected to community hall.

'St Johns Estate Community Kitchen Garden'. St Johns Estate, New North Road, hackney. Site description: large triangle of unused grass at the front of the estate.

Chapter 35

Food, homes and gardens: public community gardens potential for contributing to a more sustainable city

Carolin Mees[1] and Edie Stone[2]
[1]Institute for History and Theory of Design, Berlin University of Arts and GreenThumb, 49 Chambers Street, New York City, NY 10007, USA; [2]GreenThumb, Community Gardens Program of New York City, Department of Parks and Recreation, 49 Chambers Street, New York City, NY 10007, USA; mees.carolin@gmail.com

Abstract

Community gardens are multi-functional urban land uses that supplement traditional open spaces. They are critical for urban living in their contribution to recreation, food security and sustainable urban development. In New York City community gardens were first created by groups of residents during the 1970s; public spaces on vacated land, individually used for social gathering, to grow food and beautify neighbourhoods. These gardens have been maintained over the years in common management by resident groups, while being registered and overseen by GreenThumb, the City's community garden programme. Due to economic pressure on land the number of community gardens fluctuated during the last 30 years. Today there are about 600 gardens in the city, with over 300 gardens situated on public land. In recent years, in answer to climate change and the current economic crisis, specific elements within these gardens have been increasingly employed by gardeners to maximise food production, to conserve water or to deal with waste issues. The municipality has also begun to recognise the importance of maintaining sustainable gardens and their contribution to urban food security. Beehives, chicken coops, rainwater harvesting, and structures are examples of community garden components that were recently legalised in New York City, financially supported by NGOs or regulated in new ways. The chapter will look in detail at the employment and function of those garden components by employing methods of qualitative data analysis, discussion of secondary literature, and of participant observation to evaluate how the city's community gardens are performing in the interrelation of city and residents to prove their contribution to a sustainable 21st century city.

Keywords: land use, community gardens, open-space planning, urban density, low-income

35.1 Introduction: New York City's community gardens and urban agriculture development

Urban public gardening in New York City has a long tradition, starting with the 'top down' government organised and commonly used Victory and War Garden movements during the time of the Great Depression and the First and Second World Wars (Mees, 2010a). Community gardening activity picked up again during the economic crisis of the 1970s, then as a grassroots movement (Stone, 2009). The City of New York at the time did not object to this grassroots activism, but tolerated the voluntary efforts of local residents as a welcomed relief to the city's budget: Community gardens were started all over the city like the Garden of Happiness in the South Bronx, and run-down neighbourhoods were revived, when residents' groups regained social control over their neighbourhoods (Ferguson, 1999; Fox *et al.*,1980; Hassel, 2002; Mees, 2010; Stone, 2009) (Figure 35.1).

In 1978 New York City's Mayor Edward Koch created 'Operation GreenThumb' as a part of the Department of General Services to cope with this community garden movement and to regulate this unofficial use of public land. Utilizing Federal Community Development Block Grant money, the programme – now called 'GreenThumb' and part of the Department of Parks and Recreation – continues today as the official programme of the City that licenses public land for community gardening and provides materials and technical assistance to

Figure 35.1. Garden of happiness in the South Bronx (Mees, 2009).

gardeners. The number of public gardens in the city has changed considerably over the last 30 years: Especially during the 1990s economic boom the City administration under Mayor Rudolph Giuliani demolished many gardens in favour of housing. In response, the gardeners organised themselves better over the years – citywide and within their neighbourhoods (Ferguson, 1999; Fox *et al.*, 1980; Hassel, 2002, Mees, 2010b, Stone, 2009). Thus community garden groups surveyed by GreenThumb in 2003 reported high levels of participation in community-improvement, political and social activities not related to the garden space. In addition over 50 percent of the gardeners interviewed indicated that they were involved with block associations and community boards (Stone, 2009).

The gardeners' organisational activity forced the city government to reconsider their stance towards community gardening on public land: In September 2002 the Community Gardens Settlement was published, declaring which gardens in danger of development were now to be made 'permanent'. For the eight-year term of the settlement, this legal agreement governed all municipal regulation of community gardens (Office of the Attorney General, 2010).

Then in October 2010 the first New York City community garden rules came in effect to replace the terminating settlement. These rules guarantee the same protections for community gardens situated on public land that were granted by the 2002 Attorney General's settlement, i.e. stating that gardens under Parks Department's jurisdiction are preserved as gardens as long as they registered with and became licensed by GreenThumb. In addition the rules demand active and responsible use of community garden land, thus including a social aspect of gardening that is meant to prevent privatisation and to secure the public use of the city-owned land (New York City Community Garden Rules, 2010: http://www.nycgovparks. org/sub_about/rules_and_regulations/rr_6.html#licenses).

Expecting the population of New York City to grow up to 10 million by 2030, Mayor Bloomberg announced the PlaNYC 2030 initiative in December 2006 targeting environmental concerns. A startling omission from this version of the PlaNYC report was any discussion of food security or the provision of community gardens in every neighbourhood. PlaNYC 2030 instead focused, in a top down manner, on the provision of recreation opportunities in a public open space within 10 minutes of every residence in a neighbourhood. (PlaNYC 2030, 2011) This omission was recognised by multiple NGOs and political figures interested in food aspects of public health. This inspired the release of several reports focussing on the importance of improving food security in the city including the FoodWorks New York initiative by City Council Speaker Christine Quinn, which includes community gardens as a potential important contributor to healthy food access (FoodWorks, http://council.nyc.gov/ html/action_center/food.shtml).

In addition the New York City Community Garden Coalition (NYCCGC) made a focussed effort to make sure that gardeners participated in 'Community Conversations' held by the Mayor in every City borough to gather concerns of New York residents for the proposed update to PlaNYC. They demanded further discussion of community gardens in PlaNYC

2.0. This last update to PlaNYC 2030, published April 2011, mentions objectives including to 'review existing regulations and laws to identify and remove unnecessary barriers to the creation of community gardens and urban farms' and to 'develop a multi-faceted strategy to increase access to affordable and healthy foods and reduce the environmental and climate impacts of food production, distribution, consumption, and disposal' (A Greener, Greater New York City, 2011). Steps to achieving these objectives are outlined in the plan as well: identification of potential new urban agriculture or community garden sites on public land, planting of 129 new community gardens and creation of an urban farm on sites of the New York City Housing Authority, establishment of five additional farmers markets and an increase in the number of community volunteers registered with GreenThumb by 25% to expand support for community gardens into new underserved neighbourhoods (A Greener, Greater New York City, 2011). Therefore, for the first time in the history of community gardening and urban agriculture in New York City, there is a comprehensive plan stating the city's commitment to facilitate urban agriculture and community gardening in upcoming decades.

35.2 Discussion: a community garden's components contribute to the generation of a sustainable urban environment

According to the City's zoning resolution generally in zones allowing 'open uses', as well as 'agricultural uses', including greenhouses, nurseries, or truck gardens, (is allowed) provided that no offensive odours or dust are created, and that there is no sale of products not produced on the same lot' (New York City zoning ordinance Section 22-14 B). Each of these specific garden components are chosen by a gardening group according to the members' preferences as well as the neighbourhoods' needs. These elements are constructed over time at a garden site, so that the layout of a garden is in most cases an organic addition of these elements as needed and as resources allow rather than the result of a formal design process. In order to create more sustainable urban gardens, GreenThumb and partner NGOs offer extensive education to community gardeners about gardening techniques and the various elements available for garden use. In addition grants and assistance with construction are often provided (GreenThumb, 2011).

According to a report by the United States Department of Agriculture issued in November 2009, 14.6% of American families were food insecure at least some time during the year 2008, and prevalence rates of food insecurity and very low food security were the highest recorded since 1995 when the first national food security survey by the United States Department of Agriculture was conducted. (Nord *et al.*, 2008) At the same time in New York City 275 of 311 community gardens surveyed by GreenThumb and situated on public land, i.e. 84% of public gardens were growing fruits and vegetables (Survey of GreenThumb community gardens programme, New York City's Department of Recreation, 2009). The majority of the gardens surveyed reported that at least 50% of the garden's area was devoted to food production. Of the about 600 community gardens in New York City today, about 400 are situated on public land (unpublished data collected as a part of GreenThumb garden registration, 2011) (Figure 35.2).

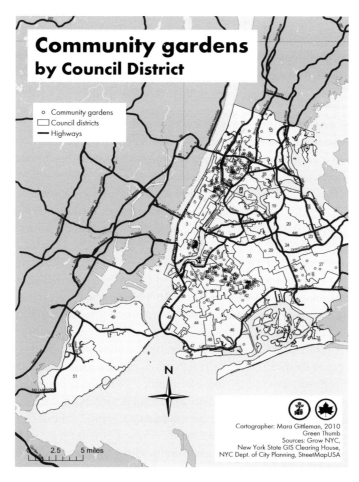

Figure 35.2. GreenThumb community gardens map.

An accurate calculation of the number of acres of public land devoted to community gardens and urban agriculture has yet to be determined due to inconsistent record keeping methodology and mapping techniques across the different City agencies holding these land parcels. A calculation can be made, however based on Parks Department managed community gardens that are accurately mapped by the agency. Parks' community gardens occupy 45.8 acres. Based on the 50% figure reported by community gardeners as area devoted to food production, one could estimate that 22.9 acres of public land under Parks management are devoted to food production (unpublished data provided by Department of Parks and Recreation Parklands Division). Based on unpublished data collected by GreenThumb staff and other NGOs, the percentage of gardens under other City agencies jurisdiction devoted to food production is likely to be even higher than 50%.

This is not a lot of acreage on a commercial agricultural scale, but by utilising intensive growing practices such as square foot gardening – a practice of planning small but intensively planted gardens by which harvests per foot (0.3 m) of garden are increased due to the rich soil mixture, well-spaced plants, and prevention of weeds – and by employing other elements in order to make more effective use of a small growing space, community gardens are producing a surprisingly large quantity of food (Bartholomew, 1981).

Yields of the food production in New York City's community gardens are currently being quantified by the Farming Concrete project supported by GreenThumb.

> In 2010, 110 community gardeners weighed their harvests and crop inventory was conducted in 67 community gardens as part of Farming Concrete's efforts. We used the average yields from the gardeners who weighed their harvests to estimate that in those 67 gardens, 87,700 lbs of fresh produce was grown on just 1.7 acres, worth more than $200,000 (farming concrete, 2011).

This citizens' science project is the first of its kind in New York City (farming concrete, 2011) and its results are mapped and made available to the public continuously (Garden Maps, 2011, http://gardenmaps.org). As this was the first year of a pilot project, the 2010 yield data are approximated; therefore the actual amount of fruits, vegetables, eggs, honey, herbs and other food products on land used for community gardening is currently not known. The farming concrete project plans to collect data using the same methodology for the next two years in order to generate more accurate results (farming concrete, 2011). GreenThumb and other NGO staff report a general increase in public desire for urban food growing space, demonstrated by the average 5 persons daily inquiring at GreenThumb's office looking for a space to garden in the city, mainly citing the desire to grow food (unpublished data).

The vegetables grown most often in community gardens registered with GreenThumb are tomatoes, which are grown in 94.7% of about 150 gardens. Herbs like basil (87.3%) and mint (80%) are favoured slightly over collard greens (78.9%), lettuce (77.6%), sweet peppers (77.2%), beans (76%), eggplant (73.1%), parsley (71.5%) and cucumbers (69%). The fruits most often grown in GreenThumb's community gardens were apples and peaches: There were 128 apple trees and 102 peach trees in about 70 community gardens (Gittleman *et al.*, 2010).

The location of available food sources is often conversely related to dense urban population areas. In low-income areas government subsidised housing towers prevail in built environments and an expectation of low profitability in these areas results in development of few supermarkets. Community gardens, which are more prevalent in these low-income areas due to availability of land, provide additional and accessible sources of food. The significance of this relationship is supported by the fact that the New York City Housing Authority is currently seeking Federal funding to promote the creation of food producing gardens at the housing developments it manages (Figure 35.3).

It appears from initial results generated by farming concrete, that community gardens are an important component of New York City's food security network in the context of 'Community food security (CFS) (which) exists when all community residents obtain a safe, culturally

Figure 35.3. East New York farm (Just Food, www.justfood.org).

acceptable, nutritionally adequate diet through a sustainable food system that maximises community self-reliance and social justice' (Bellows and Hamm, 2003). The gardens contributions to food security, as recently recognised in FoodWorks, PlaNYC, and other New York City government documents, are particularly significant due to the geographic concentration in low income, food insecure neighbourhoods (FoodWorks; PlaNYC). This concentration is likely a result of a combination of factors, including increased need in these areas and perhaps more significantly the restrictions placed on GreenThumb's funding, which require that the programme can only operate in census districts where a large percentage of residents fall under the federally determined poverty line (United States Department of Housing and Urban Development, 2010).

35.3 Soil: safety and management

In order to use public land for gardening, a group of at least ten residents has to apply at GreenThumb for a license. If the land is available and granted, the first step after licensing it for use as a garden is to make sure – for both the protection of both the gardeners and the City – that the soil is free of lead and other contaminants. Because most public garden sites in New York were once built up with residential buildings that often contained lead paint, GreenThumb assumes that most garden soil is contaminated. In order to facilitate safe food production, the programme provides clean topsoil and lumber to create raised beds. Ironically the construction of raised beds is less expensive than testing the soil and remediating it in most cases (unpublished data). In order to help determine the extent of contamination and the efficacy of using raised beds to counter potential exposure to contaminants, GreenThumb has participated since March 2010 in a four year National Institute of Health (NIH) funded study to assess the extent and effects of soil contamination in urban gardens. The goal of the 'Healthy Soils Healthy Communities' study is to provide programmes and resources to more fully address the questions and concerns that gardeners and others have about interpreting soil test results, managing gardens and soils, and protecting public health (Healthy Soils Healthy Communities, 2010). Beds are usually built to be no more than four feet (1.2 m) wide to facilitate access to plants from the sides without stepping into the planted area. At a length of between eight and twelve feet (2.4-3.6 m) a typical gardening bed allows one person to produce (when applying an intensive method of cultivation), a high amount of produce in a relatively short time. Before starting cultivation, the soil should be amended with compost or manure, collectable at a garden site, to increase its organic matter content for improved fertility to produce more food during a season (Figure 35.4). Research has also demonstrated that increasing organic matters can bind heavy metal particles and reduce uptake of contaminants into food plants (Puschenreiter *et al.*, 2005). Initial results of the NIH study indicate that utilisation of raised beds can significantly reduce exposure to lead (Lang, 2010).

Figure 35.4. Composting in Garden of Union, Brooklyn (Rasheed Hislop, 2010).

35.4 Composting and waste stream reduction

The average New York City household discards two pounds of organic waste each day, adding up to more than one million tonnes of organic material a year (New York City Department of Sanitation). This 'waste', when composted into new soil, is a potential resource rather than an economic liability for the city. The New York City Department of Sanitation also collects fall leaves for composting citywide and has set up some pilot projects to recycle additional organic material through composting. During the Department's 'Compost Giveback' events New York City residents can get unlimited amounts of free, finished compost. Still, in order to independently improving the quality of their garden's soil, the city's community gardeners usually include a composting area in their garden situated away from housing or sitting areas because of odours. Training is available for low cost through the city's two Botanical gardens, which enables community gardeners and others to become 'master composters'. As a part of this training, participants must provide several hours of community service helping others to develop or improve composting areas in gardens or at schools. Out of 137 GreenThumb community gardens recently surveyed, 65.6% had a composting area in the garden. (Gittleman *et al.*, 2010) Unfortunately since the solid waste management is considered the purview of the Department of Sanitation and community gardens are recognised by the city primarily as recreational land uses provided by the Parks Department, studies of the potenital contribution to waste stream reduction by composting in community gardens have not been conducted.

35.5 Storage and shelter = the garden house

Even though there are various official regulations regarding the employment of specific garden elements in community gardens, the City has always been especially interested in regulating the construction of self-built structures. Since the first GreenThumb leases were issued in 1978, gardeners have had to agree to observe certain rules. For example, the garden lease could in the past 'be terminated if there is an 'illegal structure' on the property, an ambiguous term defined as 'any enclosed structure' (Sciorra, 1996).

There are however small garden houses on a great number of community gardens, especially on those where a majority of the members a garden group have a cultural background in the Caribbean, Latin America or Puerto Rico. They call their garden houses 'casitas' and use them in the way the word implies, i.e. translated, as little houses. 'Casita interior space is furnished with many of the comforts of home; a table and chairs, a couch, and even a television. A number are outfitted with a small but operational kitchen complete with a refrigerator, running water and a working stove' (Mees and Stone, 2009; Sciorra, 1996).

To the City's administration these uses of the 'casita' make the small building not only a symbol of urban poverty (Sciorra, 1996: 76), but also a sign of the privatisation of public space. City officials consequently are interested in imposing rules in regard to the use of community garden land and the elements placed on the land. Their argument is based on the safety and liability standards applied to non-community managed parkland. These standards are often not met in a community garden – and in fact often cannot be met due to the nature of community gardens' grassroots and self-funded development.

More than ten years after, 'as a result of increased media attention, city hall pressured GreenThumb to develop an officially-sanctioned, standardised, open-air structure ...' (Sciorra, 1996) the New York City Building Department's Technical Affairs and Borough Commissioners issued in February 2006 guidelines concerning the construction of structures in community gardens. These garden structure regulations can be found online in the *GreenThumb Gardeners' Handbook* (GreenThumb Gardeners' Handbook, 2011). They specify that the roofed area of a structure has to be a maximum of 150 square feet (about 14 m²), the height is restricted to a maximum height of 10 feet (about three metres) and that a distance of six feet (about 1.80 metres) from all property lot lines must be maintained. In addition, the enclosure has to be optically permeable. GreenThumb added additional guidelines of items that cannot be kept in garden structures like for example space heaters, kerosene stoves or any other heating unit, household furniture such as couches, televisions or non-portable electronic devices like stereo equipment, barbecue grills and refrigerators permanently connected to electric power, etc. – i.e. any item that indicates possible use of the structure as a dwelling space.

To facilitate self-built and self-designed casita-type structures remaining a part of the urban landscape of New York City, GreenThumb developed with casita builders and architects a prototype of a self-buildable and low cost structure meeting building department guidelines. The building typology of the new wooden shed or 'Gardenhaus' is based on the structural traditions found in New York City's community gardens to reveal the cultural background of community gardeners. Other building typologies incorporated into the Gardenhaus' design are the stage and the gazebo, which are both common elements in a community gardens' public landscape (Figure 35.5). The Gardenhaus has been designed to allow adaptation to various sites and individualisation by the gardeners.

A guidebook instructing gardeners how to build acceptable structures by depicting the process of a Gardenhaus' construction, showing drawings as well as a material list can be downloaded from GreenThumb's webpage (www.greenthumbnyc.org). GreenThumb provides most of the materials needed for construction of garden houses, and encourages self-reliance and skill building in the gardens by providing construction plans instead of labour. The importance of having this type of structure in a community garden for storage, shelter in bad weather and community gatherings is illustrated by the fact that of 208 community gardens surveyed by GreenThumb and GrowNYC in 2009, over forty percent reported having a gazebo/casita in their garden. (Gittleman *et al.*, 2010)

35.6 Electricity and solar panels

Since New York City based commercial electricity providers do not normally provide electricity for garden sites, it has been an illegal but common practice since the 1970s to

Figure 35.5. Gardenhaus prototype, Vogue Garden, Bronx (Mees, 2008).

tap a light pole in the vicinity of the garden in order to obtain electricity (unpublished GreenThumb data). Some gardeners are able to make a deal with an adjacent neighbour to obtain electricity. There are not many community gardeners making use of solar panels due to the fact that the acquisition and installation of solar panels with adequate capacity cost an average of about $400, often too expensive for low-income gardening groups. In addition GreenThumb and other community garden support groups do not have funds yet to provide gardens with grants for solar power. The lack of funding is partially due to the former 'interim' status of all community gardens as a land use, which obviously raises the question of the sense in investing in expensive, long-term improvements to temporary gardens.

Nevertheless, in more gentrified neighbourhoods like the East Village, some gardening groups have invested in solar power since the 2002 Settlement granted their gardens protected status under Parks jurisdiction. For example the 6BC Botanical Garden at East Sixth Street between Avenues B and C, features a waterfall powered by a total of six solar panels – four mounted on a grape arbour and two on a garden shed's roof. The solar facility is not only moving water but allows for charging garden power tools, such as grass trimmers, and to power a loud speaker and lighting when holding cultural events. Although in this case the solar powered pond is used for ornamental fish and plants could just as easily be used to support aquaculture for food production.

35.7 Water access and rainwater harvesting systems

Community gardeners in New York City at first illegally accessed municipal fire hydrants for water. Today GreenThumb is in charge of distributing hydrant access permits approved by the Department of Environmental Protection and handing out hydrant wrenches and hose adaptors (GreenThumb Gardeners' Handbook, 2011). Some community gardens pay a fee to owners or tenants of adjacent buildings to make use of the neighbours' water. When a drought in the summer of 2001 caused the Department of Environmental Protection to restrict community gardeners' access to hydrants, the non-profit organisation GrowNYC and GreenThumb founded the Water Resources Group to promote the idea of water conservation through the installation of rainwater harvesting systems in community gardens (Figure 35.6). Today there are 55 of these systems in gardens citywide. With New York City receiving an average of about 40 to 50 inches of rain a year, some gardens are able to collect and store over 7,000 gallons of water (GrowNYC, 2010).

Because the topography of different gardens is so variable, each collection system has to be custom designed to suit the individual site. In general the tanks are installed next to an adjacent building or a shade structure or garden house for the collection of runoff rain water from the roof, and fixed with metal strapping to remain stable in a storm. Prices vary according to the design and size of the tank. The size of the garden determines the minimum size of the tank: To water all plants in a garden that is smaller than the average garden size of fifty feet (15m) by one hundred feet (30m) may only need a one hundred and sixty five

Figure 35.6. Rainwater harvest system, Garden of Happiness (Mees, 2008).

gallon (625 l) tank, and if the garden lot is larger in size a thousand gallon (3,785 l) tank can be installed. The minimum standard to collect rainwater in community gardens is a fifty five gallon (208 l) rain barrel. Recycled barrels once used for shipping olives and other food products are often utilised to save money (GrowNYC, 2010).

35.8 Season extension = the greenhouse

It is common knowledge that the practice of starting vegetables and flowers in greenhouses in late winter and early spring, and then transplanting outside as soon as the weather warms, is especially useful for growing traditional and culturally important plants that are native to, or common in warmer climates. In addition crops grown in greenhouses are protected from excesses of heat or cold, inclement weather and pests. Light and temperature control in greenhouses allows for easily turning unproductive into arable land. Hydroponics can be used in greenhouses to create additional independence from the quality of the local soil. For many community gardeners however a greenhouse is too costly and also requires a high level of maintenance (Figure 35.7).

Although GreenThumb has at times provided greenhouses to community gardens, the cost of their acquisition and manpower required for installation limited the number distributed to under 20 citywide (unpublished GreenThumb data). The closed environment of a greenhouse has its own unique requirements, compared with outdoor production. Pollination of plants has to be directed for example, so that often bumblebees are introduced as pollinators. Pests

Figure 35.7. Greenhouse in the Garden of Happiness (Mees, 2009).

443

and diseases, and extremes of heat and humidity, have to be contained, and irrigation has to be provided year-round. In addition the temperature and humidity of greenhouses must be constantly monitored to ensure optimal climatic conditions for the plants cultivated, particularly with winter production of warm-weather vegetables.

A small greenhouse known as a cold frame is widely used in community gardens because it takes up less space than a greenhouse and is easy to install and low cost since it can be constructed from found materials such as discarded windows. In New York City cold frames are employed by many community gardeners to grow lettuces and other early spring crops before the danger of frost has fully passed.

35.9 Adding proteins via the chicken coop

Even before the renewed interest in combating climate change through local cultivation of food, New York City residents with cultural backgrounds in Puerto Rico, the Dominican Republic, or other places in the Caribbean as well as from the Southern part of the United States traditionally kept chickens in their gardens for reasons of subsistence and food security (Interview with Jose Soto; Mees, 2010). In general, keeping chickens in a community garden enhances the food supply with meat and fresh eggs and improves the quality of the soil. Chickens aerate soil through scratching, remove pests they consume, and recycle kitchen

Figure 35.8. Chicken keeping in New York City (www.justfood.org).

and garden scraps. Today many community gardeners are attending workshops provided by GreenThumb to learn on how to build chicken coops and how to keep chickens (Just Food, 2010: http: //www.justfood.org) (Figure 35.8).

Keeping chickens in the City of New York is legal. According to the Health Code of the City of New York roosters, ducks, geese and turkeys are not allowed in any built-up portion of the city (New York Health Code § 161.01b[11] and 161.19) – hens, however, are classified as pets and can be kept. For community gardeners adding chickens requires obtaining the consent of GreenThumb and the garden's neighbours before constructing the coops, which must be located more than 25 feet from an inhabited building. By law chicken coops have to be kept clean and are to be whitewashed or otherwise treated in a manner approved by the Department of Health at least once a year (New York Health Code § 161.01b[11] and 161.19).

New York State and Federal Department of Agriculture laws apply to the sale of eggs and meat. The number of chickens that can be kept in New York City is almost unlimited – only producers with 3,000 birds or more, or anyone packing eggs from hens other than their own for sale have to register with the Federal government (U.S. Code Title 21, Chapter 15: Egg Products Inspection Act Section 1044, 1970). Still, the more chickens that are raised in a garden the more likely it is that flies and vermin are attracted, foul smells or excessive noise created, i.e. 'nuisance conditions', which again are illegal (New York Health Code § 161.11 Prevention of nuisances; cleaning).

In general in residential districts and commercial districts sale of agricultural products that were produced on your lot is permitted (New York City Zoning Resolution Sections 22-14, 32-13, 42-11). According to regulations outlined in the community garden license issued by New York City agencies through GreenThumb, sale of eggs and other agricultural products directly to consumers from the garden site is allowed as long as profits from sales are reinvested in garden maintenance. Because of State and Federal requirements that meat for sale must be processed in licensed slaughterhouses, however, meat from community garden raised chickens may be consumed by the gardeners but not sold. To make keeping chickens in New York City easier, the non-profit organisation Just Food set up a 'City Chicken Project' to provide training in raising chickens and in building coops (Just Food, 2010: http: //www. justfood.org). Still, out of 208 community gardens surveyed in 2009, only 3.8% have chicken coops (Gittleman et al., 2010).

The construction of a chicken coop in a community garden is again regulated by the New York City Building Department's guidelines concerning the construction of structures in community gardens (GreenThumb Gardeners' Handbook, 2011). Thus a chicken coop can be built without obtaining a building permit as long as it is a temporary structure and complies with the rules in regard to dimensions and distance from lot lines. In addition coops should be covered on top to protect against wild animals, contamination with avian flu by wild birds, and vandalism. Healthy chickens however require access to the outside. Construction should provide for ventilation, should be equipped with perches and nesting boxes and has to be easy to walk into and clean. The coop needs to be cleaned and water, food and bedding changed about once a week (Just Food, 2010).

Keeping chickens for egg production in urban areas has implications for human heath since it is providing adequate nutrition in areas underserved by food sources. In the US, 97% of eggs produced (60%world-wide) are estimated to come from battery cages (typically six birds housed together in a small 'battery cage', usually about eight by eight inches (20 cm) (Berger, 2010). The current recall of over 300 million eggs due to Salmonella infection illustrates a simple point that food safety advocates have been making for years – diseases can spread much more easily between animals crammed into tight quarters like battery cages (Neumann, 2010).

The Humane Society of the U.S. says that nine studies in the last five years have found that salmonella is anywhere from three to 50 times more likely in caged-hen facilities as opposed to those where the hens are cage-free (Berger, 2010). But it is not just the egg-laying hens that are consolidated – it is also the production facilities. Five states produce half of the U.S's eggs, a fact that explains how a problem at a facility in one county in Iowa has made people all over the country sick. The salmonella-tainted eggs, for instance, while originating from just two farms in Iowa, were distributed by at least 36 different brands to over 14 states (Berger, 2010). Food security will be increased in urban areas if keeping of chickens in community gardens is promoted and if acceptance of potential related nuisances increases among urban residents, thereby removing the main obstacle to pro-livestock legislation.

35.10 Pollinators and added value products via the beehive

The perspective of community gardeners on the one hand and the government of New York City on the other on increasing food production by keeping animals is still very much at odds in most cases, but as the example of development of bee keeping proves, there has been gradual change towards more acceptance of the gardeners actions.

In general, keeping bees in a garden benefits food production, since bees are necessary for the pollination of plants (Figure 35.9).

The production of honey per year varies according to a number of factors like the number of bees, their, the availability of nectar-producing flowers, and weather, but in a typical year, it is possible for one hive to produce around 50 pounds of honey in a good season (British Bee Keepers Association). The addition of honeybees adds to the variety of food provided by a garden and also allows the production of wax candles and soaps, all of which can be sold with high profitability at local farmers markets. The income provided by selling these products can be used to supplement the typically low-income gardener's family food budget or be reinvested in garden infrastructure.

Figure 35.9. Beehive in Joseph Daniel Wilson Memorial Garden Harlem (Eric Frey, 2010).

But, even though it is an obvious improvement to food security in the urban environments, under the Health Code of New York City beekeeping in the New York City was illegal for more than ten years, from 1999 until March 2010. In 1999, former Mayor Rudolph Giuliani introduced a new law that banned the keeping of wild animals. In this context honeybees were listed together with hornets and wasps as venomous insects in the health code's list of prohibited and wild animals, considered to be 'naturally inclined to do harm' (New York City City Health Code's §161.01). The city's Health Department fine for keeping bees in the city was $2,000. Hundreds of New Yorkers however defied the ban and maintained hives, founding the New York Beekeepers Association (http://www.britishbee.org.uk/faq.php) to advocate for legislation once more legalising the now outlawed hives.

After beekeepers in the US reported 30 to 90% losses of bees, a phenomenon called Colony Collapse Disorder, during the winter 2006-2007, the US Environmental Protection Agency officially acknowledged the value of beekeeping. 'Honeybees pollinate about one-third of the human diet, and pollination is responsible for $15 billion in added crop value' (US Environmental Protection Agency, 2010). The New York City Department of Health held a public hearing on February 3, 2010, regarding its proposed amendments to Health Code Article 161.01, which included a revision that would allow the legal keeping of honeybees. Nearly 80 individuals and organisations submitted testimony in support of lifting the ban (Just Food, 2010). On March 16, 2010, the Board of Health voted in favour of the Department's proposed amendment, and the revised code made beekeeping legal once more in New York City. Beekeepers currently have to file a notice with the Department of Health and Mental Hygiene including their contact information and their beehive's location. In addition they have to adhere to appropriate beekeeping practices including taking care that their beehives are placed in a way that prevents the bees from attacking passers by or neighbours.

35.11 Landscape architecture and community design in the layout of the garden

In New York City community gardens on public land are registered mainly through GreenThumb at the Department of Parks and Recreation, but are managed by any group of a minimum of ten residents. That means the gardeners decide themselves on how to make use of the land, on the garden's design and set-up. The placement of different modules in a garden thereby determines the intensity of its spatial use for social and agricultural purposes, i.e. the benefits and the effectiveness of this garden for the neighbourhood and the city in general. But the gardeners are not only interested in the effective food production and social context, but also in expressing their individual and cultural identity through the creation of their garden. The design of community gardens is consequently a participatory design process.

When a group of residents starts a community garden, first the site should be measured in order to make a simple, to-scale site map as a basis for discussion. Usually two or three garden design meetings should be held to generate ideas and to visualise the design by defining zones for placement of different garden modules. During the meetings these zones can be moved around on the map as the group discusses various possibilities to start their garden's layout. When the group's decisions are recorded it is ensured that decisions made can be communicated to others, so that the progress will not be slowed. In general a garden plot is assigned to each community gardener and placed in the sunniest area of the garden.

But the placement of the other modules is part of a design process. With the changing members of a gardening group the design preferences of the group are altered over time, so that the placement of a module will be discussed several times in different group constellations, before its actual construction begins.

In addition the economic situation of the neighbourhood alters over time, so that the demand for space for the production of food or for recreation and regeneration is more or less strong over the years. Consequently a community garden has to be and remain flexible in its layout in order to provide a privately usable public open space for diverse groups of residents over time, i.e. for gardeners of various cultural backgrounds and preferences as well as for different social and economic requirements.

35.12 Garden layout and productivity = the community garden and community farm

There are many examples of community gardens primarily used for agricultural purposes. One example is the Taqwa Community Farm (Figure 35.10).

This garden was started in 1992 and is situated at W 164 Street in the Highbridge neighbourhood in the Western part of the Bronx. A hydroponic system – a method of growing plants using mineral nutrient solutions, in water, without soil – was installed in some areas of the garden to cultivate fruit and vegetables independently from the condition of the soil. On the two-acre garden lot some 30 to 35 gardeners cultivate an average more than 2,000 pounds of food per year. Consequently more is produced than consumed by gardeners themselves and the surplus is sold at an on-site farmer's market as well as donated to the 'Grow and Give table' where fresh fruits and vegetables are offered to those in need.

In addition to a large area with individual and community-managed garden beds this garden contains about 40 fruit trees, a chicken coop with a flock of 13 chickens as well as a seating area overlooked by a children's mural. In addition rainwater is collected from the roof of the

Figure 35.10. Taqwa Community Farm (Rasheed Hislop, 2009).

adjacent three-story apartment building and stored in a 1000-gallon (3,785 l) polyethylene tank. The gardeners built the system under the supervision of GrowNYC staff to collect water from the adjacent building. They estimated that the system has the potential to collect 17,000 gallons for the season from 1000 square feet of roof collecting area (http://www.grownyc.org/openspace/rwh/taqwa).

35.13 Conclusion: the context of housing, food production and commonly managed open spaces for sustainable urban development

To sustain urban residents during an economic crisis, quality of life must be maintained, especially in densely populated low-income districts. Commonly managed and privately used public open spaces, in short community gardens, provide a low cost and effective means of providing open space thereby improving quality of life in neighbourhoods. It is equally important however to recognise the contribution that maximising food production on urban land makes to maintaining quality of life via fresh food access, hunger reduction and resulting improvements to public health and well being. For this purpose, garden modules, as discussed above, are commonly introduced in New York City's community gardens by the gardeners themselves.

If a garden's components are to match the needs of the neighbourhood, the modules must be installed based on the collective decision of the gardening group that commonly manages the public land as a garden. Therefore it is necessary to allow for components to be installed, rearranged, omitted and edited as the changing composition and needs of the group determines and alters the land use within the garden over time.

Despite the demonstrated benefits of this 'organic' design approach, the way New York City's administration – and some of the private non-profit gardening organisations – are dealing with community gardens in general and with garden elements like 'casitas' specifically, reveals the on going tendency of top down garden management groups and governments to gradually transform community gardens from privately used, public grassroots-landscapes into more formally constructed and used public open spaces.

The recent recognition by segments of New York City government of the import role community gardens play in sustaining people during a weak economy may help to alleviate the inherent tension between the municipal need to control the use of its land and the increase in garden sustainability resulting from a more 'organic' design approach. Ideally this will be followed by the preservation of the gardens as community managed self-sustainable infrastructural facilities. Whether this can be accomplished by for example the creation of a specific community garden zoning, needs to be proven through further research. Studies need to further consider the connected nature of community gardens in neighbourhoods and as citywide coalitions as well as the effect of their concentration in economically underprivileged city districts.

For now, the prevailing argument in the discussion of community gardens in government circles is that they are too small in area to provide a significant contribution to food security, environmental or social justice or to be an essential component of the sustainable 21st century city. Current on going research, which acknowledges these connections and reviews community gardens' contributions as a whole will provide important data towards disproving this unsubstantiated assumption.

References

Bartholomew, M., 1981. Square foot gardening: A new way to garden in less space with less work. Rodale Press, Emmaus, PA, USA 352 pp.

Bellows A.C. and Hamm, M.W., 2003. International origins of community food security policies and practices. United States Critical Public Health, Special Issue: Food Policy. 13: 107-123.

Berger, M.O., 2010. US: Salmonella outbreak tied to factory farming. Inter Press Service, August 27, 2010. Available at http://ipsnews.net/newsTVE.asp?idnews=52643.

Farming concrete, 2011. Harvest 2010 Report. Available at http://farmingconcrete.org/2011/04/19/2010-harvest-report.

Ferguson, S., 1999. The death of little Puerto Rico. New Village Journal 1. Available at www.newvillage.net/Journal/Issue1/1sacredcommon.html.

Fox, T., Koeppel, I. and Kellam, S., 1980. Struggle for space: The greening of New York City, 1970-1984. Neighborhood Open Space Coalition, New York, NY, USA, 165 pp.

Gittleman, M., Librizzi, L. and Stone, E., 2010. New York City community gardens survey – Results 2009/2010. GrowNYC and GreenThumb (Eds.). Available at http://www.greenthumbnyc.org/pdf/GrowNYC_community_garden_report.pdf.

GreenThumb, 2011, GreenThumb Gardeners' Handbook. City of New York, Department of Parks and Recreation, New York, NY, USA. Available at http://www.greenthumbnyc.org/pdf/gardeners_handbook.pdf.

GrowNYC, 2010. Rainwater harvest system at Taqwa farm. GrowNYC, New York, NY, USA. Available at http://www.grownyc.org/openspace/rwh/taqwa.

Healthy Soils, Healthy Communities, 2010. A research and education partnership with urban gardeners. Cornell Waste Management Institute. Cornell University, Ithica, NY, USA. Available at http://cwmi.css.cornell.edu/healthysoils.htm.

Lang, S., 2010. The dirt on urban gardens: Some contamination but help is on the way. ChronicleOnline, Cornell University, Ithica, NY, USA. Available at http://www.news.cornell.edu/stories/Dec10/NYCSoils.html.

Mees, C. and Stone, E., 2009. Preserving community gardens in NYC: Strategy in public space development? Booklet of the Conference 'X-LArch III, Landscape – Great Idea!', University of Natural Resources and Applied Life Sciences, Vienna, Austria, pp. 82-87.

Mees, C., 2010a. A public garden per resident? The socio-economic context of homes and gardens in the inner city. Acta Horticulturae 881, International Society for Horticultural Science. Leuven, Belgium.

Mees, C., 2010b. Community gardens in New York City: Privat-gemeinschaftlich genutzte öffentliche Gärten für innerstädtischen Wohnraum im Freien. In: Gröning, G. and Hennecke, S. (Eds.) Kunst-Garten-Kultur. Dietrich-Reimer Verlag, Berlin, Germany.

Neumann, W., 2010. Egg recall expanded after Salmonella outbreak. New York Times, August 18, 2010. Available at http://www.nytimes.com/2010/08/19/business/19eggs.html.

Nord, M., Andrews, M. and Carlson, S., 2008. Measuring food security in the United States. United States Department of Agriculture, Economic Research Report. Available at http://www.ers.usda.gov/publications/err83/err83.pdf.

Office of the Attorney General, 2010, Community Gardens Settlement. Office of the Attorney General, Albany, NY, USA. Available at http://www.ag.ny.gov/media_center/2002/sep/sep18a_02.html.

PlaNYC 2030, 2011. A greener, greater New York City. Available at http://www.nyc.gov/html/planyc2030/html/home/home.shtml.

Puschenreiter, M., Horak, O., Friesl, W. and Hartl, W., 2005. Low-cost agricultural measures to reduce heavy metal transfer into the food chain – a review. Plant Soil Environment 51: 1-11.

Sciorra, J., 1996. Return to the future: Puerto Rican vernacular architecture. In: King, A.D. (Ed.) Re-presenting the City: Ethnicity, capital, and culture in the 21st-century metropolis. New York University Press, New York, NY, USA, pp. 60-84.

Stone, E., 2009. The Benefits of community-managed open space: Community gardening in New York City. In: Campbell, L. and Wiesen, A. (Eds.) Restorative commons: creating health and well-being through urban landscapes. United States Forest Service Northern Research Station, New York, USA, pp. 122-137.

Stone, E., 2000. Community gardening in New York city becomes a political movement. Paper presented at the conference 'Perspectives of small-scale farming in urban and rural areas – about the social and ecological necessity of gardens and informal agriculture', Berlin, Germany.

United States Department of Housing and Urban Development, 2010. Community Development Block Grant Program. Available at http://www.hud.gov/offices/cpd/communitydevelopment/programs/.

United States Environmental Protection Agency, 2010. Pesticide issues in the works: Honeybee colony collapse disorder. Available at http://www.epa.gov/opp00001/about/intheworks/honeybee.htm, accessed 16 May, 2011.

Von Hassel, M., 2002. The struggle for Eden. Community gardens in NYC. Bergin and Garvey, Westport, Connecticut, USA, 183 pp.

Chapter 36

How food secure can British cities become?

Howard C. Lee
Hadlow College, Hadlow, Tonbridge, TN11 0AL, United Kingdom; howard.lee@hadlow.ac.uk

Abstract

An accompanying chapter in this book (Van de Valk and Broekhof) has reviewed the gulf between alternative (local) food production systems and global industrial agriculture. This chapter considers the pressures for urban and peri-urban food security and reviews the potential for establishing reliable yield baselines. The implications for localised urban land use strategies are critically assessed.

Keywords: urban, peri-urban, agriculture, planning

36.1 Introduction: what are the pressures for greater food security in Britain?

36.1.1 Food and fuel

It has been estimated that the country can currently feed about 65% of its citizens from domestic production (UK Agriculture, 2010: http://www.ukagriculture.com/statistics/farming_statistics.cfm), but that this could easily decline if fossil based inputs are compromised as postulated for peak oil depletion models (Brandt, 2007; Campbell, 2006; Greene *et al.*, 2006; Guseo *et al.*, 2007). Restricted fuel supplies and associated price rises, are also likely to affect the transportation of goods: 'As conventional oil supplies run down… rising prices…could put transport in conflict…with other energy demands…' (Woodcock *et al.*, 2007: 1083). This projection is rapidly becoming fact: currently (September 2011) the prices of petrol and diesel have increased to new heights. If such costs continue to rise there will come a point when diesel-fuelled freight will become unviable, inevitably moving the focus of food production and consumption to a relatively local level.

36.1.2 Where we live

Approximately 80% of European citizens, including those in Britain, live in towns and cities (Antrop, 2004) and if transport becomes limited, most of their food will have to be produced nearby in urban and peri-urban sites by means of urban agriculture (UA). The challenge is to determine what reliable yields might be obtained from UA, the projected area (the 'food footprint') that will thus be needed to feed urban citizens and how this might be achieved efficiently *via* a re-assessment of localisation strategies.

36.2 Food footprints for cities

A food footprint is considered as 'The land required around a town to feed its population' (Geofutures, 2010). Very few examples of such determinations exist to date and those, such as that undertaken by Geofutures and which are based upon rather unclear calculations, cannot be considered sufficiently stringent. Ultimately, the accurate determination of food footprints depends upon an appreciation of the likely yield baseline (guaranteed minimum yield) per unit area that might be achieved in urban and peri-urban zones.

36.3 Yield baselines

If UA is to be optimally developed then it will be vital to attempt a quantification of the factors liable to affect yield and whether baselines can be reliably generated. It is suggested that this depends upon an understanding of UA in terms of: (1) yield constraints; (2) management options; and (3) recent technologies and ecological alternatives.

36.4 Yield constraints in urban agriculture

There have been some publications on the potential yield constraints of UA and, in a review by Eriksen-Hamel and Danso (2010: 86), these are listed as: '...water availability, nutrient supply, soil degradation, pests and soil pollution.' A review of urban allotments by Cook *et al.* (2005) supports these constraints and also emphasises the important effects of choice of crop species and cultivar (cv). Access to information about the yield potential of cvs shows considerable variation, often as a result of differing disease and pest resistance characteristics (NIAB, 2010): thus, the constraints of unwise species and cv choice should not be underestimated. Cook *et al.* (2005) further emphasise the potentially negative effects of soil pollution on yield, a relatively hidden hazard unless a full history of urban land use is known: there are case studies in UA where human health has been threatened by consuming vegetables grown on contaminated soils (e.g. Nabulo *et al.*, 2010).

36.5 Management options for urban agriculture

There is a need to focus on how the management of crops in UA might achieve reliable yield baselines. What are the farming options and management implications?

It is worth starting a food security review of yields from UA by attempting to gain insights from the most accurate published models. These are mostly focused on field based agricultural yields for arable crops in rural areas, such as a recent review by Reilly and Willenbockel (2010). Such authors emphasise that when considering agriculture for food security, 'the food system [must be seen as] multi-dimensional and includes social, economic, biophysical, political and institutional dimensions. Using a model as a proxy to this system raises ontological and epistemological issues...' (p. 3050). The implication from their review is that predicting crop yields is fraught with uncertainties, especially:

- technical uncertainties concerning the quality of data available to calibrate the model and to determine input assumptions;
- methodological uncertainties because sufficient technical knowledge may be lacking;
- epistemological uncertainties – referring to changes in human behaviour and values, the randomness of nature, technological surprises and so-called high impact, high uncertainty events (such as Extreme Weather Events, e.g. Beniston 2007 for Europe and Mirza 2003 for developing nations), all of which are almost impossible to reliably predict.

Most if not all these uncertainties are likely to apply to UA when attempting to establish yield baselines, though the majority of UA has and will doubtless be dominated by vegetable husbandry, due to the greater year-on-year yield potential. For example, Briar *et al.* (2010), demonstrates tomato yields under polythene tunnel cropping at about 42 kg m^{-2} whilst the author's own experience is that optimal yields of field grown cereals such as wheat (produced using relatively high fossil-based inputs) would be the equivalent of about 1 kg m^{-2}. Clearly there are major differences of dry matter and nutritional content between tomatoes and wheat grain but crudely, 42 kg tomatoes and 1 kg wheat (as wholemeal flour) contain 7,140 kcals and 3,100 kcals, respectively (McGovern, 2001). This difference is clearly in favour of vegetable production and would be further enhanced if wheat yields (with reduced inputs post peak oil) were factored in: post-peak yield effects for high-input arable crops are currently unclear but likely to be negative and profound.

There are several management options for growing vegetables, which can be applied to urban sites: protected (glass, plastic) and open, cultivation.

36.5.1 Protected cropping – glass

Growing vegetables in glasshouses is very well documented, and authoritative practical guides are available, such as Beckett (1999). The technology for controlling environmental growing conditions is also well established (e.g. Timmerman and Kamp, 2003). Though none of the above specifically relates to urban sites, there is clear applicability. Academic studies of crop growth models in glasshouses are well established, though again not in urban areas. Challa (2002) compares field and glasshouse grown crops and, whilst acknowledging that '... the physiological principles involved in the growth of glasshouse and field crops are not basically different...' emphasises that 'many important [glass]house crops are multi-harvested...' '... have high water content and...are sold fresh....' Challa also states that '...modern [glass]house production systems provide the grower with a highly advanced, but expensive system for controlling the aerial and root environment of the crop.' (p. 47). Such relatively intensive management to maximise growth for optimal yield is supported by Lee (2010) who reviews the need to generate the most favourable root zone conditions for glasshouse grown tomatoes. Reliable examples of glasshouse grown vegetable yields for Britain can be found in Anonymous (2011), where the following estimated marketed yields for 2009 are given (their Table 1, p. 4) in t ha^{-1}yr^{-1}: tomatoes (*Solanum lycopersicum,* round, vine, plum, and cherry) 414; cucumbers (*Cucumis sativus*) 497; celery (*Apium graveolens* var *dulce*) 75; sweet peppers (*Capsicum annuum*) 262. Growing Lettuce (*Lactuca sativa)* under glass is thought likely to yield about 33 t ha^{-1}yr^{-1} (A. Harvey, personal communication).

36.5.2 Protected cropping – plastic

There are many practical guides to growing vegetables under plastic (e.g. Salt 2001), and which include fleeces, films, cloches and tunnels. Morales *et al.* (2010) also present a good strategic review. Urban vegetable production under plastic has been studied by Briar *et al.* (2010) who reported on a peri-urban study to achieve optimal organic production of tomatoes. They demonstrated a clear potential yield benefit of protected (covered) cropping of vegetables:

> High tunnels were constructed as plastic enclosed structures measuring 6.4m×14.6m
> which provide partial control over environmental variables, chiefly temperature,
> moisture as rain or snow, and solar radiation. As a result, vegetable cropping is possible
> in high tunnels over a wider portion of the calendar year compared to open field settings
> (Briar *et al.*, 2010: 2).

They concluded:

> ...that maintaining the soil...in a biologically-active state during the cold period of
> the early spring months...using high tunnels may have increased N availability for the
> subsequent tomato crop. Thus, the observed increase in tomato yield...was most likely due
> to the combination of enhanced N availability and season extension (Briar *et al.*, 2010: 7).

The maximum yields obtained in this study were about 23 kg tomatoes per plant across the growing season (see Figure 2 in Briar *et al.*, 2010), which at a spacing between plants of 0.45×1.22 m equates to approximately 42 kg m^{-2}. As an annual yield per unit area this is quite close to that reported above, under glass.

Hydroponic production under glass or plastic is commercially widespread in parts of Britain (e.g. Thanet Earth, 2010: http://www.thanetearth.com/) and might be adapted for small scale urban production, though detailed yields figures in such a situation are currently lacking.

Overall, research supports the fact that some form of protection for crops, usually polythene or glass, considerably increases yield. Cantliffe *et al.* (2001) state that 'yields from [glass]house crops are generally 10 times more than comparable field-produced crops.' (p. 195) and this is supported for wider studies, such as in China (Qiu *et al.*, 2010). For the latter, this has led to widespread changes in crop management:

> [Glass]house vegetable production has played an important role in increasing farming
> incomes during the last two decades. Accordingly, many farmers have shifted from
> conventional cereal cropping to [glass]house production systems. In Shouguang county,
> Shandong province, more than 65% of the arable land is now used for intensive [glass]
> house production... (Qiu *et al.*, 2010: 81).

However, the benefits of protected cropping for yield become complicated when climate change is factored in. Olesen and Bindi (2002) somewhat ambiguously state that: 'The main effects of a climatic warming anticipated for protected crops are changes in the heating and cooling requirements of the housing' (p. 250).

36.5.3 Open, unprotected cropping

The other option for UA is field grown vegetables. Before that is reviewed, it is considered helpful to review non-UA, commercial, unprotected cropping of vegetables. Two examples are shown in Table 36.1.

A similar range of yields for organic unprotected vegetables can also be seen (Lampkin and Measures, 2001). Lettuce is the only crop in Table 1 (average 22.5 t/ha/yr), which can be compared with protected cropping (estimated at 33 t/ha/yr) a difference which is not ten times greater, as reported above but which is still considerable. However, yield data shown here are insufficient: further conclusions will not be possible until a much more comprehensive

Table 36.1. Yields (t/ha/yr) for open, unprotected vegetable production in Britain (Agro Business Consultants, 2010; Nix, 2008).

Vegetable species		Yield (t/ha/yr)	
English name	Latin name	Nix (2008) (pp. 48-49)	Agro Business Consultants (2010) (p. 433)
Carrots maincrop	Daucus carota ssp. sativus	65	
Dry bulb onions	Allium cepa	41	
Brussel sprouts	Brassica oleracea var. gemmifera	13	14
Winter cabbage	Brassica oleracea var. capitata	33	45
Spring cabbage	Brassica oleracea var. capitata		12
Spring beans	Phaseolus vulgaris	11	
Cauliflower	Brassica oleracea var. botrytis	14	13
Broccoli	Brassica oleracea var. italica	14	
Calabrese	Brassica oleracea botrytis cymosa	10	10
Sweetcorn	Zea mays var. saccarata	13	
Lettuce	Lactuca sativa	23	22
Parsnips	Pastinaca sativa	27	
Leeks	Allium ampeloprasum var. porrum	20	
Rhubarb	Rheum rhabarbarum	47	55
Vining peas	Pisum sativum		4
Courgettes	Cucurbita pepo ovifera		30
Asparagus	Asparagus officinalis		2

data base has been developed. Climate change is also a factor to consider when predicting the yields of unprotected urban cropping: a useful review by Collier *et al.* (2008) consider Extreme Weather Events (EWEs), mentioned earlier, for such crops. They conclude that predicting yields will be confounded by interactions between EWE stresses such as drought and the irregular and potentially aggressive responses of pests and diseases.

Yield data for vegetable production in urban areas are also challenged by a limited range of published figures. This is in spite of a long history of vegetable husbandry on British urban allotments (Cook *et al.*, 2005). The most available yield data are shown in Table 36.2 for open, unprotected urban cropping. Yields for Table 36.2 compare quite promisingly with those for commercial production in Table 36.1.

36.6 The controversy of recent technologies and ecological alternatives

In addition to basic production for UA reported above, it is necessary to consider all possible means to enhance yields, including new technologies. Nationally and across other parts of western Europe there have been studies to secure optimal food security in general, such as that by The Royal Society (2009) and Conway and Waage (2010). This has led to a proposed agricultural policy of 'sustainable intensification (SI),' considered to be the '...challenge of raising productivity without compromising the flow of valuable ecosystem services' (Omer *et al.*, 2010: 1926). Within this definition the prospect of 'raising productivity' is thought to require an integration of existing and emerging developments (Anonymous, 2010), such as nanotechnology (Joseph and Morrison, 2006) and recombinant DNA procedures (Conway and Waage, 2010). The principles of SI have been embraced by the British Government's Department of the Environment, Food and Rural Affairs (Defra): for example see Defra (2010). This can be contrasted with production systems which eschew high-technology solutions, the most well known being organic farming. Organic regulations do not permit the use of genetically modified organisms and a recent paper on the implications of GM for organics (Azadi and Ho, 2010) concludes that the environmental risks of such technologies weigh against any potential yield benefits and are a continuing cause for concern. Scientists

Table 36.2. Review of possible yields from UA sites in Britain (Thornhill (2010), who adapted data from Butcher (2009), Brown (2010), Horrocks (2010), Tompkins (2006)).

Source/site	Crop category	Yield (t/ha/year)
Allotment grower, Scotland	Fruit and vegetables (1:4)	21.1
Market gardens, Hackney	Salads and leafy greens	21.6
Ministry of Agriculture 1941	Allotment produce	27.2
NSLAG Survey 2009	Fruit and vegetables	30.0
1975 RHS Trials	Salads, vegetables, soft fruit	31.3
1970's Which Magazine Trials	28 cvs vegetables (note: most production incorporated mineral fertilisers)	40.0

who support Agroecology (the production of food using ecological criteria and with optimal support for smallholder famer livelihoods) are even more sceptical (e.g. Altieri and Rosset, 1999). There have, however, been counter arguments against such a view: (Trewavas, 2008) argues that eco-scientists and the lay public who are wary of novel agricultural technologies are misguided 'amateurs' whose opposition is a potential delay to widespread technological innovation, which can offer increased yields and strengthen food security.

36.7 Confidence of achieving yield baselines

So, can reliable yield base lines be confirmed for urban food production? Although no data are available here for protected UA, general experience indicates that protected cropping would seem likely to accrue higher yields than the unprotected yields shown in Table 36.2. Also, better micro climate management is likely under glass, which assists when attempting to develop more consistent yield forecasts. However, all the above cropping is vulnerable to additional yield-threatening risks: unexpected EWEs (especially important for unprotected crops), contaminated soil, disease and pest attack, the unclear availability of nutrients from composts and other sources, and the uncertain supply of irrigation water. Additionally the adoption or not of recent technologies and management regimes reviewed above will affect likely yields.

There are considerable implications for food footprint calculations, if yield base lines in UA are so uncertain. Planning for sufficient space to feed citizens in and around towns and cities will be frustrated unless a more reliable yield data base can be accumulated. The ongoing development of community allotment projects by the Transition Town movement and other NGOs, suggest that more practical experience about likely yields in UA will soon be accumulated. However, the classic response of any scientist is to suggest that further high quality assessments are required: and this author has to agree with that premise. Whilst accumulated practical data are invaluable, they need to be combined with well structured, replicated experiments. Ultimately an applied agronomy guide for UA will be needed, to assist in planning for reliable food footprint determinations.

36.8 Localisation strategies

Localisation is discussed here in relation to food networks. What key issues emerge from the academic debate?

36.8.1 Organic and/or local?

The importance of organic management has already been discussed and its application for UA at local level is debatable. Although there has been optimism that organic production offers a viable alternative to high-tech farming in Britain (Rigby *et al.*, 2001) and across Europe (Lee *et al.*, 2008) the priority in local UA, organic or not, seems to be that of the 'food citizen' as defined by Lang *et al.* (2009), who will take a more active role in supply chains. The local food citizen in urban areas can be seen as developing knowledge about agricultural production by means of mutual support and by taking more power over decisions (Carr, 2006), which is referred to by Seyfang (2006) as 'ecological citizenship.' McCullum *et al.* (2005) contend that community food security will depend upon: (1) the creation of multi sector partnerships and networks, and (2) participatory decision making and policy development. Local decision making is further developed by Gray (2007) who terms it 'bioregionalism' and argues that it relates to the '...small scale society that is socially and ecologically responsive' (p. 790). Furthermore, it is suggested that such local groups will need to encourage in-migration of skilled people to encourage entrepreneurship, which is seen as vital for the progression of agrarian-based ecological modernisation and better sustainability (Cowell and Parkinson, 2003; Jarosz, 2008; Kalantaridis, 2010; Marsden and Smith, 2005).

So, can a small scale, local, entrepreneurial society with food production and nearby consumption be considered as best for security of supply? Some argue that this is so (Blanc, 2009; North, 2010; Seymoar *et al.*, 2010) while others have mixed views (Edward-Jones *et al.*, 2008; Hinrichs, 2003; Smithers *et al.*, 2008; Sundkvist *et al.*, 2001). In terms of economics, local, self-reliant food producing societies are thought to be counter to global expansionist scenarios and part of a new self sustaining economic paradigm (Curtis, 2003; DuPuis and Goodman, 2005; North, 2010). Overall, agro food politics is neatly summarised by Guthman (2008) as having four themes: (1) consumer choice; (2) localism; (3) entrepreneurialism; and (4) self improvement. Some important concluding points from Curtis (2003) for food eco-localism are pertinent, that:

1. *Sustainable use and preservation of ... varied eco-systems requires locally adapted knowledge, communities, products, cultures and practices.*
2. *Sustainability requires locally adapted and symbiotic forms of social, physical, human and financial capital. Business enterprises, networks, education, money, banking, and investment all need to be locally oriented.*
3. *By uniting production and consumption within eco-local boundaries, both positive and negative externalities of the production and use of goods and services are localised. This creates pressure to reduce pollution and resource use and to increase positive externalities such as ecological restoration and community building.*
4. *The relatively small scale of production in decentralised local economies may be efficient as the goal is not simply to maximise a single output relative to its inputs, but to do so in a particular social context with multiple goals. Efficiency is [thus] redefined* (pp. 98-99).

36.8.2 Population pressures and zoning the compact city

Population pressure emerges as a consistent key issue for many studies of urban food security: clearly the denser an urban area the greater the pressure to feed citizens, combined with reduced green areas for cultivation (Hopkins *et al.*, 2010; Pauleit *et al.*, 2005). Urban green spaces are seen as especially important not just for growing food but as a contributor to human comfort, sense of place and general quality of life (Gomez *et al.*, 2001; Millard, 2004;

Whitford *et al.*, 2001). Some studies report that inner city living has stimulated citizens to seek to move to less populated suburban areas (e.g. Howley, 2009 for Dublin). Such movements out of cities are surely unsustainable: the concept of the high density 'compact city' (or city of short distances) will be vital if goods and services are to remain reliably available (Howley, 2009; Condon *et al.*, 2010). A good example is the famous Portland case study in the USA (North West Environment Watch, 2002), where

> *...growth management softened the impact of rapid population increase in the metropolis. Portland Metro's urban growth boundary restrained suburban sprawl, slowed the loss of rural land and open space, and provided better transportation alternatives by channelling development into compact neighbourhoods that use land and urban infrastructure more efficiently* (p. 6).

Maintaining compact cities and limiting future suburban sprawl seems vital if food security is to be achieved.

What are the planning implications for designing food footprints across compact cities? It is suggested that the following is required:

- Urban mapping, to determine what areas are available for UA (e.g. Pauleit *et al.*, 2005 for Merseyside; Gill *et al.*, 2008 for Manchester; Owen *et al.*, 2006 for the west Midlands).
- Zoning around cities, such as that by Condon *et al.* (2010) at the interface between urban and agricultural land. Brown (2010) proposes a more sophisticated zoning for urban, peri-urban and rural hinterland (zones 1, 2 and 3, respectively).
- The implementation of planning which facilitates an eco-localist approach, where the food citizen is empowered in the development of UA. As Watson (2009) discusses, the traditional master plan used by planners needs to evolve towards a more strategic spatial approach: planning which is '... transparent, participative and inclusive...' (p. 167).

What proportion of the food footprint of a city might be produced within urban food zones? Brown (2010) refers to: '...human-scale, organic, mixed farms located in and around urban areas which are directly connected to the urban communities they feed and which enable those communities to source increasing amounts of food from close to where they live,' but does not quantify yields in (her) zones 1, 2 and 3. As reported above, Condon *et al.* (2010) study the urban-agricultural interface (in essence the peri-urban zone 2 of Brown) and recommend that:

> - *...two-thirds of this land [be used] exclusively for agriculture...;*
> - *such agricultural land comes under the ownership of the associated municipality, referred to as Community Trust Farming;*
> - *[and that this land] is leas[ed] (very favourably) ... to agricultural entrepreneurs [with a stipulation that it be farmed] exclusively for local/regional markets...* (Condon *et al.*, 2010: 109).

This author supports the recommendations by Condon *et al.* and is currently working on a food footprint for Maidstone, the county town of Kent, UK and with a premise that zones 1,

2 and 3 might provide 10, 70 and 20%, respectively of the food needs of citizens. This project is developing the mapping techniques shown by Gill *et al.* (2008) and will ultimately seek to engage with other scientists and local stakeholders for participative discussions and feedback. It is intended that a jointly agreed, urban sustainable land use strategy to ensure food security in Maidstone is ultimately developed.

36.9 Conclusions: scenario planning and organisational ambidexterity

Urban food production and security depend upon an ability to look ahead into an uncertain future and plan accordingly. The risks of not doing so will be food shocks, food insecurity and famine (Dilley and Boudreau, 2001). The procedure of flexible forward planning is referred to as 'scenario planning' (Chermack, 2005) and considered as: (1) a means of making sense of impending complexities and ambiguities and; (2) embracing strategic change by exploring the limits of possibilities for the future. Success in scenario planning is thought to depend upon 'organisational ambidexterity', i.e. the ability to balance deliberate (pre-planned) and emergent (flexible, adaptive) approaches (Bodwell and Chermack, 2010). The implication is that horticulturalists, agriculturalists, ecologists, economists, social scientists, architects and urban planners must work together to consider towns and cities holistically, as urban agroecosystems and, after objective studies, develop protocols which are realistic and achievable. We need to plan to be prepared, essentially, for the unexpected.

Acknowledgements

The author would like to thank final year honours students for their helpful inputs: Nina Thornhill, Jeff Conway, Richard Casey, Diana Martineau and Sean Clements (BSc Sustainable Land Management) and Richard Masson and Sam Jackson (BSc International Agriculture). He would also like to thank colleagues: Caroline Jackson and Alan Harvey for valuable critical inputs.

References

Agro Business Consultants, 2010. The agricultural budgeting and costing book. Agro Business Consultants Ltd., Leicestershire, Uk, 474 pp.

Altieri, M. and Rosset, P., 1999. Ten reasons why biotechnology will not ensure food security, protect the environment and reduce poverty in the developing world. Agricultural Biological Forum 2: 155-162.

Anonymous, 2011. Greenhouse Yearbook and Buyers Guide. ACT Publishing, Suton, UK, 75 pp.

Anonymous, 2010. How to feed a hungry world. Nature 466(7306): 531-532.

Antrop, M., 2004. Landscape change and the urbanization process in Europe. Landscape and Urban Planning 67: 9-26.

Azadi, H. and Ho, P., 2010. Genetically modified and organic crops in developing countries: A review of options for food security. Biotechnology Advances 28: 160-168.

Beckett, C., 1999. Growing under glass. The RHS Encyclopedia of Practical Gardening. Royal Horticultural Society, London, UK, 191 pp.

Beniston, M., 2007. Linking extreme climate events and economic impacts: Examples from the Swiss Alps. Energy Policy 35: 5384-5392.

Blanc, J., 2009. Family farmers and major retail chains in the Brazilian organic sector: Assessing new development pathways. A case study in a peri-urban district of São Paulo. Journal of Rural Studies 25: 322-332.

Bodwell, W. and Chermack, T.J., 2010. Organizational ambidexterity: Integrating deliberate and emergent strategy with scenario planning. Technological Forecasting and Social Change 77: 193-202.

Brandt, A.R., 2007. Testing Hubbert. Energy Policy 35: 3074-3088.

Briar, S.S., Miller, S.A., Stinner, D., Kleinhenz, M.D. and Grewal, P.S., 2010. Effects of organic transition strategies for peri-urban vegetable production on soil properties, nematode community, and tomato yield. Applied Soil Ecology 47: 84-91.

Brown, J., 2010. 'Growing communities' Manifesto for feeding the city. Taking our food system back.' Available at http://www.transitiontownbrixton.org/wordpress/wp-content/uploads/2010/01/manifesto.pdf.

Butcher, J., 2009. Investigating the potential for the expansion of urban agriculture in the city of Edinburgh. Ecology and Conservation Management dissertation, University of Edinburgh, Edinburgh, Midlothian, UK. Available at www.cityfarmer.info/2009/10/04. Accessed 21 January 2011.

Campbell, C.J., 2006. The Rimini Protocol, an oil depletion protocol: heading off economic chaos and political conflict during the second half of the age of oil. Energy Policy 34: 1319-1325.

Cantliffe, D.J., Secker, I. and Karchi, Z., 2001. Passive ventilated high-roof greenhouse production of vegetables in a humid, mild winter climate. In: Fernandez, J.A., Martinez, P.F. and Castilla, N. (Eds.) Production of Crops in mild winter climates: current trends for sustainable technologies. Acta Horticulturae 559: 195-201.

Carr, E.R., 2006. Postmodern conceptualizations, modernist applications: Rethinking the role of society in food security. Food Policy 31: 14-29.

Challa, H., 2002. Crop models for greenhouse production systems. In: Baker, J. and Lieth, H. (Eds.) Models for plant growth and control in greenhouses: Modeling for the 21st century – Agronomic and Greenhouse Crop Models. Acta Horticulturae 593: 47-53.

Chermack, T.J., 2005. Studying scenario planning: Theory, research suggestions, and hypotheses. Technological Forecasting and Social Change 72: 59-73.

Collier, R., Fellows, J., Adams, S., Semenov, M. and Thomas, B., 2008. Vulnerability of horticultural crop production to extreme weather events. In: Halford, N., Jones, H.D. and Lawlor, D. (Eds.) Effects of Climate Change on Plants: Implications for Agriculture. Aspects of Applied Biology 88: 3-13.

Condon, P.M., Mullinix, K., Fallick, A. and Harcourt, M., 2010. Agriculture on the edge: strategies to abate urban encroachment onto agricultural lands by promoting viable human-scale agriculture as an integral element of urbanisation. International Journal of Agricultural Sustainability 8: 104-115.

Conway, G. and Waage, J., 2010. Science and Innovation for Development. UK Collaboration on Development Services, London, UK. Available at www.ukcds.org.uk.

Cook, H.F., Lee, H.C. and Perez-Vazquez, A., 2005. Allotments, Plots and Crops in Britain. In: Viljoen, A. and Howe, J. (Eds.) CPULs: The productive urban landscape. Architectural Press, London, UK, pp. 206-216.

Cowell, S.J. and Parkinson, S., 2003. Localisation of UK food production: an analysis using land area and energy as indicators. Agriculture, Ecosystems and Environment 94: 221-236.

Curtis, F., 2003. Eco-localism and sustainability. Ecological Economics 46: 83-102.

Department of Environment, Food and Rural Affairs, 2010. Food 2030. HM Government, UK, 81 pp.

Dilley, M. and Boudreau, T.A., 2001. Coming to terms with vulnerability: a critique of the food security definition. Food Policy 26: 229-247.

DuPuis, E.M. and Goodman, D., 2005. Should we go "home" to eat?: toward a reflexive politics of localism. Journal of Rural Studies 21: 359-371.

Edwards-Jones, G., Mila i Canals, L., Hounsome, N., Truninger, M., Koerber, G., Hounsome, B., Cross, P., York, E.H., Hospido, A., Plassmann, K., Harris, I.M., Edwards, R.T., Day, G.A.S., Tomos, A.D., Cowell, S.J. and Jones, D.L., 2008. Testing the assertion that 'local food is best': the challenges of an evidence-based approach. Trends in Food Science and Technology 19: 265-274.

Eriksen-Hamel, N. and Danso, G., 2010. Agronomic considerations for urban agriculture in southern cities. International Journal of Agricultural Sustainability 8: 86-93.

Geofutures, 2010. Food footprints. Available at http://www.geofutures.com/2009/07/food-fooprints-re-localising-uk-food-supply/.

Gill, S.E., Handley, J.F., Roland Ennos, A., Pauleit, S., Theuray, N. and Lindley, S.J., 2008. Characterising the urban environment of UK cities and towns: A template for landscape planning. Landscape and Urban Planning 87: 210-222.

Gómez, F., Tamarit, N. and Jabaloyes, J. 2001. Green zones, bioclimatics studies and human comfort in the future development of urban planning. Landscape and Urban Planning 55: 151-161.

Gray, R., 2007. Practical bioregionalism: A philosophy for a sustainable future and a hypothetical transition strategy for Armidale, New South Wales, Australia. Futures 39: 790-806.

Greene, D.L., Hopson, J.L. and Li, J., 2006. Have we run out of oil yet? Oil peaking analysis from an optimist's perspective. Energy Policy 34: 515-531.

Guseo, R., Dalla Valle, A. and Guidolin, M., 2007. World oil depletion models: Price effects compared with strategic or technological interventions. Technological Forecasting and Social Change 74: 452-469.

Guthman, J., 2008. Neoliberalism and the making of food politics in California. Geoforum 39: 1171-1183.

Hinrichs, C.C., 2003. The practice and politics of food system localization. Journal of Rural Studies 19: 33-45.

Hopkins, R., Thurstain-Goodwin, M. and Fairlie, S., 2010. Can Totnes and district feed itself? Exploring the practicalities of food relocalisation. Available at www.transitionnetwork.org.

Horrocks, P., 2010. What is your plot worth? National Society of Allotments and Leisure Gardeners Limited. Available at http://www.nsalg.org.uk/page.php?article=658&name=What+is+your+plot+worth%3F.

Howley, P., 2009. Attitudes towards compact city living: Towards a greater understanding of residential behaviour. Land Use Policy 26: 792-798.

Jarosz, L., 2008. The city in the country: Growing alternative food networks in Metropolitan areas. Journal of Rural Studies 24: 231-244.

Joseph, T. and Morrison, M., 2006. Nanotechnology in Agriculture and Food. Institue of Nanotechnology. Available at www.nanoforum.org.

Kalantaridis, C., 2010. In-migration, entrepreneurship and rural-urban interdependencies: The case of East Cleveland, North East England. Journal of Rural Studies 26: 418-427.

Lampkin, N. and Measures, M., 2001. 2001 Organic Farm Management Handbook. University of Wales, Organic Advisory Service (EFRC), Aberystwyth, UK, 182 pp.

Lang, T., Barling, D. and Caraher, M., 2009. Food Policy. Integrating Health, Environment and Society. Oxford University Press, Oxford, UK, 313 pp.

Lee, A., 2010. Improving tomato fruit quality. Reducing or eliminating fruit physiological disorders with correct root zone management. Practical Hydroponics and Greenhouses May/June 2010: 53-59.

Lee., H.C., Walker, R., Haneklaus, S., Philips, L., Rahmann, G. and Schnug, E., 2008. Organic farming in Europe: a potential major contribution to food security in a scenario of climate change and fossil fuel depletion. Landbauforschung – vTI/Agriculture and Forestry Research 58: 145-152.

McCullum, C., Desjardins, E., Kraak, V.I., Ladipo, P. and Costello, H., 2005. Evidence-based strategies to build community food security. Journal of the American Dietetic Association February 2005: 278-283.

McGovern, S.G., 2001. The Healthy Eating Plan. Geddes and Grosset, Scotland, UK, 288 pp.

Marsden, T. and Smith, E., 2005. Ecological entrepreneurship: sustainable development in local communities through quality food production and local branding. Geoforum 36: 440-451.

Millard, A., 2004. Indigenous and spontaneous vegetation: their relationship to urban development in the city of Leeds, UK. Urban Forestry and Urban Greening 3: 39-47.

Mirza, M.M.Q., 2003. Climate change and extreme weather events: can developing countries adapt? Climate Policy 3: 233-248.

Morales, A., Marin, A. and Pacheco, V., 2010. Plastics in the present agriculture. Use advantages. Plasticulture 8: 69-79.

Nabulo, G., Young, S.D. and Black, C.R., 2010. Assessing risk to human health from tropical leafy vegetables grown on contaminated urban soils. Science of the Total Environment 408: 5338-5351.

National Institute of Agricultural Botany, 2010. National Lists of Crop Cultivars. Available at http://www.niab.com/pages/id/106/National_Listing.

Nix, J., 2008. Farm Management Pocketbook. 38th edition. Imperial College London, London, UK, 255 pp.

North, P., 2010. Eco-localisation as a progressive response to peak oil and climate change – A sympathetic critique. Geoforum 41: 585-594.

North West Environment Watch, 2002. Sprawl and smart growth in metropolitan Portland: Comparing Portland, Oregon, with Vancouver, Washington, during the 1990s. Sightline Institute, Seattle, WA, USA. Available at http://www.sightline.org/research/sprawl/sprawl-and-smart-growth/sprawl_smart_port/portlandgrowth.pdf/view.

Olesen, J.E. and Bindi, M., 2002. Consequences of climate change for European agricultural productivity, land use and policy. European Journal of Agronomy 16: 239-262.

Omer, A., Pascual, U. and Russell, N., 2010. A theoretical model of agrobiodiversity as a supporting service for sustainable agricultural intensification. Ecological Economics 69: 1926-1933.

Owen, S.M., MacKenzie, A.R., Bunce, R.G.H., Stewart, H.E., Donovan, R.G., Stark, G. and Hewitt, C.N., 2006. Urban land classification and its uncertainties using principal component and cluster analyses: A case study for the UK West Midlands. Landscape and Urban Planning 78: 311-321.

Pauleit, S., Roland, E. and Yvonne, G., 2005. Modeling the environmental impacts of urban land use and land cover change – a study in Merseyside, UK. Landscape and Urban Planning 71: 295-310.

Qiu, S.J., Ju, X.T., Ingwersen, J., Qin, Z.C., Li, L., Streck, T., Christie, P. and Zhang, F.S., 2010. Changes in soil carbon and nitrogen pools after shifting from conventional cereal to greenhouse vegetable production. Soil and Tillage Research 107: 80-87.

Reilly, M. and Willenbockel, D. 2010. Managing uncertainty: a review of food system scenario analysis and modelling. Philosophical Transactions of the Royal Society B 365: 3049-3063.

Rigby, D., Young, T. and Burton, M., 2001. The development of and prospects for organic farming in the UK. Food Policy 26: 599-613.

Salt, B., 2001. Gardening under plastic. Batsford, London, UK, 127 pp.

Seyfang, G., 2006. Ecological citizenship and sustainable consumption: Examining local organic food networks. Journal of Rural Studies 22: 383-395.

Seymoar, N-K., Balantyne, E. and Pearson, C.J., 2010. Empowering residents and improving governance in low income communities through urban greening. International Journal of Agricultural Sustainability 8: 26-39.

Smithers, J., Lamarche, J. and Joseph, A.E., 2008. Unpacking the terms of engagement with local food at the Farmers' Market: Insights from Ontario. Journal of Rural Studies 24: 337-350.

Sundkvist, A., Jansson, A.M. and Larsson, P., 2001. Strengths and limitations of localizing food production as a sustainability-building strategy – an analysis of bread production on the island of Gotland, Sweden. Ecological Economics 37: 217-227.

The Royal Society, 2009. Reaping the benefits: science and the sustainable intensification of global agriculture. The Royal Society, London, UK, 73 pp.

Timmerman, G.J. and Kamp, P.G.H., 2003. Computerised Environmental Control in Greenhouses. PTC, Ede, the Netherlands, 267 pp.

Tomkins, M., 2006. The Edible urban landscape: an assessment method for retro-fitting urban agriculture into an inner London test site. MSc Thesis, Architecture: Advanced Environment and Energy Studies, University of East London, London, UK. Available at www.cityfarmer.org/MikeyTomkins_UA_thesis.pdf.

Trewavas, A., 2008. The cult of the amateur in agriculture threatens food security. Trends in Biotechnology 26: 475-478.

Waggoner, P.E., 2006. How can EcoCity get its food? Technology in Society 28: 183-193.

Watson, V., 2009. 'The planned city sweeps the poor away…': Urban planning and 21st century urbanisation. Progress in Planning 72: 151-193.

Whitford, V., Ennos, A.R. and Handley, J.F., 2001. 'City form and natural process' – indicators for the ecological performance of urban areas and their application to Merseyside, UK. Landscape and Urban Planning 57: 91-103.

Woodcock, J., Banister, D., Edwards, P., Prentice, A.M. and Roberts, I., 2007. Energy and transport. The Lancet 370: 1078-1088.

Chapter 37

Food seams: planning strategies for urban borders in New Orleans

Brittney Everett
201 St. Charles Avenue Suite 4318, New Orleans, LA 70170, USA; brittney.everett@gmail.com

Abstract

Since Hurricanes Katrina and Rita in 2005, food insecurity has increased exponentially in New Orleans. Many supermarkets have not re-opened since these disasters, and the majority that have are concentrated in the more affluent communities. Residents in lower income neighbourhoods, many of which do not own cars, have been forced to rely on local corner stores where healthy food options are limited or they must use public transportation to travel long distances across the city to purchase affordable food. These circumstances have left a formidable gap in the socioeconomic fabric of the city. This chapter, which is a culmination of academic and independent research, proposes a food planning approach for New Orleans in which urban borders between communities are transformed into socially and culturally rich urban food seams that provide healthy, affordable food retail to underserved residents. By identifying the weaknesses in New Orleans current approach to food security and exploring food initiative case studies in other United States cities, this research will develop recommendations for how food seams can expand food security in New Orleans.

Keywords: food security, farmers markets, low income, socioeconomic, hurricane katrina

37.1 Introduction

Although it is a city internationally renowned for its food culture, New Orleans, like many other cities, struggles to provide residents with access to sufficient, safe, and nutritious food. The farmers markets and mobile food carts that made healthy food options widely available in he early 20th century have been replaced by big box supermarkets that are inaccessible to many residents. The damage from Hurricanes Katrina and Rita in 2005 has exacerbated the issue of food security – only fifty percent of the supermarkets that existed prior to the storms have re-opened (New Orleans Food Policy Advisory Committee, 2008). Nearly half of these supermarkets are located in the Garden District and Uptown, where the average income is considerably higher than most of the city as illustrated in Figure 37.1. Many households do not have access to grocery stores because they do not own cars and public transportation routes are limited (Greater New Orleans Community Data Center, 2010). Residents in low income neighbourhoods, such as Central City, Desire, and Treme, rely heavily on corner stores that offer limited healthy food options, focusing on more lucrative, high-calorie snacks and alcohol. Many of these stores do not stock fresh fruit and vegetables and stores that do are few and far between (Custer, 2009).

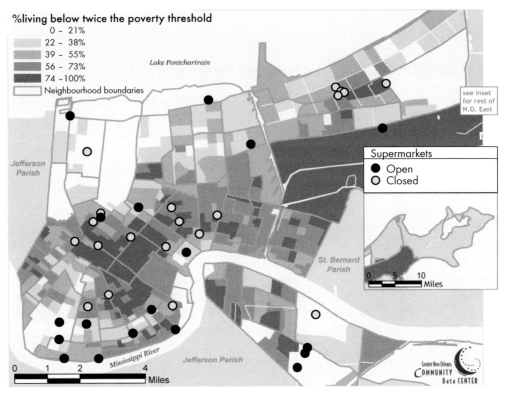

%living below twice the poverty threshold
- 0 – 21%
- 22 – 38%
- 39 – 55%
- 56 – 73%
- 74 –100%
- Neighbourhood boundaries

Lake Pontchartrain

Jefferson
Parish

see inset
for rest of
N.O. East

Supermarkets
- ● Open
- ○ Closed

St. Bernard
Parish

0   5   10
Miles

0   1   2   4
Mississippi River   Jefferson Parish
Miles

Greater New Orleans
COMMUNITY
Data CENTER

Figure 37.1. Supermarkets in New Orleans post-hurricane Katrina vs. percent living below twice the poverty threshold (based on Tulane University of Medicine, 2010).

While many food organisations in New Orleans – including the New Orleans Food Policy Advisory Committee (FPAC) (2008), the New Orleans Food and Farm Network (NOFFN) (2010) , and the Louisiana Public Health Institute (LPHI) – are dedicated to creating farmers markets and increasing food security, they have yet to establish tangible food planning strategies for the city. The New Orleans Community Food Charter of 2007 demonstrates residents' support of local food production, food education, and the development of food infrastructure but does not provide planning strategies for *how* to increase food access. In January of 2010, the New Orleans City Planning Commission approved Goody Clancy's 'Plan for the 21st century: New Orleans in 2030' which totals more than 500 pages yet dedicates only a few bullet points to food access issues. These recommendations – provide fresh produce retail within walking distance to all residents, support urban agriculture, and commission a market analysis to locate feasible sites for supermarkets – offer a basic vision for food security in New Orleans but lack specific, implementable design strategies for how to increase food security.

This research explores how the transformation of urban borders into urban 'food seams' can expand food security in New Orleans. Food seams are fresh food retail venues located at urban borders between neighbourhoods and their primary mission is to provide healthy food to the underserved. Urban borders are often plagued by racial, ethnic, and economic tensions because of their location between socio-economically distinct communities (Everett, 2010). Jane Jacobs berates borders as 'zones of low value' and 'dead ends of use,' suggesting that by increasing social activity and using these edges more productively, 'the curse of the border vacuum' can be ameliorated (Jacobs, 1961). Kevin Lynch concurs, '...an edge may be more than a simply dominant barrier if some visual or motion penetration is allowed through it – if it is, as it were, structured to some depth with the regions on either side. It then becomes a seam rather than a barrier, a line of exchange along which two areas are sewn together...' (Lynch, 1960).

Food seams offer solutions for two problems: the problem of food insecurity and the problem of urban borders based on the notion that food is a shared value and experience among people on all sides of borders.

This chapter develops a theory for food seams by comparatively analysing food security projects in New Orleans and other food projects in the United States. The strengths and weaknesses of these projects will be investigated within four major themes: food diversity, programming, accessibility, and affordability. Conclusions drawn from this analysis will prove the benefits of locating food projects at borders between communities and inform food planning recommendations for the city of New Orleans.

37.2 Food (in)security in New Orleans

The following case studies provide an overview of food projects in New Orleans, focusing specifically on the challenges that limit the impact of these projects on food security. Figure 37.2 illustrates the location, frequency, and accepted methods of payment of each venue in relation to citywide income levels

In 2008, the New Orleans Food and Farm Network (NOFFN: http://www.noffn.org) established the Hollygrove Market and Farm (http://hollygrovemarket.com) to provide fresh, local produce to neighbourhood residents. The farm includes expert gardens for growing demonstrations and community garden space in which residents can grow their own food. The market is one of few in the city that has a permanent, indoor facility. Hollygrove Market offers a small variety of fresh produce, dried goods, and dairy as well as a seasonal produce box for USD 25.00. For a small fee, the market's delivery service will bring produce boxes to customers that cannot attend. The greatest challenge for Hollygrove Market is to attract neighbourhood residents. Many Hollygrove residents are elderly or have severe health issues which limit their ability to attend the market. As a result, only 10 to 15% of the market's patronage is from Hollygrove (Vance, 2010). Furthermore, the market only operates two days a week and is located deep within this neighbourhood at the western edge of Orleans Parish. Transportation options are limited – patrons must drive, bike long distances, or take the bus, which has limited routes in that area – ultimately making Hollygrove Market and Farm inaccessible for most New Orleanians.

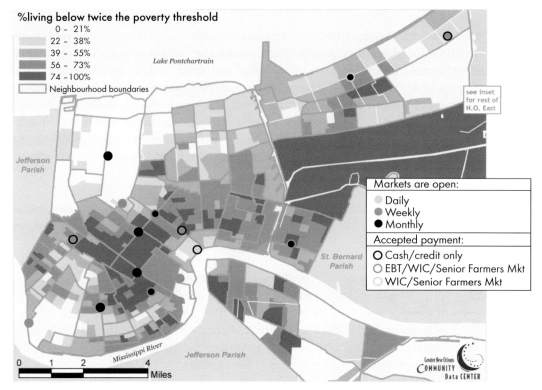

Figure 37.2. New Orleans farmers markets vs. percent living below twice the poverty threshold (based on Tulane University of Medicine, 2010).

The Crescent City Farmers Market (2010: http://www.crescentcityfarmersmarket.org) provides three weekly markets located in three different neighbourhoods. These markets have not significantly impacted food security because these market sites are located in predominantly white, affluent communities. These markets are inconvenient for the low income, most of who rely on public transportation. Operation hours and food options are limited which is inconvenient for many residents. Marketumbrella.org, founder of the Crescent City Farmers Market, tackles financial affordability through its Market Match programmeme, which matches up to USD 25.00 of what patrons spend in EBT food stamps. The anticipated increase in patron diversity has been limited because these markets are located far from residents who use food stamps. Marketumbrella.org openly admits that they primarily attract shoppers who have other healthy food options (Marketumbrella.org, 2010: http://www.marketumbrella.org). The question remains: why hasn't the Crescent City Farmers Market established markets at sites that are more accessible to underserved residents?

The mission of the newly completed Healing Centre is to provide a holistic, sustainable centre that heals and empowers residents by emphasising physical, nutritional, intellectual, and spiritual well-being (New Orleans Healing Center, 2011: http://neworleanshealingcenter.

org). The project is located on the St. Claude Avenue commercial corridor, one of the most visible socioeconomic borders in the city, where the mixed race Bywater neighbourhood, the affluent Marigny, and the predominantly black, lower income St. Roch and Upper Ninth Ward neighbourhoods intersect (Greater New Orleans Community Data Center, 2010: http://www.gnocdc.org). The Healing Centre's vibrant, colourful building is highly visible in the community and conveniently located on a main public transit line. It also has a diverse portfolio of programmes including a cooperative grocery store, organic restaurant, yoga studio, fitness centre, and art gallery. Many of these programmes favour more affluent residents, which can be a deterrent to lower income residents patrons. This can also reinforce the stigma that healthy grocery stores are only for affluent residents (Coffman, 2010). The food cooperative's ability to alleviate the food desert in which it is located will likely increase if it can provide programmes that attract lower income patrons as well.

While these food projects and many others like them have increased fresh food access for some New Orleanians, their overall impact on food security has been limited because these efforts lack strategic planning and analysis at the urban scale. These food node initiatives have resulted in small-scale, disjointed markets buried within neighbourhoods that lack the visibility, reliability, accessibility, and affordability they need to benefit the underserved population. Alternatively, the food seam initiative proposes locating food programmes along the edges of low income communities so they are accessible to the underserved and also highly visible to anyone passing through the area. Food seams offer a holistic and integrative approach to food security that collectively establishes an urban network of food retail throughout the city as illustrated in Figure 37.3.

37.3 Threading the needle

The following case studies illustrate the benefits of locating food projects near urban borders. By considering food diversification, programme diversification, accessibility, and economic affordability, these projects have made significant impacts on food security in their respective cities.

37.3.1 Findlay Market

In Cincinnati, Findlay Market's commitment to providing diverse food options that meet the varying budgets of its patrons has enabled it to attract diverse people from the neighbourhood as well as people from all over the city. The market is convenient and affordable to the underserved, low income population of the Over-the-Rhine neighbourhood in which the market is located. Findlay also acts an urban destination for the affluent because of its large scale and unique offerings of ethnic food, gourmet spices, and distinctive homemade goods. Vibrant, food-themed murals enhance the visual connection between the market and Central Parkway, the main vehicular thoroughfare that forms the border between Over-the-Rhine and the West End, as shown in Figure 37.4. The market's proximity to this main street makes it accessible by both car and public transportation.

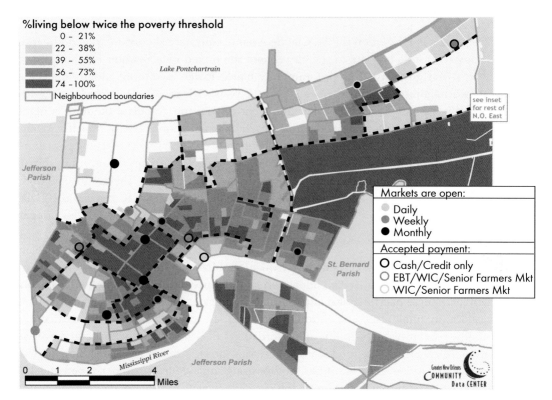

Figure 37.3. New Orleans farmers markets vs. urban borders (based on Tulane University of Medicine, 2010).

Findlay Market is able to offer diverse food options because it has both permanent and flexible infrastructure. The indoor market provides each vendor a stall equipped with electrical hook-ups, cold and dry storage, a preparation area, and a refrigerated display case. In these areas, vendors sell meat, seafood, dairy, and prepared foods. The pedestrian street that surrounds the indoor market provides informal space where vendors can set up tables to sell non-food items. The produce market consists of a covered promenade that extends to the parking lot like a peninsula, enabling farmers to conveniently park their produce trucks adjacent to their sale tables. Findlay Market is able to attract more patrons because it is surrounded by mid-rise buildings that have additional ground level food retail.

Findlay Market is a thriving retail food district because of the diverse programmes it offers. The market provides tables and seating areas where patrons can enjoy ready-to-eat foods and beverages sold at the market. The market's ambitious schedule of concerts, parades, and festivals has established it as a social destination. Findlay Market's Arts in the Market programme enables at-risk youth to create art installations that distend colour and optimism into the blighted Over-the-Rhine neighbourhood. These art projects extend the market's identity to the nearby playground and baseball field where children can play while their parents shop at the market.

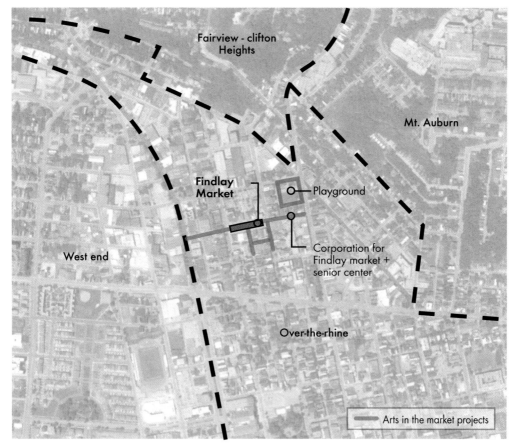

Figure 37.4. Findlay Market map (based on Tulane University of Medicine, 2010).

37.3.2 32nd Street farmers market

For 30 years, the 32nd Street Market in Baltimore (2010: http://www.32ndstreetmarket.org) has maintained a loyal clientele because of its convenient location at a highly trafficked, transit-accessible border. The market venue, which consists of tents and tables organised in a large parking lot, is located among a moderate-income African American neighbourhood, a low income Korean community, an affluent neighbourhood, and Johns Hopkins University (Fisher, 1999). Nearly 30% of people walk to the market (Rey, 2010). Additionally, the market's strategic location at the edge of the Greenmount Avenue Commercial District – a large retail destination that includes restaurants, supermarkets, clothing shops, bookstores, hair salons, and other services – makes it an urban destination for residents from all over the city. Visitors can use one of five public transit lines that run within two blocks of the market as shown in Figure 37.5.

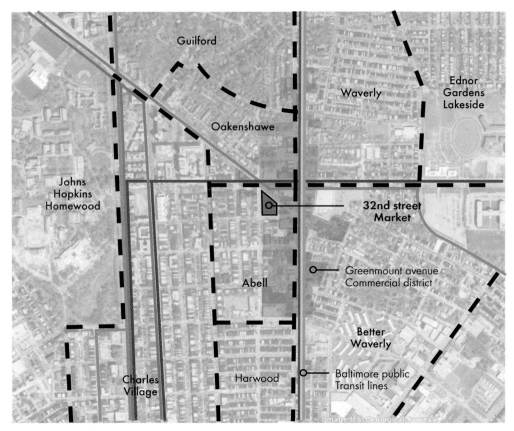

Figure 37.5. 32nd Street market map (based on Tulane University of Medicine, 2010).

The 32nd Street Market offers diverse, affordable food options that meet the cultural and economic needs of the surrounding communities. The market offers fresh produce, baked goods, meats, and dairy and also a variety of prepared ethnic foods, including Asian, Ethiopian, Indian, and Mexican cuisine, that represent the diverse cultural backgrounds of the market's patrons. The vendors are also representative of local neighbourhood demographics, which helps foster a collective sense of belonging among patrons. Wireless EBT machines help make groceries more affordable for lower income residents. The market also offers additional money to these patrons for the purchase of fresh produce. The market's large scale – nearly 50 vendors attend weekly – enables it to function as a one-stop-shop for residents, limiting supplemental trips to the local supermarket to fulfil a week's grocery needs. The market's strict attendance policy for vendors and year-round Saturday schedule helps ensure its reliability (Rey, 2010).

In addition to offering food retail, the 32nd Street Market acts as a venue for community events. Cooking demonstrations, activities for children, and weddings are just a few of the special events that help attract patrons. The market uses leftover funds from the previous year to provide grants to local schools, choirs, and churches. It also provides community space in which nonprofits and community organisations can disseminate information and people can come together to discuss community issues.

37.3.3 City Farm

City Farm, an urban farm and market in Chicago, demonstrates how urban agriculture can simultaneously provide food to local residents and mitigate blight. The mission of City Farm is to transform vacant properties throughout the city into productive green spaces and provide highly nutritious food to diverse neighbourhoods (Resource Center Chicago, http://www. resourcecenterchicago.org). The pilot site is located at the border between the low income Cabrini-Green neighbourhood, a large former public housing project, and the affluent Gold Coast neighbourhood as illustrated in Figure 37.6. The farm sells its produce to chefs at local restaurants and residents from both neighbourhoods at its on-site market stand. The project is designed to be mobile; once land is farmed, its compost can be moved and reused to start other projects throughout the city.

City Farm demonstrates the ability of urban agriculture to create jobs for local residents. Thousands of people have visited City Farm to learn about sustainable farming practices and receive workforce training. The project also offers advanced training for those that want to convert their own land into food gardens. City Farm intends to create a permanent training facility so that it can continue to expand its workforce training programme.

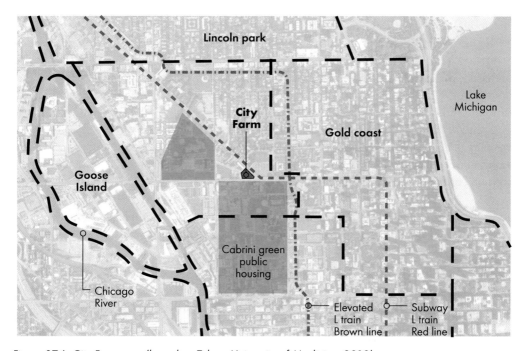

Figure 37.6. City Farm map (based on Tulane University of Medicine, 2010).

37.4 Seaming New Orleans

The following section identifies the current challenges faced by food security initiatives in New Orleans and provides recommendations for how food seams can mitigate these issues based on lessons learned from the Findlay Market, 32nd Street Market, and City Farm case studies.

Challenge: the small scale and infrequency of New Orleans' markets hinder their reliability.
Recommendation: food seams must be large-scale initiatives with permanent facilities and extensive food offerings in order for them to be efficient, reliable alternatives to traditional supermarkets.

If residents must supplement farmers market visits with visits to the local supermarket, farmers markets will become inconvenient and obsolete. Hence, food seams must have sufficient food options so that they can function as one-stop food shops. In order to offer a variety of food, these sites require permanent facilities equipped with electrical access, refrigeration, cooking surfaces, preparation areas, and storage. Ready-to-eat foods and beverages are critical components of food seams because they attract people who are looking to eat a meal but not necessarily purchase food that needs preparation. They also encourage people to linger, which increases social activity at the food seam.

Challenge: New Orleans' farmers markets lack a culturally and economically diverse patronage.
Recommendation: food seams must offer diverse food options to attract diverse patrons.

Since food seams are located at borders between neighbourhoods, it is imperative that they provide diverse food offerings that are valued by residents in all surrounding communities. Food seams can attract economically diverse patrons by providing both affordable and more expensive food retail as demonstrated by the Findlay Market case study. Like 32nd Street Market, food seams with ethnically diverse food options and market vendors are more likely to attract ethnically diverse patrons and foster a sense of belonging among them.

Challenge: New Orleans' farmers markets have low attendance.
Recommendation: by incorporating diverse programmes into food seams, these sites can attract more visitors and become socio-cultural anchors in the community.

Food seams can become social venues by hosting music performances, festivals, cooking demonstrations by local chefs, parades, community meetings, and special events for children.

Recreation space, such as playgrounds and athletic fields, can encourage physical fitness. Food seams can also provide an opportunity to engage youth in art and food education projects. Since borders are oftentimes comprised of neglected, blighted properties, these art projects can create vibrancy and optimism in such areas. Food seams can also partner with existing community organisations to strengthen their presence in neighbourhoods. City Farm demonstrates how urban agriculture can transform borders into productive landscapes. Food seams can include urban farming projects to help create jobs and foster a sense of community among otherwise disparate neighbourhoods.

Challenge: many of New Orleans' farmers markets are inaccessible to the underserved populations of the city.
Recommendation: for maximum visibility and accessibility, food seams should be located on main thoroughfares that have access to public transportation.

Many of New Orleans' markets are buried deep within neighbourhoods and are not highly visible. These food nodes tend to attract people from one particular community. Food seams at borders, on the other hand, will put food retail within walking distance of multiple communities rather than just one. By developing food seams on or near public main streets, they are likely to receive more exposure and be near existing public transportation routes. If food seams are also near existing retail, they feed off patrons from those venues to increase their own patronage.

Challenge: New Orleans farmers markets are not affordable for the underserved populations.
Recommendation: food seams must be located among economically diverse communities.

By providing both affordable and expensive food retail at food seams, revenue from affluent patrons that purchase the more expensive, gourmet food can help financially stabilise the market, enabling it to provide affordable food options for lower income patrons. Food seams can also develop finance programmes, like 32nd Street Market and the Crescent City Farmers Market, in which the market matches a certain US Dollar amount of what patrons spend in EBT/WIC food stamps. This amount can then be spent on fresh produce to encourage low income residents to make healthy food selections.

Challenge: many of New Orleans' farmers markets lack the funding they need to be large-scale projects.
Recommendation: food seams must seek opportunities for public and private investment.

By combining financing from public entities such as local government and advocacy groups with that of private banks and foundations, food seams can receive the funding they need to become large scale projects (The Reinvestment Fund, The Food Trust, and Greater Philadelphia Urban Affairs Coalition, 2009). Advocacy organisations like the Food Trust in Pennsylvania can help procure support for these projects by creating task forces that can educate the public, business leaders, and policy makers about the need for food retail development.

37.5 Conclusion

In order to address the problem of food insecurity in New Orleans, a food master plan should developed for the city that identifies potential food seam sites and strategies for how to make these sites attractive, accessible, and affordable for low income residents. The aforementioned recommendations provide guidelines for where and how to develop these food seams. The next step will be to further analyse each selected site and its surrounding neighbourhoods to determine the appropriate food offerings and programmes, assess accessibility, and identify funding opportunities. The development of food seams in New Orleans can help the city become internationally renowned for not only its food culture but also its innovative approach to food security.

References

Custer, S., 2009. Healthy Corner Stores for Healthy New Orleans Neighborhoods. Congressional Hunger Center. Available at http://healthycornerstores.org/wp-content/uploads/resources/NOLA_Healthy_Corner_Stores_Toolkit.pdf.

Everett, B., 2010. Katrina the Seam[STRESS]: Rebuilding through flood architecture. In: ReBuilding: Proceedings of the 2010 98th Annual Meeting in New Orleans. ACSA Press, Washington DC, USA, pp. 212-220.

Fisher, A., 1999. Hot peppers and parking lot peaches: evaluating farmers' markets in low income communities. Community Food Security Coalition, Portland, OR, USA. Available at http://www.foodsecurity.org/pubs.html.

Goody Clancy, 2010. The Plan for the 21st century: New Orleans 2030. Available at http://www.nolamasterplan.org.

Jacobs, J., 1961. The Death and life of great American cities. Vintage, New York, NY, USA, pp. 257-269.

Lynch, K., 1960. The image of the city. MIT Press, Cambridge, USA, 194 pp.

New Orleans Food Policy Advisory Committee, 2008. Building healthy communities: expanding access to fresh food retail. Available at http://www.sph.tulane.edu/ PRC/pages/FPAC.htm.

The Reinvestment Fund, The Food Trust, and Greater Philadelphia Urban Affairs Coalition, 2009. Pennsylvania Fresh Food Financing Initiative. The Reinvestment Fund, Philidelphia, PA, USA. Available at http://www.trfund.com/resource/downloads/Fresh_Food_Financing_Initiative_Comprehensive.pdf.

Interviews

Coffman, D. Personal interview. 2 Aug 2010.

Rey, V. Telephone interview. 12 Aug 2010.

Vance, A. Telephone interview. 28 Jul 2010.

Chapter 38

The CPUL City Toolkit: planning productive urban landscapes for European cities

Katrin Bohn[1] and André Viljoen[2]
[1]*Fachgebiet Stadt & Ernährung, Technische Universität Berlin, Fakultät VI – Planen Bauen Umwelt, ILAUP – Institut für Landschaftsarchitektur und Umweltplanung, Sekretariat EB12, Straße des 17. Juni 145, 10623 Berlin;* [2]*School of Architecture and Design, Faculty of Arts, University of Brighton, Mithras House, Lewes Road, Brighton BN2 4AT, United Kingdom; katrin.bohn@tu-berlin.de*

Abstract

Cities across the world seek policy guidance, good practice examples and further evidence for the impact of urban agriculture, and its relationship to a viable and sustainable food policy. In Europe, the potential environmental and socio-cultural benefits of introducing productive landscapes into cities are now widely acknowledged, although not (yet) to the extent that they are manifest as essential urban infrastructure. This chapter explores ways in which designers and planners can continue to play a significant role in conceiving, advocating and delivering the integration of sustainable food systems into the urban fabric. The authors will summarise 10 years of design and research work on Continuous Productive Urban Landscape (CPUL), and will review their evolving *CPUL City* concept in the context of two European cities: Berlin and London. The chapter will focus on the historic lessons, current practices and future strategies learned from these cities and present a first summary of specific proposals for guidance on implementing productive urban landscapes. This guidance – the *CPUL City Toolkit* – aims to provide an overview of the key steps necessary when planning and implementing urban agriculture as part of coherent productive urban landscape strategies. Four methods of action defining the Toolkit will be introduced: Action U+D (Bottom Up + Top Down), Action VIS (Visualisation), Action IUC (Inventory of Urban Capacity), Action R (Design Research). The chapter concludes with a reflection on the rapidly evolving practice and policy in Berlin and London related to the CPUL City Toolkit, as CPUL components begin to move 'out of the gallery' and into everyday urban infrastructure.

Keywords: Continuous Productive Urban Landscape, urban agriculture, sustainable urban planning and design, urban and regional infrastructure, urban food systems

38.1 CPUL: essential infrastructure

Cities across the world are seeking policy guidance, good practice examples and further evidence for and about the impact of urban agriculture. Since 2005, the authors, for example, have been asked to present their CPUL City concept to public and professional audiences in Canada, Cuba, Denmark, France, Germany, Italy, Ireland, the Netherlands, Norway, Portugal,

Spain, Sweden, Switzerland, the UK and the US. Additionally, invited articles about the concept have been published widely, including in China, Korea and Russia.

Central to the CPUL concept is the coherent integration of urban agriculture into interlinked, multi-functional – productive – open urban space networks that complement and support the built environment. CPUL advocates such productive landscapes as essential elements of sustainable urban infrastructure (Viljoen and Bohn, 2009, 2005a). CPUL is a physical and environmental design strategy and the concept proposes that urban agriculture can contribute to more sustainable and resilient food systems while also improving the urban realm. The *CPUL City* concept provides a strategic and associative framework for the theoretical and practical exploration of implementing productive landscapes within existing and emerging cities (Bohn and Viljoen, 2010) (Figure 38.1).

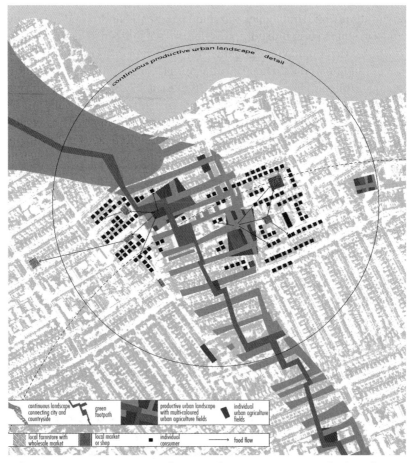

Figure 38.1. The CPUL concept. Green corridors provide a continuous network of productive open space containing foot paths and cycle ways. Fields for urban agriculture and other outdoor work and leisure activities are located within the network and serve adjacent built-up areas (Bohn&Viljoen Architects, 2002).

Within the CPUL concept, urban agriculture refers in the main to fruit and vegetable production, as this provides the highest yields per square metre of urban ground (see Chapter 36 by Lee). Key features of CPUL are food growing, leisure and commercial outdoor spaces shared by people, natural habitats, ecological corridors and non-vehicular circulation routes. Its network connects existing open urban spaces, maintaining and, in some cases, modifying their current uses.

CPUL impacts on a city qualitatively with respect to citizens' experience and quantifiably with respect to reduced negative environmental impact (Viljoen and Bohn, 2005b). The concept recognises that each site and city will present a unique set of conditions and competing pressures informing the final shape and extent of its productive landscapes. It envisages a 'mixed economy' of growers practicing urban agriculture: projects for the community and by the community, small scale and large scale, commercial and communal (Figure 38.2).

The CPUL concept grew out of the authors' design research exploring the role of urban agriculture within urban design and was first designed for and then defined by Bohn&Viljoen Architects respectively in 1998 and in 2004 (Viljoen and Bohn, 1999; Viljoen *et al.*, 2004; Viljoen and Tadiveau, 1998).

These studies, as well as the research of statistical, mostly UK-centred data, resulted in the CPUL City concept being underpinned by a number of interrelated social, environmental, economic and design arguments for what would amount to a radical change in the configuration and programming of open urban space. The overarching desire was to find more self-sustaining ways of living (Viljoen and Bohn, 2000) (Figure 38.3).

The CPUL concept has benefitted from favourable comment from activists including Rob Hopkins, founder of the Transition Towns Network (Hopkins, 2006), and is cited by academics and practitioners (Cultivate Kansas City 2011; Hodgson *et al.*, 2010; Mougeot 2005; Smit 2005; Taylor Lovell and Johnston 2009).

Figure 38.2A,B. Imagine a CPUL as an open urban space where intensive urban agriculture and convivial outdoor places for residents complement each other and are designed and built into a coherent infrastructural landscape (Figure 2A, Bohn&Viljoen Architects 'Cuba: Laboratory for urban agriculture', 2002; Figure 2B, Bohn&Viljoen Architects 'The Continuous Picnic', 2008).

Figure 38.3. Unlocking Spaces project in Brighton: a public event showing how a CPUL might transform a mono-used public space. The event was designed and run by Bohn&Viljoen Architects in collaboration with local residents and the University of Brighton and funded through a 'Creative Campus Initiative' award (Jonathan Gales, 2010).

38.2 The role of planning and design in raising awareness for food-productive urban landscapes

For urban agriculture, a solid body of literature exists since the 1990s. This concentrates on urban agriculture's positive impact with respect to food security, public health and income generation in places with high levels of social and economic deprivation (Cruz Hernández and Sánchez Medina, 2003; Egziabher *et al.*, 1994; Koc *et al.*, 1999; Mougeot, 2005).

The publication in 1996 of the book *Urban agriculture: food, jobs and sustainable cities* (Smit, 1996) was a landmark in defining an international role for urban agriculture and may be considered seminal to a sequence of publications, academic and popular. While planning for urban agriculture had already been on the development agenda, the publication in 2005 of *CPULs* (Viljoen, 2005) was the first time a book was devoted to presenting a design strategy for the coherent integration of urban agriculture into cities.

Within design disciplines, the dissemination of new ideas takes place as much through the medium of exhibitions as through the publication of academic papers. In these disciplines, a rapid increase in interest, exploration and dissemination of ideas about designing urban space for productive landscapes/urban agriculture is evident. In Europe, the breakthrough in the exploration of design consequences and possibilities arising from urban agriculture was reached in 2007, when the Netherlands Architecture Institute (NAi) Maastricht curated an exhibition titled 'De Eetbare Stad/The Edible City' (Anonymous, 2007). This brought together an international group of leading architects, artists and designers all, at that time, exploring urban agriculture within their work. Since then, the number of similar exhibitions and 'public works' hosted by leading international design institutions has continued to increase (Figure 38.4).

The subject's closeness to low-energy and sustainability debates, its ability to synthesise seemingly unconnected issues and the fascination with scenarios for an urban future, may be the reason for the notable presence of architects in the early moments of this 'movement'.

Today, the CPUL (City) concept is complemented by other concepts for integrating urban agriculture into contemporary Western cities. Often these start from an interest different to CPUL and result in a different set of proposals, but all have begun to explore the design possibilities of growing food within the urban realm. Most notably, these are Carolyn Steel's *Sitopia* (Steel, 2008), Dickson Despommier's *Vertical Farms* (Despommier, 2010) and CJ Lim's *Smartcities* (Lim and Liu, 2010).

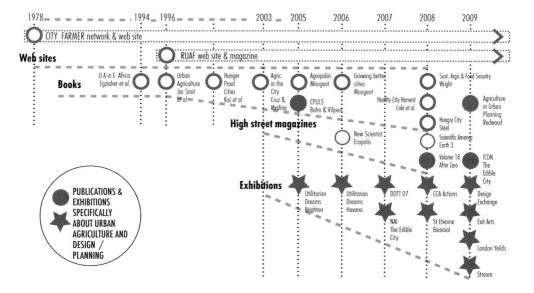

Figure 38.4. The increasing number of exhibitions about urban agriculture and CPUL hosted by arts and architecture institutions and galleries indicates how these subjects are entering the international architectural and urban design discourse (the chart is not exhaustive, but reflects trends evident to the authors in their practice) (Bohn&Viljoen Architects, 2009).

38.3 The CPUL City Toolkit

The lack of policy and design guidance on urban agriculture has not prevented the establishment of successful initiatives like the Prinzessinnengarten (Nomadisch Grün GmbH, 2011) in Berlin or the Capital Growth (Capital Growth, 2010) programme in London. In many respects, practice is outstripping policy, but development can be very contingent and often lacks coherence. In some cases, with Detroit being the prime example, different development strategies and approaches can result in a highly contested environment, where issues of food sovereignty, political and economic approach can polarise opinion (Gallagher, 2010).

The reasons for this uneven development are complex and intertwined and include:
• the complex nature of urban food systems;
• different local contexts for countries, cities and individual sites, including their different food cultures, physical and logistical site conditions and trading patterns;
• diverse agricultural practices and organisational structures for urban agriculture operations;
• lack of long term experience with urban agriculture projects, other than for allotments and community gardens;
• the current lack of evaluation of comparable projects and the inconsistent dissemination of transferable knowledge;
• competition with commercial developers for valuable urban land within expanding cities and the lack of resources for infrastructure projects within shrinking cities;
• scepticism regarding urban agriculture's legitimacy as an urban land use.

With this in mind we are developing a CPUL City Toolkit as a planning and design guide for implementing more localised urban food systems. The Toolkit aims to meet the demand for systematic, practical, graphically descriptive and transferable know-how. Some of these aspects are covered elsewhere, for example the Transition Towns' *Local food* book has systematic, practical and transferable information aimed at communities with engaged activists (Pinkerton and Hopkins, 2009).

The CPUL City Toolkit, however, will work within a larger framework referencing existing best practice where it exists, while also describing a spatial urban design strategy. Whilst the concept can be thought of in terms of 'localisation agendas', it explicitly recognises the need for coherent municipal planning frameworks to manage its infrastructural implementation. The Toolkit therefore aims to address activists and the design, planning and governance professionals who can provide a managed and strategic overview at the level required for implementing productive urban landscapes as urban infrastructure.

Our work has led us to conclude that four distinct methods of action are most relevant to the architectural, urban design and planning professions. The methods acknowledge a need for inter- and trans-disciplinary action, while also helping to define particular tasks within the competency of an individual. These four methods of action define the CPUL City Toolkit, and what we have provisionally titled the 'CPUL Clover Methods' (Figure 38.5).

38.3.1 Action U+D = Bottom Up + Top Down

Infrastructural projects such as CPUL need parallel top-down and bottom-up initiatives. An urban agriculture project will have the best chance of long-term success, when it can rely on local initiators and supporters and when these enter negotiation processes with their local authorities or municipalities. Cuba's organoponicos (Viljoen, 2005) and New York's Green Thumb initiative (see chapter by Mees and Stone) are good examples. A spectrum of bottom-up motivations can be identified ranging from community-led to entrepreneurial initiatives. Within each category, further differentiation can be noted, for example community-led programmes that are driven by imperatives of empowerment and ex-/inclusion, e.g. Growing Power in Milwaukee (2011), Berlin's Intercultural Gardens Gardens (Stiftung Interkultur, 2011), or those in more affluent areas that aim for broader educational and lifestyle choices, e.g. Fortis Green Community Allotments Trust (2010) in London. Entrepreneurial-led projects similarly range from those advocating small-scale, but individually viable market gardens, e.g. the social enterprise Growing Communities (2011) in London to larger scale 'corporate' approaches, e.g. Hantz Farms in Detroit (Hantz Farms, 2009). What is required now, is the evolution of policy to support diverse bottom-up initiatives and accommodate these within a coherent framework that adds to urban experience, urban resilience and the quality of urban space.

38.3.2 Action VIS = visualising

The qualities of CPUL and urban agriculture need visualising to raise public awareness and influence decision makers. This is one of the primary skills of architects, planners and designers. To further the case for urban agriculture, this role widens to include alongside the design of productive urban spaces also the public and visually descriptive dissemination of ideas, data and best-practice examples in the form of exhibitions, installations, talks, websites and publications. Here, the professional becomes the agent of change, which carries on a long, and at times problematic, tradition of the architectural manifesto as a herald of future change and challenges. Within this action, the CPUL City concept has aimed from the outset to underpin its 'vision' with a concrete body of research.

Figure 38.5. The CPUL Clover Method. Four tools of action for use to enable the successful implementation of an urban agriculture project (Bohn&Viljoen Architects, 2010).

38.3.3 Action IUC = inventory of urban capacity

An 'inventory of urban capacity' is necessary, especially of spatial, stakeholder and managerial capacities. At the beginning of the relatively short history of the urban agriculture movement, planning emphasis was given to identifying and mapping available urban space (soil quality, pollution, water, exposure, adjacency to markets and compost) as shown, for example, in the city of Portland's *Diggable City* report (Balmer *et al.*, 2005) or the *Elephant & Castle Study* for London (Tomkins, 2009). In recent years, it has become clear that stakeholder and managerial/maintenance capacity is as important. Evidence for a few different approaches is becoming available, e.g. the increase of local growing capacity through active community inclusion work by the Bankside Open Space Trust in London (B.O.S.T., 2011) or the increase of maintenance capacity when shared between council and urban farmer as practised by Lichtenberg council and the Agrarbörse in Berlin (Agrarbörse, 2011). While available space is finite (although often under-estimated), stakeholder and managerial capacity can be increased. One of the top-down approaches that has proven successful is the funding of extension programmes focussed on developing agricultural and managerial skills (business and social enterprise), most notably in Cuba (Viljoen, 2005).

38.3.4 Action R = design research

Constant research, development and consolidation of the CPUL concept is needed to adapt it to changing circumstances. Social and environmental conditions can change rapidly, locally and globally. To keep pace with such developments, but also to scrutinise the achievements of concepts such as CPUL, these strategies have to undergo continuous evaluation and evolution. Theory and practice need to be able to accommodate change. Applied design research is needed to develop different procedural, spatial and business models for different scales of production.

For example the prototyping and *in situ* testing of suitable growing techniques are two of the most efficient ways of evaluating design options. We have found that exhibitions and installations provide a good initial testing ground for what seem fairly straightforward propositions: Our designs for an 'urban agriculture curtain' and a 'growing balcony' (Bohn and Viljoen, 2009), both utilising hydroponic systems, have indicated that subtle variations in design, location and maintenance can have a significant impact on yield and ease of utilisation (Figure 38.6). Similar lessons can be expected with respect to business models and different scales of production.

38.4 London and Berlin: first steps in testing the Toolkit

London and Berlin provide the following early examples of different approaches to applying Action U + D:

Figure 38.6. The Urban Agriculture Curtain. Working prototype by Bohn&Viljoen for a vertical productive urban landscape as part of the exhibition 'London Yields'. The system developed with Hadlow Agricultural College utilises industry standard hydroponic components and produces fortnightly crops for use in the Building Centre's restaurant (Bohn&Viljoen Architects, 2009).

38.4.1 Cultivating the Capital – initiating a debate top-down

In January 2010, the Greater London Authority's Planning and Housing Committee published its report *Cultivating the Capital: food growing and the planning system in London* (London Assembly, 2010). It based its findings on a detailed consultation period with about 50 farmers, growers associations, government departments, food systems experts, government departments,including the authors of this chapter and an extended review of literature and best practice.

The report makes nine policy recommendations and calls for changes to the planning system to exploit the capital's potential to integrate commercially viable food growing in the city. It highlights the need for amendments to the *London Plan* and local authority planning policies to encourage food growing in London.

38.4.2 Capital Growth – a framework for supporting bottom-up initiatives

In 2009, in parallel with the preparation of *Cultivating the Capital*, a partnership initiative between London Food Link, the Mayor of London Boris Johnson, and the Big Lottery's Local Food Fund, established the 'Capital Growth' programme. Capital Growth is administered by the food-charity 'Sustain' and aims to establish 2012 new community-based urban food growing initiatives by the year 2012. The scheme builds capacity by offering 'practical help, grants, training and support to groups wanting to establish community food growing projects as well as advice to landowners' (Capital Growth, 2011).

38.4.3 The London Plan 2011

The London Plan, sets the planning framework for greater London, and in 2011, influenced by both *Cultivating the Capital* and *Capital Growth*, it included specific policy commitments to incorporate 'Land for food'. (Greater London Authority, 2011). This significant policy commits the city to: (a) support agriculture in particularly in the Green Belt (peri-urban agriculture); (b) encourage the use of land for food near to urban communities (urban and peri-urban agriculture); (c) require boroughs to identify potential land for food growing, may include roofs (urban agriculture).

38.4.4 Croydon's 2011 core strategy planning document

The London Plan is implemented by local authorities, and we can identify the first adopters of policy to encourage productive landscapes. The London Borough of Croydon is notable in having explored and consulted widely on urban agriculture (U + D), and has included Productive Landscapes as a specific land use category in its final draft core strategy planning document, scheduled for ratification in late 2011 (London Borough of Croydon, 2011).

38.4.5 Berlin's 'green vision' – a top-down responsive mode

Berlin's Senate Department for Urban Development publicly presented its draft for a 'green vision' (*Grünes Leitbild*) for the German capital in September 2010. In 2009, a team of two landscape architecture practices started working on the draft strategy supported by a series of expert think tanks. Following public presentation, Berlin residents were invited to comment on the draft.

'Natural. urban. productive' is the draft strategy's subtitle mirroring the three main directions in which the city seeks to channel its future open space planning. The draft does not demand urban agriculture, but explicitly includes it as a recognised use under the heading 'productive'.

> *"Productive' is defined as 'an urban interpretation of the cultured landscape, of open space, that is generated not only through its designers, but equally through its users.'*
> (Bohn and Giseke, 2010).

This inclusion of productive spaces into the draft strategy is also the result of campaigns by numerous groups of Berlin activists who advocate the innovative, uncomplicated and productive use of the many brownfield sites within the city. Most notable are '*urbanacker e.V.*' (2011) and '*AG Kleinstlandwirtschaft*', who make up the two strongest members of the Berlin grass-root network of gardening activists. They grew out of actual conflicts over public space uses which sharpened both, their and the council's view of issues around urban agriculture (Figure 38.7).

Whilst the two reports, *Cultivating the Capital* and *Grünes Leitbild*, might be considered the most important stepping stones in terms of the planned integration of food producing landscapes into London and Berlin respectively, the processes leading to their formulation highlight the different natures of the urban agriculture movement in each city.

The London and Berlin examples suggest that the Toolkit's method of action 'Bottom Up + Top Down' is beginning to be employed and is resulting in measurable outcomes. London already benefits from having the 'London Food Board' (top-down) and the food charity 'Sustain: The Alliance for Better Food & Farming' (bottom-up). Sustain has a national remit, but is very active in London, both with respect to advocacy and delivery of food-related projects (Sustain, 2011). The London Food Board 'is an advisory group of independent food

Figure 38.7. Urban Agriculture in Berlin. In Berlin, social productivity is an important driver of food growing initiatives. This map shows food growing sites with a strong social/communal aspect (Stadt und Ernährung, 2011)

policy organisations and experts which oversees the implementation of The Mayor's *Food Strategy: healthy and sustainable food for London*, published in 2006, and coordinates work and leads the debate on sustainable food issues in the Capital' (London Assembly, 2011). Both organisations are active in the policy arena and in raising municipal and institutional awareness of sustainable food systems and are now in the position to push for the implementation of urban agriculture projects.

Berlin does not have a Food Board or Food Policy Council, but early communication between the Senate (as expressed in the draft green strategy) (top-down) and gardening activists' network 'urbanacker' (bottom-up) resulted in a very promising experiment, executed on the site of the former Tempelhof Airport, which was closed in 2008.

Having been allocated space by the Senate, *'urbanacker'* is creating its most ambitious public communal urban agriculture project yet, the *'Allmende Kontor'* (The Common's Office). Described as a platform for urban agriculture, *Allmende Kontor* is now promoted by the Senate as an example of a successful Pioneer Project in the high-profile temporary uses at Tempelhof (Figure 38.8).

The London and Berlin examples both represent the start of a process. The report *Cultivating the Capital* has resulted in London adopting policy to support 'land for food' and in some cases specifically the concept of productive landscapes; Berlin's 'Green Vision' is on the way to being adopted as official policy for the city. So far the projects directly supported by each city are community-based (Capital Growth and Allmende Kontor), and both have a significant awareness raising component allied with good public relations potential for local politicians.

38.5 Conclusion and what happens next

London and Berlin are not unique as cities actively exploring the impact of productive urban landscapes and urban agriculture. As with other cities, the approaches they have taken lead to different results within their respective urban contexts, especially with regard to the type of urban agriculture initiatives supported and the role of design and the planning system. What is common to both is that their activities have already involved food growers, the

Figure 38.8. The Allmende Kontor. Six months after its opening in May 2011, the 5,000 m² site, one of the biggest in the 'Pioneer Project Scheme', is fully occupied by local food growers (Bohn&Viljoen Architects, 2011).

architecture/design/planning professions and local government authorities. It can also be seen that the projects cited directly involved city authorities, advocacy groups and non-commercial community-based food groups. London is examining commercial growing and the planning system, but direct results are not yet evident. Commercial initiatives do exist in each city, and it is likely that in the future commercially focused local food projects will be integrated into the larger planning system. Although beyond the scope of this chapter, this is already beginning to occur in other cities, e.g. Milwaukee and New York.

With their work on planning guidance documents including urban agriculture, Berlin and London have joined other cities worldwide by entering a new territory. These relatively broad, city-wide discussions and their position within the cities' planning departments represent a change in municipal attitude towards urban agriculture.

In both cases, the willingness of the local governments to engage with food growing issues has lead to increased interest in their work by local citizens: Berlin's activists are participating in a constructive exchange with the Senatsverwaltung für Stadtentwicklung, and Londoners have experienced greater support for the establishment of community based food-growing.

However, it would be naïve to conclude, that the role of the architect or planner has diminished with the worldwide interest in urban agriculture. Most of the strategic and urban design work still has to be done. There is a difference between recognising the need to change the current urban food system and to have identified, designed and implemented meaningful, extensive, appropriate and sustainable solutions.

The CPUL City Toolkit with its condensed list of actions aims to be a method for finding such design solutions and critically evaluating them. It provides a framework of actions to generate debate, disseminate ideas and facilitate implementation; its use and success will be tested and depend on the degree to which its actions are adopted. It is not unlikely that the cases quoted will soon be overtaken by more rigorous, successful or unique examples of European urban agriculture. But we believe that the 4 methods of action presented are robust enough to be able to accommodate new insights for quite some time.

The architectural, design and planning professions have the specific skills to design space or design into space. They become agents of change through inter-disciplinary work with both, community-led and local government initiatives. At no time, will they replace community-led food-growing initiatives. However, it might be through these professions that the processes of implementing spatial design solutions for urban agriculture can be managed in a sustainable, inter-disciplinary, spatially coherent and long-lasting manner.

References

Agrarbörse, 2011. Agrarbörse Deutschland Ost e.V. Available at: http://www.agrar-boerse-ev.de/.

Anonymous, 2007. De Eetbare Stadt. naim/bureau europa. Available at http://www.bureau-europa.nl/tentoonstellingen/archief/352-de-eetbare-stad.

Balmer, K., Gill, J., Kaplinger, H., Miller, J., Peterson, M., Rhoads, A. Rosenbloom, P. and Wall, T., 2005. The diggable city: Making urban agriculture a planning priority. prepared for the city of Portland. Nohad A. Toulan School of Urban Studies and Planning, Portland State University, Portland, OR, USA. Available at http://www.diggablecity.org/.

Bohn&Viljoen Architects, 2009. The Urban Agriculture Curtain and Growing Balconies. Available at http://www.bohnandviljoen.co.uk.

Bohn K. and Giseke U., 2010. Reiche Ernte für die Stadt. Garten und Landschaft 12/2010: 26-28.

Bohn K. and Viljoen A., 2010. Continuous productive urban landscape (CPUL): Designing essential infrastructure. Landscape Architecture China 2010: 24-30.

B.O.S.T., 2011. Bankside Open Spaces Trust. Available at http://www.bost.org.uk/.

Cruz Hernández, M. and Sánchez Medina, R., 2003. Agriculture in the city: A key to sustainability in Havana, Cuba. Ian Randle Publishers, Kingston, Jamaica, 210 pp.

Capital Growth, 2011. What's the big idea? Available at http://www.capitalgrowth.org/big_idea/.

Cultivate Kansas City, 2011. Urban agriculture and urban planning & design. Available at http://www.cultivatekc.org/resources/planning-design.html.

Despommier, D., 2010. The vertical city farm: feeding the world in the 21st century. Thomas Dunne Books, St Martin's Press, New York, NY, USA, 305 pp.

Egziabher, A., Lee-Smith, D., Maxwell, D., Mernon, P., Mougeot, L. and Sawio, C., 1994, Cities Feeding People: An Examination of Urban Agriculture in East Africa, International Development Research Centre, Ottawa, Canada.

Fortis Green, 2010. Welcome to the home of the Fortis Green Community Allotments Trust. Available at http://www.fortisgreenallotments.co.uk/.

Gallagher, J., 2010. Reimagining Detroit: Opportunities for redefining an American city. Painted Turtle Book, Detroit, USA, 166 pp.

Greater London Authority, 2011. The London Plan. Greater London Authority, London, UK, pp. 240.

Growing Communities, 2011. *Growing Communities*: Hackney's organic farmers' market and box scheme. Growing Communities, London, UK. Available at http:/ www.growingcommunities.org.

Growing Power, 2011. Growing Power, Inc. Available at: http://www.growingpower.org/.

Hodgson, K., Caton Campbell, M. and Bailkey, M., 2011. Urban agriculture: Growing healthy, sustainable places. (PAS 563), American Planning Association, Planning Advisory Service, Chicago, IL, USA.

Hopkins, R., 2006. Review of CPULs – Continuous productive urban landscapes. Designing urban agriculture for sustainable cities. Transition Culture. Available at http://transitionculture.org/essential-info/book-reviews/cpuls/.

Hantz Farms Detroit, 2009. Introducing Hantz Farms™. Available at http://www.hantzfarmsdetroit.com/introduction.html.

Koc, M., Macrae, R., Mougeot, L. and Welsh, J. (Eds.), 1999. For hunger-proof cities sustainable urban food systems. International Development Research Centre, Toronto, Canada.

Lim, C. and Liu, E., 2010. Smartcities + eco-warriors. Routledge, Oxfordshire, UK, 253 pp.\

London Assembly, 2010. Cultivating the Capitol: Food growing and the planning system in London. The London Assembly, London, UK.

London Assembly, 2011. What is the London Food Board? The London Assembly, London, UK. Available at http://www.london.gov.uk/london-food/general/what-london-food-board.

London Borough of Croydon, 2011. Croydon Development Plan Document – Proposed Submission for Publication. London Borough of Croydon, UK 60 pp.

Mougeot, J.A. (Ed.), 2005. Agropolis: The social, political and environmental dimensions of urban agriculture. Earthscan, London, UK, 12 pp.

Nomadisch Grün GmbH, 2011. Prinzessinnengarten, Urbane Landwirtschaft. About us. Available at http://prinzessinnengarten.net/about/.

Pinkerton, T and Hopkins, R., 2009. Local food: how to make it happen in your community. Transition Books an imprint of Green Books, Totness, UK, 216 pp.

Smit, J. (Ed.), 1996. Urban agriculture: food, jobs and sustainable cities. United Nations Publications, Blue Ridge Summit, PA, USA, 328 pp.

Smit, J., 2005. The ethics of urban agriculture. The urban agriculture network, Washington DC, USA. Available at http://www.cityfarmer.org/EthicsUA.html.

Steel, C., 2008. Hungry city: how food shapes our lives. Chatto & Windus, London, UK, 400 pp.

Stiftung Interkultur, 2011. Übersicht aller Interkulturellen Gärten in Berlin. Available at: http://www.stiftung-interkultur.de/berlin.

Sustain, 2011. About Sustain. Available at http://www.sustainweb.org/about/.

Taylor Lovell, S. and Johnston, D.M., 2009. Creating multifunctional landscapes: how can the field of ecology inform the design of the landscape? Frontiers in Ecology and the Environment 7: 212-220.

Tomkins, M., 2009. The elephant and the castle; towards a London edible landscape. The Urban Agriculture Magazine 22: 37-38

Urbanacker, 2011. urbanacker.net. Available at http://urbanacker.net/.

Viljoen, A. and Tardiveau, A., 1998. Sustainable cities and landscape patterns. In: Proceedings of Passive and Low Energy Architecture (PLEA) 98 Conference, Lisbon, Portugal, PLEA. pp. 49-52.

Viljoen, A. and Bohn, K., 1999. Elasticity. In: Europan 5 – Europan Results, Paris, France, pp 285.

Viljoen, A. and Bohn, K., 2000. Urban Intensification and the Integration of Productive Landscape. In: Proceedings of the World Renewable Energy Congress VI, Part 1, Pergamon, Oxford, UK, pp 483-488.

Viljoen, A., Bohn. A. and Pena Diaz, J., 2004. London Thames gateway: Proposals for implementing CPULs in London Riverside and the Lower Lea Valley. Report for the Architecture and Urbanism Unit of the London Assembly. University of Brighton and CUJAE Havana, Brighton, UK.

Viljoen, A. and Bohn K., 2005a. Don't forget the food – rural agriculture and urban agriculture. Littoral Rural Design Forum Conference "Architecture and Agriculture", DEFRA Conference Centre, York, UK.

Viljoen, A. and Bohn K., 2005b. Continuous Productive Urban Landscape – Urban agriculture as essential infrastructure. Urban Agriculture Magazine, 15: 34-36.

Viljoen, A. and Bohn K., 2009. Continuous Productive Urban Landscape (CPUL) Essential Infrastructure and Edible Ornament. Open House International 34: 50-60.

Viljoen, A., 2005. Continuous Productive Urban Landscapes: Designing Urban Agriculture for Sustainable Cities. Architectural Press, Oxford, UK.

Chapter 39

Designing multifunctional spatial systems through urban agriculture: the Casablanca case study

Christoph Kasper, Undine Giseke and Silvia Martin Han
Technical University Berlin (TU Berlin), School VI, Planning Building Environment, Chair of Landscape Architecture, Open Space Planning, Straße des 17. Juni 152, 10623 Berlin, Germany; christoph.kasper@tu-berlin.de

Abstract

One of the most powerful drivers of global change is urbanisation and the complex interactions of urban areas with their physical environment. With this in mind, in 2005 the German Federal Ministry of Education and Research established a research programme on *Future Megacities*. One of these research projects, Urban Agriculture Casablanca (UAC), investigates to what extent urban agriculture can make a relevant contribution to the design of a sustainable city. This project seeks to include the provision of open space within integrated and sustainable urban growth in the emerging megacity Casablanca, Morocco. The conceptualisation of urban agriculture is a strategy that could offer a way to integrate green productive infrastructures (GPI) to food planning. The spatial system is to be understood as a working and living environment for a portion of the urban population, meaning that urban farmers become spatial producers of open space and, at the same time, of a new urban milieu – the *rurban*. The specific structure of megacities and the discussion about resource-efficient and climate-optimised urban development also raises the issue of how urban-rural linkages within an urban region should be dealt with and what positive interactions between the city and agriculture there might be. These urban-rural linkages have the potential to create improved livelihoods in combination with spatial integration: to develop climate-optimised, multifunctional spatial systems. Megacity dynamics require the linking of top-down and bottom-up strategies, as well as the development of tailor-made solutions. Thus, the execution of integrated case studies in peri-urban spaces and the implementation of four pilot projects were of central importance in the research work. Based on the action research approach, the pilot projects specify the synergy potentials between city and agriculture in connection with other relevant sectors of the social and economic system.

Keywords: urban rural linkages, rurban, productive landscape, action research, green productive infrastructure

Christoph Kasper, Undine Giseke and Silvia Martin Han

39.1 Introduction: productive landscapes as a constructive urban element in the emerging megacity Casablanca

One of the most powerful drivers of global change is urbanisation, including socio-economic transformations and the complex interactions of urban areas with their physical environment. With this in mind, in 2005 the German Federal Ministry of Education and Research (BMBF) established a research programme on *Future Megacities* in order to develop energy- and climate-efficient structures in urban growth centres in newly industrialising and less developed countries.[41]

The challenges faced by future megacities, to list but a few, include considerable – and partially uncontrollable – spatial growth, fragmented spaces, substantial population growth, the increasing divide between rich and poor, problems of providing adequate housing, the guaranteeing of adequate environmental and living standards, and the maintenance of a technical infrastructure – particularly for transportation – as well as the challenges posed by looming climate change. Within the framework of this programme, the BMBF's ten research projects concerning the future of urban growth centres around the world cover many of these topics. This is, however, the first time that agriculture is being discussed as an integrative factor of urban development, as it is here in the Casablanca project. And thus urban research is putting more emphasis on the previously neglected issue of productive open space systems, as well as on the question of methods of urban food production.

The discussion of potential open space systems for urban growth centres is thus being given a new dimension: It no longer merely concerns their contribution to recreation and ecological compensation, but also to the issue of feeding the city. The fact that productive landscapes can make a contribution to feeding cities in the sense of a food planning system and that they should be integrated into the planning process is self-evident. And yet the specific structure of megacities and the discussion about resource-efficient and climate-optimised urban development also raises the issue of how urban-rural linkages within an urban region should be dealt with and of what positive interactions between the city and agriculture there might be. This is exactly what the Casablanca project presented in this chapter focuses on.

According to UN estimates made in 2004 with regard to population statistics in agglomerations in emerging and developing countries, Casablanca can expect to have considerable population growth from 3.6 million residents in 2003 to 4.5 million in 2025 (Royaume du Maroc, 2008: 271). Casablanca therefore belongs to the category of emerging megacities that the BMBF's research programme is emphasising.

---

[41] Federal Ministry of Education and Research Funding Programme (2005-2013): *Research for the sustainable development of the megacities of tomorrow*, Focus: energy and climate-efficient structures in urban growth centres. UAC – Urban Agriculture as an Integrative Factor of Climate-Optimised Urban Development, Casablanca / Morocco.

Due to accelerated demographic change that has led to slower population growth, the five million inhabitant mark will probably not be met until 2025. Nevertheless, the latest statistics show that the Casablanca region is the largest urbanised region in the Kingdom of Morocco and the 13th largest city in Africa (UN-HABITAT, 2010: 53).

Casablanca – in Arabic *Dar el Beida* – continues to serve as the country's economic motor. 60% of Moroccan industry is concentrated in Casablanca, which is also North Africa's largest harbour. At the same time, deindustrialisation, the growth of the tertiary sector, and above all the failure to provide housing coupled with climate-led increases in rural migration to the cities, are triggering profound transformational processes and producing new and dynamically changing land use patterns (Figure 39.1). According to the megacity types of Hall and Pfeiffer (2000) and regarding different indicators, Casablanca can be classified more in terms of dynamic growth than informal hyper-growth: it is a specific type of city that can be learned from.

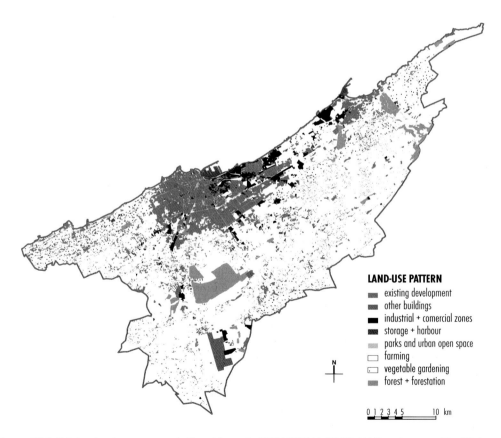

LAND-USE PATTERN
- existing development
- other buildings
- industrial + comercial zones
- storage + harbour
- parks and urban open space
- farming
- vegetable gardening
- forest + forestation

0 1 2 3 4 5   10 km

Figure 39.1. Existing land use patterns in Casablanca in 2004 (SDAU, 2008; UAC project graphics/Christoph Kasper).

39.2 The conceptualisation of urban agriculture as a multifunctional open space system

This project is conceptually based on the thesis that it makes sense to develop an open space system for a mega-urban area by using existing agricultural space. These areas have a special potential to become multifunctional spatial systems that can provide a variety of answers to today´s urban challenges, including the ability to provide a city with food. Thinking about the potential of urban open space in this way means overcoming the modern monofunctional idea of open space and viewing open space systems as a productive green infrastructure (PGI) that interacts with the city on different – ecological, social, and cultural – levels, thus creating new synergies.

This means – as an overall goal of the project – converting agricultural areas into a productive and green infrastructure by establishing a more varied system of agricultural cultivation and gaining co-benefits through strengthening its multifunctionality. According to PRIEMUS, multifunctional urban land use means the implementation of more functions in a determined place in a determined period of time (Priemus *et al.*, 2004: 270). The character of multifunctionality is to create synergies by diversifying and interweaving the urban system within a demarcated area. Urban agriculture in itself can thereby already be seen as a conceptualisation of multifunctionality in the urban context, as it combines both the urban and rural spheres.

The following sub-goals have been identified in a common future search workshop and through the process of scenario writing. A productive green infrastructure (PGI):
- should contribute to the supply of urban food;
- should provide recreational and leisure opportunities;
- should contribute to resource efficiency and urban recycling management;
- should contribute to regulative services with regard to climate change;
- should integrate living-space functions and infrastructure functions;
- and should be beautiful.

One point within this approach should be given special emphasis: This spatial system is to be understood not merely as an urban infrastructure but also as a working and living environment for a portion of the urban population, meaning that urban farmers will become spatial producers of both open space and of a new urban milieu – the *rurban*. The previous rural form of living therefore becomes an integrated factor in urban development, generating *the rurban* as a new urban milieu (with specific spatial, functional, and social interconnections). This term has been generated within the UAC project and is the result of a different point of view when compared to other projects that focus on the European context (Esparcia and Buciega, 2005, Vanden Abeele and Leinefelder, 2007) with regard to matters of different planning tools but not in the sense of a new urban milieu. Another example of the concept of *rurban* can be found in the *Goa 2100 Project*, an entry in the international competition on Sustainable Urban Systems Design (Revi *et al.*, 2006). This concept follows an ecosystematical approach, but compared to the UAC approach components like economical practices and socio-cultural practices are missing.

To date five sub-concepts have been derived from these goals:

- Sub-concept 1: urban agriculture as a space for regional food production
  This sub-concept is concerned with the question of to what extent urban agriculture can contribute to the sustainable provision of nutrition for the city and, therefore, how much it can promote climate-optimised diets and urban-regional nutritional sovereignty. This question is primarily examined from a nutritional science, social, and cultural perspective, but above all in terms of its spatial impacts. The concept of urban agriculture is not only that agricultural production contributes to the generation of income and to household strategies of the population involved in the agricultural economy. It simultaneously poses the question of to what extent regional domestic production can contribute to nutritional sovereignty, not only for individuals but for the overall urban population, and therefore implicitly supports sustainable nourishment and in a wider sense the alleviation of poverty through the generation of income and value chains, and the establishment of green industries.

- Sub-concept 2: urban agriculture as beautiful, productive, and recreational space
  The provision of sufficient open space for leisure and relaxation is an important goal in current urban development policies, as it is in Casablanca. Urban agriculture supplements the spectrum of traditional urban open space such as parks, which were largely created during the process of colonial city expansion and are concentrated in the inner city. We examine the question of how urban agriculture can be productive in a classical sense while simultaneously serving as urban leisure and recreational space in a special cultural context. For example, one of the pilot projects deals with the questions of how peri-urban farmers – besides street-vending local products – can also offer regional tourist facilities to weekend visitors (weekend picnics are a widespread leisure activity in Morocco) to Casablanca, thus increasing their income.

- Sub-concept 3: urban agriculture and resource-efficient urban-rural cycles
  This conceptual building block concentrates on establishing interactive and resource-efficient cycles between agriculture and the city, and thus on the generation of co-benefits and synergies between the urban and the rural. One focal point is to use treated wastewater from urban areas for agriculture (Figure 39.2), which would allow for a considerable increase in irrigated farming *vis-à-vis* the current levels. Constructed wetlands can, in addition, be integrated into agricultural landscapes. This subsidiary concept is linked not only to specific technological solutions, but is accompanied by processes that have to be established over the long-term, such as the introduction of new forms of agricultural production and the acquisition of new cultural practices in interface management.

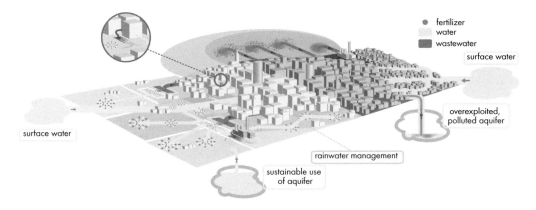

Figure 39.2. The idealised resource-efficient city vs. business as usual (UAC project graphics).

- Sub-concept 4: urban agriculture and climate regulative services
  Set against the background of climate variations and climate change, our project is particularly interested in the services that agriculture can provide regarding the regulation of the ramifications of climate fluctuations and extreme conditions, i.e. measures which would contribute to integrated risk management. Urban agriculture should, in addition to the mere production of food and open space, be systematically developed to a point that it becomes a regulative provision in adapting to climate change. In terms of the analysis to date, the focus here is on flood protection, flood management, alleviating the consequences of drought, and contributing to urban cooling.

- Sub-concept 5: urban agriculture as rurban living space
  In this sub-concept the perspective of the inquiry is directed toward the transformation of agricultural areas into space for living. In the first instance urban agriculture can be seen as the carrying out of agricultural production in an urban context, but at the same time it can also serve as a living area for some of the megacity's inhabitants. These persons make their living through their permanent activity in the productive landscape. In other words: they live and work in a rural sphere within an urban space that can be characterised as the afore-mentioned *rurban,* and the *rurbanite* becomes a new form of living (in the sense of farmers who adopt an urban lifestyle). Urban agriculture can thus create a new type of green urban infrastructure that is – contrary to parks – multifunctional and inhabited. The result will not be the anti-urban romanticisation of the rural common to the large western urban conurbations of the last few centuries, but instead new coexistences and synergies of the rural and the urban for a sustainable city. This mutual absorption creates new forms of coexistence and allows for the emergence of new long-term synergies, values, living strategies, and spatial structures.

39.3 Designing a spatial model for urban agriculture on a regional scale

The second basic approach of the UAC project is that the concept of an urban-regional open-space structure based on urban agriculture is a viable response to changing spatial patterns in urban growth centres, matched by a conscious integration of urban-rural linkages in a polycentric city. As opposed to previous concepts that concerned themselves with the interaction between the city and the countryside – such as Frank Lloyd Wright's Broadacre City (Wright, 1958) – the aim here is not a spatial and scattered weaving together of the two structures. Agricultural areas are to be developed as a productive green infrastructure for a compact city in which synergies between rural and urban structures are specifically generated at various spatial, functional, and socio-cultural levels. Until now urban agriculture as an integrative developing strategy for an open space system in mega-urban growth centres has not existed.

Within the project, the identification of potential spaces for productive landscapes and multifunctional space systems has thus been undertaken from several different perspectives. An important basis for understanding the mechanism of spatial production is knowledge of agricultural spaces, their different inhabitants, their practices, productivity, and soil qualities, and equally, how they are perceived and valued within the region.

In the spatial model different types of agricultural landscape correspond to the urban context and extend from inner-city micro areas in single buildings and districts to expansive areas of intense production on the urban periphery. The project's main focus, however, is on today's peri-urban spaces.

The quantitative orientation framework of the current master plan (Royaume du Maroc, 2008) forms the basis for the first modelled approaches to the development of a green productive infrastructure based on urban agriculture. An important success within the first phase of the project is the fact that agriculture has been considered an urban planning category in the current development of the master plan, and that nearly 60,000 ha within the urban region's administrative borders have been designated as agricultural land. Figure 39.3 shows the spatial distribution of nine identified categories of urban agriculture as they are linked to urbanised areas.

The goal is to develop specific strategies and action plans to qualify the different categories of rural-urban linkages by a dual-track urbanism. An urban planning and open space planning research competition carried out in September 2010 has already made an important contribution to this (Giseke, 2011).

39.4 The implementation of pilot projects as a first result of the action research approach

Besides the concretisation of the conceptual approach and the in-depth sector analyses, the execution of integrated case studies in peri-urban spaces and the implementation of four pilot

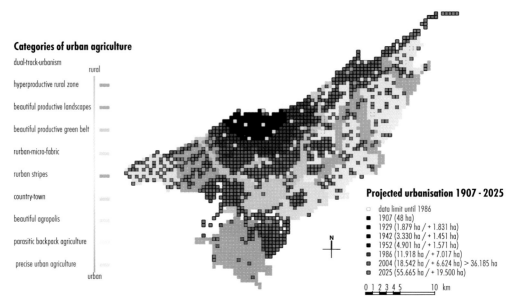

Categories of urban agriculture

Figure 39.3. Nine categories of urban agriculture linked to urbanised areas (SDAU, 2008; UAC project graphics/Christoph Kasper).

projects was of central importance in the research work done up to now. Here the focus will be on the pilot projects.

In the preparatory phase different potential synergies between the city and agriculture were discussed, of which four were considered to be very fruitful and were thus subsequently chosen for more extended investigation; a process reinforced by the choice of suitable locations for the pilot projects (Figure 39.4). Because the four pilot projects are arranged in such a manner that the question of urban agriculture is always addressed in connection to other relevant sectors of the social and economical system, various different groups and stakeholder constellations were identified using specific data and knowledge inventories. Moreover, based in turn on a communicative-participatory procedural process (workshops, work meetings, reports and monitoring, site trips, field research) a working organisation was created within the project itself.

Pilot project 1, Industry and Urban Agriculture, demonstrates potential synergies between agricultural and industrial sectors, which normally act as competitors for water use. Pilot project 2, Informal Settlement and Urban Agriculture, concentrates on the question of if and how the inhabitants of informal settlements in peri-urban spaces could exploit urban agriculture to improve their living standards. In pilot project 3, Peri-urban Tourism and Urban Agriculture, we devoted ourselves to the question of the linkage between urban agriculture and peri-urban tourism in a scenic landscape valley near Casablanca. Pilot project 4, Healthy Food Production and Urban Agriculture, is aimed at securing the potential

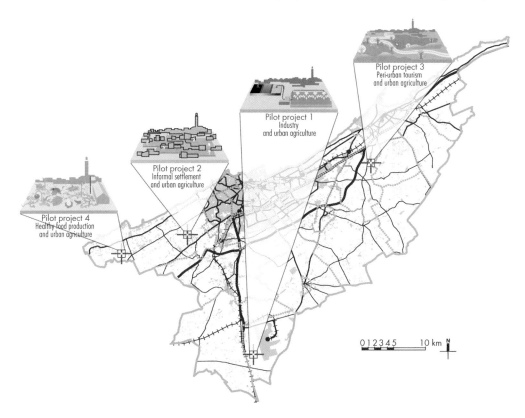

Figure 39.4. Location of the pilot projects (UAC project graphics/Christoph Kasper).

for a better income for peasant farmers living in the peri-urban area and at teaching them organic farming methods (UAC, 2010).

39.5 Challenges and opportunities

This inter- and transdisciplinary research project is being executed jointly by researchers and practitioners from Morocco and Germany. It is managed by the Technical Universiy Berlin – Department of Landscape Architecture and Open Space Planning. That the organisation and coordination of such a complex inter- and transdisciplinary project was taken over by landscape architects did not happen by accident. Because they are used to dealing with the complex system 'landscape', they are well-suited to investigate and assess the various dimensions of the research topic at hand and to work together with a variety of partners to generate transdisciplinary and integrated results.

The project places the three dimensions of urban agriculture, urban development, and climate change together in a new perspective framework whilst simultaneously placing all three in a new operational constellation under the heading of governance and specific technical support. Taken together they form a so-called *Project Rhombus* (Figure 39.5). The Project Rhombus is a methodological tool for linking the four topics central to the project's fundamental research question (based on the sub-questions) in order to arrive at integrated results.

The project's fundamental trans-disciplinary alignment is already evident in the composition of the team. Cooperation on both the German and the Moroccan sides involves the working together of very different scientific disciplines – planners and designers, agricultural scientists, climatologists, sociologists, ecologists, process engineers, and nutritionists.

Of special note is the fact that the relevant authorities – the city planning, regional planning, and regional agricultural authorities – could be won over as permanent project partners through an intensive process of communication, the innovative project approach, and favourable timing. Moreover, various civil society associations and local farmers are project partners, or have been directly tied into project work by partners in the pilot projects.

The project explicitly combines an actor-related approach with a land area-related approach. It does not confine itself to intra-urban practices but instead encompasses the potential of existing agricultural areas as the basic structure with which to create a multifunctional open space system in an urban region, with a focus on peri-urban spaces. Gaining acceptance for the topic of 'urban agriculture' in this sense has been a first success of the project's work. Agriculture has never been associated with a modern urban context in Morocco, but this attitude is now considered to have been successfully transformed. The intensive debate in Casablanca about possible synergies of the urban and agriculture opened up new avenues, allowing for an inverse connection to pressing questions of urban development. Thus, the substantial updating of the master plan for the Casablanca region (scheduled since 2006, adopted in 2010) offers a window of opportunity for integrating the project's initial results. Stimulated by the project, agriculture has been declared an urban land use category and has been pegged to open space development targets.

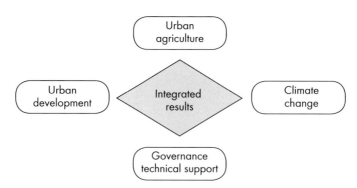

Figure 39.5. Project Rhombus (UAC project graphics).

In addition to the generation of different scientific results on the various sectoral levels, the aim of this inter- and trans-disciplinary action research project focuses (with the help of the project rhombus) on a combination of scientific knowledge and practical knowledge. The goal is to find answers to the project's overall research questions until the end of the project in March 2013:

- To what extent can urban agriculture play a significant part in adapting to the consequences of climate change?
- To what extent is urban agriculture an innovative strategy for the sustainable conservation of urban open space in tomorrow's megacities?
- To what extent can urban agriculture contribute to the struggle against poverty?
- How can urban agriculture be integrated as a crucial element of urban development in accordance with local conditions?

The preliminary design of a spatial model now serves as a basis for further investigation into how agriculture and urban development can be integrated and contributes to climate-optimised urban development. The initial development assumptions have been further concretised, rejected, or verified by means of in-depth sectoral and sub-area analyses, the step-by-step development of multifunctional modules, and the evaluation of the results of the pilot projects. One important question that will be looked at in depth involves the contribution the local production of food will make in terms of feeding the urban region and shortening food-chains. Simultaneously, an attempt will be made to understand the mechanism of current space production and the dynamics of the structural development of the urban region by, among other things, identifying patterns of spatial appropriation. Urban and peri-urban agriculture involves complex interactions between social, economic, and environmental phenomena in locations that are undergoing extremely rapid change. In the Casablanca region agricultural areas also serve as a reserve of land for future construction. What has already become clear is that the spatial question of preserving land is linked to both the question of establishing new, economically attractive, and more resource-efficient methods of agricultural production, distribution and refinement, and the learning of new social and cultivation practices. This requires examples of good practice, (i.e. the pilot projects), awareness-raising, and the kind of network-building that occurs in public events and at round tables, and, most importantly, it will require time.

References

Esparcia, J. and Buciega, A. 2005. New rural – Urban relationships in Europe: A comparative analysis. Experiences from the Netherlands, Spain, Finland and France. Local Development Research Institute, University of Valencia, Spain, 56 pp. Available at http://www.rural-urban.org/files/5fb1d5fe03b938775 f1c535bc705d63b.pdf.

Priemus, H., Rodenburg C.A. and Nijkamp P. (Eds.) 2004. Multifunctional urban land use: a new phenomenon? A new planning challenge? Built Environment 30: 269-349.

Giseke, U., 2011. Urban agriculture Casablanca. Design as an integrative factor of research. Technische Universität Berlin. special Topos insert –The International Review of Landscape Architecture and Urban Design, issue 74, Germany, 48 pp.

Hall, P. and Pfeiffer, U., 2000. Urban Future 21. A global agenda for twenty-first century cities. E & FN Spon, London, UK, 365 pp.

Revi, A., Prakash, S., Mehrotra, R., Bhat, G.K., Gupta, K. and Gore, R., 2006. Goa 2100: The transition to a sustainable RUrban design. Environment and Urbanization 18: 51-65.

Royaume du Maroc, 2008. Plan de Développement Stratégique et Schéma Directeur d´Aménagement Urbain de la Wilaya du Grand Casablanca (SDAU). Projet de Diagnostic et Enjeux du Developpement, Rapport Justificatif, Ministere de l'Interieur, Agence Urbaine de Casablanca, 394 pp.

UAC (Urban Agriculture Casablanca), 2010. The four pilot projects. Available at http://www.uac-m.org/pilot-projects/.

UN HABITAT, 2010. The State of African Cities 2010. UN HABITAT, Nairobi, Kenia, 268 pp.

Van den Abeele P. and Leinfelder, H. 2007. New Alliances for a rurban region. Available at http://www.ifhp2007copenhagen.dk/Components/GetMedia.aspx?id=9a3665d8-f234-4df2-a493-fe1248e79b7d.

Wright F.L., 1958. The living city. Horizon Press, New York, NY, USA, 255 pp.

Chapter 40

Good planning for good food: mechanisms within the English planning system to support sustainable food and farming

Harriet White and Suzanne Natelson
Sustain: the alliance for better food and farming, 94 White Lion Street, London N1 9PF, United Kingdom; harrietwhite@googlemail.com

Abstract

The planning system shapes our urban and rural areas by guiding and regulating the use of land, and through the making of place and space. This chapter explores the mechanisms within the planning system in England that provide an opportunity to support sustainable food systems at a local level. The chapter is based on findings in a recently published report 'Good Planning for Good Food' that investigates the policies and practices planners are taking locally to support and encourage a range of sustainable food initiatives such as supporting local production and distribution of food, promoting and protecting non-commercial growing spaces, and defending retail diversity and easy access to local shops that meet day-to-day food needs.

Keywords: spatial planning, local food systems, UK

40.1 Introduction

The planning system shapes our urban and rural areas by guiding and regulating the use of land, and through the making of place and space. The planning system is inherently intertwined with our food system. It influences the availability of land for commercial and non-commercial agriculture, shapes the retail environment, and it outlines options for the management of food waste. Thus it can undermine, or support and promote a more sustainable food system. This chapter outlines research undertaken by *Sustain* on the English Planning System. It highlights components of the national policy framework that shape how the planning system intersects with the food system, and explores how local level policy could, and is being used to, positively shape local food systems.

Whilst this chapter does not define a sustainable food system (this is explored in detail elsewhere, including Lang 1999), it is based on the principle that the food system should be ethical, equitable and environmentally-sound in meeting society's needs. These principles are reflected in Sustain's Sustainable Food Guidelines (Sustain, 2010b) and this is the working definition of a sustainable food system for the purposes of this chapter.

Sustain is an alliance, representing around 100 national public interest organisations, that advocates 'food and agriculture policies and practices that enhance the health and welfare of people and animals, improve the working and living environment, enrich society and culture and promote equity'[42]. In recent years, Sustain has looked at how planning can influence and better support sustainable food and farming objectives. This work has involved responding to planning policy consultations to promote the inclusion of food in planning policy, identifying positive practical steps planners can take, and exploring how food businesses and community organisations can use existing planning mechanisms to help them meet their objectives (e.g. increasing allotment provision), and influence planning policy development so that policy better addresses issues such as food poverty (Sustain, 2004). Recently, Sustain has produced a report (*Good planning for good food*, White and Natelson, 2011) that aims to inform planners about practical steps they can take to support sustainable food and farming objectives through their work.

This chapter summarises some of the findings of the *Good planning for good food* report. The report's findings and recommendations are the result of primary research (communication with local planning authorities, other governmental bodies such as Primary Care Trusts, and third-sector organisations) and secondary research (including of national and local planning policy frameworks) to identify examples of best practice. As such, this chapter does not detail academic research but outlines examples of what could be, and is currently being done through the English Planning System to support a sustainable food and farming system. It also touches on what opportunities and barriers exist to expanding this role. Given the existing statutory changes being made to the English Planning System that will 'thin' the strategic policy framework and place greater emphasis on local level planning, Sustain's emphasis on what local planning practitioners can do is particularly pertinent.

40.2 Planning reform: the fertile ground of spatial planning

The English planning system is guided by a national planning policy framework, and includes the development of local and neighbourhood plans and decision-making on individual planning applications. In the UK spatial plans and planning policies establish visions for the development of localities and guide, alongside political and economic forces, place-based development decisions, from the location of new housing and infrastructure, to the conversion of a single shop into an office. Over the last decade, the English Planning System has undergone reform, changing it from a traditionally land-based system, to one that seeks to address the many factors that shape spaces and make places.

Planning in England developed in reaction to the 'evils of the nineteenth-century city' (Hall, 2002) – notably issues of health and housing, and the acknowledgement that overcrowded and unsanitary urban areas came at a financial cost and with the potential for social unrest. Public health legislation at the end of the nineteenth century and the first planning legislation

---

[42] Sustain website: http://www.sustainweb.org/about/.

at the beginning of the twentieth century sought to address sanitary conditions, including how the built environment influenced these. Ebenezer Howard played an important role with his publication in 1898 of Garden Cities of Tomorrow. This outlined a vision of 'garden cities' where the benefits of both the town and country could be enjoyed, with agricultural self-sufficiency being a central component (Hall, 2002). In the post-war period, planning in England evolved as a land-based system: it both enabled and restricted the development of land, with the nationalised right to develop land (1947) being the central tenet of this.

Over the last decade, this land-based system has undergone reform. The 2001 Green Paper[43], *Planning: delivering a fundamental change* (ODPM) articulated aspirations for a reformed system: planning that would 'deliver sustainable development' (ODPM, 2001: i). To achieve this, planning needed to evolve from something that was predominantly land-based and procedural (ODPM, 2005) to embracing a broader place-shaping role (Lyons, 2007: i); hence the new title of *Spatial Planning*. This would require it to be better integrated with other public services, to coordinate policy action, and to reach out to partners – private, public and voluntary – to help realise a vast array of community aspirations and government objectives (Healey, 2006; Morphet *et al.*, 2007). There was also a strong 'community' thread running through planning reform that envisioned a planning system that would 'fully [engage] people in shaping the future of their communities' (ODPM, 2001: 1).

Whilst there is debate about the success or otherwise of this 'break with the past', planning professionals nonetheless have new policy tools and political rhetoric supporting this new multi-dimensional and collaborative role.

40.3 The food and farming system: opportunities for planning

This section examines where the planning system intersects with the production (non-commercial and commercial) and retail components of the food system within both rural and urban areas, and what opportunities there are at these points for local-level planners to support sustainable food and farming objectives. The authors have made a distinction between 'commercial' agriculture – food that is grown for the purpose of sale – and 'non-commercial' – food that is produced for personal consumption, or primarily for educational or community cohesion purposes. This distinction is made because the 'needs' of commercial and non-commercial food production predominantly differ (although access to land is recognised as a common need), as do the interfaces between these forms of production and planning (i.e. the points at which planning influences them). It is recognised that there are important opportunities in other components of the food chain, including in food waste disposal, but due to space constraints these are not explored here.

---

[43] In the UK, the production of a Green Paper is a tentative proposal by the Government but without a commitment to action. It is often the first step towards changing the law.

40.3.1 Commercial food production

The *UK Foresight Programme*, which reports directly to the Cabinet Office and Chief Scientific Advisor, produced a report in 2010 on the challenges that will face land management in the UK over the next 50 years (Government Office for Science, 2010). This recognised the potential need to increase the amount of food produced, including maintaining a critical capacity of high-quality farmland and the physical infrastructure that supports it. The report explores various incentives for planners, developers, local communities, farmers and government, including for example how to support farmers to produce more food (tackling the issue of spiralling land values) and how businesses, local government and residents can work together to decide on land use.

As outlined in Planning Policy Statement[44] (PPS) 7 *Sustainable development in rural areas,* one of the government's objectives for rural areas is

> *to promote sustainable, diverse and adaptable agriculture sectors where farming achieves high environmental standards, minimising impact on natural resources, and manages valued landscapes and biodiversity; contributes both directly and indirectly to rural economic diversity; is itself competitive and profitable; and provides high quality products that the public wants* (CLG, 2004).

In line with national planning policy and the findings of the Foresight *Land Use Futures* report, planning departments should ensure the continued protection of Best and Most Versatile (BMV) agricultural land, to safeguard the most fertile land for agricultural production[45]. Farming in the Green Belt can be promoted and enabled through the planning system as a way of helping agricultural production that is closer to the consumer. Local planning policy[46] can also establish criteria for what farming-related development would be acceptable and supported on Green Belt land, as well as on greenfield sites more broadly. This would help to ensure planning plays a positive role in supporting farmers' commercial viability. Planning policy and decisions about planning applications could be sensitive to the growing need for farms to diversify and modernise (e.g. by developing on-farm processing facilities and farm shops). Conditions can be attached to planning permission to ensure that developments support sustainability objectives. So, for example, planners could require a percentage of produce sold in on-site farm shops is from the farm or the local area, and that a further percentage is from within the region.

---

[44] In England, there are 21 National Planning Policy Statements (PPS) which outline the principles, priorities and targets which guide how areas should be developed and land used. They cover a range of topics, such as climate change, transport, open spaces and coastal planning.

[45] Whilst the contribution that planning can make in supporting sustainable commercial food production may be more limited compared with non-commercial growing, there are still some key areas in which planning can play a supportive role.

[46] Local Authorities produce a portfolio of planning documents (a Local Development Framework), guided by national policy, which outline the Authority's vision for the development and use of land in their locality. One document is the Core Strategy, which contains a range of local planning policies.

40.3.2 Non-commercial food production

Whilst non-commercial food production does not currently contribute significantly to production overall, community growing projects can provide invaluable educational, health, environmental and social benefits (Garnett, 1999; Iles, 2005). Non-commercial food production is also growing in popularity – an estimated 10% of Britain's population is currently engaged in urban food growing initiatives (Iles, 2005). Protecting existing growing spaces, and identifying locations for new allotments and community gardens represent key opportunities for local planners to support the production of sustainable local food. New spaces can be supported through: urban regeneration schemes; new housing and commercial development; green infrastructure strategies; and new town developments.

Development plans are statutorily required to make provision for allotments where there is an identified requirement (Allotments Act 1925). In addition, Planning Policy Guidance (PPG) 17 *Planning for Open Space, Sport and Recreation* (CLG, 2006), states that local authorities should ensure adequate provision of open space (this includes allotments, community gardens and city farms), including setting standards for this provision, based on local assessments of need and opportunities. PPG17 also states that existing open spaces may only be built on where an assessment shows the land to be 'surplus to requirements' (CLG, 2002: para.10). The benefits of community growing projects can be subtle and difficult to measure, but qualitative and quantitative evidence could be identified to support policy proposals to retain or establish such growing spaces. Infrastructure Delivery Plans, which form part of the Local Development Framework (LDF), are an important opportunity for local authorities to provide growing space in response to local need.

Enfield Borough Council's Open Space and Sports Assessment states that 'all residents within the Borough should have access to an allotment garden within 800m of home' (Enfield Council, 2006: xv). The assessment identified areas of allotment deficiency, highlighting opportunities in these areas to create new allotments, and stated that 'opportunities to bring forward allotment space may arise through new development' (Enfield Council, 2006: xxvi). However, it is recognised that statements in LDF documents, such as the latter, are often solely 'statements of support' rather than specific policy commitments (White, 2010 personal communication). In reality, for a new development, issues such as land value, profit margins, and aesthetics will all shape development priorities and therefore what land uses take precedence.

To encourage actual, as well as theoretical support for proposals for non-commercial food production, the link to specific objectives and targets should be articulated and quantified where possible in a range of policies. For example, all local authorities are required to sign up to National Indicator (NI)[47] 56 *Obesity in primary school age children in Year 6.* The

[47] National Indicators are how central government manages local government. http://www.communities.gov.uk/publications/localgovernment/finalnationalindicators (note: this system is currently under review (2010)).

role that community food growing can play in addressing such a target has been recognised by a national agency (Food Standards Agency, http://www.food.gov.uk/multimedia/pdfs/enforcement/laani056.pdf). Collaboration with other local authority departments and non-governmental bodies, such as health professionals and food growing organisations, can also support the creation and maintenance of food growing spaces. These relationships can be identified in Infrastructure Delivery Plans or Green Infrastructure Strategies so that responsibility for implementation is clearly articulated and monitored.

40.3.2 Food retail: access and choice

A sustainable food retail environment is one that is accessible, affordable, healthy, diverse, and incorporates localised and environmentally benign food systems. Although the forces shaping the food retail environment are multi-faceted and complex, the planning system can influence the nature of the food retail environment and support this sustainable vision.

PPS4 *Planning for sustainable economic growth* states that local authorities should 'identify deficiencies in the provision of local convenience shopping and other facilities which serve people's day-to-day needs' and to assess whether provision 'allows genuine choice to meet the needs of the whole community' (CLG, 2009: EC1.3 and EC1.4). Concurrently, PPG13 Transport states that local authorities should, when preparing development plans and considering planning applications, 'locate day to day facilities which need to be near their clients in local centres so that they are accessible by walking and cycling' (CLG, 2006: 6). Thus the national policy framework should promote access and choice in the retail environment.

Local planning policies could therefore seek to ensure provision of shops that meet day-to-day needs, and protect existing shops of this type, providing detail of the type of shops this includes. Hackney Borough Council Core Strategy's Policy 13 *Town Centres* provides such detail: 'Shops that provide essential day-to-day needs for the local community such as a baker, butcher, greengrocer, grocer, specialist ethnic food shop...in the borough's town, district and local shopping centres as well as shopping parades and corner shops will be protected from changes of use away from retail' (Hackney Council, 2010).

Importantly local planning policies can protect 'sole shops' (i.e. the only shop in that neighbourhood, or the only shop in that neighbourhood that sells particular goods, such as fruit and vegetables). *Town Centre First* policies, which make a presumption against out of town development, are also key in supporting the vitality and viability of local centres, and in protecting food shops that are accessible by walking, cycling or public transport (CLG, 2009).

It is recognised that planning cannot influence *what* is sold in local shops; other mechanisms are important here. The Buywell Retail Project (Sustain, 2010a), for example, has successfully promoted sales of fresh fruit and vegetables in convenience stores in some of London's most deprived areas. However, ensuring retail outlets do not change use where they meet a day-today need or where they are 'sole shops' is a crucial step in ensuring people have easier access to healthy and sustainable food.

Accessibility standards and methods for measuring improvement can be useful in developing solutions to food access problems. Such standards can be included in development plan documents and supplementary planning documents (SPDs). The Department of Health for the West Midlands is developing a 'physical accessibility standard for healthy food', to be used by local authorities across the region (Department of Health West Midlands, 2009). As part of this, a region-wide mapping exercise has been undertaken to show the location of outlets likely to sell fruit and vegetables (the proxy for 'healthy' food). These maps will be used by the region's local authorities to gauge 'hot spots' where there are potential problems with access to healthy food, and then develop and implement policies to address such problems.

As well as affecting shops, planning can promote street and farmers markets. These markets have multiple benefits: they can increase the diversity of the food available for people to buy; they can strengthen the local community and enhance the vitality of localities; they can provide a market for small producers, including community food growing projects; they provide an additional source of revenue for farmers; and they can provide an arena in which producers and consumers are brought closer together. Protecting existing markets and creating new ones can be done through local planning policies, regeneration projects and master planning processes.

However, as with shops, it is recognised that it is difficult to ensure that produce sold at farmers markets is both affordable for local residents and sold at a fair price for the farmers and producers (Crewe, 2004). Again, non-planning measures play an important role here. Market managers need to ensure stall costs are kept to a minimum and that there is diversity in the types of stall-holder, with affordable basic foods being available alongside value-added processed foods (Friends of the Earth, 2000).

Finally, planning can restrict the over-concentration of hot food takeaway shops (classified as A5 units in the Town and Country Planning (Use Classes) Order). Restrictions can include: buffer zones around sensitive places, such as schools and youth facilities, within which new A5 units will not be permitted; standards for the number of A5 units (as a percentage of overall unit number) allowed in an area; or standards for the distance between A5 units. The London Borough of Waltham Forest (2009) has an adopted SPD, which details criteria for the restriction of A5 units. As a result, Waltham Forest has already turned down 12 applications for A5 units (White 2010, personal communication) demonstrating the direct impact such an approach can have. However, planning can only influence whether more A5 units are given permission; it cannot affect existing units. Thus, as with shops and markets, non-planning mechanisms are important for shaping what is sold in fast food outlets.

40.4 Some barriers faced by planners

Whilst the national planning policy framework contains statements of support for sustainable food and farming, there is less awareness more generally among both planners and non-planners about the importance of these issues, and there is weak political support to enable change that is wide-spread and 'third order' (Lang *et al.*, 2009). In addition, the onus is on local planners to use the national policy framework to facilitate positive change on the ground, and this is not routine practice; planners that do proactively support positive change are considered to be visionary or 'going against the grain'.

A key issue for planners is balancing diverging demands and needs; the use of space is increasingly contested among a wide range of different sectors of society. The desire to protect a local landscape vies with the drive towards a lower-carbon society, which vies with changing lifestyle choices that are more resource intensive. Local-level activities interact with international drivers of change: local people's preferences with international economic trends. Planners and other policy makers are tasked with balancing conflicting public attitudes, and mediating between the different preferences of individuals and those of communities, and with the growing need for societal resilience in the face of future change.

What is often seen in many decision-making processes is that the balance between different sectors has historically tipped towards the more powerful actors. The planning system is a mechanism through which a more equal balance can be established by enabling less powerful sectors of society to have an influence. This is not to say that all of the less powerful actors will agree with each other. With any development decision, a range of interests, stakeholders and policies (often contradictory) are involved, and needs have to be balanced and compromises made to reach agreements.

Steps are being taken by some local authorities to use existing planning mechanisms to support more sustainable food systems. However more work needs to be done to raise awareness amongst planners and policy makers of the importance of supporting sustainable food and farming through planning policy and practice. Political support also needs to be strengthened to encourage this action at the local level.

References

Crewe, K., 2004. Planners and equitable food distribution. Progressive Planning 158: 2.

Department of Communities and Local Government, 2004. Planning Policy Statement 7: Sustainable Development in Rural Areas. Available at http://www.communities.gov.uk/publications/planningandbuilding/pps7.

Department of Communities and Local Government, 2009. Planning Policy Statement 4: Planning for Sustainable Economic Growth. Available at http://www.communities.gov.uk/publications/planningandbuilding/planningpolicystatement4.

Department of Communities and Local Government, 2011. Planning Policy Guidance 13: Transport. Available at http://www.communities.gov.uk/publications/planningandbuilding/ppg13.

Department of Health West Midlands and JMP Consultants Ltd., 2009. Developing a physical accessibility standard for healthy food in the West Midlands. Available at http://www.foodwm.org.uk/resources/Exec_Summary_Final.pdf.

Enfield Council, 2006. Enfield open space and sports assessment. Available at http://www.enfield.gov.uk/downloads/file/1229/volume_1_final_assessment_of_open_space_needs-august_2006.

Friends of the Earth, 2000. The economic benefits of farmers' markets. Friends of the Earth Trust, London, UK. Available at http://www.foe.co.uk/resource/briefings/farmers_markets.pdf.

Garnett, T., 1999. City Harvest – The feasibility of growing more food in London. Sustain, London, Uk.

Government Office for Science, 2010. Land use futures: Making the most of land in the 21st century. Available at http://www.bis.gov.uk/assets/bispartners/foresight/docs/land-use/luf_report/8507-bis-land_use_futures-web.pdf.

Hackney Council, 2010. Hackney core strategy development plan document. Available at http://www.hackney.gov.uk/core-strategy.htm.

Hall, P., 2002. Cities of cities of tomorrow: an intellectual history of urban planning and design in the twentieth century. Blackwell Publishing, Oxford, UK, 553 pp.

Healey, P., 2006. Territory, integration and spatial planning. In: Tewdwr-Jones, M. and Allmendinger, P. (Eds.) Territory, identity and spatial planning. Routledge, Abingdon, UK, pp. 64-80.

Iles, J, 2005. The social role of community farms and gardens in the city. In: Viljoen, A. (Ed.) Continuous productive urban landscapes: designing urban agriculture for sustainable cities. Architectural Press, Oxford, UK, pp. 82-88.

Lang T., 1999. Food policy for the 21st century: can it be both radical and reasonable?. In: Koc, M. MacRae, R., Mougeot, L. and Welsh, J. (Eds.) For hungerproof cities: sustainable urban food systems. IDRC, Ottowa, Canada, pp. 216-224.

Lang, T., Barling, D. and Caraher, M., 2009. Food policy: integrating health, environment and society. Oxford University Press, Oxford, UK, 336 pp.

London Borough of Waltham Forest, 2009. Waltham forest supplementary planning document: Hot food takeaway shops, spatial planning unit. London Borough of Waltham Forest. Available at http://www.walthamforest.gov.uk/spd-hot-food-takeaway-mar10.pdf.

Lyons, M., 2007. Lyons Inquiry into Local Government: Places-shaping: a shared ambition for the future of local government. The Stationary Office, London, UK, 44 pp.

Morphet, J., 2007. Shaping and delivering places for tomorrow. Town and Country Planning 76: 303-305.

ODPM (Office of the Deputy Prime Minister), 2001. Planning: delivering a fundamental change. Available at http://www.communities.gov.uk/archived/publications/planningandbuilding/planningdelivering.

ODPM, 2002. Planning Policy Guidance 17: Planning for Open Space, Sport and Recreation. Department for Communities and Local Government, London, UK, 15 pp. Available at http://www.communities.gov.uk/documents/planningandbuilding/pdf/ppg17.pdf.

ODPM, 2005. Planning Policy Statement 1: Delivering Sustainable Development. Available at http://www.communities.gov.uk/documents/planningandbuilding/pdf/planningpolicystatement1.pdf.

Sustain, 2004. How London's planners can increase access to healthy and affordable food. Available at www.sustainweb.org/publications.

Sustain, 2010a. Buywell Retail Project. Sustain, London, UK. Available at http://www.sustainweb.org/buywell/buywell_shops/.

Sustain, 2010b. 7 Principles of Sustainable Food. Sustain, London, UK. Available at http://www.sustainweb.org/sustainablefood/.

White, H. and Natelson, S., 2011. Good planning for good food: How the planning system in England can support healthy and sustainable food. Sustain, London, UK, 39 pp.

Chapter 41

The Food Urbanism Initiative

Craig Verzone[48]
Verzone Woods Architectes Sàrl – paysage, urbanisme, architecture, Route de Flendruz 20, 1659 Rougemont, Switzerland; fui@vwa.ch

Abstract

The domains of agriculture and urbanity have traditionally been perceived as mutually exclusive despite their extreme interdependence. In recent years, a burgeoning grass-roots movement has emerged with the aim of re-integrating agriculture into the life and health of the city. The time is ripe for yet more innovative spatial solutions to the possibilities surrounding food and urbanism. The Food Urbanism Initiative (FUI) aims to examine the overall impact of food on urban design and to study the potential of new architectural and landscape strategies for the integration of food production, processing, distribution and consumption in the contemporary city. This essay outlines a methodology for generating strategies to facilitate urban development that integrate both city life and food production cycles into a more harmonious coexistence that is socially, economically, and environmentally responsible. Using Switzerland as a point of departure, FUI explores the potentialities of Food Urbanism through its recent history and contemporary context. Four primary research groups comprise the multi-disciplinary team to explore the broad and wide-ranging issues inherent to the subject. The project aims to establish both innovative and feasible proposals for the purpose of grafting the domains of city and farm. The FUI methodology relies upon the development of a set of essential tools including a webportal, public opinion survey, urban mapping, prototyping and a series of pilot projects corresponding to multiple scales of proposed intervention. Food Urbanism proposals, programmes and projects are sprouting up across the globe and are providing valuable insight. FUI hopes to advance these efforts by thoroughly researching the movement and actively developing solutions to help reinforce the liaison between theory and practice, between architecture and agriculture, between city and farm.

Keywords: urban design, architecture, agriculture, landscape architecture, public space

41.1 Introduction

Food systems have historically played a vital role in the design of cities, though that role is not always acknowledged by designers. In her book *Hungry City,* Carolyn Steel argues that the ability to supply the city with food was the single-most important control to urban growth in pre-industrial societies (Steel, 2008). Modern innovations in logistical networks and the

---

[48] www.foodurbanism.org, www.vwa.ch.

industrialisation of food continue to alter the spatial and living patterns of the contemporary city, especially as the world's urban population continues to increase at a rapid rate. The Food Urbanism Initiative (FUI) examines the unique and complex relationship between farming and the city, investigates the impact of food on urban design and develops new architectural and landscape strategies for the integration of food production, processing, distribution and consumption in the contemporary city. FUI outlines the fundamental intentions of this movement as well as examining strategies meant to facilitate urban development that integrate both city life and food production cycles into a more harmonious coexistence that is socially, economically, and environmentally responsible (Figure 41.1).

This chapter presents Food Urbanism in relationship to the Swiss context. In addition, the chapter describes the Food Urbanism Initiative, its team organisation and methodology as well as the first phases of FUI research that include the launching of a Public Survey and globally-aimed Webportal. In 2010, FUI was awarded a three-year Swiss National Science Foundation Research Grant in the National Research Programme 65 entitled 'New Urban Qualities'. The research programme

> aims at (further) developing concepts and strategies for new urban quality and at testing the feasibility of the research findings. The concepts and strategies will demonstrate innovative ways to achieve urban development, urban redevelopment, and urban planning in Switzerland that are realisable in the medium term (2030) and long term (2050) (Swiss National Science Foundation, 2011).

Figure 41.1. Urban vision.

'Food Urbanism', a term used occasionally in the recent past by a number of different projects (Food Urbanism, 2007[49]; Grimm, 2009), posits that spaces of food production, distribution, and consumption share the ability to structure urban form in the contemporary city, and that by considering food in the broader sense, the health of both the city's residents and its physical environment can be improved. Food Urbanism encompasses more than urban agriculture, realising that the relationship between food and the consumer has the opportunity to create more complex interactions than those implied by the mere presence of agriculture as a single, disconnected and peripheral land use in the broader urban metropolis. If Food Urbanism is to become a viable model for structuring or restructuring the contemporary city, serious challenges and objections must be addressed, and potentials need to be carefully mined.

FUI opens research under the hypothesis that urban design strategies that integrate well conceived and designed food production models and facilities will provide new urban quality. In testing this hypothesis, the FUI team addresses two fundamental research questions.
1. How can urban design and public space making integrate, encourage and facilitate urban food production?
2. What types of urban food production affect urban design and improve urban quality?

The primary objective of the FUI research is to uncover and develop the synergies between urban design and urban food production so as to achieve new urban quality. The first phase of research is marked by defining FUI parameters, establishing the FUI target audience and initiating the collection and assessment phase. Within this period a Webportal sets in motion the congregating of a world-wide research community and web-based archive of projects, a Public Survey reveals the scope of public acceptance, Urban Mapping outlines the potentials for food urbanism, and the FUI Agronomic Criteria define the parameters of urban food production possibilities. See Figure 41.2.

41.2 Switzerland and sustainable agriculture

Switzerland is an ideal place for the study of the integration of urban life with food production, and for developing models that can be implemented both locally and beyond. More so than most industrialised nations, Switzerland has already implemented policies to protect rural modes of life and food production as well as rural land resources. High quality, locally produced food resources have long been seen as part of the nation's patrimony, stemming from a belief that resource independence can help secure political independence, and that the unique beauty of Switzerland's landscapes is closely tied to the aesthetic appeal of land devoted to agricultural production. Nonetheless, the pressure exerted on Switzerland from the European Union, globalisation and/or open markets provides Swiss citizens with difficult choices in the supermarket, often reduced to the 'best buy' or 'made in Switzerland'.

---

[49] See http://www.nfp65.ch/E/portrait/Pages/default.aspx.

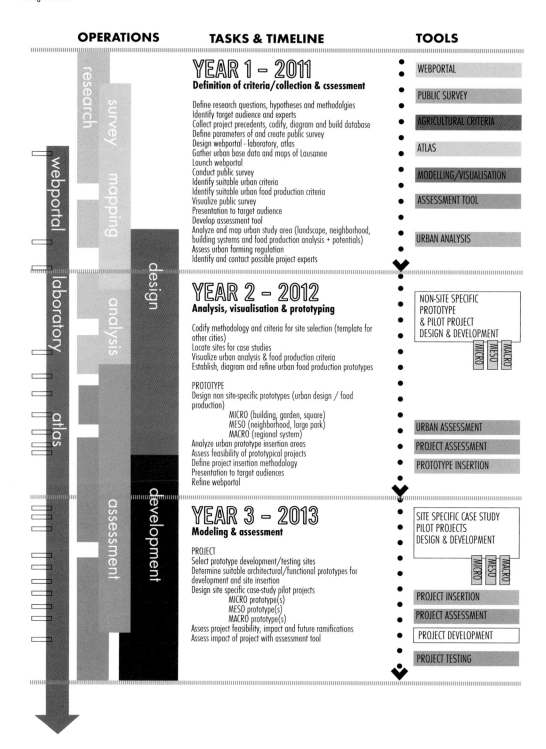

OPERATIONS   TASKS & TIMELINE   TOOLS

research
survey
mapping
webportal
laboratory
atlas
design
analysis
assessment
development

YEAR 1 – 2011
Definition of criteria/collection & cssessment

Define research questions, hypotheses and methodolgies
Identify target audience and experts
Collect project precedents, codify, diagram and build database
Define parameters of and create public survey
Design webportal - laboratory, atlas
Gather urban base data and maps of Lausanne
Launch webportal
Conduct public survey
Identify suitable urban criteria
Identify suitable urban food production criteria
Visualize public survey
Presentation to target audience
Develop assessment tool
Analyze and map urban study area (landscape, neighborhood,
building systems and food production analysis + potentials)
Assess urban farming regulation
Identify and contact possible project experts

WEBPORTAL
PUBLIC SURVEY
AGRICULTURAL CRITERIA
ATLAS
MODELLING/VISUALISATION
ASSESSMENT TOOL
URBAN ANALYSIS

YEAR 2 – 2012
Analysis, visualisation & prototyping

Codify methodology and criteria for site selection (template for
other cities)
Locate sites for case studies
Visualize urban analysis & food production criteria
Establish, diagram and refine urban food production prototypes

PROTOTYPE
Design non site-specific prototypes (urban design / food
production)
          MICRO (building, garden, square)
          MESO (neighborhood, large park)
          MACRO (regional system)
Analyze urban prototype insertion areas
Assess feasibility of prototypical projects
Define project insertion methodology
Presentation to target audiences
Refine webportal

NON-SITE SPECIFIC
PROTOTYPE
& PILOT PROJECT
DESIGN & DEVELOPMENT
MICRO MESO MACRO

URBAN ASSESSMENT
PROJECT ASSESSMENT
PROTOTYPE INSERTION

YEAR 3 – 2013
Modeling & assessment

PROJECT
Select prototype development/testing sites
Determine suitable architectural/functional prototypes for
development and site insertion
Design site specific case-study pilot projects
          MICRO prototype(s)
          MESO prototype(s)
          MACRO prototype(s)
Assess project feasibility, impact and future ramifications
Assess impact of project with assessment tool

SITE SPECIFIC CASE STUDY
PILOT PROJECTS
DESIGN & DEVELOPMENT
MICRO MESO MACRO

PROJECT INSERTION
PROJECT ASSESSMENT
PROJECT DEVELOPMENT
PROJECT TESTING

Figure 41.2. FUI project timeline.

At the same time, Switzerland has invested heavily in efficient modern infrastructures for connecting city centres to the countryside, mitigating much of the physical and at times cultural segregation between urban and rural areas. The often clear delineation between relatively compact urban areas and the surrounding countryside allows for greater interaction between urban and rural components than that available in the sprawling megalopolises of Asia or the Americas. Switzerland is not immune, however, to the many problems that have accompanied urban population expansion in the rest of the world. Despite only modest population growth, for example, the footprint of urbanisation in the nation has more than doubled in the last 50 years (Swiss Statistics, 2011), implying a trend towards lower density development at the periphery rather than a densification of urban centres. Such single-use development provides few of the benefits of either urban or rural living.

> In the last centuries, the Swiss landscape has been the result of evolving agricultural production systems and their changing processes have inevitably left their marks on the terrain. These modifications have reflected the changing needs for food, fibre and energy and the associated policy measures. Until 1850 Swiss agriculture was dominated by subsistence farming, focused upon diverse production. With the emergence of the steam engine and the corresponding transport facilities, wheat and other crops were imported to Switzerland. Price relationships between crops and livestock changed significantly. Agricultural production switched from crops to milk and meat production making grassland the dominant land-use. Economic crises and two world wars led to the emergence of a highly regulated agricultural sector. Since WW2 crop production has been supported in order to secure food availability in times of need which has led to a re-emergence of a wider crop variety (Huber and Lehmann, 2009).

> The legitimacy of promoting multifunctional agriculture in Switzerland is largely derived from Article 104 of the Swiss constitution that ensures a market-based agricultural sector that takes into account natural resource conservation and a decentralised settlement pattern. It is based on the assumption that farming provides public services that are not remunerated by the market and therefore justify government intervention. Direct payments (ensuring a farm income above the national average) and eco-payments (giving additional support to farmers that apply certain agro-environmental measures) are especially designed to compensate for these public services. The expected loss of farm income due to lower prices are compensated through more direct subsidies (Aerni et al., 2008).

Another area of interest for urban agricultural initiatives concerns public perception as some agricultural uses are considered unsightly or have been deemed a nuisance. The issue of public perception, however, presents far more opportunities than obstacles, as most agricultural uses are considered favourable to creating a sense of place, even in urban areas. In and around the city of Lausanne, Switzerland, for example, the presence of vineyards is considered highly desirable and links the city to its past. More often than not, the desirability and necessity of agriculture is deemed worthy of subsidy, not only to help guarantee the food supply, but also for its tangible and intangible aesthetic qualities. The government and its people regard agricultural subsidies supporting heritage agriculture as auspicious and go as far as to consider it a preservation tactic defending the nation's most referential landscapes. (Verzone and Woods, 2006) After all, the character of the Swiss Alps would change dramatically if its cows were to disappear.

General public perception concerning urban agricultural initiatives, however, provokes anxiety about how contemporary urban activities can coexist with the traditional operations of agriculture. In urban and peri-urban areas, opportunities to protect agricultural heritage and active terrains of production are considerably more rare. Aesthetic and environmental aspects of agriculture need to improve tools measuring their credibility to be accepted in any cost-benefit equation. A parking lot may be an economically viable use for a particular parcel of land, but the surrounding properties might benefit much more, if only in terms of real-estate value, from planting mixed bands of lettuces and berries. Urban agriculture should be considered as a land-use with ramifications that extend much beyond the core economics of its crops (Appelbaum, 2009; Rankin, 2009) (Figure 41.3). While regulatory frameworks often seek to separate uses, neighbourhoods rich with overlapping uses and users expresses a stronger, richer civic realism than those without (Rowe, 1997).

Landscape management tools provide yet another set of opportunities needing refinement if urban food production is to become viable. Subsidy structures and regulatory frameworks, for example, can be used to leverage urban agricultural projects for public benefit. The Swiss government currently compensates farmers for developing beneficial landscape typologies

Figure 41.3. Urban vision.

that provide valuable ecosystem services. Farmers receive compensation for establishing and maintaining Surfaces for Ecological Compensation SECs such as extensively farmed grasslands, hedgerows, fish breeding ponds, trees in fields, wetlands, etc. and can receive extra subsidies if such interventions are proven to be part of a greater ecological network (Swiss Federal Office of the Environment, 2011). Once established that agriculture in the city is itself a positive contributor to the urban ecological network, and that professional farmers are willing to become partners in maintaining this landscape, then subsidies might become available to support such changes (likely in the areas that suffer most from neglect).

FUI is manifested during a period of both political and public local support towards the subject in Lausanne. Heightened sensitivity is seen often in academic journals, the popular press (Wiles, 2009) and within the administrational agendas. Some of most important of local authorities are supporting FUI through direct feedback and assistance. Municipal authorities in the department of urban design, parks and gardens as well as agriculture and forestry, already following creative initiatives of their own (Les Plantages Lausannois, 2011), have reviewed the initiative and have offered support. To help launch the assessment of the case study area physical parameters, they have provided the FUI team with the municipal urban cartographic database.

Professionally, the planning community around Lausanne has begun to acknowledge the importance of Food Urbanism at a regional scale and is keen to modify typical procedures in the planning process such as those occurring in the Projet d'agglomération Lausanne-Morges PALM and Schèma Directeur du Nord Lausannois SDNL studies. Academically, the University of Lausanne UNIL and the School of Landscape, Engineering and Architecture HEPIA in Geneva organised a conference around the topic in October of 2010. Furthermore, the 2011 Congress of the International Federation of Landscape Architects in Zurich chose the theme Food Urbanism (chaired by the author) as well as Peri-Urban Agriculture for four of its twenty-eight sessions, marked by the presentation of more than sixteen different papers, thus emphasising the gravity and importance of the subject (IFLA, 2011).

41.3 Agriculture to digital culture: a multidisciplinary approach

The Food Urbanism Initiative is piloted by urban designers, architects, and landscape architects and draws on the expertise of researchers in fields ranging from agriculture, our first culture, to digital culture, our latest. The initiative is conducted by an interdisciplinary team composed of four primary research groups (see Figure 41.4). Verzone Woods Architectes VWA, a design studio working across the domains of urban design, architecture, and landscape architecture pilots the endeavour. The EPF Lausanne Media and Design Lab LDM brings its specific knowledge in the synthesis of visual and spatial information with quantitative data so as to create formal and systems organisational models. The Agri-food & Agri-environmental Economics Group AFEE of the ETH Zurich approach FUI from an agro-economic viewpoint, weighing economic properties of urban land costs, consumer preferences (their willingness to pay) and the expense of production on small parcels with non-economic motivation or public

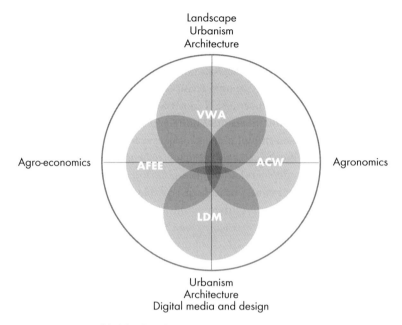

Figure 41.4. Team expertise and fields of study.

interest value, namely leisure, knowledge and social benefits. Agroscope – proficrops ACW, a federal research laboratory and institution extensively connected to the Swiss agricultural sector is responsible for identifying the optimal composition of crops, infrastructure and activities suitable for the city and for assisting to insert them into appropriate FUI projects.

The design studies will be informed by techniques from the biological sciences, the social and the economic sciences and meet in the spatial realm of the contemporary city. The team is balanced between research and practice. Incorporating representatives of the agriculture and economic sectors and their academic branches into the FUI team as well as from both the academic and professional realms insures the preliminary step of implementation: connecting research to its application through feasible proposals (see Figure 41.5).

41.3.1 foodurbanism.org

FUI creates and incorporates the Weportal www.foodurbanism.org as a vital tool for research and implementation, one critical to the success of the short-term, intermediate and long-term goals of the initiative. Foodurbanism.org establishes a dynamic, mesh-like connection with the world of urban agriculture and its many actors working in the area of spatial development. It is a mechanism to give identity to the Food Urbanism Movement, to feed the research process by facilitating and expanding data collection, to assemble and build an on-line community, to catalyse pluri-disciplinary interaction and to disseminate the research in the field of urban agriculture.

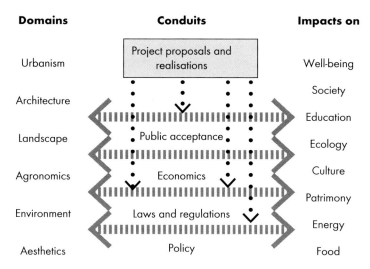

Figure 41.5. Domains of research activity, conduits of change and project impact.

41.3.2 Atlas: how can the database grow into a richer source of content including detailed descriptions and illustrations of projects?

The Food Urbanism Atlas germinates from foodurbanism.org, its genesis. An atlas as defined by the Concise Encyclopaedia Britannica is a 'collection of maps or charts, usually bound together' and 'often containing pictures, tabular data, facts about areas, and indexes of place-names keyed to' specific locations. Pertaining to FUI, the Atlas is a collection of thoroughly documented and carefully catalogued and sorted projects, efforts and initiatives that infuse strands of our food system into the urban domain. It is an environment where authors of projects across all disciplines, geographies and cultures contribute their work while feedback, input and insight is also elicited. The Atlas becomes not only a repository of information but a tool for the exchange and dissemination of ideas and information enriching the knowledge and dynamism around the projects and the urban agriculture movement.

The Atlas is a growing illustrated database of Food Urbanism project precedents. Open to the public, the Atlas provides access to the entire range of scales and typologies of Food Urbanism related projects from around the world. It is organised around and categorised by a series of topics conditioning the subject; urban context, scale, relationship to the ground, access, purpose/intention, agronomic process as well as author and location (see Figure 41.6). Projects are tagged, grouped and searchable by topic. The Atlas postings also include events and discussions, also tagged, grouped and searchable by topic. By organising and archiving project precedents, displaying new events and topical discussions, the Atlas operates as a dynamic, educational and extensive conduit into the domain. Although publicly visible, Atlas entries are uploaded by website members who require inscription thus offering the Weportal a means to control its content and communicate with its group.

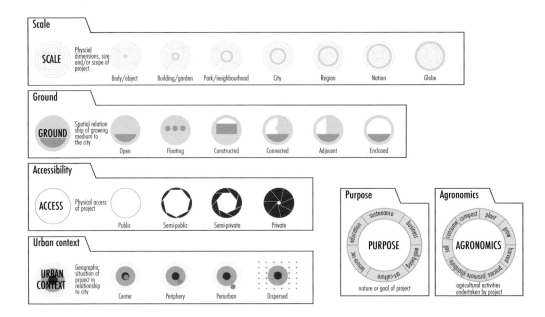

Figure 41.6. Atlas characteristics templates.

41.3.3 Laboratory: what are the appropriate tools to facilitate connections between disparate entities preparing similar research in the field of urban agriculture?

The establishment of a knowledge base supporting FUI reveals a wide array of recent project examples scattered across the globe (City Farmer News, 2011: http://www.cityfarmer.info/; RUAF, 2011: http://www.ruaf.org/). On one hand, this shows that the movement is active and wide-spread. On the other hand, however, it uncovers the fact that the knowledge-base is disparate and the researchers, with little to no contact amongst one another, are spear-heading overlapping efforts. The determination to create a forum for discussion, feedback and active collaboration regarding ongoing projects and the complex range of Food Urbanism issues propels the development of the online Laboratory found on the Weportal.

The Laboratory is organised around publicly visible team pages that act as team identity portals communicating team events, discussions, new projects, etc. Furthermore, each team has access to their own private pages visible only to their own inscribed members. Apart from being customisable, this privatised workspace allows teams to communicate internally via discussions, events posting and link sharing (projects, articles or videos). Teams are given the choice to post/share communications freely within the boundaries of their team Laboratory, with other selected research teams or with the all the research team members of foodurbanism.org.

Essentially, the Laboratory facilitates contact between teams by keeping members abreast of their fellow researchers' work through its discussion and event panels thus stimulating more profound research, inquiry and testing. Enhancing the Laboratory performance, a platform of project-review is offered to its inscribed teams. Providing identity to teams, keeping team members abreast of current research, facilitating access and communication between teams and encouraging peer review, the Weportal designed within the scope of the FUI aims to modify the mode of research within the growing field of Food Urbanism.

41.3.4 Public survey

Urban dwellers play prominent roles as consumers, users and citizens of their cities. Without their assistance, investment in time, money and political support, urban agricultural initiatives are unlikely realisable. Therefore, the assessment of the case study region does not only focus on physical parameters, but also includes social ones, such as the habitants' expectations and wishes. A comprehensive public opinion survey distributed by postal mail to a random sample of 2,500 inhabitants of the city of Lausanne was launched in March 2011. The final response rate corresponds to 35.6% (889 respondents).

The Public Survey identifies the market potential of certain agricultural products and the demand for leisure gardening opportunities. Likewise it measures the potential support or opposition with respect to public interests in aspects of Food Urbanism and Urban Quality and generates criteria that help define and drive a series of parametric models. These results act as the foundation for the mapping and criteria determination project phases and provide a base for decision making for the prototype and pilot project generation.

The Public Survey inquires about the willingness to pay WTP for 'urban' fruits and vegetables and about their perceived characteristics. Preferences regarding leisure time, public space use, gardening and cultivation are examined. Additionally, public interests are considered, such as the possible spatial, aesthetic, ecological and social effects of agriculture. The survey based evaluations of the market products, leisure opportunities and the public interest are put into the larger context of the respondents' perceptions regarding topics such as food security, local food supply and the preservation of natural and cultural resources. Separating the survey data into socio-demographic pools additionally helps to clarify personal motivations.

Apart from the testing of hypotheses (such as about a majority's approval of urban agriculture), the survey provides rankings of different options regarding the organisation, design and conceptualisation of urban agriculture projects. Additionally it allows for characterisation of specific groups within the population (e.g. persons who could be interested in participatory projects or the potential consumers). Thus the survey results feed the design of prototypes and pilot projects. The FUI survey results allow the generation of criteria (e.g. choice of plants, preferred planting sites, population's participation, social functions) that help define and drive a series of FUI parametric models. Those who support the initiative also reveal its specific potentials (i.e. market products, leisure opportunities or public good) thus shaping design parameters to meet an existing demand and increase feasibility. The survey also provides rankings of different options regarding the organisation, design and conceptualisation of urban agriculture projects. Additionally it allows for characterisation of specific groups within the population (e.g. persons who could be interested in participatory projects or the potential consumers).

41.4 Modelling and prototyping

Parametric modelling refers to a computer interface that can be used to iteratively generate design solutions based on parameters related to the site and its physical and cultural characteristics (Woodbury, 2011). The parametric approach consists in determining which parameters are most relevant to the favourability of a given site for urban agriculture, and the analysis of sites within the study area based on these parameters. (Cache, 2004) In the early FUI project development parametric engines will be used as a means of visualising survey data; then as an analytical tool for the evaluation of urban sites in terms of their potential for agriculture; and finally as a means of supporting the generation of design.

Parametric modelling was selected over alternative technologies for visualisation, geographic data analysis and design development for a number of reasons. First, the capacity of the parametric engine to iteratively generate multiple scenarios in the analysis and design phases of the project offers the ability to consider a wider range of possible solutions and quickly determine which parameters have the greatest impact on the agricultural potential and urban character of the sites under consideration (Provoost, 2009; Rocker, 2002). Among the parameters that will be considered for the purposes of analysis and design include the built neighbourhood density, population density, ratio of built to open space, proximity to transportation and other amenities, availability of roofs suitable for gardening, as well as climatic and environmental factors (solar orientation, soil quality, shading, wind exposure, elevation).

Another benefit of parametric modelling is the possibility to develop time-based scenarios that reflect anticipated change in the relevant parameters. Rather than envisaging static solutions, the parametric engine offers the ability to develop staged solutions that have the potential to evolve as the site itself changes over time. It is also expected that the parametric engines developed in the visualisation, analysis and design phases will play a role in the communication of the results and in consensus-building among project participants as well as the target audience. As a means of quickly visualising relationships inherent in complex datasets (such as the survey responses or the available environmental data) the engines will provide a basis for discussion among neighbourhood residents, stakeholders and other interested parties. Similarly, during the analysis and design phases the engines will allow the rapid testing of scenarios in response to feedback from the project's target audience, and the refinement of specific scenarios.

In the final phases of FUI, it is anticipated that the parametric engines developed for data visualisation and analysis will support the process of design development, offering a means of generating multiple possible design solutions that each respond to criteria selected in the analysis phase of the project. The parametric modelling helps formulate prototypes of urban farming in specific contexts that are used to identify and spatially describe systems of sustainable micro farming in conjunction with the city and to provide a genetic code that can be readily adapted to diverse contexts. Through urban, architectural and landscape design and planning, new solutions for incorporating small-scale agriculture into urban contexts are explored (Figure 41.7). This is accomplished by the automated generation of multiple designs through the variation of site and building parameters, and the evaluation of these designs using a simulation of the resulting conditions for agriculture. The parametric engine also facilitates the automated adaptation of a single design for incorporation on multiple sites, each with unique requirements and characteristics.

The FUI is unique in marrying disparate data from various disciplines to generate design solutions using parametric modelling. Within the team, parametric modelling is a medium for consensus-building, a mode of communication for discussion and project results. The models also visualise complex relationships becoming a basis for discussion among all stakeholders. Agricultural and economic analysis sews the extensively developed knowledge within the Swiss agricultural sector to its new urban context.

41.5 Conclusion

The FUI will draw from and in turn inform many promising new architectural and urban design initiatives, including hybrid architectural typologies (Andraos and Wood, 2010) and

Figure 41.7. Urban vision.

building components that incorporate food cycles as significant programme elements. Cities around the world, and especially in Europe, are incorporating green roofs and facades into existing and new construction with proven environmental and aesthetic benefits (Da Cuhna *et al.*, 2009) and using these technologies for added productive purposes may be desirable but more research is needed to make it credible. Switzerland, as an industrialised nation that continues to grow while also recognising its need to protect its traditional rural and agricultural processes, becomes an ideal location for such research and design. Pilot projects supporting urban agriculture are in desperate need of support, development and testing. The Food Urbanism Initiative hopes to fertilise the new and promising field of urban agriculture by propelling current research and defining new ground for its implementation.

Acknowledgements

This text stems from work prepared for a Swiss National Science Foundation Grant within the National Research Programme 65 entitled New Urban Quality. Excerpts from this text were composed with the assistance of Joseph Claghorn and Dr. Anna Crole-Rees (general text), Dr. Therese Haller (Public Survey), Dr. Jeffrey Huang, Mark Meagher and Trevor Patt (Modelling and Prototyping).

References

48th IFLA World Congress, 2011. Sessions List. 48th IFLA World Congress, June 27-29, 2011, Zurich, Switzerland. Available at http://www.ifla2011.com/schedule/sessions.html.

Aerni, A., Allan, R. and Lehmann, B., 2008. Nostalgia versus pragmatism? How attitudes and interests shape the term sustainable agriculture in Switzerland and New Zealand. Food Policy 34: 227-235.

Anonymous, 2011. Les Plantages Lausannois. City of Lausanne, Olympic Capital. Available at http://www.lausanne.ch/view.asp?docId=35740&domId=65390&language=E.

Andraos, A. and Wood, D., 2010. Above the Pavement – the Farm!: Architecture & Agriculture at Public Farm 1. Princeton Architectural Press, New York, NY, USA, 206 pp.

Appelbaum, A., 2009. Growing with the crops: Nearby property values. New York Times, July 1, 2009.

Cache, B., 2004. After Jean Prouvé: Non-standard folding software. In transcripts of the colloquium devices of design. Canadian Centre for Architecture (CCA) and the Daniel Langlois Foundation for Art, Science, and Technology (CCA). Montreal, Canada, pp. 52-57.

Campbell, L. and Wiesen, A., 2009. Restorative commons: creating health and well-being though urban landscapes. USDA Forest Service, Newton Square, PA, USA, 278 pp.

Carrot City, 2011. Designing for urban agriculture. Available at http://www.ryerson.ca/carrotcity/.

Da Cuhna, A. and Université de Lausanne (Eds.), 2009. Urbanisme végétal et agriurbanisme. Urbia – Les Cahiers du développement urbain durable N° 8. University of Lausanne, Lausanne, Switzerland.

Grimm J., 2009. Food Urbanism – A sustainable design option for urban communities. Senior thesis in landscape architecture, College of Design, Iowa State University, Ames, IA, USA. Available at http://johnsonlinn-localfood.webs.com/Planning%20Resources/Food%20Urbanism_Grimm.pdf.

Huber, R. and Lehmann, B., 2009. WTO agreement on agriculture: Potential consequences for agricultural production and land- use patterns in the Swiss lowlands. Geografisk Tidsskrift-Danish Journal of Geography 109: 131-145.

Provoost, M. (Ed.), 2009. New towns for the 21st century the planned vs. the unplanned city. Sun, Amsterdam, the Netherlands, 256 pp.

Rankin, B., 2009. Local food is not always the most sustainable. Harvard Design Magazine 31 Vol. II: 101-104.

Rocker, I., 2002. Versioning: In-forming architectures. AD Architectural Design 72: 10-17.

Rowe, P., 1997. Civic realism. MIT Press, Cambridge, MA, USA, 266 pp.

Steel, C., 2008. Hungry city: how food shapes our lives. Chatto & Windus, London, UK, 400 pp.

Swiss National Science Foundation SNSF, 2011. New Urban Quality – National Research Programmeme NRP 65. Available at http://www.nfp65.ch/E/portrait/Pages/default.aspx.

Swiss Statistics, 2011. Federal Administration. Available at http://www.bfs.admin.ch/bfs/portal/en/index/themen/01.html.

Swiss Federal Office of the Environment OFEV, 2011. Compensation écologique. Available at http://www.bafu.admin.ch/landschaft/00522/01649/01650/index.html?lang=fr.

Viljoen, A. (Ed.), 2005. CPULs. continuous productive urban landscapes: designing urban agriculture for sustainable cities. Elsevier, Oxford, UK, 280 pp.

Verzone, C. and Woods, C., 2006. Cadrages – Paysage et aménagement du territoire. Volume 1: Etat de lieu. Report of the initiative of the Service de l'aménagement du territoire SDT, Service de l'économie, du logement et du tourisme SELT, Rougemont, Switzerland.

White, M., Przybylski, M. (Eds.), 2010. Bracket – on farming – Almanac 1. Actar, Barcelona, Spain, 252 pp.

Wiles, W., 2009. Farming has become Fashionable. Icon-Magazine 072: 70-78.

Woodbury, R., Gün, O.Y., Peters, B. and Sheikholeslami, M., 2010. Elements of parametric design. Taylor and Francis, Oxford, UK, 300 pp.

Chapter 42

Room for urban agriculture in Rotterdam: defining the spatial opportunities for urban agriculture within the industrialised city

Paul. A. de Graaf

Paul de Graaf Ontwerp & Onderzoek, Aelbrechtskade 10-a, 3022 HK Rotterdam, the Netherlands; info@pauldegraaf.eu

Abstract

The research presented here provides a top-down perspective of the Rotterdam urban landscape and its opportunities for urban agriculture. This perspective aims to be instrumental in bottom-up entrepreneurial driven realisation of urban agriculture projects in cities in industrialised countries in general and Rotterdam in particular, firstly by uncovering and mapping opportunities for urban agriculture and secondly by showing their potential through design case studies. The chapter will focus on the definition of opportunities and its underlying rationale. First, promising types of urban agriculture are defined based on international best practice and expert judgement of the local context. Secondly, a set of criteria is formulated that defines opportunities based on the demands and on the benefits for the city these types have spatially, environmentally and socially. It is argued that the definition of spatial opportunities and a corresponding typology for urban agriculture should be investigated in relation to the role it can play in making the city more sustainable both in a social and an environmental sense and the potential of urban agriculture to function as a system that is more than the sum of its parts, making the city more resilient to changes caused by climate change or other environmental problems, such as depletion of finite resources, loss of biodiversity etcetera.

Keywords: typology, mapping, urban planning, systems thinking

42.1 Introduction

42.1.1 Context

Internationally there is a growing interest in urban agriculture. This is also true for the Netherlands but unlike other places this hardly leads to initiation of concrete urban agriculture projects, at least not projects that go beyond the scale of community gardens, allotment gardens, school gardens and art projects. With all due respect to these manifestations of urban agriculture, what is missing are projects that fulfil the potential for an urban agriculture that is economically viable and that uses the specific opportunities of the urban environment to produce food and non food products. According to Eetbaar Rotterdam ('Edible Rotterdam'), an independent expert group that wants to promote, initiate and investigate such urban agriculture in Rotterdam, this is partly due to the different stakeholders being unaware of these opportunities.

The project 'Room for Urban Agriculture in Rotterdam' wants to change this by mapping opportunities for the realisation of urban agriculture in urban space, focusing on the existing urban fabric and showing through design case studies how these opportunities can be realised in these spaces. It is part of the efforts of Eetbaar Rotterdam to investigate viable strategies for urban agriculture in the Dutch context (Graaf and Van der Schans, 2008).

The project is formulated by Paul de Graaf Research & Design, on behalf of Eetbaar Rotterdam and funded by the 'The Netherlands Architecture Fund'. The research is conducted by the author with input from experts from Eetbaar Rotterdam, Delft University of Technology and Wageningen University, including disciplines such as architecture & urban planning, care, food retail, industrial ecology and agriculture, as well as city officials.

41.1.2 Focus of this chapter

Mapping opportunities is not just about geographical notation but just as much about definition of what to map: trying to understand what agriculture can mean in an urban context and imagining the possible relations agriculture, and the urban context as we know it, can engage in. When agriculture and urban conditions mutually influence each other, new types of agriculture evolve. Therefore traditional criteria for finding suitable space for agriculture are not transposable to the city. All kinds of spatial, socio-cultural and environmental conditions have to be taken into consideration: its built and green milieus and the social diversity of their inhabitants, its zoning and its underlying waste (water) and energy infrastructure. Together these conditions form a landscape that can be mapped. The resulting map is not a road map for an existing practice but an invitation for a practice to evolve that is beneficial to the city. As such the research is a form of exploratory planning based on facts and expert opinion from a diversity of disciplines.

In the tradition of the work by Bohn and Viljoen (Viljoen, 2005) and other urban design and planning professionals that followed in their footsteps (e.g. Bhatt and Kongshaug, 2005; Grimm, 2009; Hohenschau, 2005) its results are aimed to be of scientific interest in its method and analysis and of practical merit in communicating the opportunities of urban agriculture to stakeholders, in this case, in Rotterdam.

42.2 Definitions and methods

42.2.1 Urban agriculture

In this project urban agriculture is defined as agriculture inside and adjacent to the city, that is shaped by interaction with local urban conditions and that produces, processes and distributes food and non-food for the local urban market. Although most Dutch agriculture is urban or peri-urban by international definitions, this research limits itself to types of agriculture that differ from common Dutch agricultural and horticultural practice. Of specific interest are types of urban agriculture that close loops locally either internally or as part of the local food system, and fulfil urban needs that can be spatial, socio-cultural and environmental on the basis of viable economic models.

42.2.2 Opportunities and services

According to Luc Mougeot of the International Development Research Centre (IDRC): 'Urban agriculture is typically opportunistic.' This is meant in a positive way:

Its practitioners have evolved and adapted diverse knowledge and know-how to select and locate, farm, process, and market all manner of plants, trees, and livestock. What they have achieved in the very heart of major cities, and dare to pursue despite minimal support, and often in the face of official opposition, is a tribute to human ingenuity (Mougeot, 2006).

Urban agriculture has to find its place in the existing city and its ingenuity in making use of opportunities clearly sets urban agriculture apart from its rural counterpart. The IDRC research focuses on developing countries, and there are obvious differences between examples in e.g. Ghana, Johannesburg or Rosario and the examples from United States and Canada, but this seems to be a recurring characteristic of urban agriculture. The urban farmer is an entrepreneur that sees these opportunities and 'designs' his business accordingly. As these opportunities often lie in fulfilling urban needs and solving urban problems, urban agriculture can be defined, not just as producing its food, but also performing other services as diverse as education, job training, water reuse or treatment, cooling and beautification (Figure 42.1).

42.2.3 Types and forms

The opportunistic nature of urban agriculture results in a diversity of forms that make classification difficult. Existing examples are shaped in response to local circumstances, as well as the preferences of the farmer. Therefore it is helpful to distinguish between types and forms, in parallel with the difference between typology and morphology that is used in urban planning and design (e.g. Komossa, 2003). A type is the abstract scheme with generic characteristics and a form is the site-specific interpretation and realisation of a type (or a combination of types). This approach – together with the specific Dutch context – distinguishes the typology developed here from other proposed typologies, such as the one made for the city of Ames, Iowa (Grimm, 2009).

Types of urban agriculture are defined as cultures or combinations of cultures (poly-cultures) that can be classified along characteristics such as: medium, nutrient management, use of knowledge, labour, capital, energy and technology, degree of integration in buildings. The type and its characteristics define the criteria for opportunities.

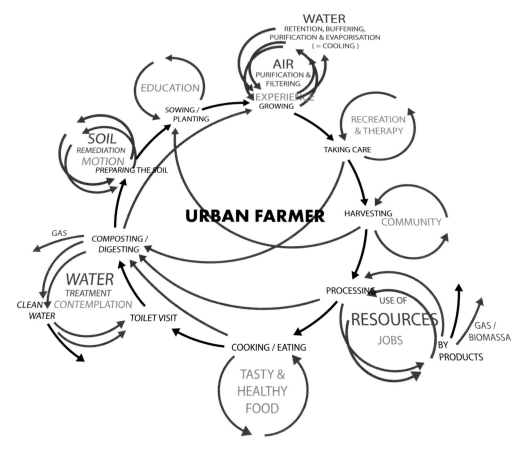

Figure 42.1. Urban agriculture in service of the city. The urban food (nutrient) cycle as and its potential social and environmental spin-off.

Forms are the local, site and time specific interpretations of a type or combination of types in a design, in which the farmer determines which opportunities are seized and how. Scale, spatial lay-out and exact combination of types are formed in the interplay between the farmer and the location. Site specific interpretation include the way incoming flows from local sources are incorporated, the optimal use of available space in relation to e.g. micro-climate, sunlight as well as access and visibility and the inclusion of services for the neighbourhood. These forms come also with specific forms of organisation in terms of labour, capital, time (i.e. a business plan).

42.2.4 Mapping

Because of its opportunistic character urban agriculture is resistant to top down planning. However planning can play a role in identifying opportunities for individual practices and, secondly, assessing their role in the functioning of the city as a whole, as part of the urban food, waste, water and energy system. The mapping of opportunities is a way of investigating to what extent urban agriculture can benefit the city, what its scale and impact could amount to. By making the benefits apparent and tangible it can create a context in which the urban agriculture that is most beneficial to the city is recognised as such and can be stimulated.

Secondly mapping can be a communication tool for opening the eyes of stakeholders to what urban agriculture can mean to the city. The aim is to map and show Rotterdam as a landscape of opportunities to people that are or want to become involved in urban agriculture professionally and to the people that these urban farming entrepreneurs will meet on the way to realising their ideas: civil servants, developers, investors.

In this context to identify promising locations plot by plot is beside the point. It is too detailed and too precise and therefore quickly outdated. Instead the map will indicate in which areas of the city one is likely to find space and opportunities for (specific types of) urban agriculture on the basis of more general spatial, socio-cultural and environmental characteristics of the areas, such as density, average income and the surplus of storm water run-off. For this reason it is chosen not to work with GIS but with a number of more general maps in combination with expert opinion and on-the-ground observation to deduct, compose and draw a map.

42.3 Defining an urban agriculture typology for Rotterdam

42.3.1 Promising types of urban agriculture

As shown in Table 42.1 both city and agriculture have their needs that the other can (partly) fulfil. In determining promising types the first step is to determine what types are suitable for the city (in how far the city can satisfy its agricultural needs) and secondly which of these suitable types are most beneficial to the city (in how far it can fulfil the needs of the city).

According to a revised version of the Von Thunen model crops are identified that would have the greatest competitive advantage over other crops in the city. These are crops that have a high yield on a relatively small area, that by their quality justify a higher price than supermarket products, e.g. because there is no need to artificially prolong shelf life with cooling, and crops that in a smart way tap in to local resources of (waste) energy, water, nutrients and materials (Graaf and Van der Schans, 2008).

Typically this includes vegetables and dairy (because of shelf-life), but also mushrooms and micro-stock such as fowl (double purpose) and raised fish can be viable options. Growing these crops and raising these stocks in the city however is only possible if it can be done profitably on a small scale and if the way the crop is grown can be adapted to make optimal use of available space and resources, without being a nuisance to the inhabitants.

From this follows a range of types that have in common that they are flexible in their use of space and that they can function at a relatively small scale. The Southeast False Creek Urban Agriculture Strategy prepared for the City of Vancouver gives a good overview of this

Table 42.1. Matrix of supply and demand of agriculture and the city.

Demand	Supply
Agricultural needs	Urban supply
sunlight/daylight	plenty of sun-exposed surface
nutrition/fertiliser	waste flows (nutrition, irrigation, substrate, heat)
irrigation	micro-climate
soil/substrate	vacant space
micro-climate/environment	niche space
space	temporary space
loading capacity (integrated in buildings)	underused constructive capacity
labour (intensive/extensive)	labour force (employees)
market	customers
Urban needs	Agricultural supply
public green design & management	aesthetics
ecosystem services	relative biodiversity
education (nature,	experience of seasons
food production, life skills)	hands-on learning/work experience
therapeutic work	therapeutic work
appropriate jobs	skilled and unskilled labour
water storage	water intake & evaporation
climate control (cooling / heating) at building and neighbourhood level	evaporative cooling
improvement water, soil and air quality	purification of water, soil and air
waste treatment and management	organic waste treatment

range (Holland Barrs Group, 2002). The types differ in their relation to the soil and the built environment, their exchange with the essential flows of the city and in the impact they have on public space socially and aesthetically. Thus they offer different benefits to the city and most likely will also respond to different opportunities. For this research promising types were ordered in a matrix along two axes: control versus self-organisation and soil-bound versus building integrated. The field defined by these two axes can be further characterised by a number of keywords and shows difference in the needs these types have (and thus in their potential). In this matrix roughly four types in different varieties can be distinguished (see Figure 42.2).

On the one end of the matrix presented in Figure 42.2 there is forest gardening, which is a soil-bound and largely self-organising outdoor garden. On the other end are highly controlled, indoor installations such as hydroponics and aquaponics. In between there is soil-based cultivation which is how many people grow their own vegetables, and is the most accessible because it requires less knowledge and capital then the other two. These four types (Figure 42.3), that will be described more in depth below, supplement each other in the

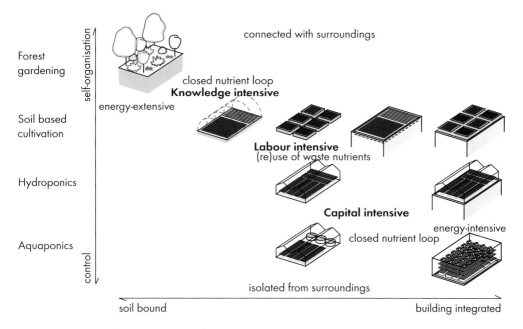

Figure 42.2. Matrix of promising types of urban agriculture.

products and services they offer, in the diversity of needs they have, and in the way they serve spatial, socio-cultural and environmental needs of the city.

Type 1: forest gardening

Forest gardening is a way of building up a garden as an ecosystem. It is also a spatial model working with 4-7 vertically organised layers of trees, shrubs, plants, rhizomes etcetera. It is soil based, and when managed correctly largely self-sustaining. It uses mostly perennials and resembles both in functioning and appearance a natural forest edge. Forest garden is a poly-culture that offers products, such as fruits, roots and nuts that supplement those typical of

Figure 42.3. Four promising types of urban agriculture: forest gardening; small-plot soil-based cultivation; rooftop hydroponics; indoor aquaponics.

other types of urban agriculture, as well as medicinal plants, bamboo and other non-food products (Jacke and Toensmeier, 2005).

It potentially offers a radically different, potentially viable model for public green maintenance. In contrast to other types of urban agriculture it provides a green outdoor environment throughout the whole year and it optimises output (produce) in relation to the energy (mostly labour) put in. The city of Rotterdam has large areas of public green (especially in postwar reconstruction housing areas) that are being maintained at minimum cost, with the consequence that these areas are underused and degraded. Here, as well as in parks and botanical gardens or zoos, forest gardening could offer a viable alternative. The forest garden has the educational potential to show how nature closes loops as an inspiration for closing loops in the built environment.

Typically the forest garden will not make the city greener but rather improve the quality of existing green areas, both aesthetically and ecologically. Its role in the city metabolism is limited, with hardly any exchange of streams, except for a considerable capacity for keeping and evaporating storm water. Finally creating a forest garden takes knowledge and experience as well as a considerable period of time to build up (approx. 10 years), which requires long-term investment.

Type 2: small-plot soil-based cultivation

Soil-based cultivation of predominantly vegetables on small plots taking advantage of the extended growing season in the city. This urban agriculture type can be said to be the most common type of urban agriculture as it has been practised in allotment gardens, community gardens and school gardens for a very long time, mostly for self-sustenance and educational purposes.

What makes this type relevant for Rotterdam is the way Small-plot Intensive Farming (SPIN farming) developed by Wally Satzewich has optimised this type as a business model that allows urban farmers to earn an income off small private plots of land that are available in cities, such as back yards and rooftops (Satzewich and Christensen, 2005). Because of the comparatively high density of Dutch cities and the smaller backyards the potential for growing vegetables is smaller than in Canada or the USA; a significant part of the backyards in Rotterdam do not catch enough sun to profit from the extended growing season. For cultivating rooftops constructive capacity and accessibility are the main issues, limiting this type to heavily constructed buildings especially from the 50's and 60's. When situated in dense urban areas with little outdoor green, rooftop farms can offer an oasis, reducing heat-stress, floating fine dust and CO_2, as well as beautifying the roof landscape.

From a socio-cultural point of view this type has the potential to provide local jobs, as a lot of the skills are present amongst citizens, particularly immigrants. Cheap access to land and the lack of expensive machinery make this type easily accessible for people wanting to start their own business either part-time or full-time.

Type 3: rooftop hydroponics

Rooftop hydroponics systems grow vegetables and fruits on substrate in greenhouses, that run on photovoltaic cells, capture and use rainwater and can use exhaust air from the building underneath for additional heating and possibly even CO_2. Further sustainability claims include the cooling effect of green on the roof, less food miles and reduction of use of pesticides. With the right design they could function as educational and even therapeutic spaces although in this sense hydroponics offers a more limited environment than e.g. Forest gardening.

The potential for hydroponics lies in its high yield combined with its light weight (70-120 kg/m2 – comparable to an extensive green roof). A quick scan showed that experts estimated that a large part of the flat roofs in Rotterdam can have an extensive green roof without additional constructive measures (Voll, 2006). Another potential lies in the integration of knowledge and innovation from the Dutch greenhouse sector, in the close-by Westland area, such as the Energy Producing Greenhouse. With this technology rooftop hydroponics could use heat-exchange to store summer heat in the ground (cooling the city) and use it for heating in the winter. Rooftop hydroponics could help to close the nutrient cycle inside the city. If nutrients from household waste water would be recaptured e.g. through anaerobic digestion, and made available for use in hydroponic systems, they would solve the current dependency of hydroponics on artificial fertiliser that depends on finite resources.

Type 4: indoor aquaponics

Aquaponics systems can be set up in urban settings using a limited amount of technology in order to keep investment costs down and operation and maintenance relatively simple. This makes these systems more flexible in scale and operation and allows them to operate in the limited space available in the city, e.g. using vacant buildings. Aquaponic systems are more diverse and complex than hydroponics and have a more advanced cycling of nutrients. Hydroponics' need for fertiliser is solved by adding a step in the circulation of water. Rearing fish adds organic waste (fish excrement) to the water feeding the plants. More steps can be included, such as composting with the use of black soldier fly larvae that can be used as fish food. It is this innovative thinking in terms of cycling combined with the low-tech do-it-yourself attitude that makes this type of urban agriculture very promising, both as a means of turning waste into a resource and as an educational model. In its basic form a self-contained system it has the potential to engage with different urban waste streams. The diversity of crops and stocks, with each their different needs, e.g. amount of sun/daylight and temperature, makes it possible to make optimal use of different conditions inside vacant buildings. Its diversity in produce, including fish, makes it supplemental to food production from other types of urban agriculture.

Despite the relative low-tech approach, capital investment is still considerable; higher than other types of urban agriculture but lower than traditional horticulture. Also investment in knowledge is needed.

42.3.2 Mapping opportunities

On the basis of the Rotterdam urban agriculture typology, spatial, socio-cultural and environmental criteria are defined that can be mapped (see Figure 42.4). Spatial criteria are mapped through an urban typology of milieus (divided in three main categories: housing, industry and green), that are distinguished by the conditions they offer agriculture (in terms of micro climate, sun exposure, constructive capacity and accessibility of underused space). Some socio-cultural and environmental characteristics can be attributed to these milieus as well (e.g. proximity of restaurants, availability of household nutrient stream). On the basis of their qualities, each of these milieus can be judged more or less suitable for each of the four types.

Closely related to the spatial layer are zoning restrictions and the subsoil (which in a city like Rotterdam is influenced by spatial development rather than vice versa). Additionally, environmental maps locate sources of (waste) water, nutrients, energy, heat, e.g. areas where excess storm water is a problem.

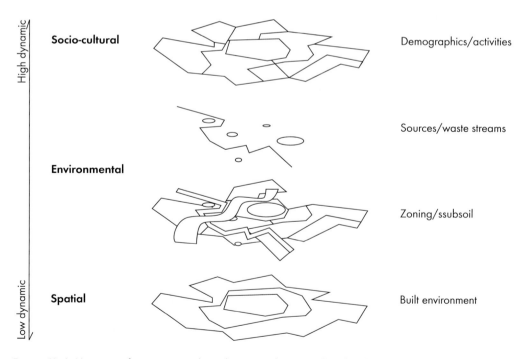

Figure 42.4. Mapping of opportunities based on spatial, socio-cultural and environmental criteria.

Socio-cultural criteria are the most difficult to map. This layer is much more dynamic than the spatial layer and to a lesser extent the environmental layer as it consists of people, their background and motivation. Except through the filter of urban milieus and its social characteristics, this layer is not further represented by maps as this takes an effort and expertise beyond the scope of this research.

As the next step opportunity maps have been made by super-positioning maps of criteria. The criteria were weighed and adjusted in relation to the characteristics of the types. A further distinction was made between Small-plot Soil-based Cultivation (Type 2A) and Soil-based Cultivation on Rooftops (Type 2B) as these have very different requirements. The process of mapping and weighing is not further discussed here, but resulted in distinctive opportunity maps for each type. These opportunity maps have then been combined in a policy map for urban agriculture in Rotterdam (Figure 42.5), that shows which type is most useful to the city in what area. The results of the research – the maps and a number of design case studies illustrating their use – will be presented to and discussed with stakeholders. If this results in commitment from stakeholders the actual testing of the map can begin. This phase will be largely dependent on entrepreneurs willing to start up an urban farming business. At this

1. Forest gardening   2a. Small-plot soil-based cultivation   2b. Soil-based cultivation on rooftops

3. Rooftop hydroponics   4. Indoor aquaponics   Policy map

Figure 42.5. Opportunity maps for each urban agricultural type (1-4), and their integration into a policy map (lower right corner) showing which type is most suitable and beneficial where, based on a sustainable planning perspective.

point it is still a question as to who has the combination of farming expertise and affinity with the city to become these urban farmers.

42.4 Some concluding remarks

Planning policy can be based not just on the urban agriculture initiatives that are already there, but on the possibilities specific types of urban agriculture offer for solving problems a city is dealing with. As a form of exploratory planning, mapping opportunities for urban agriculture helps with making evidence based decisions about where to allow and support particular types of urban agriculture. Thus planners, by playing a pro-active role in stimulating urban agriculture, can co-create a sustainable city.

The areas of opportunity that were mapped for each of the four promising types of urban agriculture can be said to represent the areas where they are most likely to flourish, or their 'preferred habitat'. When overlapping these maps in a combined map questions about the design and planning of the city are raised. In case of competing claims which type is to be preferred? And where do conflicts of interest with other types of public green (e.g. urban nature areas) occur? The planner has to weigh spatial (aesthetic), socio-cultural and environmental considerations against each other. The combined opportunity map is a first attempt to answer these questions and consider the sustainable contribution of urban agriculture from an integrated, holistic viewpoint, but it is by no means a definitive map.

The next step is to get the interest and involvement of planning officials for a pro-active approach to urban agriculture. Urban agriculture is on the political agenda in Rotterdam and this opens up possibilities, but there are a lot of different and sometimes conflicting ideas about what urban agriculture is. It is hoped that this research will inform the municipal process of defining an official standpoint and policy on urban agriculture. Here the opportunistic nature of the approach helps as it is complementary to other spatial planning policy on e.g. community gardens and allotments, green roofs, public green maintenance, etcetera.

It is encouraging that parallel to this research the first professional urban farm is being realised with the enthusiastic support of some dedicated municipal officials. Hopefully 'Room for Urban Agriculture in Rotterdam' will further stimulate the wider realisation of a professional urban agriculture. When more professional urban farms are operational systemic organisational questions about nutrients, energy and water flows can be more fully assessed and addressed.

References

Bhatt, V. and Kongshaug, R. (Ed.), 2005. EL – making the edible landscape. A study of urban agriculture in Montreal. Mc Gill University, Montreal, Canada. Available at http://www.mcgill.ca/mchg/pastproject/el/.

De Graaf, P. and Van der Schans, J.W., 2008. Integrated urban agriculture in industrialised countries. Design principles for locally organised food cycles in the Dutch context. In: SASBE09 Conference Proceeedings. Available at http://www.sasbe2009.com/papers.html#papers.

Grimm, J., 2009. Food Urbanism. A sustainable design option for urban communities. University of Iowa, Ames, USA. Available at http://www.database.ruaf.org/ruaf_bieb/upload/3129.pdf.

Hohenschau, D.L., 2005. Community food security and the landscape of cities. University of British Columbia, Vancouver, Canada. Available at http://www3.telus.net/public/a6a47567/DLHohenschau.pdf.

Holland Barrs Planning Group, 2002. Southeast false creek urban agriculture strategy. Holland Barrs Planning Group, Vancouver, Canada. Available at http://vancouver.ca/commsvcs/southeast/documents/pdf/urbanagr.pdf.

Jacke, D. and E, Toensmeier, 2005. Edible forest gardens. Vol. 1: Vision & Theory. Chelsea Green, White River Junction, USA, 654 pp.

Mougeot, L.J.A., 2006. Growing better cities. International development research centre, Ottawa, Canada, 97 pp.

Satzewich, W. and Christensen, R., 2005, SPIN Farming. How to grow commercially on under an acre. Spin Farming LLC, Canada. Available at http://spinfarming.com/.

Viljoen, A. (Ed.), 2005. Continuous productive urban landscapes. Designing urban agriculture for sustainable cities. Architectural Press, Oxford, England, 280 pp.

Voll, L., 2006. Rotterdam groen van boven. Toepassing van groene daken in Rotterdam. Gemeente Rotterdam, Rotterdam, the Netherlands. Available at http://www.eetbaarrotterdam.nl.

Chapter 43

Progress through process: preparing the food systems report for the Chicago Metropolitan Agency for Planning GoTo2040 Plan

Lynn Peemoeller
Food Systems Planning, Wollinerstraße 2, 10435 Berlin, Germany; Lynn@foodsystemsplanning.com

Abstract

How will we continue to produce food and feed our population in 2040 while planning for population growth, transportation, homes, and commerce in the Chicago Metropolitan region? This is the question that frames the Chicago Metropolitan Agency for Planning 2040 Food Systems Report, which was developed over the course of 9 months through a participatory community planning process with over 130 food systems stakeholders from the urban and peri-urban areas that make up the Chicago Metropolitan Region. The report was the first report of its kind for the region. It required a collaboration of stakeholders who had not previously worked closely together to define existing conditions, a vision for 2040, a set of recommendations, and a list of indicators. This chapter examines the process by which the report was produced using the community planning strategy and the challenges and opportunities it presented.

Keywords: regional planning, community food systems, stakeholder process

43.1 Introduction

Planners in the US have begun to incorporate food systems into their work in a number of ways, from regional, systemic long-term approaches to specific, focused approaches like food access, healthy lifestyles, value chain development, and reduction of food waste among others.

The American Planning Association, the United States Department of Agriculture and other organisations like the Community Food Security Coalition have helped facilitate the development of food systems policy with a growing number of support resources and publications. It is clear that food systems have a seat at the table among the many other sub-disciplines of urban planning.

As food systems ideas and networks have grown throughout the US, Canada and the E.U., so has the evolution of the work. Professional, scholarly and government efforts are shifting from more targeted programmatic work to more comprehensive and multi-level approaches to food systems planning. As planners and policy makers know, regional planning requires a coordination of a complex set of land use activities, infrastructure, and settlement growth across a significantly large area. As this becomes more defined, parallels between food systems and the framework for regional planning have begun to emerge and become integrated. Recent examples of this in the U.S. include the 'Greater Philadelphia Food Systems Study' (2010) a multi-year, multi dimension effort by the Delaware Valley Regional Planning Commission. And 'Planting Prosperity and Harvesting Health' (2008) a report by the Institute of Portland Metropolitan Studies.

The Food Systems Report (for the Chicago Metropolitan Planning Agency GoTo2040 Plan), commissioned by the Chicago Community Trust (CCT) in 2009 is one such effort that takes a place among several dozen other local reports to examine potential strategies for implementing a regional vision for the future in the official Go To 2040 Comprehensive Regional Plan, led by the Chicago Metropolitan Agency for Planning (CMAP), the regional Metropolitan Planning Organization for the greater Chicago area, the third largest metro region in the United States. The Go to 2040 plan is based on nearly three years of research, deliberation, and public input. The plan recommends action in four broad themes: Livable Communities, Regional Mobility, Human Capital, and Efficient Governance. Within this, the Food Systems Report is incorporated.

As the first effort to develop a regional food systems plan for the Chicago metropolitan region, this chapter will shed light on the outcomes and the process by which the Food Systems Report was produced. The outcome of the applied community planning strategy will be discussed and the challenges and opportunities it presented in bringing together diverse stakeholders, creating a shared vision, and producing an authentic report reflecting that process.

43.2 Putting it together

GO TO 2040 is the Chicago Metropolitan region's official comprehensive plan, intended to help the many communities of the seven county region face and implement recommendations in order to guide sustainable prosperity by preparing to meet its needs in 2040 in light of predicted population growth, demographic shifts, and climate changes.

Food systems was recognised as a topic of special consideration as one of 12 quality of life issues such as Health, Emergency Preparedness, Safety, Education, Arts, Culture and so on. In 2008, the Chicago Food Policy Advisory Council (CFPAC) and the City of Chicago Department of Zoning and Land Use Planning were chosen by CCT, the funder, to co-convene and lead an advisory council to develop a vision of the regional food system for 2040, define existing conditions, generate a list of indicators, and develop a set of recommendations in a report for submission in 2009.

The partnership between the CFPAC a non-profit volunteer advocacy group and the City of Chicago Department of Zoning and Land Use Planning was strategically chosen to balance resources that neither group had on its own. The author of this chapter participated and managed the project in affiliation with CFPAC.

Chicago Food Policy Advisory Council Resources:
* an active network of people representing communities throughout the city of Chicago and in some surrounding areas;
* a reputation for citizen engagement;
* a diverse network representing people in a range of ethnicities, ages, and income levels;
* partners willing to convene community based meetings.

City of Chicago Department of Planning and Development Resources:
* paid, skilled and knowledgeable staff;
* administrative support;
* access to wide range of data;
* existing partnerships with government and other relevant stakeholders.

As soon as the lead agencies were established, an advisory committee of both urban and rural residents of the seven-country region was assembled. This included over 30 individuals representing a range of stakeholders in the regional food system. Participants were representatives from city, county and state government, private business, farmers, urban advocacy groups, rural advocacy groups, health advocates, educators, funders, non-profit organisations, and policy specialists.

The idea of assembling an advisory committee was to invite a wide range of participants into the process. This stems from basic principles of community planning or in other words a bottom-up approach to developing the plan. Indeed, food systems are often framed as 'community food systems' and 'community food security', which utilise methods of democracy, and citizen engagement for the process of making visible the relationships between the actors in the food systems such as the producers, processors, distributors, and consumers and creating relevant, accessible place based solutions (Born *et al.*, 2008). Public opinion was a fundamental part of this project, as it is able to provide CMAP and CCT with a baseline of information for what issues are of importance and priority to diverse communities.

In this example, the goal was to represent as much ethnic and geographic diversity from the Chicago metropolitan region as possible. This presented several challenges. The first was that not all people invited were able to participate in the meetings. Barriers to participation included the location of the meetings, which were held in downtown Chicago. Expensive parking, travel time and traffic were all challenges for those that had to travel far. The timing of the meetings was during working business hours. This was difficult for those whose participation was not supported by their job. Another challenge for some was the unfamiliarity with the topic, the lead agencies, and the relevance to their work. In the end, the advisory committee was represented by a group of dedicated individuals who had the resources and desire to participate.

As it became clear that geographic and ethnic diversity were going to be challenges in the representation of the advisory committee, a second level of outreach was developed.

The CFPAC developed a strategy to partner with community organisations in different areas of the city of Chicago to insure that an inclusive process was followed to identify needs, priorities and assets at the community level. A total of over 60 people attended four meetings.

Additionally the lead agencies worked together to partner with a rural advocacy group to assemble a rural meeting of over 20 farmers from the seven county region to discuss issues of importance related to the agricultural community. Over the course of the project, input was also gathered from smaller privately arranged meetings with various stakeholders such as those involved in land use regulations, environmental regulations, business interests, hunger advocates, spiritual communities, and workforce development specialists in the region.

In the end, over 130 individuals and organisations were represented in the development of the vision and the writing of the report. This was a satisfactory outcome for the effort put into the process.

43.3 Methodology

As food systems continue to be defined in the field of planning, one of the challenges for practitioners is deciding what methodologies should be used to approach the work. There are a number of utilised models in the United States that often take a bottom-up, community based or adaptive learning process approach to planning (Reed and Mark, 2006).

A 'Community Food Assessment' is a methodology used to determine the existing resources in a community and is often used to measure and guide community food security. The United States Department of Agriculture (2002) created a toolkit and guidelines for the Community Food Assessment process that are meant to provide a standardised set of measurement tools for assessing various indicators of community food security.[50]

A more complex hybrid incorporating similar methodology is the 'ommunity Food Security Assessment Toolkit', which according to the Community Food Security Coalition (1997), includes information on access to food; hunger, nutrition, and local agriculture data; in addition to the inventory of community food resources; and policy perspectives.

Several community food assessments have been undertaken in the United States at different scales, from the regional and state level, to the county and town level. There are many examples. A few include the San Francisco Collaborative Food Systems Assessment

---

[50] The USDA Community Food Security Assessment Toolkit was developed through a collaborative process, initiated at the community Food Security Assessment Conference sponsored by ERS in June 1999. It is designed for use by community-based nonprofit organisations and business groups, local government officials, private citizens, and community planners.

(2005), Toledo Community Food Assessment (2007), and the Burlington Community Food Assessment (2004).

In 2008 the Northeastern Illinois Community Food Security Assessment was released. This included a baseline study of access to healthy, culturally appropriate food broadly in the region and within in six communities in-depth. This area largely corresponded to the seven county CMAP region. Results from this report were referenced and utilised in the Chicago Food Systems Plan.

As experienced through practice, a range of factors can play a part in application of methodology including access to information, training, resources, scale of the project, clients, partners, funding, and more. It would be a valuable contribution to the professional or academic field to survey different food systems planning methodologies and compare outcomes. This would help to further establish guidelines for the practice of food systems planning and expectations from the process.

The methodology of the Food Systems Report is an example of a hybrid of approaches. It did not rely on an intellectual or academic construction of process. Rather, the approach and methods were based on concepts of community based planning and derived from the following set of circumstances:
- expectations and pre-set framework for deliverables from CCT as the funder;
- specific needs from CMAP, the Metropolitan Planning Organization partner;
- expertise and resource capacity of the lead agencies;
- timeframe and funding;
- wide representation of community input from diverse stakeholders;
- participatory and collective input and development of the report.

Additionally, in Chicago, it is important to mention that this was not the first attempt to study the local food system. There were several reports & studies previous to the Food Systems Plan besides the Northeastern Illinois Assessment that informed the content and approach of the report. Significantly, two were produced by the lead agencies. The 'Eat Local Live Healthy' report released by the Department of Zoning and Land Use Planning (2007) and the report 'Building Chicago's Community Food Systems' released by the Chicago Food Policy Advisory Council in 2008. Also at that time a statewide task force was completing a Local Food Farm and Jobs study. So while there were active citywide recommendations and statewide recommendations for local food systems planning, nothing had been defined at the regional level at that point.

The challenge of how to build the work plan and approach the methodology of the project created some tension among the lead agencies. While individuals had worked together before there had been no history of collaboration between the two agencies. Both agencies had different missions and members of the project team had a variety of professional training. Additionally the agencies had the pressure of making the process relevant to the entire region when both were Chicago based. The effect of these circumstances created an internal process that required a significant amount of up-front time to get everyone in agreement and to develop the framework for the project.

43.4 Building the vision

One of the first exercises was to develop a shared vision. The lead agencies coordinated a facilitated visioning workshop with the advisory committee to begin to imagine what the regional food system would look like in the year 2040. The committee produced a list of principles, goals, and strategies that covered topics such as; access, fairness and justice, environment, community, economy, and health. Although the level of participation from advisory committee members in the visioning exercise was high, by the end of the meeting, the results were only loosely organised. This presented a challenge for the lead agencies to further organise and represent the outcome as a cohesive vision.

The lead agency team searched for an organisational system to represent the wide range of feedback received from the advisory committee. After lengthy consideration, the decision was to utilise the concept of sustainability to frame the vision. This was chosen for several reasons; firstly because it resonated throughout the feedback from participants, and secondly because its framework is consistent with ongoing work of CMAP in the region.

The vision (excerpt from report)

In 2040 we will have a regional food system that nourishes our people and the land. The food system will:
- achieve economic vitality by balancing profitability with diversification in all sectors;
- preserve farmland and enhance water and soil quality in closed loop systems;
- contribute to social justice through equal access to affordable, nutritious food;
- support vibrant local food cultures based on seasonality and availability.

Excerpt continued

> *The regional food system strategy proposes to incorporate all of these ideas into one model: The Four Pillars of Sustainability. Sustainability is often referred to as a three legged stool with economic, environmental, and social components. A newer, more innovative model of sustainability seeks to include a fourth prong, culture. "This model recognises that a community's vitality and quality of life is closely related to the vitality and quality of its cultural engagement, expression, dialogue, and celebration. It increasingly recognises the need to incorporate culture and creativity in sustainable plans and strategies" (CECC, Cultural Research Salon, 2006). Here, culture encompasses every day choices that we make, inherent in our habits and traditions. In order for the food system to change, attitudes must change.*

The vision was the driver for the report. After it was created it was utilised throughout the duration of the project to engage individuals and to create an opportunity for feedback. While the vision did not change after its initial development by the lead agencies, the response to specific points were incorporated into the report in the appendix of stakeholder feedback.

43.5 Writing the report

The framework for the report outcomes were developed by CCT and CMAP prior to the project start. This required the following to be incorporated; a vision for the region, a set of indicators on how to find food systems data and track the progress of the region over time, the existing conditions of the region, and recommendations for the region on how to achieve the vision.

Writing the report was the culminating challenge for the group. The lead agencies led the process with a voluntary writing committee of 20 people over the course of approximately three months. A number of factors contributed to challenges; the short time frame for research, limited complete data sets available at the regional level, the cross section of different government structures involved in the regional level, lengthy debate on key issues like conventional agriculture versus sustainable agriculture, how to contextualise urban agriculture, and how to define the food shed in the regional context. With the framework for an open participatory process perhaps the most difficult task was organising the report in light of the many different contributors and varying perspectives.

Excerpt continued

> *In the beginning phases of preparing this report we quickly realised that it is impossible to separate what goes on at the most local level of the food system without acknowledging the national and global food system that we are linked to as consumers and producers. Worldwide, a number of trends are affecting our food system, including what and where food is grown, and at what cost. Global issues include climate change, the unstable price and supply of oil, the challenges of supply and demand based on the limits of arable land, increased population growth, water and air pollution, loss of biodiversity, and changing markets. Regional decisions must be made with full knowledge of the larger context in which they reside.*

Achieving consensus was not easy in the structure and development of the report, and at times it created less than favorable working conditions, which perhaps resulted in an end product that was not entirely satisfactory to all the participants. However, there are several points to this process, which are worth considering:

- A diverse group of core stakeholders stepped up to the challenge. Many of these people were not involved from the beginning of the process, but became dedicated to the outcome. The open, participatory process allowed this.

- A range of food systems data such as pertaining to farmland, local food production, food access, the food service sector, public health and diet related disease, and others were investigated and analysed for the first time in regional context. (A separate report and working group was funded by the CCT to develop a study and report on the topic of Hunger in the region. This was cross-referenced.)
- The writing of the report gave the various stakeholders an opportunity to come together to share information and debate topics of importance to the regional food system. This also established new networks within the group and strengthened existing ones.
- The outcome of the report allowed participants to come away with a working framework to understand how to approach the regional food system.

Ultimately, the resulting report released in October 2009 took a conservative rather than action oriented approach to the regional food system. This is reflected by the focus on research, very broad goals and recommendations, and no well-defined action plan.

Excerpt continued

> While it would be encouraging to reach the end of the process that resulted in this report and be able to say that there is a broad consensus on obvious regional policies that must be implemented, food systems planning in the Chicago region is not yet at that stage. The more modest outcome for this report is likely to be that planners and policymakers working at different government levels in the seven county region, and in close collaboration with the private and nonprofit sectors, will be able to determine which avenues may prove most fruitful in impacting the food system to ensure that the growing population of the Chicago region is supported by a food system that is economically, socially, culturally, and environmentally sustainable.
>
> The recommendations, while broad, do point to the strong need for a greater gathering of data and increased public understanding. The outcome of this report may well be that many of the same players who served on the advisory committee will join with new colleagues met during the planning process and decide which questions must be researched first on the path to good decisions and creation of corresponding policies. What elements of the food system make sense to address first? What steps logically follow? Having the research from this report in hand to consider these questions is a critical step.

43.6 Lessons learned

This chapter reflects an ambitious effort to define and discuss the future of the food systems in the Chicago metropolitan region through an open participatory process. As a professional participant leading this process, there are several observations that can be valuable to the planning field.

Due to the conditions mentioned previously, maintaining the cohesion of the process required a great deal of time and energy from the lead agencies. The open participatory process resulted in frequent polemics about the issues like how define food systems regionally, this led to constant re-organisation of information in the report and in the end, time was lost in this process.

Although previously defined food systems work was acknowledged and referenced, it was not universally utilised in the preparation of the report. Rather than cohesive buy-in to shared ideas, individual knowledge resources were defended and incorporated sectionally.

In the end, we can reflect on the value of the report to the stakeholders and its contribution to the field of food systems planning and policy development. It clearly provided value to the funder and the metropolitan planning agency and has been officially incorporated into the GO TO 2040 regional plan. In this way it acts as an effort to educate and inform regional citizens and policy makers that food systems (more important than 'food systems' as a whole are the relevant components) have a place in the greater regional planning process.

However for the lead agencies and the stakeholders in the process, many of whom are involved in food systems work on a daily basis, it is not clear that a high value in the outcome of the report is shared by all. Several factors contribute to this; this report took place among several other concurrent food systems defining efforts at the state and community levels, the lack of action plan did not provide a way for participants to stay engaged, leadership of the lead agencies and other stakeholder groups changed during the process- leading to a lack of successive cohesion, and in the end, general discord about the strength of the positioning of the report.

To come away from this effort focused on the challenges it presented is to say that there were no lessons learned and no progress was made. What can be seen through all this is the significant contribution it made to a process that had not previously been attempted at the regional scale. A vision for the future was created. Success will be defined in the long-term. From those discussions and debates what emerged is a new network of food systems ideas fitted to the regional scale. Since the goal is 2040, this plan has hopefully set a framework from which a slow crystallisation of ideas can grow to create a more sustainable food system for the Chicago metropolitan region.

References

Aubrun, A., Brown, A. and Grady, J., 2006. Conceptualizing US food systems with simplifying models. Frameworks Institute, Washington DC, USA. Available at http://www.frameworksinstitute.org/.
Block, D., Chavez, N., Birgen, J., 2008. Finding food in Chicago and the suburbs. The Report of the Northeastern Illinois Community Food Security Assessment. Available at: http://www.csu.edu/nac/neil_community_food_security.htm.

Centre of for Policy Studies on Culture and Communities, 2006. Culture: The Fourth Pillar of Sustainability. Centre of for Policy Studies on Culture and Communities, Vancouver, BC, Canada. Available at http://www.cultureandcommunities.ca/resources_sustainability.html.

Community Food Security Coalition, 1997. Community food security: A guide to concept, design, and implementation. Available at http://www.foodsecurity.org/pubs.html.

Delaware Valley Regional Planning Commission, 2010. The greater Philadelphia food systems study. Available at http://www.dvrpc.org/food.

Institute of Portland Metropolitan Studies, 2008. Planting prosperity and harvesting health. Available at http://www.pdx.edu/ims/food-systems-forum-2008.

Samina, R. and Born, B., 2008. Planners guide to community and regional food planning. APA Press, Chicago, IL, USA, 111 pp.

Pothukuchi, K., 2004. Community food assessment: A first step in planning for community food security. Journal of Planning Education and Research 23: 356-377.

Pothukuchi, K. and Kaufman, J., 2000. The food system: A stranger to urban planning. Journal of the American Planning Association 66: 113-24.

Reed, M.S., 2006. An adaptive learning process for developing and applying sustainability indicators with local communities. Journal of Ecological Economics 59: 406-418.

The United States Department of Agriculture, 2002. Economic research service community food security assessment toolkit. Available at http://www.ers.usda.gov/Publications/EFAN02013/.

Chapter 44

Slow briefs: slow food... slow architecture

Joanna Crotch
Mackintosh School of Architecture, Glasgow School of Art, Glasgow G3 6RQ, Scotland, United Kingdom; j.crotch@gsa.ac.uk

Abstract

We are moving too fast… fast lives, fast cars, fast food… and fast architecture. We are caught up in a world that allows no time to stop and think; to appreciate and enjoy all the really important things in our lives. Recent responses to this seemingly unstoppable trend are the growing movements of Slow Food and Cittaslow. Both initiatives are, within their own realms, attempting to reverse speed, homogeny, expediency and globalisation, considering the values of regionality, patience, craft, skill and longevity. The analogy between Slow Food and Slow Architecture are embraced at the Mackintosh School of Architecture where in the third year of the undergraduate programme, student design briefs are planned to address many practical issues with sustainability at the core; through a number of 'food centric' projects students are encouraged to consider how materiality and construction contribute to a sustainable architecture where craft, sensuality, and delight are explored, with consideration given in both micro and macro contexts. Working on the premise that speed driven architecture can result in a visually dominant architecture, one in which the spaces created are viewed rather than felt; the projects required a 'Slow' haptic response. Adopting the Slow Food Movement and its principles as the metaphor for 'Slow Architecture', our aim was to design a programme to embrace these key principle and encourage students to investigate a more holistically sensual approach to their architecture, with the aim to encourage investigations of proposals that would extend beyond the 'visual'. A further ambition was to test if such projects would encourage a dialogue into how architecture can respond to the wider current issues of energy and food production and produce intelligent sustainable and appropriate proposals.

Keywords: sustainability, slow food movement, Cittaslow, multi-sensory, architectural education

44.1 Slow briefs

> *Architecture has the capacity to be inspiring, engaging and life-enhancing. But why is it that architectural schemes which look good on the drawing board or the computer screen can be so disappointing "in the flesh"?* Eyes of the Skin (Pallasmaa, 2005).

Pallasmaa's statement challenges the domination of the visual understanding of architecture over the other sensory experiences. His words are key to the direction and the formulation of studio design briefs for third year undergraduate students at the Mackintosh School of Architecture. The programme's intentions are to question the prioritisation of visual experience in the conception and making of buildings through investigations into the multi-sensory realms of architecture. Pallasmaa's book is one of the key texts on the book list along side text by Peter Zumthor (2006b), a known advocate of Slow Architecture. Both argue that the dominance of the visual realm in today's culture has suppressed any thorough engagement with our other senses. The question that Pallasmaa poses has fuelled much discussion regarding how we as teachers can encourage students to recognise that architecture is a phenomenon that extends beyond the visual. As teachers we have to impart practical information that is critical for students of architecture to acquire as they move through their training, giving both breadth and depth to their growing knowledge. The danger is that the delivery of this information can be theoretical and abstracted from reality. Our educational system does not support a slow approach that allows a 'hands on' course structure as class cohorts grow to groups of 100 students and more, with reduced budgets and limited time the possibility of connecting architecture and sensuality is not an easy task, knowledge of materials, their properties and aesthetics are more efficient to teach in theory rather than hands on explorations of the kinaesthetics of the materials.

Sustainability is another subject area that is more efficient to teach in the lecture theatre where, for example, passive solar gain is represented by a series of red and blue arrows flowing through a diagram, and the full integration of embedded environmental concepts do not extend beyond these figures. Our ambition was to enable these key principles to be considered and applied by the students to their own work, thus making connections and understanding theories that are no longer abstract but grounded in a level of reality, be it via a theoretical proposal.

The situation of being 'withdrawn' from sensory reality is aggravated when one considers the context and culture in which the majority of students entering a course of architecture live. A fast lifestyle where almost anything and everything is available 24/7.... at the push of a button. They are members of a digital and electronic era where the computer rules. Are individuals within this fast and sanitised lifestyle de-connected with the sensual aspect of life?

Zumthor in his book *Thinking Architecture* (2006a) states: 'The strength of good design lies in ourselves and in our ability to perceive the world with both emotion and reason. A good architectural design is sensuous. A good architectural design is intelligent.' And whilst discussing teaching and architecture he goes on to say: 'All design work starts from the premise of this physical, objective sensuousness of architecture, of its materials. To experience architecture in a concrete way means to touch, see, hear and smell it'

How can this 'fast electronic life style' embrace the less tangible but crucial qualities that make good design, and also consider relevant current issues? It would be foolish to imagine that removal of these digital key tools would either be possible or sensible; more appropriately, how can they help to facilitate explorations that will produce architectural proposals that have truly considered the experiential qualities that embrace these 'fast' mediums along side a more craft driven approach. Another factor that should be considered and tapped into is our own individual experience of architecture and our memory of those experiences. This is a great resource, although one that students seem less able to access or value. The experience we have from our childhood and our youth; our memories of space, light, smell and touch that are imbedded in our psyche; the house in which we grew up, places and events that we have experienced, the special, the memorable, the stuff that we don't realise that we know... subconscious knowledge.

As well as looking back and reflecting, students are also encouraged to look to the future and carefully consider the impact of their designs in today's....and tomorrow's world and make intelligent and thoughtful proposals to engage with the needs of our society and the unsustainable demands that we are currently putting on our planet.

44.2 Slow food – slow architecture

> *A man without memories loses his past and with that his future. He can no longer relate to his environment nor with himself* (www.slowarchitecture.com).

With these questions at the forefront of our minds we embarked on writing a series of studio design briefs that would unlock experiences and allow for sensual explorations. With the inclusion of hands on workshops that gave the opportunity to explore materials and how their application might enhance the experiential qualities of proposals, alongside practical requirements. The application of discoveries from workshops were put in place to help direct and facilitate the students through a more holistic design process and complimented the more 'traditional' tools, hand drawings, physical models and computer renders. Our starting point was food, more precisely Slow Food (http://www.slowfood.com). Slowness being a critical response and reaction to our fast lives, fast architecture and fast food. Food is something that we can all connect with, our need for food as living beings is essential to our survival, but also an element in our lives that touches our senses so directly, gives joy and contentment and evokes memories of past experiences. The ambition for the year's work was to utilise Slow Food as our metaphor. It would help to enable students to make a direct relationship with a multi sensory proposal; cooking and eating being a thread that would continue through all the projects and provide the accessible link for students to embrace, explore and develop the more difficult concepts relating to the atmospheres of their buildings through sight, smell, touch and sound. It was also hoped that the issues raised would begin a conversation about the social, environmental and economic issues relating to food in terms of growing, eating, trading and sustainability.

The connection between food and architecture is in many ways clear; proportion, form, shape and arrangement alongside selection of ingredients or materials, and thereafter composition of the parts. The result is a product that will give personal feelings and sensations. Food and cooking is a shorter and more transient art than architecture but no less sensually powerful and can be equally as memorable.

The Slow Briefs set design projects that embraced the principles of Slow Food, a movement founded in 1986 by Carlo Pertrini in Italy. Slow Food strives to preserve and promote regional cuisine and food growth. This movement has initiated a slow subculture where travel, shopping and design are areas to which similar considerations have been applied. Cittaslow (2009) is one such movement, with its aims to improve the quality of life in towns by resisting globalisation, homogenisation and celebrating the diversity of culture and specialities of a town and its surrounding hinterland. Engagement with the 'Slow' Movements' gave further depth to our projects where the consideration of regionality and uniqueness would be considered.

Reflecting on the programmes aims and utilising the Slow Movements principles as our vehicle we devised a series of studio design briefs for the year. A sequence of projects that would embrace all that we believed would nurture the connections with sensuality and atmosphere, regionality and sense of place, sustainability and community and embed the learning outcomes of the third year within them. Two main projects were designed, the first challenge was the design of a small but complex building placed sensitively and appropriately within a rural context, with both public and private elements, and specific environmental requirements. A deepening of the proposals came from the technical requirements with the design for specific environmental conditions and an understanding of construction and structure. The second project complimented this and began with the analysis of a small urban settlement located close to the first site. Following interrogation of the place, the challenge was to design a strategy for regeneration of the town, within which a small public building would be placed; programmatically linked to the proposal from the first brief. Both projects dealt with programme, context and technology at different scales. The design briefs were supported with hands on workshops and lectures by cooks, food specialists and architects, together with cooking and feasting (Figure 44.1). A visit to Barcelona between the two main projects reinforced the importance and relevance of 'food spaces' by directly connecting the theory with experiences explored, investigated and enjoyed first hand.

44.3 The briefs

44.3.1 One day in june

The year commenced with a summer project; titled 'One day in June'. This project was designed to act as a preamble to the design briefs and an opportunity to develop an understanding of indoor environments through a series of real world investigations. The students were asked to plan an experiment that would take place on the 21st June, the summer solstice, the longest day. The experiment required them to select an existing interior space, and by appropriate means register a salient aspect of a very particular environment that they had identified. The design of a methodology was required together with a device to measure or register the selected sense. This first foray with the sensual produced a diverse array of experiments

Figure 44.1. '[P]lan; after the feast' (William Knight; 2010).

and results, from the daily range and depth of smells within a spice shop in Morocco, to the movement of air in a cow shed in Orkney (Figure 44.2), and drawings of sounds within Glasgow's Kelvingrove museum.

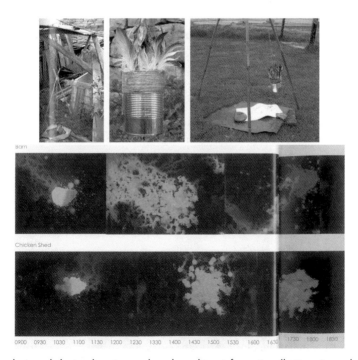

Figure 44.2. The designed device (top images) and resulting information (bottom image) collected for the 'One day in June' experiment (Andrew Wilkinson, 2010).

44.3.2 Cookhouse

The design of a 'Cookhouse' followed. The students were introduced to the philosophy of the Slow Food Movement and Slow Architecture, where the principles; connection to place, sustainability, regionality, craft and uniqueness were discussed, and immediately they were asked to consider architecture beyond the visual – an architecture that would address all the senses. The purpose of the food metaphor was presented as an overt tool with which to aid the design of a building that would enhance and celebrate life and food in a careful, crafted and sustainable way, and also address all the senses. Student's proposals should embrace and carefully respond to the uniqueness of the context in which it would be anchored.

This new building type would bring together a community of 15 individuals to live from the land for a year. The proposal for a cook house and its supporting landscape would be a place where slow food is grown, cooked and eaten. The ambition for the building was to provide a place to 'change gear', a place for its residents to jump off the treadmill and reflect holistically on their lives, an opportunity to take stock of priorities that are so often neglected due to the fast pace that many of us lead our lives.

The proposals considered the uniqueness of place and engender all the sensual qualities that the Slow Architecture/Slow Food Movement promotes. Sustainability is an essential aspect of the Movement, and therefore the students were explicitly asked to address this in both architectural terms but also in terms of lifestyle and community. Proposals were 'off grid' and so had to be self sufficient in respect of energy, water, waste as well as food, transport and community. Through careful partnership of architecture and food, the house would actively facilitate the enjoyment of growing, cooking and eating and provide a positive and sensitive environment within which these activities could be enjoyed.

This brief required the consideration of how proposals would resonate with all of the inhabitants senses, necessitating the consideration of light, scale, materials, colour, and texture, sensing through not only the eye, but also having an awareness of temperature, smell and sound. A further requirement was to investigate modes of presenting these less tangible, but no less critical elements of the proposed buildings. This was done through experimenting with materials, working at full size, written descriptions and other modes, including recordings and film to capture, explore and communicate their ideas.

Careful selection of a site in order to support the ideas was important. Perth and Kinross was chosen, as it is Scotland's first, and currently only Cittaslow member. Perth has also adopted the Slow Food Charter. We sought a site with an historical resonance and real sense of place.

Ardoch Fort near the small settlement of Braco on the banks of the River Knaik became the project's touchstone. On the edge of the broad and fertile valley of Strathearn, 2000 years ago, Ardoch became home to one of the earliest and most northerly outpost of the Roman Empire.

The Romans selected their sites with great care. Proximity to water and good agricultural land being fundamental to the choice of place and the subsequent sustainability of the settlement in what was a hostile country. The choice for the location of the student's proposals was similar; the proximity of a watercourse and good quality arable and grazing land was fundamental to the prolonged existence of the facility (Figure 44.3).

A thorough site analysis was undertaken with students recording both the physical and the sensual qualities of the site. Using their bodies to explore the site – walking the extent of the policies, running the ramparts, wading in the river and sheltering in the walled garden. A grasp of local farming and the seasonal nature of crops and livestock were required to gain an understanding of the agricultural possibilities that the site could provide over an annual growing cycle, thus validating proposals for growing and rearing of livestock. Students located their proposals within a selected area with care and sensitivity towards the existing marked landscape, views, topography, meteorology and orientation.

The preparation of food and the celebration of eating were the core activities to be housed. It was made clear that this was not a commercial kitchen – but a place where home grown and reared food would be caringly prepared and eaten. The spaces were to be designed to recognise and celebrate this fact. An additional area for weekend dining was also a requirement of the brief. This is where the local community would be invited to enjoy the fruits of the land and Cook House, not as a restaurant, but a place for guests to eat with the chefs in a homely environment – a response to the 'anti-restaurant' movement that is currently building, and in keeping with the sustainable culture of the Cook House.

Figure 44.3. Ardoch Fort.

Support spaces for food storage were also required to allow the building envelope to provide the desired environmental conditions, through a daily and annual cycle – stability and control of humidity and temperature were key elements. To compliment the communal spaces a residential area was to be included; a place of retreat from the busy group spaces; private bedrooms and living spaces for the residents to call home during their stay in the Cook House. Throughout the project students were encouraged to explore both the haptic and the ephemeral, utilising physical models, hand drawings and computer renders (Figure 44.4).

Figure 44.4. Explorations of atmospheres and qualitative experiences (Nathan Cunningham, Stefano Belingardi, Jack Hudspith and Hugo Corbett, 2010).

44.3.3 'Supra-market'

The second major project took the students to Creiff, a near neighbour to Ardoch Fort. Creiff is a town with an embedded history of trade and local industry. The town was granted a charter in the middle ages; in the 18th century it became the centre for Scotland's cattle trade; in later times its factories produced paper and lace, and with the development of rail links with Central Scotland it became a thriving Victorian holiday spa town. Creiff has evolved and morphed through these various different identities and now finds itself as an ailing rural town suffering from the effects of a declining and ageing population. A new community campus school and an out of town supermarket have taken facilities and life away from the town's core; Creiff is now a town struggling to define its identity in the 21st century.

Students carried out extensive observations and analysis of the town, recording its history and growth, identifying and understanding core, periphery and morphology, becoming familiar with what has made Creiff the town it now is. Recognising the potential of markets as social and community regenerators, as discussed by Steel in Hungry City (2009), a market building was proposed. The project demanded students to design a strategy for new growth and further more to design a 'supra-market' and new external public space which would sit as part of this strategy. The proposed facility transcends the notion of a farmers market – becoming a permanent venue for the current and nomadic monthly market, a public foothold for the Cook House and various other social facilities. It was also required to have a true sustainable agenda in terms of the local community, building again on the concepts of the Slow Food Movement. The establishment of such a proposal would have the ambition to act as a catalyst to revive the town's centre and reverse the current trend of deterioration. Again Slow Food was the tool for students to build their proposals upon. Supported by the Cittaslow charter that the region has embraced, students were confronted with real issues. The connection with Slow Food with this second brief utilised food as a pragmatic tool to engage students with the possibility of how a successful building within a designed framework could act as a catalyst for rehabilitation and sustainability, extending in a scale far greater than the that of the building itself. Recognising what the town had to offer; a farmers market and a range of local producers; and working within the existing built context, food was a realistic tool for solving the challenges posed to them. (Figure 44.5).

44.5 Conclusion

Each year students are confronted with design challenges that will enable them to meet the aims and objectives of the degree programme at the Mackintosh School. The 'Slow Briefs' were specifically designed to move our ambitions for the student's proposals beyond the visually dominant results that have been produced through more traditional project briefs. With food as our metaphor for architecture we hoped that all students would relate to the analogy at some level, and allow engagement more easily with concepts beyond the visual. Students were encouraged to utilise a multitude of design tools to enrich their investigations, along side visits to kitchens and restaurants. Growing was addressed with an allotment garden

Figure 44.5. A new community 'supra-market' in Crieff (Kugathas Kugarajah and Eu Jin Lim, 2010).

established on the roof terrace of the department and this was complimented with organised group cooking and feasting (Figure 44.6). The hands-on workshops were designed to give students the opportunity to explore the rituals and celebration of food through growing, cooking and eating; lectures by chefs, food specialists, material experts and architects engaged in the food industry all supported the briefs.

A study visit to Barcelona focused on markets, community buildings, and public spaces providing an opportunity to smell, touch and experience a plethora of atmospheres, the aim was to provided further richness to their proposals.

The resulting outcomes have been mixed, and as with an analysis of proposals presented by any student cohort there have been varying levels of success particularly relating to our ambitions for this years work. The following comments are based on observations made of the work as scrutinised during the intensive assessment period and the degree show that followed.

Our belief that the students would embrace the concept of food with ease and allow a more immediate connection to the sensual aspects of the project than a more traditional brief might give, was not immediately evident. Work produced by some students did show an engagement with expressing atmospheres beyond the visual with experiences gained from the practical workshop clearly feeding into their design proposals. Conveying and communicating these explorations and the concluding designs with the use of the computer and physical models did result in some sophisticated outcomes with photo realistic visualisations created on the computer. These were complemented with full scale castings in concrete; knowledge of food growing, harvesting and cooking, and an understanding of integrated energy strategies. The design of objects beyond their architecture was also explored from recepies to furniture to cutlery (Figure 44.7). This success was recognised with two students from the cohort being short listed for the RIBA Bronze medal 2010, and Jack Hudspith going on to be awarded this prestigious award. The judges made specific comments relating to how successfully the proposal demonstrated an awareness of the relationship of internal and external spaces and of how these spaces were inhabited.

Figure 44.6. Cooking with concrete at the Mac 2010.

In respect of the entire year what was particularly successful and demonstrated by the majority of the students was the application of environmental and construction strategies. Embedding and testing these essentially lecture based subject into a design brief has produced some thoughtful and innovative proposals. Environmental strategies in the most part are integrated and not 'attached bells and whistles'. An understanding and application of sustainability, beyond the scientific, concerning lifestyle and community was clearly demonstrated in the majority of proposals. Careful location, orientation and efficient building envelopes were evident across the cohorts proposals. Materiality was also generally more holistically approached, utilising the experience gained from timber and concrete workshops, at which the students had worked at full scale with these materials (Figure 44.6). The impact of a more informed material selection moved beyond a visual choice, to one that also considered the qualitative aspects of the spaces created by the materials and the resulting environmental performance.

In the Cook House project student's engagement with food and cooking was disappointing in some cases; with students defaulting to re-planning their designs and more conventional

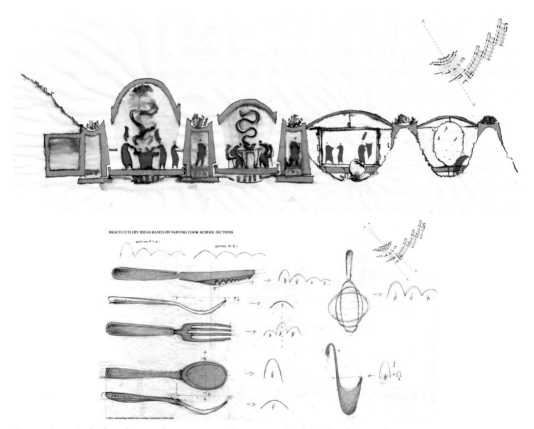

Figure 44.7. Ardoch cutlery and environmental response (Jack Hudspith, 2010).

explorations rather than investigations of texture and ambience, and addressing the bigger issues relating to sustainability that we hoped would be explored. In contrast to this the 'supra-market' proposals demonstrated how students embraced the notion of utilising redundant spaces within a town 'master plan', developing appropriate strategies for community and food development and production, supporting their more formal public space and building proposals (Figure 44.8).

In conclusion it is difficult to confirm if third year undergraduate students are able to grasp and explore these additional, more abstract layers of information that we so keenly desire. At this relatively early stage in their training does solving the practical become the default and the more challenging and elusive qualities of their architecture remain undiscovered? At what

Figure 44.8. Creiff 'supra-market' (Kugathas Kugarajah and Helen McCormack, 2010).

stage should this line of enquiry sit hand in hand with the technical and practical issues that a brief poses– where a truly holistic mindfulness of their ideas is explored and communicated?

Perhaps it is akin to the development of our palette....it just takes time and lots of 'tastes' before an appreciation and enjoyment of such can be valued and experienced.

References

Pallasmaa, J., 2005. The eyes of the skin: Architecture and the senses. Wiley, Chichester, UK, 59 pp.

Steel, C., 2009. Hungry city: How food shapes our lives. Vintage, London, UK, 400 pp.

Zumthor, P., 2006a. Thinking architecture. Birkhauser, Berlin, Germany, 96 pp.

Zunthor, P., 2006b. Atmospheres. Birkhauser, Berlin, Germany, 75 pp.

Chapter 45

GreenEngines, a pedagogic tool on sustainable design and productive landscapes

Ana M. Moya Pellitero and Josué da Silva Eliziário
Principals StudioMEB, Rua Vicente Vaz das Vacas 41, 5E, 8500-74 Portimão, Portugal;
moya@studiomeb.com

Abstract

Since 2009, the office StudioMEB (architecture, planning and landscape), is developing action research in the field of sustainable landscape planning and sustainable food planning design. The office created the research platform GreenEngines, with the aim of exploring the potentialities of productive landscapes to generate a sustainable territory that is respectful to the existing local environment with a multifunctional character, community involvement, heritage and cultural identity. GreenEngines develops as a pedagogic tool in the education of multidisciplinary teams, involving students of architecture, geography, landscape architecture, design, urban planning and environmental studies among others. Through education, students can understand the complexity involved in strategic landscape planning for the preservation, creation or re-invention of productive landscapes, with the objective to reach a sustainable equilibrium between economy, society, culture, the environment, and food production. Our work methodology takes into consideration landscape urbanism, landscape planning, environmental planning and theory of complexity principles. Our premise is the academic and professional collaboration, at an international level, with experts, to create a transversal platform of exchange of knowledge, among different disciplines (geographers, environmentalists, landscape architects, architects, town planners, and designers) that integrate their knowledge in a common goal. For each case study, we establish contact with local governments, town halls and local institutions that collaborate with us, providing information about the real local problematic, and the expertise of the local professionals involved. The research objective is to draw a strategic landscape plan for the future development of the case study site, taking into account flexible dynamics, scenario thinking, actors involved, and processes over time, which relate with changes and re-adaptation. The present chapter describes how sustainable food planning studies fit into the present debate on landscape urbanism and productive landscapes. Landscape urbanism, in our research approach involves social analysis, cultural and community patterns. The chapter explains in detail the research structure of GreenEngines as a pedagogic tool and the response of European schools to our teaching methodology, evaluating obstacles and drivers in this initiative.

Keywords: productive landscape, landscape urbanism, education, landscape planning strategies, food planning

Ana M. Moya Pellitero and Josué da Silva Eliziário

45.1 A sustainable productive landscape

We state that a productive landscape is any natural, rural, coastal or urban environment used and exploited for agricultural, industrial, business or touristic activities. In the case of the rural territory, the shift, in recent years, towards the construction of solar, or wind power plants, together with the production of industrial agriculture for bio-fuels, bio-mass, and economies of scale, has transformed many rural landscapes into technological and productive ecological deserts, expelling society from their environment. The rural landscape is not built any more on culturally productive patterns adapted to climatologic and geographic conditions, and maintained along the centuries by the work of local family farming. Globalisation and liberal markets ask for a rural territory with strong competences, adaptability and quick innovation. However, the rural dynamics can never adapt to merchandise policies that are destroying slowly local traditions, heritage, culture, biodiversity and habitat. They are forcing people to migrate to urban areas, creating social uncertainty. These necessary eco-technological and industrial measures distance the rural landscape from being a truly sustainable productive environment. In the case of the urban context, urban conurbations are also productive landscapes that aim to attract business, industry and tourism. Cities suffer processes of development that are temporal and discontinuous based on intermittent economical global interests. Within countries themselves, inequality coexists between cities that belong to global networks and those other cities and regions that shrink and decay. A process of economical growth, and urban development, implies the de-urbanisation of other cities and regions, the degradation and erosion of weak territories, particularly those ones based on local economy. Cities with a high rate of unemployment and few work opportunities start suffering processes of forced shrinkage due to migration and population loss. During the 20th century, the number of shrinking cities has increased, and in 1990 more than a quarter of all large cities in the world have already experienced population losses (Rieniets, 2006). The moving of industries offshore, due to differences in wage costs, affects cities in developed countries. Only cities that are the home to the players of globalisation enjoy the privilege of having stable growth and urban development (Müller, 2006). To avoid inequality, poverty and migration, a sustainable urban territory is needed. Sustainability, based on a local economy is necessary for the economic survival of cities, mainly those ones that run outside the global network.

The research platform GreenEngines was created in 2009, at the office StudioMEB, inside a collaborative network between Universities, practitioners and local governments with the concern and urge to explore the potentialities of local productive landscapes to generate a self-sustained territory. It addresses the research of strategic future alternatives for specific case studies inside rural, urban, natural and coastal environments. Which new planning strategies and transformative processes could guide changes and improve self-sustained productive local geographies? We state that any productive landscape entitled to be called sustainable should accomplish the following: First, it belongs to a cultural construction, which adapts to the cultural landscape and the local environment, with a clear strategy of preservation and maintenance of the cultural values and the identity of the territory, including the revitalisation of the palimpsest of traditions, heritage (built and natural) and collective memory. Second, the landscape is multifunctional with different actors involved in the same space (energy and food production, industry, tourism, education, leisure, culture, nature, health, housing, commerce). Third, it takes into account social participation, involving the self-maintenance and self-organisation of the space. It encourages individuals to interact with their close environment through participatory processes and a close physical experience. Fourth, it values the phenomenological qualities of the space. The territory is acknowledged by sensory experiences within the parameters of space and time. It is experienced by emotions, memories, and mental bonds. The phenomenal richness of the landscape is present in the social imaginary, the collective memory, the desires, the sensorial and the poetic experience of the inhabitants. Fifth, it considers new models of mobility thinking in alternatives to the car, and betting for intermodal ways of transportation (pedestrian, bike, bus, train).

For the attainment of self-sustained rural and urban productive landscapes, GreenEngines believes that both of them need to reach a hybrid character. In the moment that the rural territory urbanises, and the urban territory ruralises, the hybrid cohesion between both worlds enlarges the capacity of local development, strengthens the cultural landscape and enlarges new emotional linkages with the territory. With a rural urbanity, there is an interaction between the rural environment with new urban services and infrastructures (new alternative industries, neo-rural social movements, tourism, and education). In the case of urban rurality, the city takes advantage of the proximity of peri-urban rural areas, and the creation of rural parks and green rural networks, integrating food gardens in the urban structure, as an ecological lung and source for local food production, self-organised activities and ecological food education. The challenge embodied in Agenda 21 is to achieve a social commitment for sustainability through collective participation, social debate, and educational diffusion. The introduction of urban agriculture within the city allows for social education about ecological, seasonal and organic farming. It allows growing food for local consumption, managed by the local community, renting the land from the municipality for private use inside the same local framework. However, urban agriculture will never be sufficient to sustain the entire food needs of a city, but it allows a social interaction within the environment based on a close physical experience, helping to educate new generations, introducing new sustainable habits.

45.2 Research methodology

In our action research, GreenEngines establishes a methodology of work; taking into consideration disciplines such as landscape urbanism, landscape planning, environmental planning and complexity theory.

Landscape Urbanism[51] is a hybrid practice that emerged in North America and Europe in the late 90's as a new design discipline to respond to the conditions of sprawl under the phenomena of post-industrialisation of the urban territories. That is when landscape emerged as a model for contemporary urbanism, especially in the context of complex natural and urban environments and in that sense 'landscape supplants architecture's historical role as the basic building block of urban design' (Waldheim, 2006: 37). Elements within each of the design practices – landscape architecture and urban planning – have moved towards a shared form of practice that is involved in the conceptual scope of landscape design and planning. In other words the 'capacity to theorise sites, territories, ecosystems, networks, and infrastructures, and to organise large urban fields' (Corner, 2006: 23). The discipline of Landscape Urbanism is still an emerging field of study because it has been subject to different interpretations. Therefore, it still needs to build a body of research that concretises and enlarges divergent standpoints. In the postgraduate courses on Landscape Urbanism at London's Architectural Association, run since 2000, landscape is interpreted as a 'machine'. According to C. Najle, co-editor of '*Landscape Urbanism: a manual for the machinic landscape*', landscape is a cybernetic universe with its own laws (Najle, 2003: 141). Fluxes and processes, inherent to the constant time-space evolution of the landscape, are mapped using dynamic systems. They employ the science of complexity and emergence to create new organisational planning models (development prototypes). They gather information (social, ecological, economical, spatial), and with specific software, they decode the information, synthesising it and systematically processing it, to create a design process capable of accommodating change and indeterminacy. This type of approach gives importance to the constant appropriation and actualisation of data, giving shape to self-representational structures that distance from the real physical context. Contemporary practitioners of Landscape Urbanism also experiment with the use of infrastructural and ecological systems as ordering mechanisms in shaping urban settlements. Landscape then starts offering strategies for urban design. A historical reference was the competition of Parc de la Vilette (1982). Both the winning project of Bernard Tschumi, and the second entry by OMA, used the landscape as a medium to give shape to the city. They set up a practice that articulated layers and players, being at the same time flexible and strategic, including urban infrastructures, unplanned relationships and events.

Landscape planning also helps to rethink the variables for a dynamic sustainable territory, in which economic growth supports social progress and respects the environment. A team

---

[51] Landscape urbanism was anticipated in the Symposium 'Landscape Urbanism' in 1997, organised by Charles Waldheim. Later, he, as an editor, collected essays by practitioners in all the mentioned disciplines in the book 'Landscape Urbanism Reader' (2006). In America stands out the recollection of articles, in the format of a summary textbook, with the works of professionals that have paved the way towards this current theory, in 'Center 14: On Landscape Urbanism' (2007) edited by Dean Almy with the collaboration of Center of American Architecture and Design and the University of Texas at Austin School of Architecture. Landscape Urbanism has been also a topic for publication in Europe in *Landscape urbanism: a manual for the machinic landscape* (2003) edited by Mohsen Mostafavi and Ciro Najle, with articles by different authors completed by projects developed within the framework of the landscape urbanism programme at the Architectural Association, London. It has also been discussed in round tables and workshops, like in the Colloquium 'Articulating Landscape Urbanism' organised by the Chair of Landscape Architecture in Wageningen University in 2007, the Netherlands.

of researchers, headed by Prof. Carl Steinitz, in the department of Landscape Architecture and Planning at Harvard University, have provided, since the 1990s, a modelling strategy for planning assessment. The model, of an analysis scenario-based study for alternative futures, considers which are the actors and issues responsive to policy and planning decisions. In the work *Alternative futures for changing landscapes* (2003), the team of investigators guided by C. Steinitz, propose an approach that follows the typical decision-making processes and choices that shape the future of a region. It identifies a simultaneous set of policies and planning decisions applied in the context. A scenario is created that reflects choices selected among the possible options for each policy. Steinitz defines the word scenario as an 'outline of events, typically the plot of a story, play, or film' (Steinitz C, 2003). This scenario generates an outline or hypothetical plot for a future. When choices are made, it results in a scenario that is tested and assessed for its consequences in the development of the territory. The individuals play a role in the creation of urban space and they have the potential to reflect and respond critically to their environment and develop themselves through learning. A landscape planning strategy establishes those designed actions that will trigger processes of change and social involvement with the environment. When this social involvement is self-organised there is an interaction between the formal and the informal, appearing a complex system that evolves over time with new stages that are not predictable by a linear causality but by a circular one that is constantly readapting (Portugali, 1999). According to Juval Portugali in '*Self-Organization and the city*' (1999), self-organisation is defined as an autonomous action based on complex and uncertain relationships happening between organisations and individuals.

In our action research, GreenEngines creates a methodology of work that integrates the four themes involved in the practice of Landscape Urbanism and outlined by James Corner in 'Terra Fluxus' (2006). First, it 'considers urban processes over time', second 'anticipates strategic scenarios and operational logics through a wide range of scales', third 'reconsiders representational and operative techniques', and fourth 'takes into account the phenomenal richness of physical life (social imaginary, collective memory, desires, the tactile and the poetic)' (Corner, 2006). The planning of a sustainable strategy for a specific case study contemplates the research by design at different scales. In each scale level (large, medium, and detail), it is possible to discover different phenomena, processes and relationships affecting the planning of a territory and its landscape, together with an approach to detailed architectural and urban design interventions. What differentiates us from other approaches in landscape urbanism is that we distance ourselves from strategic prototypes, and formal representations of dynamic data and visual modelling. Instead, we give priority to scenario thinking, actors involved, and processes over time, which relate to changes and re-adaptation, reflecting a particular view of society and the groups that compose it. We look at the networks of interactions and multifunctional aspects of the actors involved through a scenario-based analysis. We divided the research work into five stages, which correspond to different work scales: analysis, strategy, tactics, actions, and evaluation. In the first stage, the analysis of the location helps to detect the potential, qualities and problems of the site at large scale. It listens carefully to the existing cultural and ecological landscape. It searches for the potentials, qualities and problems of the territory. In the second stage, the strategy aims to draw the big picture of a planning proposal.

In this stage, it is important to consider the time implementation of the planning decisions. It is also important to value what already exists and to reconsider the potentialities of the space. The landscape operates in a certain way and it is essential to evaluate which parts of it should change in order to arrive to a sustainable scenario. The strategy answers the question why. Why do we choose a specific location for a specific purpose? Why do we modify an aspect of the existing landscape? Why do we preserve certain features and we erase others? The result of this stage is the mapping of the main operational logics and scenario thinking, taking into account the number of actors involved and the time-strategies (Figure 45.1 and 45.2). The third stage involves the tactics. It centres in the middle scale in planning, testing a sustainable model using exploratory and inductive logic. One decision brings to the next. It answers the question how. How are we going to arrive to a result in a specific area of the plan? How different programmes and actors interact with each other in the design approach? The result

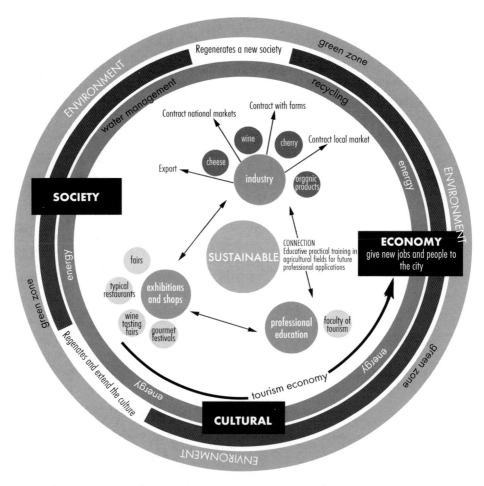

Figure 45.1. Planning a strategy for Cova da Beira, Portugal. Diagram showing the relationship among actors (education, industry, facilities and tourism).

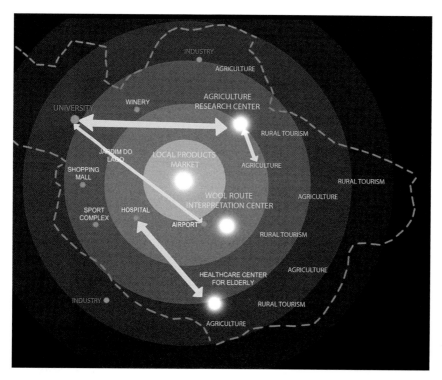

Figure 45.2. Map of Cova da Beira's case study with the strategic location of players.

of this stage is middle scale plans, sections and photomontage images of the area. The fourth stage, implies the search for design actions. It involves the small scale, where decisions are materialised into the detail of specific urban design, architecture, landscape architecture or landscape design detail interventions. It answers the question what. What type of landscape we will have as a result? What type of road system? What type of urban-rural park? What type of farming and network of farms?, etc. The result of this stage are plans, sections and details of design actions, including perspectives of the view and atmosphere of the landscape. The evaluation stage helps to draw conclusions about the different approaches based on the same methodology and the same objectives. The evaluation looks at the strengths and weaknesses of each proposal. It also compares results and draws conclusions about common results and divergences.

We include in our educative research work social analysis specific to each case-study. Any planning strategy should consider ecology and community. It is the social involvement with the close environment and the understanding of potential self-organised processes of the community that can generate designed interventions that trigger processes of change evolving over time. In this respect, we take into consideration principles of environmental planning, being aware that any alteration of the nature of the landscape, no matter how small, has deep implications for the ecological processes of the immediate area and the larger region. It also involves the respect for the heritage, the cultural identity and the historical context.

45.3 GreenEngines as a pedagogic tool

GreenEngines has already carried out two international summer workshops. The first one took place in the Faculty of Architecture of Barcelona, University Polytechnic of Catalonia, on July 2009, with the topic 'Barcelona Tres Turons Park, a case study'. StudioMEB coordinated the event with Istanbul Kültür University (Istanbul, Turkey), with the collaboration of Yildiz Technical University (Istanbul, Turkey), and the participation of Delft Technical University (the Netherlands), University Autonomous of Barcelona (Spain), Polytechnic University of Barcelona (Spain), Elisava School of Design Barcelona (Spain), Consultancy and Engineering DHV, Department of Environment and Transportation, Eindhoven (the Netherlands), and the Department of Urban Planning from Barcelona City Council. The area of 'Tres Turons' is a strategic green space in the northern part of the city of Barcelona, with great environmental, historical and social values. Inside the area, there are already three consolidated urban parks. The original landscape of the hills, once agro-forest, was transformed by human activity during the nineteenth and twentieth century. It contains old quarries, a civil war anti-aircraft battery, and informal self-constructed post-war settlements erected during the rural migrations, together with shanties, wastelands and sightseeing public platforms. The case study involved 122 ha, mainly in public ownership, with 94 ha of green areas. The workshop aimed to test, inside a student design studio, the implementation of a self-sustained urban agro-ecosystem for the area. The landscape planning strategy had to consider the main catalyst processes for the generation of a community self-organised productive landscape, where urban agriculture (food gardening) integrates inside a location with a heterogeneous character, embracing different historical sites. The students were introduced to GreenEngines research methodology (Figure 45.3). The workshop was structured in five stages (analysis, strategy, tactics, actions, and evaluation). Each day the students were given the goals that were to be achieved in each stage, in order to help them to handle the complexity of moving through different scales. The idea was to learn by progress, testing day by day the results. The students were divided in three multidisciplinary groups, mixing different levels in expertise and education. Each group was guided by assistants and invited reviewers, experts in the field of landscape architecture, urban planning, architecture, visual arts and geography.

The second international summer workshop took place in Covilhã, Portugal, on July 2010, with the topic 'Covilhã, landscape of change, a prototype for a new integrated rural-urban growth model'. A prototype for a new integrated rural-urban growth model' was developed with the collaboration of the University of Beira Interior (UBI, Portugal), Department of Civil Engineering and Architecture, and the guest participation of Istanbul Kültur University (IKU,), Yildiz Technical University (YTU) and Istanbul Technical University, with a total of 33 students. Six multidisciplinary groups were created. Invited guests participated with us from the Wool manufacturing Museum, the Association of rural development of Serra da Estrela, and University of Brighton (England). Our aim, in the second edition of GreenEngines, was to discover how to regenerate a peri-urban territory, with a rural character, achieving a new rural-urban model of sustainable development (Figure 45.4). Inside Cova da Beira, the case study had a perimeter of 14 km and a total of 1000 ha. The students had to develop

Figure 45.3. Team of students at work in GreenEngines Workshop, UBI, Covilhã, Portugal, 2010 (GreenEngines).

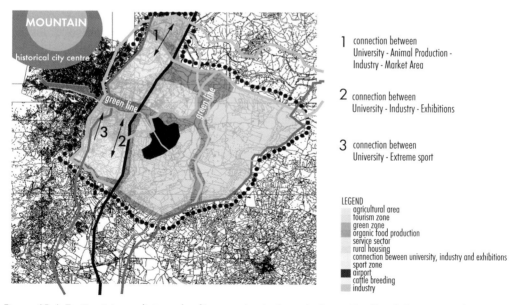

1 connection between
   University - Animal Production -
   Industry - Market Area

2 connection between
   University - Industry - Exhibitions

3 connection between
   University - Extreme sport

LEGEND
   agricultural area
   tourism zone
   green zone
   organic food production
   service sector
   rural housing
   connection beween university, industry and exhibitions
   sport zone
   airport
   cattle breeding
   industry

Figure 45.4. Tactics: intermediate scale of intervention in Cova da Beira, Covilhã. A Green Line along water streams surrounds the different programmatic areas.

a landscape planning strategy in which economical, social, cultural and environmental objectives were interlinked. Their goal was to create different scenarios with actors such us education institutions and research centres; local industry adapted to an ecological model of local production of organic food and energy; clean renewable energies; and housing areas for elderly, tourists, students and working families. The goal of the planning strategy was also to bring together different social groups, thinking in the maintenance and self-organisation of rural-educational activities, food and energy production, all-season tourism attractions, health therapies and the natural and rural landscape. It was also important for the students to consider the rehabilitation of the rural landscape heritage. The valley then becomes a cultural and historical park; platform for education, leisure, health and sports. Another premise was the preservation of the water streams as important ecological corridors, with the recuperation of their fringes and humid areas. These streams were in the past an important part of the cultural landscape of Covilhã because all the historical textile factories were located in their fringes.

45.4 Evaluation and conclusion

In the teaching of strategic landscape planning we apply a work methodology that integrates the practice of Landscape Urbanism (processes over time, scenario thinking, new operative techniques, and the social imaginary) and Landscape Planning (scenario-based analysis), with sustainable food planning systems. The postgraduate courses on Landscape Urbanism at the Architectural Association London, approach the study of landscape urbanism mainly under the point of view of the design of time-based urban processes with the goal to control morphological, spatial and data transformations of the territory. Our work method, instead, gives a great importance to the planning of self-sustained productive landscapes, where any intervention in the territory considers social analysis, understands the evolution of cultural community patterns, and preserves the cultural historic heritage and identity of the territory. Each planning strategy includes sustainable food planning systems combined with other different scenarios and actors involved (education, health, tourism, industry, or leisure among others). A self-sustained productive landscape is always a complex multifunctional space adapted to a specific cultural environment, considering always the social involvement of the community (Figure 45.5).

In the first workshop in 2009, all the group proposals were respectful to the context and the location of study. The students understood the importance to make visible processes that were already occurring in the site. They guided these processes by urban design and architectural interventions, limited in specific and strategic locations that were studied carefully based on site analysis. The design strategy was the core of the intervention. Each group took different strategic decisions based in the same analysis (Figure 45.6). None of the three groups gave to their landscape proposals a specific scenic design. They simply let the site be and evolve by itself, without any formal design of a park. All three groups agreed that in order to plan a sustainable space, a study of the relation between different actors was needed, such as education, tourism, recreation, and commerce. They also agreed how important it is

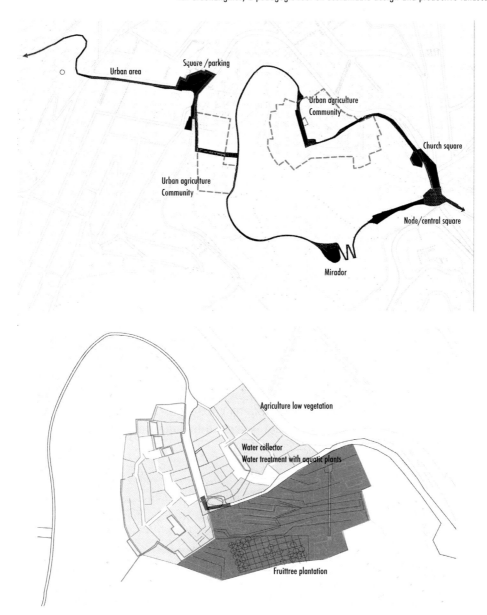

Figure 45.5. Tactics: intermediate scale of intervention in El Carmel, Tres Turons, Barcelona. Circular circuit around a housing community.

to transmit and to map complex processes of interactions and relations between actors, and how difficult is to represent them in legible maps. The progression through different scales within a process thinking method helped the students have clear strategies, and at the same time, to be able to understand the physical and tactile nature of the site, within which people move and interact.

Figure 45.6. (A) Design action: access node for parking area and elevators in La Rovira. Structure covered by a technological skin of solar cells and humidity collectors. (B) Design action: community urban agriculture in the slopes of the quarry in La Rovira. Pedestrian access with a ritual ascendant path experience.

In the second workshop, in Covilhã, in 2010, the students learned that before approaching any planning decision, it was necessary to evaluate and analyse the problems and potentials of the site. All their proposals were sustained in this clarifying analysis of the existing territory. They discovered the importance of considering different actors that interact with each other at the same time, involving operational logics and scenario thinking, and working at different scales.

In both workshops, teaching students a new methodology and research aims, within the topic of sustainable food planning was a complex task. On the one hand because the programme was only one week long, with students from different disciplines and levels of study, with multicultural visions and approaches and with the large scale in planning. The methodology of work was new for all of them. But in both workshops, we succeeded to make the process of work understandable to the students by never giving the whole picture at once, but dividing and discovering the tasks and research goals, day by day. There was also a great curiosity, on the student side, in the introduction of sustainable food planning strategies and food gardening as a catalyst of social processes and community self-organisation. The greatest barrier in this initiative has been the logistics and the limited budget, in order to bring together in one week and in one location, students from different Universities, guest professionals and local experts. However, what has facilitated our task has been the great positive response to our pedagogic initiative of many Universities, Institutions and practitioners from different disciplines. Since we started in 2009, we have counted with their help and collaboration. With them, we have enlarged the multidisciplinary character of our research.

We aim to apply our research experience into new case studies, in different international contexts (urban, natural, rural, or coastal), with the participation of local experts, local governments and with the collaboration of Universities and local Institutions. From these future international case studies we can learn how sustainable productive landscapes are able to adapt to different geographic, historic, cultural and social realities. We also aim to establish new academic and professional exchanges in order to enlarge the academic and professional community knowledge on strategic landscape planning for productive landscapes (always including food planning systems), and to educate the new generations of professionals that will have to deal in the near future with local geographies disconnected from global networks.

Acknowledgements

We would like to thank the students participating in the GreenEngines International Workshop at Polytechnic University of Catalonia, Barcelona in 2009 (C. Briceño and S. Kiliç for Figure 45.6; I. Casanova for Figure 45.5), and the students participating in the GreenEngines International Workshop at the University of Beira Interior, Covilha, in 2010 (J. Pinto, I. Ince, B. Ozirisen, H. Yenicikan, F. Kaya for Figure 45.1 and 45.4; and M. Betãnia, E. Sisman, T. Duarte, E. Yilmaz, D. Karadeniz, T. Ekiz for Figure 45.2). The illustrations in this chapter were a result of their work.

References

Almy, D. (Ed.), 2007. On landscape urbanism. Center vol. 14. Center for American Architecture and Design, University of Texas at Austin School of Architecture, Austin, TX, USA, 355 pp.

Corner, J., 2006. Terra fluxus. In: Waldheim, Ch. (Ed.) The landscape urbanism reader. Princeton Architectural Press, New York, NY, USA, pp. 28-33.

Mostafavi, M. and Doherty, G. (Eds.), 2010. Ecological urbanism. Harvard University, Graduate School of Design, Lars Müller Publishers, Baden, Switzerland, 655 pp.

Müller, K., 206. Economic transformation. In: Oswalt, P. and Rieniets T. (Eds.). Atlas of shrinking cities. Hatje Cantz, Ostfildern, pp. 122-152.

Portugali, J., 1999. Self-organization and the city. Springer-Verlag GmbH & Co., Berlin, Germany, 34 pp.

Rieniets, T., 2006. Urban shrinkage. In: Oswalt, P. and Rieniets T. (Eds.). Atlas of shrinking cities. Hatje Cantz, Ostfildern, 30 pp.

Steinitz, C., Arias, H. and Shearer, A., 2003. Alternative futures for changing landscapes. The upper San Pedro river basin in Arizona and Sonora. Island Press, Washington DC, USA, 200 pp.

Waldheim, C., 2006. Landscape as urbanism. In: Waldheim, C. (Ed.) The landscape urbanism reader. Princeton Architectural Press, New Tork, NY, pp. 35-54.

Index

A

accessibility   469
 – to food   71, 111, 128, 129, 131, 134, 145, 150, 471
 – to land   388
action
 – plan   390, 555
 – research   361, 495
Actor Network Theory   116, 124
Acts of Enclosure   339
affordability   214, 469, 513
AFN   25, 26, 27, 176, 223
AFS   385, 395
Agenda 21   573
agrarian urbanism   247
agri-business   385, 394, 395
 – ecology plus   311
 – hybride   311
agricultural
 – labour   21
 – land   24, 182
agro-ecology   204, 459
agro-food industry   173
Agromere (Almere)   308
agro-park   396
agro-urban patchwork   284
Allen, Will   389
Allmende Kontor (Berlin)   490
allotment   199, 213, 215, 235, 338, 340, 342, 344, 346, 354, 511, 565
Allotments Regeneration Initiative (UK)   344
All that rot! (Netherlands)   371
Almere   309, 316, 317, 318
 – Agromere   308
alternative food
 – consumption   153, 155
 – geography   25, 27
 – movement   64, 173
 – network – *See:* AFN
 – system – *See:* AFS

American Planning Association   56, 104, 394, 547
animal
 – foods   239
 – production   191
 – welfare   190, 238
aquaponics   541
architects   56, 385
 – amateur   424
artists   380
asteraceae areas   373
auto industry   273

B

Back to Front manual   411
backyard gardening   279
Barcelona   578
 – Faculty of Architecture   578
barriers   434
 – to participation   549
bees   446
behaviour change   173, 233, 240
Berlin   479
 – Allmende Kontor   490
 – Grünes Leitbild   488
 – Prinzessinnengarten   484
 – urbanacker   490
best practice   508
biodiversity   23, 204, 358
 – year (2010)   191
bioregionalism   460
Bohn&Viljoen Architects   388, 481, 534
bonding   164
botanical gardens   439
bottom-up initiatives   84, 485, 533
Bourdieu, Pierre   151
box schemes   27
Bremen   195
Brighton and Hove
 – Food Partnership   73, 74
 – Foodshed   74
British Trust for Conservation Volunteers
 – *See:* BTCV